# Anti-Lock Braking Systems for Passenger Cars and Light Trucks - A Review

PT-29

(Selected papers through 1986)

Prepared under the auspices of the
Passenger Car Brake
Activity Committee

Published by:
Society of Automotive Engineers, Inc.
400 Commonwealth Drive
Warrendale, PA 15096

ISBN O-89883-117-2
SAE/PT-87/29
Copyright 1987 Society of Automotive Engineers, Inc.
Library of Congress Catalog Card Number: 86-63562
Printed in U.S.A.

## Editorial Advisors Preparing PT-29

J. Martin Rowell
Automotive Sector
Allied-Signal, Inc.

Paul S. Gritt
Chrysler Motors
Chrysler Corp.

# PREAMBLE

## How To Use This Book

In compiling this book, we had hoped to include trucks, buses, motorcyles, etc., but the fine material available was clearly too much. Therefore, we have concentrated on passenger cars. In each section, papers are in historical order (the first two digits indicate the year of publication).

The preface gives a general history of the development of Passenger Car ABS.

The Introduction provides a short description of the major system components, definition of some of the more common terminology and a simplified explanation of the principles upon which most systems are based. Much of the information for the introduction was drawn from the papers in this publication for which the editors are most appreciative.

The book is divided into the major sections shown below. It has not been easy to place papers into each category, and we apologize for any Inconsistencies.

1. The Background Section contains papers on basic systems configuration and theory.

2. The Regulation and Testing Section discusses a new low-friction test track and includes a recent effort to make sense out of the regulatory maze.

3. The Past ABS Section contains four papers describing specific ABS systems and their applications to some production vehicles.

4. The Current ABS Section deals with many ABS design aspects, collected from production experience and ABS systems which have recently become available.

5. The Advanced Technology Section contains papers on Algorithms for ABS, Traction Control and some initial ideas on lower cost ABS.

# PREFACE

Anti-Lock Brake Systems (ABS) have finally come of age, and are now in 1986 widely available on upper range passenger cars, trucks and buses.

The introduction of this outstanding safety feature has been sporadic over the past 20 years. Development work in the early 50's resulted in the Dunlop "Maxaret" adaptation to aircraft in 1952. Meanwhile, Ferguson Research took on development of the Maxaret to support their 4-wheel drive pioneering building 4 prototype 3 ltr. Ford Capri's in 1970.

In 1972, the Jensen Interceptor became the first series production car to offer Maxaret based ABS in conjunction with the Ferguson 4-wheel drive system using one prop-shaft speed sensor and the viscous-coupling, another development which has suddenly gained worldwide acceptance.

In the U.S., two car manufacturers became involved in what was seen as a promising new development: Ford and Kelsey Hayes produced a 1-channel, vacuum powered, rear-wheel ABS system for the 1969 Thunderbird. Chrysler, working with Bendix Corporation, launched a 3-channel 4-wheel vacuum actuated system on the 1971 Imperial. This was the first four wheel ABS in volume production. Both makers used then state-of-the-art, discrete component, analog electronics.

Perhaps due to insufficient public awareness, and un-readiness of the U.S. motorist to pay for a relatively intangible option, both systems were quietly withdrawn from the market in the mid 70's.

During this same period, however, the U.S. National Highway and Traffic Safety Administration (NHTSA) saw the potential for improving the braking performance of commercial vehicles. The performance criteria set down in standard FMVSS 121 in 1975 effectively mandated ABS for most heavy duty trucks and forced extensive redesign and up-sizing of medium and heavy truck brakes, including the fitting of front (!) brakes on many tractors for the first time.

With a compressed time table to meet the Government deadline, ABS was forced on a skeptical and unprepared marketplace, dominated by major fleets and independent truckers. Early reliability and EMI/EMC problems led to lost earnings and the trucking industry brought pressure to bear on the U.S. Government to rescind or drastically amend FMVSS 121. This event occurred in 1978 and the bad aftertaste set back the adoption of what are now clearly superior ABS systems in the USA.

Europe has meanwhile broadly welcomed ABS especially on buses and for the transport of hazardous substances. It is expected that these now-proven systems will find their way into U.S. fleets in the late 80's: the NHTSA is currently evaluating a variety of systems so that recommendations can be made.

In Europe, meanwhile, the major brake and electronic suppliers had continued to develop ABS, changing from Analog to Digital Electronics and from discrete components to hardwired integrated circuits and finally to microprocessor based computers (ECU's) and from vacuum to high pressure hydraulic power.

Planned introductions in 1974 were delayed by the first "Oil Crisis," so it was in 1978 that a Bosch developed hydraulic ABS was first used on Mercedes Sedans. This 3-channel, 4-wheel system was and remains today, of the "Add-On" type which is interposed in the existing vacuum or hydraulically assisted brake system.

Some of the Japanese brake suppliers and OEM's have adopted designs based on the Bosch System, and these can be found on several top line models in the 80's. Honda and Toyota have developed their own ABS systems in the interim, and Japan is coming to the forefront of new developments.

The next break through came with the 1985 MY introduction of the Teves "Integrated" ABS on the Lincoln Mark VII; it is also a 3-channel, 4-wheel system, the rear brakes being treated as one channel. "Integrated" implies that the ABS actuator and the hydraulic boosted foundation brakes are in one unit, displacing the vacuum booster and conventional master cylinder used with the "Add-On" ABS. The 1985 Ford Scorpio in Europe, became the first high volume vehicle to have ABS specified 100%.

As this book is published, ABS is now available on an ever wider range of passenger cars (including four-wheel drive cars) in Europe, USA and Japan, with the goal now being to introduce lower cost systems and to extend usage to medium and compact size cars and light trucks. With this in mind, Girling introduced their "Stop Control System" in Europe in 1985 which uses two hydro-mechanical actuators driven from the driveshafts of a front wheel drive vehicle to control the two diagonal channels to offer a non-electronic system costing approximately 50% of the more sophisticated ABS available today.

In the USA Kelsey-Hayes has developed a single channel Electronic Brake Control, specifically designed to be used on the rear axles of U.S. light trucks. This system was introduced by Ford in the 1987 Model year.

This brief background of the development of ABS will provide the reader with useful perspective and indicate what changes can be expected as ABS becomes available, probably on all vehicles by the mid 1990's.

In this preface, it was decided to adopt the term "ABS," for which we are indebted to Robert Bosch, who originally coined the acronym for "Anti-Blockier Schutz." It also conveniently covers "Adaptive Brake System" "Anti (Lock) Brake System," "Automatic Brake System," etc., and has clearly become an everyday automotive word.

For a future collection of papers, the scope will surely have to be broadened to include all aspects of "Tire to Road Adhesion," as Traction Control and Adaptive Suspension interact more and more, forming the need for total chassis management.

# TABLE OF CONTENTS

Introduction to ABS . . . . . . . . . . . . . . . . . . . . . . . . . . . . . . . . i

## BACKGROUND

### Theory

**Road Testing of Wheel Slip Control Systems in the Laboratory,** Edwin E. Stewart
and Lauren L. Bowler (690215) . . . . . . . . . . . . . . . . . . . . . . . . . . . . . 5
**Proportioning Valve to Skid Control — A Logical Progression,** Frederick E. Lueck,
William A. Gartland, and Michael J. Denholm (690456) . . . . . . . . . . . . . . . 15
**Electronic Anti-Skid System — Performance and Application,** P. Müller and
A. Czinczel (725046) . . . . . . . . . . . . . . . . . . . . . . . . . . . . . . . . . 25
**Design Considerations of Adaptive Brake Control Systems,** R. R. Guntur
(741082) . . . . . . . . . . . . . . . . . . . . . . . . . . . . . . . . . . . . . . . 35
**Wheel Lock Control Braking System,** Robert A. Grimm (741083) . . . . . . . . . . 53
**Influence of Antiskid Systems on Vehicle Directional Dynamics,** E. Bisimis
(790455) . . . . . . . . . . . . . . . . . . . . . . . . . . . . . . . . . . . . . . . 65
**An Analytical Approach to Antilock Brake System Design,** John W. Zellner
(840249) . . . . . . . . . . . . . . . . . . . . . . . . . . . . . . . . . . . . . . . 73
**Antiskid Systems and Vehicle Suspension,** C. Tanguy (865134) . . . . . . . . . . 87

### Systems

**Hydraulic Brake Actuation Systems Under Consideration of Antilock
Systems and Disc Brakes,** Otto Depenheuer and Hans Strien (730535) . . . . . . 95
**Introduction of Antilock Braking Systems for Cars,** Hans Christof Klein
and Werner Fink (741084) . . . . . . . . . . . . . . . . . . . . . . . . . . . . . 107
**Electronic Control Systems for Ground Vehicles,** Edward J. Hayes and
George W. Megginson (790457) . . . . . . . . . . . . . . . . . . . . . . . . . . 113

## REGULATION AND TESTING

**The Design of the MIRA Straightline Wet Grip Testing Facility,** C. Ashley,
H. C. Allsopp, V. E. Davis, F. Fielden, and K. S. MacKellar (C194/85) . . . . . . . . . 119
**Electronic Braking System,** Edward J. Hayes and George W. Megginson (780856) . . . 131
**Antilock Braking Regulations,** Paul Oppenheimer (860507) . . . . . . . . . . . . . . 135

## PAST ABS TECHNOLOGY

### Systems

**Design and Performance Considerations for a Passenger Car Adaptive Braking
System,** Thomas C. Schafer, Donald W. Howard, and Ralph W. Carp (680458) . . . . 151
**Evolution of Sure-Track Brake System,** R. H. Madison and
Hugh E. Riordan (690213) . . . . . . . . . . . . . . . . . . . . . . . . . . . . . 161
**The Chrysler "Sure-Brake" — The First Production Four-Wheel Anti-Skid
System,** J. W. Douglas and T. C. Schafer (710248) . . . . . . . . . . . . . . . . . 175
**Design and Development of a Hydraulic Powered Wheel Slide Protection
System,** Robin A. Cochrane (720031) . . . . . . . . . . . . . . . . . . . . . . . 185

# CURRENT ABS TECHNOLOGY

## Theory

**Excess Operation of Antilock Brake System on a Rough Road,** M. Satoh and
S. Shiraishi (C18/83) . . . . . . . . . . . . . . . . . . . . . . . . . . . . . 197
**The Potential and the Problems Involved in Integrated Anti-Lock Braking
Systems,** H. Leiber and A. Czinczel (C192/85) . . . . . . . . . . . . . . . 207
**Pressure Modulation in Separate and Integrated Antiskid Systems with
Regard to Safety,** W. D. Jonner and Heinz Leiber (840467) . . . . . . . . 215
**Brake Boosters Designed Specifically for Anti-Lock Braking Systems (ABS),**
Heinz Leiber and Armin Czinczel (845106) . . . . . . . . . . . . . . . . . 223

## Systems

**Antiskid System for Passenger Cars with a Digital Electronic Control Unit,**
Heinz Leiber and Armin Czinczel (790458) . . . . . . . . . . . . . . . . . 233
**Four Years of Experience with 4-Wheel Antiskid Brake Systems (ABS),**
Heinz Leiber and Armin Czinczel (830481) . . . . . . . . . . . . . . . . . 241
**Rear Brake Lock-Up Control System of Mitsubishi Starion,** Satohiko Yoneda,
Yasuo Naitoh, and Hideo Kigoshi (830482) . . . . . . . . . . . . . . . . . 249
**The First Compact 4-Wheel Anti-Skid System with Integral Hydraulic Booster,**
H.-W. Bleckmann, J. Burgdorf, H.-E. von Grünberg, K. Timtner, and
L. Weise (830483) . . . . . . . . . . . . . . . . . . . . . . . . . . . . . . 257
**Performance of Antilock Brakes with Simplified Control Technique,**
Makoto Satoh and Shuji Shiraishi (830484) . . . . . . . . . . . . . . . . . 269
**Evaluation Criteria for Low Cost Anti-Lock Brake Systems for FWD
Passenger Cars,** W. R. Newton and F. T. Riddy (840464) . . . . . . . . . . 277
**A New Anti-Skid Brake System for Disc and Drum Brakes,** Heinrich Schürr
and Adam Dittner (840468) . . . . . . . . . . . . . . . . . . . . . . . . . 289
**Cost-Benefit Analysis of Simplified ABS,** Peter Hattwig (850053) . . . . . . . 303
**Anti-Lock Brake Systems for Passenger Cars, State of the Art 1985,**
Hans-Christof Klein (865139) . . . . . . . . . . . . . . . . . . . . . . . . 311
**A New Anti-Lock Braking System for Passenger Cars and Light Commercial
Vehicles, with Integrated Hydraulic Brake Booster,** Heinz Leffler,
Erwin Petersen, and Brian Shilton (865140) . . . . . . . . . . . . . . . . 325

# ADVANCED ABS TECHNOLOGY

## Theory

**Digital Algorithm Design for Wheel Lock Control System,** Syed F. Hussain
(860509) . . . . . . . . . . . . . . . . . . . . . . . . . . . . . . . . . . . 337
**4-Sensor 2-Channel Anti-Lock System for FWD Cars,** Yasuo Kita,
Masato Yoshino, and Hideaki Higashimura (860511) . . . . . . . . . . . . 349

## Systems

**Upgrade Levels of the Bosch ABS,** Wolf-Dieter Jonner and Armin Czinczel
(860508) . . . . . . . . . . . . . . . . . . . . . . . . . . . . . . . . . . . 361
**Traction Control System with Teves ABS Mark II,** Hans-W. Bleckmann,
Helmut Fennel, Johannes Gräber, and Wolfram W. Seibert (860506) . . . . 371

**References** . . . . . . . . . . . . . . . . . . . . . . . . . . . . . . . . . . 381

**Index** . . . . . . . . . . . . . . . . . . . . . . . . . . . . . . . . . . . . . 389

# INTRODUCTION TO ABS

## The Basics of Anti-Lock

An anti-lock brake system is a feedback control system that modulates brake pressure in response to measured deceleration of the wheel so as to prevent the controlled wheel or wheels from becoming fully locked, above a pre-set minimum speed.

The majority of systems use electronic controls and this type of system will be used for purposes of illustration. (There are systems using entirely hydro-mechanical controls but the basic terminology is similar.)

## Major Components

Typical anti-lock systems consist of the following major components:

Wheel Speed Sensors: Electro-magnetic pulse pickups with toothed wheels mounted directly to the rotating components of the drivetrain or wheel hubs. These devices provide a digital signal whose frequency is proportional to wheel rotational velocity.

Electronic Control Unit (ECU): A solid state electronic device containing computer functions, sensor signal processing circuits, output device drivers for the various ABS valves and components, and failure detection logic.

Brake Pressure Modulator: An electro-hydraulic or electro-pneumatic device for reducing, holding, and restoring pressure to one or more brakes, independent of the brake pedal effort applied by the driver. Depending on design, this device may include a pump/motor assembly, accumulator and reservoir.

Wiring, relays, hydraulic tubing and connectors complete the installation by linking these devices to each other, and to the brake and electrical systems of the vehicle.

## Terminology

In discussing anti-lock systems the number of control channels, type of hydraulic circuits, and the use of "select-high" or "select-low" logic is often mentioned.

Control channel:

This refers to the portion of the brake system which the computer/modulator controls independently of the rest of the brake system. As an example, a two channel system might control the front two brakes together and the rear two brakes together.

Hydraulic circuit:

This refers to the piping arrangement of the basic brake system of a vehicle. As an example, front wheel drive passenger cars typically have two diagonal hydraulic circuits with the left front and right rear brake on one and the right front and left rear on the other circuit.

There is an important distinction between these two concepts: the hydraulic circuit configuration is only significant in the event of a loss of system pressure. The number of control channels, on the other hand, affects the basic capabilities of the anti-lock system.

For example, the typical rear wheel drive vehicle fitted with ABS has two hydraulic circuits, (front axle and rear axle) but three control channels (each front wheel and the two rear wheels).

Select-high, Select-low:

This refers to the method used by the control algorithm in the ECU to decide which wheel speed signal to use for brake pressure modulation in systems where a control channel involves more than one wheel fitted with a sensor.

Select-low means that the ECU will use the information from the slower of the two wheels and select-high means the faster of the two.

In a typical system, the two rear wheels of a vehicle might each have a sensor, but both are on the same control channel, and hydraulic circuit. A select-low principle could be employed to guarantee that both rear wheels would continue to turn, even on surfaces with different coefficients of friction under each wheel. This would ensure maximum vehicle stability.

## Basic Principles

The ability of an anti-lock system to maintain vehicle stability, steerability, and still produce stopping distances shorter than those from a locked wheel stop, on most surfaces, is a result of the shape of the mu-slip curve for the tire-to-road interface. The effective coefficient of friction between the tire and the road peaks at about 10-20% slip and falls off to 70-80% of this value at 100% slip (locked wheel). An anti-lock system attempts to keep the wheel at the peak so that the tire can still generate lateral and steering forces as well as shorter stopping distances. The performance of an anti-lock system, of a given control configuration, is directly related to how well it can hold wheel slip in this ideal 10-20% range.

The control algorithms which are used to determine when to reduce and increase the pressure to a given brake vary widely. Almost all of the control logics, however, are based on the following general approach:

The speed signal from the wheel is differentiated to produce a signal proportional to angular deceleration. If the deceleration exceeds a pre-determined maximum (such as 1.5 G's) it is assumed that the wheel is going into lock-up because the vehicle could not possibly decelerate at that level. The pressure in the appropriate control channel is reduced until the wheel stops decelerating and starts to accelerate, at which point the pressure is increased again. This cycle is repeated several times a second until the whole vehicle has slowed to approximately 3-5 KPH, at which time the system is shut off.

This brief explanation cannot begin to address the many subtle considerations which must be included in the ECU logic to produce a properly performing system, but study of some of the papers in this book will provide the reader with considerable insight into the finer points of ABS design.

# BACKGROUND

*Theory*

# Road Testing of Wheel Slip Control Systems in the Laboratory**

**Edwin E. Stewart and Lauren L. Bowler**
Engineering Staff, General Motors Corp.

**Paper 690215 presented at the International
Automotive Engineering Congress and Exposition, Detroit, Michigan, January, 1969.

THE ULTIMATE TEST of a wheel slip control system is the application of the system to the vehicle for which it is intended. However, it is difficult to develop such a system relying solely upon vehicle testing. Harned and Johnston (1)* have reported in their work that the variability in wheel slip control system performance on wet road surfaces can be as large as 15%. Such variations make system tuning virtually impossible. The effect of hour-to-hour or day-to-day changes in the vehicle and tire-road friction characteristics may completely overshadow the effect of any intentional changes made for tuning evaluation. The consistency of simulator results makes realistic tuning of wheel slip control systems possible.

When designing the laboratory simulator, as much hardware as possible was used in an attempt to minimize the car to simulator differences. The resulting design is a simulator that employs hardware for the brake system and wheel slip control system, while substituting an analog computer to replace the braked vehicle. The computer continuously solves the vehicle equations of motion.

Several possibilities exist for combining an analog computer with hardware, both open and closed loop. However, to make the operation of the simulator as nearly like vehicle operation as possible, it was necessary to operate in a closed loop fashion with information flowing continuously from the hardware to the computer and from the computer to the hardware. The link from the hardware to the computer is a d-c signal level proportional to brake pressure at the wheel cylinders, while the link from the computer to the hardware is a d-c signal level or frequency proportional to wheel velocity or acceleration.

## SIMULATOR COMPONENTS

The simulator consists of an analog computer, some standard vehicle hydraulic brake components, a wheel slip control system, and a means for controlling the simulator and for recording various simulator results. A detailed discussion of the analog computer follows in the next section. The remaining components are discussed in the ensuing paragraphs. A schematic diagram of the simulator is given in Fig. 1.

The simulator hydraulic brake system consists of a vacuum-boosted power brake master cylinder connected to a pair of rear brake shoe and drum assemblies. Since the brake drums do not rotate during simulator operation, the brake linings are greased to ensure that the shoes seat with the drums, thus resembling the action of rolling wheels.

The wheel slip control system consists of a wheel speed

---

*Numbers in parentheses designate References at end of paper.

————ABSTRACT

The use of a laboratory simulator to evaluate the performance of wheel slip control systems under controlled operating conditions is reported. It is shown how an analog computer can be interconnected with a hydraulic brake system and wheel slip control hardware to form a hybrid simulation of a vehicle installation. An analog computer can also be used to simulate vehicle dynamics and tire-to-road friction characteristics. Simulator accuracy is established by correlating laboratory results with road data. Advantages and disadvantages of using the simulator in lieu of experimental road testing are pointed out.

5

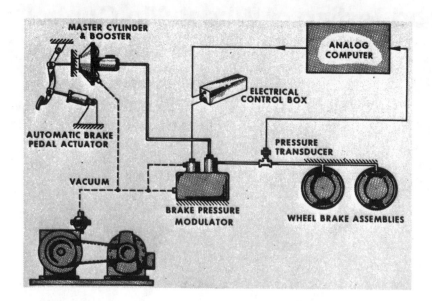

Fig. 1 - Simulator schematic

sensor, electronic control unit, and brake pressure modulator. Any type of electronic control unit that uses a wheel velocity or acceleration input and an output suitable for driving solenoid valves may be used. The brake pressure modulator is inserted in the hydraulic line between the master cylinder and brakes. It is a vacuum-powered unit of conventional design; that is, it isolates the brakes from the master cylinder during wheel slip control operation and reduces brake pressure by increasing the trapped volume from modulator to wheel cylinders. A vacuum pump and accumulator serve as the vacuum source for the brake pressure modulator and power brake booster. Vacuum levels of 5-25 in. Hg vacuum can be maintained. In addition, a pressure regulator can be used with the vacuum source to lower the apparent atmospheric pressure acting on the brake pressure modulator and power booster, thus allowing simulation of operation at altitudes to 15,000 ft above sea level.

Control of the simulator is similar to the operation of a wheel slip control system in a vehicle. The simulator is "armed" by turning on the electronic control unit and placing the computer in "OPERATE." The vehicle (simulated on the computer) will run at a constant velocity until the brakes are applied. Automatic brake application is accomplished by activating an air cylinder which applies the brake pressure (through the brake pedal mechanism) at a controlled rate to a preset maximum value. When the stop is completed, the brake pressure is removed and the computer is reset and ready for the next stop.

Real time recordings of simulator operation are made during each stop. Stopping distance and mean brake pressure may be obtained from the computer. A direct writing oscillograph is used to record the following:
1. Master cylinder pressure.
2. Modulated brake pressure.
3. Vehicle velocity (from computer).
4. Wheel velocity (from control unit).
5. Wheel acceleration (from control unit).
6. Modulator electrical control signals.
7. Miscellaneous electronic control box signals.

The pressures in the brake system are sensed by strain gage pressure transducers. Strain gage amplifiers are used to amplify these signals for recording purposes and also to send a voltage proportional to wheel cylinder pressure to the analog computer for closed loop operation. The vehicle velocity, wheel velocity, and wheel acceleration appear as voltages to the wheel slip control unit. Stopping distance and mean brake pressure are read directly from the computer digital voltmeter upon the completion of a stop. Electrical signals in the wheel slip controller are directly monitored and recorded.

## ANALOG COMPUTER AS VEHICLE MODEL

As indicated, the simulator uses actual hardware for all system components except the vehicle. For that an analog computer is used on-line to close the loop by continuously computing instantaneous vehicle and wheel motions. Fig. 2 is a block diagram of the simulator. Wheel brake pressure is the prime input to the computer models, while wheel speed is the prime output fed back to the wheel slip control unit to close the loop with the hardware. Secondary outputs from the computer models include average tire torque and vehicle stopping distance, both helpful in performance evaluation.

EQUATIONS OF MOTION FOR BRAKED WHEEL AND VEHICLE - A wheel under the influence of braking has two major torques acting on it: brake torque and tire torque. Brake torque arises from the application of brake pressure through the brake mechanism and tire torque is generated by the friction of the tire-road interface as wheel slip occurs. Tire rolling resistance and friction associated with the wheel are neglected.

Brake torque, $T_B$, is assumed to be proportional to brake pressure, $P_B$, with brake gain, $K_B$, the constant of proportionality,

$$T_B = P_B \cdot K_B \qquad (1)$$

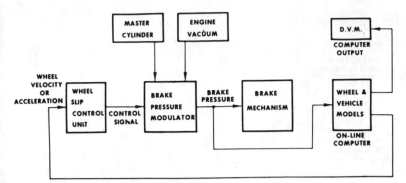

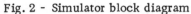

Fig. 2 - Simulator block diagram

Brake friction coefficient is a nonlinear function of wheel slip and is dependent on many variables including road surface, vehicle speed, road contaminants, tire construction, inflation pressures, and normal load. Because of these many variables, it is beyond the scope of this paper to study it. A detailed discussion is given by Harned, Johnston, and Scharpf (2). Instead, three representative curves were selected and used throughout. They are composite curves averaged for dry concrete, wet asphalt, and glare ice surfaces and are shown in Fig. 3. To obtain brake friction coefficient, instantaneous slip, $\sigma$, must be computed. Slip is defined as:

$$\sigma = [V_V - r \cdot \omega]/V_V \qquad (2)$$

where:

$V_V$ = Vehicle velocity

$\omega$ = Wheel angular velocity

$r$ = Wheel rolling radius

Once slip is calculated, the functional relation between slip and brake friction coefficient is set into a diode function generator (Fig. 3). The diode function generator then gives brake friction coefficient as an output with slip as the input.

Brake friction coefficient, $\mu$, is then related to tire torque, $T_T$, by the equation,

$$T_T = \mu \cdot N \cdot r \qquad (3)$$

where:

N = Normal load on tire

For the free body consisting of the brake, wheel, and tire the equation of motion is

$$I_W \cdot \dot{\omega} + T_B - T_T = 0, \qquad (4)$$

where:

$I_W$ = Wheel polar moment of inertia, lb-ft-sec$^2$

$\dot{\omega}$ = Wheel angular acceleration, rad/sec$^2$

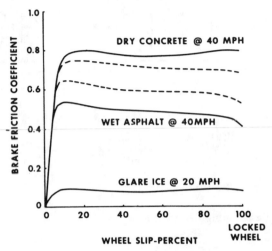

Fig. 3 - Tire brake force characteristics

Rearranging Eq. 4 gives,

$$\dot{\omega} = [1/I_W] \cdot [T_T - T_B] \qquad (5)$$

When the difference between tire torque and brake torque is positive, the wheel accelerates; and when negative, the wheel decelerates.

Once brake friction coefficient is established, vehicle acceleration, $a_V$, is obtained as follows:

$$a_V = -\mu \cdot g \qquad (6)$$

where:

g = Acceleration of gravity, 32.2 ft/sec$^2$

Vehicle velocity is then obtained by integrating vehicle acceleration,

$$V_V = \int_{t_o}^{t} a_V \, dt + V_o \qquad (7)$$

7

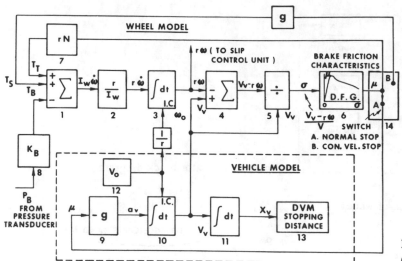

Fig. 4 - Analog computer block diagrams (wheel and vehicle models)

where:

$V_o$ = Initial vehicle velocity

$t_o$ = Time brakes are applied

$t$ = Time stop is complete

Vehicle stopping distance follows:

$$X_v = \int_{t_o}^{t} V_v \, dt + X_o \qquad (8)$$

where:

$$X_o = 0 \qquad (9)$$

COMPUTER IMPLEMENTATION OF QUARTER-CAR MODEL - Implementation of the formulas developed above results in the following computer block diagram simulating a single wheel, quarter-vehicle model. The following description refers to Fig. 4 which details how Eqs. 1-8 can be continuously solved on the computer.

Summing amplifier (1)* computes the difference between tire torque, $T_T$, and brake torque, $T_B$. This is multiplied by wheel radius over wheel inertia, $r/I_W$, (2) to obtain wheel tread acceleration, $r \cdot \dot{\omega}$, which is inturn integrated (3) to obtain wheel tread velocity, $r \cdot \omega$. Summing amp (4) and dividing circuit (5) team together to calculate the dimensionless wheel slip parameter, $\sigma$, which drives the diode function generator (6) programmed with a particular brake friction coefficient, $\mu$. Brake friction coefficient is fed back and multiplied by wheel radius times normal force, $r \cdot N$, in (7) to obtain tire torque for the wheel torque summation

_____
*Numbers in parentheses in these paragraphs refer to reference numbers in Fig. 4.

(1). Brake pressure, $P_B$, is multiplied by, $K_B$, (8) brake gain to obtain brake torque and complete the wheel torque summation.

Brake force coefficient is also multiplied by -g, the acceleration of gravity (9), to obtain vehicle acceleration, $a_v$. This is integrated (10), to obtain $V_v$, vehicle velocity, which is used in the slip calculation. Vehicle velocity is then integrated (11) to obtain stopping distance, $X_v$, for readout on a digital voltmeter (13).

To complete the simulation initial vehicle velocity, $V_o$, (12) is sent to both the vehicle (10) and wheel (3) integrators.

COMPUTER IMPLEMENTATION OF HALF-CAR MODEL - To correlate simulator performance and actual vehicle operation better, the quarter-car vehicle model can be expanded to include weight transfer and the front wheel to obtain a half-car model. The expanded model simulates an uncontrolled front wheel and a controlled rear wheel similar to the quarter vehicle model. By including weight transfer, the effect of the rear wheels on braking is decreased, reducing the improvement obtainable from wheel slip control only on the rear wheels. Thus, the simulator results will correlate better with test vehicles equipped with rear slip control units.

The half-car model is a simplified model, neglecting body pitch motions. Front and rear normal tire loads, $N_F$ and $N_R$, will be based on force and moment equations for the vehicle with weight, W, and mass, M, shown in Fig. 5.

To solve for the normal tire load at the rear, the summation of moments about the front tire patch (axis O) gives,

$$N_R \cdot (a+b) + M/2 \cdot a_v \cdot h - W \cdot a/2 = 0 \qquad (10)$$

and

$$N_R = (W \cdot a/2 - M/2 \cdot a_v \cdot h) \cdot \frac{1}{a+b} \qquad (11)$$

8

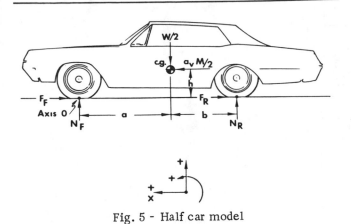

Fig. 5 - Half car model

where:

a, b, and h = Distance shown in Fig. 5

Thus, the normal force on the rear tire decreases as the car decelerates. Summing forces in the horizontal direction results in,

$$- F_F - F_R + M/2 \cdot a_v = 0 \qquad (12)$$

or

$$M/2 \cdot a_v = F_F + F_R \qquad (13)$$

$F_F$ and $F_R$, are the reaction forces that act to produce the tire torques at the front and rear wheels, respectively. Thus,

$$F_F = \mu_F N_F \qquad (14)$$

and

$$F_R = \mu_R N_R \qquad (15)$$

Substituting Eq. 13 in Eq. 11 yields,

$$N_R = \left[ \frac{W \cdot a}{2} - (F_F + F_R) \cdot h \right] \cdot \frac{1}{a+b} \qquad (16)$$

Summing the forces in the vertical direction results in,

$$- W/2 + N_F + N_R = 0 \qquad (17)$$

or

$$N_F = W/2 - N_R \qquad (18)$$

The equations of motion that apply to the braked front wheel are the same as those for the rear (developed previously). Thus, a wheel model similar to the one for the rear wheel shown in Fig. 4 could be used for the front. However, since the front wheel is not controlled and will proceed to lockup whenever brake torque is sufficiently large, a simplified model can be used. The front reaction force, $F_F$, and normal force, $N_F$, are the only front wheel parameters necessary to solve Eqs. 15, 16, and 18.

$F_F$ can be determined in the following manner. Let, $\mu_{FL}$, be the value of front wheel brake friction coefficient when the wheel is locked. Then, the reaction force when the wheel is locked is,

$$F_{FL} = \mu_{FL} \cdot N_F \qquad (19)$$

Prior to wheel lockup, the front reaction force will be less than, $F_{FL}$, and can be approximated by a force proportional to brake torque. This is due to the fact that the term, $I_W \cdot \overset{\circ}{\omega}$, in Eq. 4 is small when the slope of the brake friction coefficient versus slip curve has a positive slope. Thus, the front reaction force becomes,

$$F_F = T_B/r \quad \text{if:} \quad T_B/r < F_{FL} \qquad (20)$$

and

$$F_F = \mu_{FL} \cdot N_F \quad \text{if:} \quad T_B/r > F_{FL} \qquad (21)$$

Fig. 6 shows the functional diagram for the half-car model. It shows the quarter-car model (1)* that appeared in detail in Fig. 4 as a single element, detailing only those inputs and outputs that are required for the half-car model operation. The diagram can be used to explain how Eqs. 1, 13-16, 18, 20, and 21 can be continuously solved.

A logic switch (2) is used to determine the proper value of front reaction force, $F_F$, which is added to the rear reaction force, $F_R$, in summing amplifier (3). Rear reaction force, $F_R$, is found by multiplying $N_R$, the rear normal force, by the rear brake friction coefficient obtained from the quarter car model in (4). The sum of the reaction forces, $F_F + F_R$, is multiplied by the moment arm, h, and subtracted from the quantity half-car weight, W/2, times moment arm, a, in summer (5). The rear normal force is then found by multiplying the output of (5) by the reciprocal of the wheel base, 1/(a+b), (6). This rear normal force is fed back to (4) for the computation of the rear reaction force

---

*Numbers in parentheses in these paragraphs refer to reference numbers in Fig. 6.

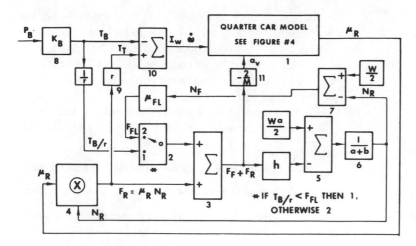

Fig. 6 - Half car model

and is subtracted from the half-car weight in summer (7) to obtain the front normal force, $N_F$. The front normal force is multiplied by the locked wheel brake friction coefficient, $\mu_{FL}$, to give the value of the locked wheel front reaction force, $F_{FL}$, for the logic switch (2). The value of brake torque over wheel radius, $T_B/r$, also sent to logic switch (2) is obtained by multiplying brake pressure, $P_B$, by brake gain, $K_B$, in (8), then multiplying brake torque by the reciprocal of the wheel radius.

The value of rear reaction force at (4) is multiplied by the wheel radius in (9) to yield tire torque, $T_T$. This quantity is then summed with brake torque for wheel torque summation (10) to give the value of wheel inertia times wheel acceleration, $I_W \cdot \dot{\omega}$, for use in the quarter-car model.

Also, the sum of the reaction forces at (3) is multiplied by the reciprocal of the half-car mass, $2/M$, in (11) to give the vehicle acceleration, $a_v$, for use in the quarter-car model.

## CLOSED LOOP OPERATION

A valuable asset of the simulator arises from the ease of setting and varying the wheel and vehicle parameters of the computer models. Thus, the simulator readily lends itself to the development of the wheel slip control system for many vehicles. After the parameters are selected for a particular vehicle, the values are entered into the computer by setting potentiometers and scaling amplifiers. Once this is done and the computer model patching is complete, the hybrid simulator with its accompanying interconnected hardware, as in Fig. 2, is ready to operate.

OPERATION IN STOPPING MODE - The operating procedure is as follows:

1. Turn on the vacuum pump for sufficient time to reach the predetermined vacuum level.

2. Turn on the electronic wheel slip control unit.

3. Switch the computer from "INITIAL CONDITION" to "OPERATE" mode. (Note: The computer will act as if it is driving at constant velocity.)

4. Finally, energize the automatic brake pedal actuator to apply full brake pressure. This simulates a repeatable driver brake application.

The simulator enacts a "panic" stop from a predetermined speed, the initial conditions set into the computer. When the computer's digital voltmeter is set to monitor the stopping distance integrator, a direct read-out of vehicle stopping distance is presented. This figure of merit for each stopping run can be used as a comparison between runs if a careful attempt is made to hold most parameters constant, save the several being studied. In this way, control unit parameters or even entire control units, modulator parameters, and vehicle parameters can be selectively optimized.

OPERATION IN CONSTANT VELOCITY MODE - In addition to stopping distance, a second and useful figure of merit for system performance is to monitor average tire torque while the simulation runs in the constant velocity mode. This is akin to pulling a trailer and braking one trailer wheel while, at the same time, applying the throttle to keep the trailer at constant speed.

If tire torque or brake friction coefficient is averaged with a low pass filter, and the simulator is set running in this constant velocity mode, then to optimize selected parameters one adjusts them until average tire torque is maximized.

To accomplish constant velocity operation, negative vehicle deceleration is added continuously to the wheel acceleration integrator, offsetting the wheel's tendency to slow down. On Fig. 4 note switch (14). When this is in position "A," the computer will operate in the normal stopping mode. In position "B," it operates in the constant velocity mode with the vehicle model "never" stopping.

## CORRELATION WITH ROAD TEST DATA

A wheel slip electronic control unit was selected for use in the correlation of the simulator results with road test results. This control unit was placed in a test vehicle and

### SIMULATOR RECORD
#### STOPPING DISTANCE 102'    LOCKED WHEEL STOP 101'

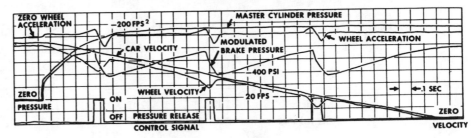

### TEST CAR RECORD
#### STOPPING DISTANCE 102'    LOCKED WHEEL STOP 101'

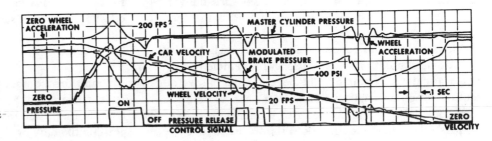

Fig. 7 - Oscillographic records showing simulator and test car results on dry concrete at 45 mph

### SIMULATOR RECORD
#### STOPPING DISTANCE =126'    LOCKED WHEEL STOP =126'

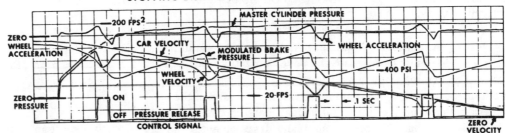

### TEST CAR RECORD
#### STOPPING DISTANCE =125'    LOCKED WHEEL STOP =126'

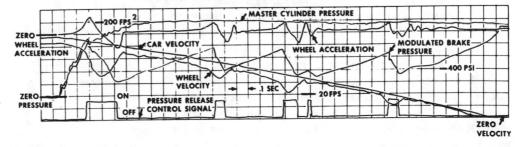

Fig. 8 - Oscillographic records showing simulator and test car results on wet asphalt at 45 mph (light water cover)

was run on dry concrete and on wet asphalt. Oscillographic records were made and average controlled and locked wheel stopping distances were obtained.

The control unit was then returned to the laboratory and installed in the simulator. Representative parameter values from the test vehicle were set on the analog computer. The dry concrete and wet asphalt road surfaces on the computer were modified, as shown by the dotted lines in Fig. 3, to bring the locked wheel stopping distances into agreement with those from the test vehicle. The half-car model was

used with front locked wheel brake friction coefficient set equal to rear. Stopping distances were recorded and oscillographic recordings were made. Figs. 7 and 8 show the oscillographic records from the vehicle and simulator for dry concrete and wet asphalt, respectively. The similarity in cycle rates, pressure levels, wheel speed droops, and wheel acceleration levels can be seen. Stopping distance comparisons for this correlation study are given in Table 1.

Each stopping distance for the vehicle installation in Table 1 is the average of five stops. The simulator stop-

| | Table 1 - Comparison of Stopping Distance | | | |
|---|---|---|---|---|
| | Dry Concrete (45 mph) | | Wet Asphalt (45 mph) | |
| | Controlled Stop, ft | Locked Wheel Stop, ft | Controlled Stop, ft | Locked Wheel Stop, ft |
| Vehicle | 102 | 101 | 125 | 126 |
| Simulator | 102 | 101 | 126 | 126 |

ping distance was repeatable to within 1% while the variation in vehicle stops on wet asphalt was ± 5% of the average value.

The simulator has been used for several studies including an altitude simulation. Correlation with vehicle test results was quite acceptable. The development of several new control system concepts and modulator changes have also been done on the simulator. Little, if any, further work was required when these modifications were applied to a vehicle installation, further indicating the validity of the simulator.

## ADVANTAGES AND DISADVANTAGES OF SIMULATOR TESTING VERSUS ROAD TESTING

Putting the simulator in proper perspective is important. It is a device built to decrease the development time of wheel slip control systems by eliminating long hours of test vehicle stopping distance data acquisition and analysis, the object being to obtain statistically significant improvements in stopping distance for the system under development. The simulator can do this because it does not have the stopping distance variability associated with it that is inherent in the car's operation. Many road surfaces may also be checked on the simulator, which would be difficult or even impossible to do with the car. For example, checking operation on hard ice during the summer months would be difficult.

The simulator is probably most effectively used in parallel with a development vehicle, so that once a desired simulator result is achieved it can be verified by setting up the same configuration in the car and comparing results. This may also help to iterate the computer simulation to a closer correlation with the real world.

In summary, the simulator makes control system development and parameter tuning fast and efficient when compared to in-car development time and cost. This results from the consistency of the simulator data, allowing either a direct comparison of runs using stopping distance, or the maximizing of tire torque while operating in the constant velocity mode.

The simulator does not eliminate the vehicle testing of the final system, which must always be done in the car. No attempt has been made to model every interaction that may occur between the vehicle and the road. Thus, care-

ful in-car evaluation will always have to be carried out concurrent with, or subsequent to, simulator development.

## CONCLUSION

Results of the work to date on the wheel slip control simulator indicate reliable correlation with instrumented vehicle testing. Due to the simulator's consistency, it makes possible the easy evaluation of small changes in systems parameters, a task difficult to perform with a road test program. Thus, the simulator has grown into a valuable tool for the design and development of wheel slip control systems.

## NOMENCLATURE

$a$ = Distance from center line of front wheel to c. of g. location, ft

$a_v$ = Vehicle acceleration, ft/sec$^2$

$b$ = Distance from center line of rear wheel to c. of g. location, ft

$F_F$ = Front reaction force, lb

$F_{FL}$ = Locked wheel value of front reaction force, lb

$F_R$ = Rear reaction force, lb

$g$ = Acceleration of gravity, ft/sec$^2$

$h$ = Distance from ground plane to cg location (c. of g. height), ft

$I_W$ = Wheel polar moment of inertia, lb-ft-sec$^2$

$K_B$ = Brake gain, lb-ft/psi

$M$ = Vehicle mass, lb-sec$^2$/ft

$N$ = Normal force, lb

$N_F$ = Front normal force, lb

$N_R$ = Rear normal force, lb

$P_B$ = Brake pressure, psi

$r$ = Wheel radius, ft

$t$ = Time variable, sec

$T_B$ = Brake torque, lb-ft

$T_S$ = Wheel synchronous torque, lb-ft

$T_T$ = Tire torque, lb-ft

$V_o$ = Vehicle initial velocity, ft/sec

$V_v$ = Vehicle velocity, ft/sec

$W$ = Vehicle weight, lb

$X_v$ = Vehicle stopping distance, ft

$\mu$ = Brake friction coefficient

$\mu_F$ = Front brake friction coefficient

$\mu_{FL}$ = Locked wheel value of front brake friction coefficient

$\mu_R$ = Rear brake friction coefficient

$\sigma$ = Wheel slip

$\omega$ = Wheel angular velocity, rad/sec

$\mathring{\omega}$ = Wheel angular acceleration, rad/sec$^2$

REFERENCES

1. J. L. Harned and L. E. Johnston, "Anti-Lock Brakes." Paper presented at General Motors Automotive Safety Seminar, July 1968.

2. J. L. Harned, L. E. Johnston, and G. Scharpf, "Measurement of Tire Brake Force Characteristics as Related to Wheel Slip (Anti-Lock) Control System Design." Paper 690214 presented at SAE International Automotive Engineering Congress, Detroit, January 1969.

# Proportioning Valve to Skid Control— A Logical Progression*

## Frederick E. Lueck, William A. Gartland, and Michael J. Denholm
Borg and Beck Div., Borg-Warner Corp.

*Paper 690456 presented at the Mid-Year
Meeting, Chicago, Illinois, May, 1969.

UP TO THIS TIME, most braking systems have been designed with a fixed front-to-rear brake ratio. This provides a fundamentally simple system with a given level of brake performance at some arbitrary loading condition.

With continuing emphasis on improved brake performance and improved driver control, this approach to brake system design is becoming outmoded. For the purposes of this discussion, brake controls will be defined as any device which intelligently modifies the driver-controlled brake system input.

It is the objective of any brake control system to maximize the stability and controllability of the vehicle while minimizing stopping distances. Brake control systems attempt to attain these objectives by making the most of the available force at the tire-to-road interface.

While skid control, hold-off valves, fixed-ratio proportioning valves, and load-sensitive proportioning valve systems attain these ends to different degrees and by different methods, the objectives of stability, controllability, and improved stopping distances are the same. For this discussion, we will confine our comments to the areas of skid control and load-sensitive proportioning valves.

LOAD-SENSITIVE PROPORTIONING VALVE

As shown in Fig. 1, as a vehicle increases its deceleration, the ratio of front-axle-to-rear-axle loading changes as a function of the vehicle deceleration, wheel base, and height of the center of gravity of the vehicle.

If we superimpose a fixed front-to-rear braking ratio over this vehicle system, it is obvious that the brake balance can only be correct for one specific vehicle deceleration. At all other coefficients of friction, either premature front-wheel or rear-wheel slide will occur.

To provide the maximum stability and minimum stopping distance, we would prefer that all four wheels lock simultaneously. When we add the variable of changing vehicle load, as shown in Fig. 2, the problem becomes even more severe in that the imbalance between the fixed brake ratio and the impressed axle load becomes even greater.

ABSTRACT

This paper discusses the development of a family of brake control devices capable of handling all vehicles from passenger cars through air-braked heavy trucks. These devices consist of: Hydraulic load-sensitive proportioning valves for passenger cars, light trucks, and vans; hydraulic load-sensitive proportioning valves for medium trucks; pneumatic load-sensitive proportioning valves for air-braked heavy trucks and tractor/trailer combinations; skid control systems for passenger cars; and skid control systems for trucks and tractor/trailer vehicles.

The paper explains how this broad approach to brake controls allows the selection of a system which can be tailored to the particular brake control needs of a specific vehicle and its duty cycle. Except for the passenger car skid control, these devices are capable of being retro-fitted on existing vehicles.

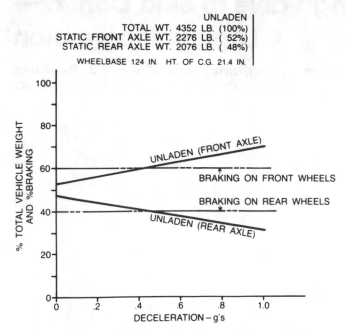

Fig. 1 - Weight transfer of a typical sedan - unladen

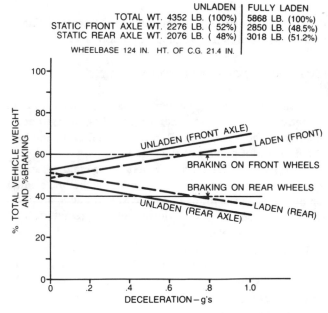

Fig. 2 - Weight transfer of a typical sedan - laden and unladen

It is the aim, then, of the load-sensitive proportioning valve to match the applied brake torque to the wheel torque generated at the tire patch.

WHAT IT THEORETICALLY
ACCOMPLISHES

If we look at this same data in a different fashion, we can observe both the effect of vehicle loading and the effect of changing coefficient of friction between the tire and road. Fig. 3 shows a plotted curve of the braking efficiency versus

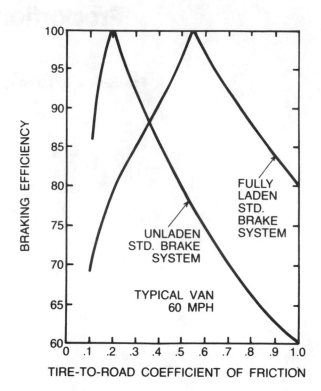

Fig. 3 - Braking efficiency versus coefficient of friction - laden and unladen

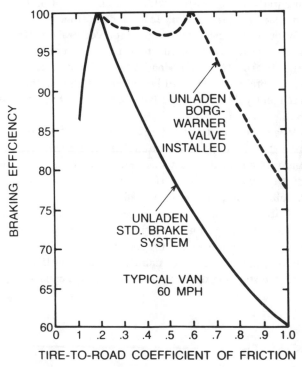

Fig. 4 - Braking efficiency versus coefficient of friction - unladen with and without load-sensitive proportioning valve

the coefficient of friction between tire and road. Braking efficiency is defined as the ratio of actual vehicle deceleration to available tire-to-road coefficient of friction.

In other words, if a vehicle is capable of attaining 0.8 g deceleration at first wheel lock on a 0.8 μ surface, the braking efficiency is defined as 100%. It is obviously the goal of any braking control system to provide maximum braking efficiency across the widest range of load and road surface.

Fig. 3 shows the brake efficiency of a typical short wheelbase van in the unladen and laden condition. Note the shift in peak efficiency between laden and unladen conditions. Rear wheel slide, which occurs to the right of the peak, exists over a wider range of coefficient of friction in the unladen condition.

Now, over the unladen curve we will overlay the performance change that can be attained by the use of a load-sensitive proportioning valve. This is shown in Fig. 4. Note that in the unladen condition, the braking efficiency has been increased over a wide range of tire-to-road friction.

## ACTUAL PERFORMANCE GAINS

To demonstrate the actual performance improvement that is attainable with a system of this type, a series of tests were conducted on this same van.

Fig. 5 shows the deceleration capability at first wheel slide at 30 and 60 mph in the unladen and laden condition. As would be expected, the major improvements are in the unladen condition, where the brake balance is most compromised in a system without a load-sensitive proportioning valve.

Fig. 6 depicts the performance of a medium truck (24,000 lb gross) under various load conditions. Here, we have plotted stopping distance with first wheel lock. You will note again that the stopping distance is significantly reduced in the unladen and half-laden conditions. This indicates that the basic brake balance was set up for the maximum load condition.

Fig. 7 depicts the same performance characteristics with load-sensitive proportioning valves applied to the driving wheels and the trailer wheels of an air-braked tractor/trailer rig (32,000 lb gross combination weight). Again, since the normal combination rig has its brake balance optimized for the loaded condition, there is little difference with and without the valve when laden. Major performance gains occur at no load or partial load.

## HOW PROPORTIONING VALVES
## ACCOMPLISH THIS

Because of the wide variety of displacement requirements for various brake systems and different operating media, a family of valves is required to cover the entire line of vehicle applications.

Fig. 8 shows a basic installation of the hydraulic load-sensitive proportioning valve for passenger car, van, and light truck. One end of the torsion bar linkage is attached to the rear axle of the vehicle. The other end is connected to the proportioning valve, which is fastened to the chassis. The torsion bar twists in relation to the load applied on the axle. The torsional reaction, which is taken by a piston on the valve, is then the load-sensitive input.

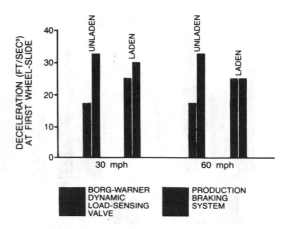

Fig. 5 - Deceleration versus speed and loading - production brake system versus load-sensitive proportioning valve system, 1969 production van

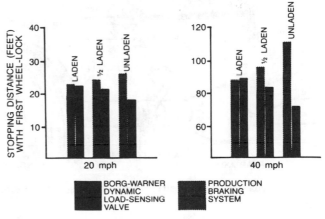

Fig. 6 - Stopping distance versus speed and loading - production brake system versus load-sensitive proportioning valve, 2 1/2-ton truck

Fig. 9 shows a schematic of the valve. The torsion bar input to the valve applies a load to the main valve piston. The torsion bar load plus the return spring load tend to keep the valve piston away from the sealing ring. When sufficient inlet pressure is generated, the inlet pressure on the large diameter of the valve is sufficient to cause the main piston of the valve to move to the left against the combined spring load.

At this point, the valve shuts off against the sealing ring; and the split point of the valve is reached. Above this point, proportioning at a fixed ratio occurs. This ratio is a function of the relative areas of the piston head and piston stem.

A typical performance curve for this valve is shown in

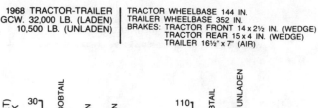

1968 TRACTOR-TRAILER
GCW. 32,000 LB. (LADEN)
10,500 LB. (UNLADEN)

TRACTOR WHEELBASE 144 IN.
TRAILER WHEELBASE 352 IN.
BRAKES: TRACTOR FRONT 14 x 2½ IN. (WEDGE)
TRACTOR REAR 15 x 4 IN. (WEDGE)
TRAILER 16½" x 7" (AIR)

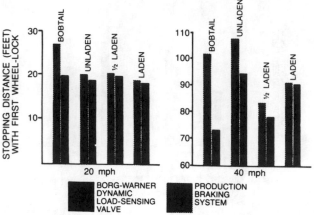

Fig. 7 - Stopping distance versus speed and loading - production brake system versus load-sensitive proportioning valve, tractor/trailer

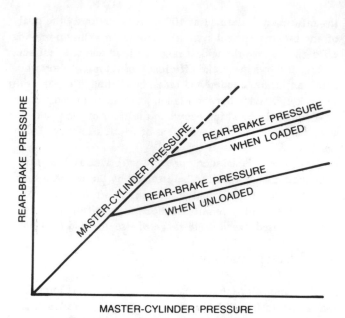

Fig. 10 - Dynamic load-sensitive proportioning valve performance

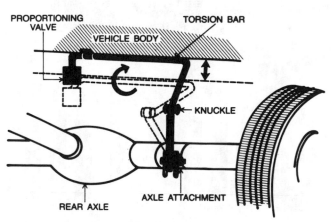

Fig. 8 - Dynamic load-sensitive proportioning valve installation

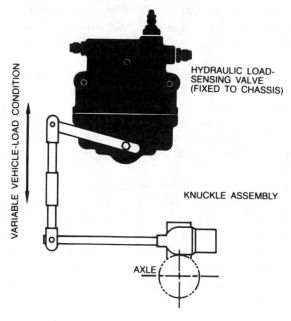

Fig. 11 - Hydraulic load-sensitive proportioning valve installation

Fig. 10. By adjusting the internal ratio of the valve and the rate of the torsion bar, the performance of the valve can be adjusted to the requirements of a particular type of vehicle.

While providing the same function, the hydraulic valve for a medium truck, as shown in Fig. 11, accomplishes this by a different means. This unit, which is fixed to the bed or chassis of the truck, is linked to the axle with a lost motion device or knuckle assembly to prevent over-travel in the valve. In this way, the vehicle axle load is sensed and transmitted to the valve.

As shown in Fig. 12, hydraulic braking pressure from the master cylinder is applied to the load-sensing valve and

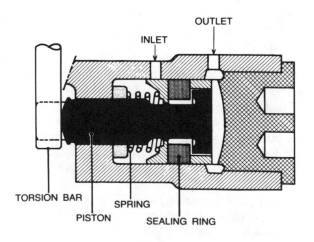

Fig. 9 - Dynamic load-sensitive proportioning valve

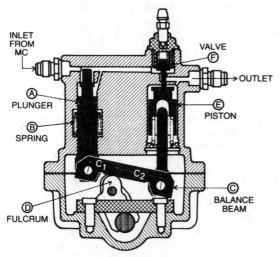

Fig. 12 - Hydraulic load-sensitive proportioning valve

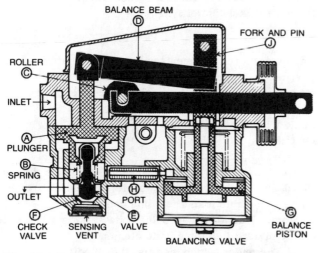

Fig. 13 - Pneumatic load-sensitive proportioning valve

acts upon plunger A, forcing it down against the load in spring B. The plunger then acts on one arm of the balance beam C, which pivots on fulcrum D. The position of fulcrum D is determined by the position of the frame relative to the axle; that is, it is a function of the axle load.

On the other arm of beam C is piston E, which moves valve F to open or close the passage between the inlet port and the outlet port. Valve F can isolate the inlet from the outlet. Balancing of this system occurs when the outlet pressure on piston E multiplied by the lever arm $C_2$ equals the inlet pressure on plunger A multiplied by the lever arm $C_1$.

Since the length of the arms is governed by the fulcrum position, which is dependent on axle load, the brake proportioning is determined by axle load. The pneumatic valve shown in Fig. 13 is similar in installation and in principle to the hydraulic valve, except that valve E acts both as an isolation valve and an exhaust valve to release the compressed air to atmosphere on brake release.

## SUMMARY

In summary, the load-sensitive proportioning valve has the ability to increase vehicle deceleration before first wheel lock to improve vehicle stability by preventing premature wheel lock. In articulated vehicles, it can minimize the possibility of jackknifing by preventing premature locking of the tractor drive wheels (1)*.

Devices of this type cannot eliminate skidding, but do significantly reduce the hazard by providing greater latitude and margin for driver error.

## SKID CONTROL

As mentioned earlier, proportioning valves -- regardless of their level of sophistication -- can only minimize the possibility of skidding by providing greater latitude for driver error.

The true skid control system seeks to prevent skidding by eliminating protracted locking of the wheels. The ideal skid control system, by preventing wheel lock, can improve vehicle stability, vehicle controllability, and stopping distances under almost all driving and road conditions.

At the present "state-of-the-art," this theoretical performance level is not attainable.

## REAR-WHEEL VERSUS FOUR-WHEEL SKID CONTROL

The ultimate skid control system must understandably be a four-wheel system if all theoretical advantages are to be obtained. However, before a choice between rear-wheel and four-wheel systems can be made, performance and economic considerations must be taken into account.

It is obvious that the application of a rear-wheel skid control system cannot provide the ultimate objective of vehicle controllability since, by its very nature, the rear-wheel system allows the front wheels to lock. In this condition, no steering forces can be generated (2). For this reason, four-wheel skid control systems are actively under development.

The rear-wheel system, however, can and does provide significant improvement in vehicle stability by assuring straight-line stops under almost all conditions. It also offers significant improvement in vehicle stopping distance. This can be attained at a price which is less than half that required for a four-wheel system of the same concept.

## PRINCIPLES OF SKID CONTROL

The basic phenomena which permits a skid control system to function is best described by the family of $\mu$-slip curves shown in Fig. 14 (3). These curves demonstrate the principle that the maximum retarding force is generated when the tire

---

*Numbers in parentheses designate References at end of paper.

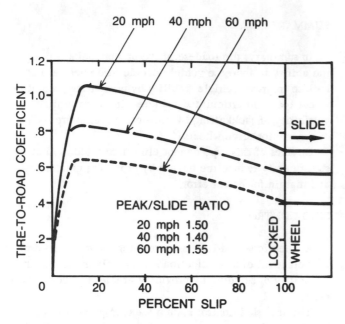

Fig. 14A - Typical μ-slip curves - wet concrete

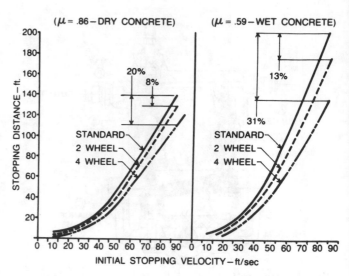

Fig. 15 - Theoretical stopping distance versus velocity for a standard brake system, rear-wheel skid control system, and four-wheel skid control system

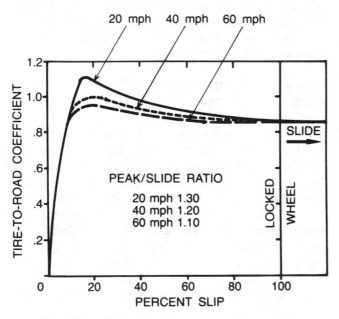

Fig. 14B - Typical μ-slip curves - dry asphalt

is rotating with a peripheral velocity less than the corresponding translational velocity of the vehicle.

The coefficient of friction at 100% slip (or locked wheel) represents the maximum deceleration that can be obtained on a locked-wheel stop. Therefore, the ratio of the peak coefficient of friction to the locked-wheel coefficient of friction represents the improvement in stopping distance that can be attained with an optimum skid control system.

Fig. 15 plots the theoretical improvement in stopping distances for two-wheel and four-wheel skid control systems on dry pavement and on a wet concrete surface. It is readily apparent from these curves that significant improvement can be obtained on lower coefficient surfaces, while less

improvement (on a percentage basis) can be attained on dry pavement. These improvements are a direct result of the ratio of peak coefficient to locked-wheel coefficient.

REAR-WHEEL PASSENGER
CAR SKID CONTROL

Since the present "state-of-the-art" has made the rear-wheel skid control system attainable, we would like to describe a workable rear-wheel skid control system.

From a functional standpoint, today's skid control systems can be divided into three basic elements:

1. The sensor, which monitors the velocity or deceleration of the wheels.

2. A logic device, which processes this information into a usable signal.

3. An actuator, which modulates the brake pressure in some fashion on command from the logic device.

A typical system schematic is shown in Fig. 16.

To move from the theoretical system to a real system, the sensor shown in Fig. 17 is mounted in each of the rear wheels and monitors the deceleration of the rear wheels. When a preselected triggering level is reached, a signal is sent to the logic module (Fig. 18), which processes the signal and provides a command function to the actuator (Fig. 19).

The actuator then reduces the pressure to the rear brakes by increasing the volume of the rear brake system. This reduction in pressure causes a reduction in brake torque and allows the wheel to accelerate. The pressure is then re-applied by the actuator, and the cycle is repeated five to eight times per second.

Typical performance of such a system is shown in Table 1. Note that as predicted, major performance improvements occur on the lower coefficient surfaces where the higher ratio of peak-to-sliding coefficients between the tire and road provide greater potential for improvement.

Table 1 - Performance Summary of the Passenger-Car
Skid Control System (1968 Sedan)

| Vehicle Speed (mph) | Apparent Coefficient of Friction | Stopping Distance (Feet) | | Percent Improvement |
| --- | --- | --- | --- | --- |
| | | Standard Brake System | Borg-Warner Skid Control | |
| 10 | .09 | 41.6 | 39.3 | 6.0 |
| 10 | .21 | 15.6 | 15.2 | 3.0 |
| 10 | .24 | 14.4 | 13.5 | 6.0 |
| 20 | .19 | 71.8 | 62.2 | 15.0 |
| 20 | .63 | 21.6 | 21.4 | -- |
| 30 | .66 | 46.8 | 44.2 | 6.0 |
| 40 | .71 | 75.2 | 68.7 | 10.0 |
| 40 | .76 | 72.1 | 68.6 | 5.0 |
| 60 | .71 | 170.2 | 164.0 | 3.5 |
| 60 | .93 | 129.3 | 123.4 | 5.0 |

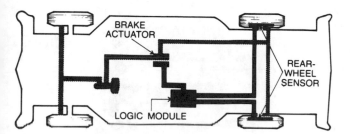

Fig. 16 - Schematic of a rear-wheel skid control system

Fig. 17 - Rear-wheel skid control system sensor

Fig. 18 - Rear-wheel skid control system logic module

SKID CONTROL FOR COMMERCIAL
VEHICLES

For trucks and tractor/trailers, skid control can provide the same advantages as for passenger cars. In addition, on articulated vehicles jackknifing can be minimized by preventing locking of the driving wheels on the tractor (4).

A schematic of a typical air-operated skid control system is shown in Fig. 20. Three components make up this system:

Fig. 19 - Rear-wheel skid control system actuator

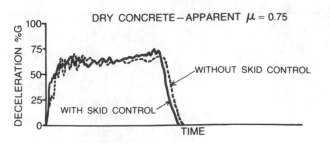

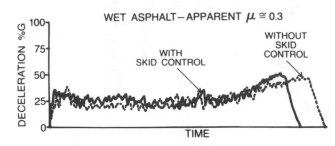

Fig. 21 - Vehicle deceleration versus time - with and without skid control - 30 mph, 32,000 lb GCW, tractor/trailer

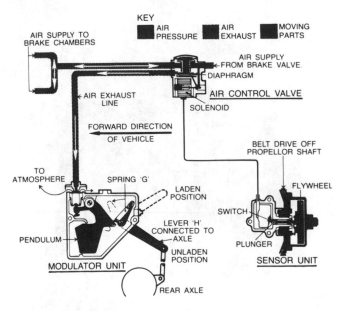

Fig. 20 - Schematic of a pneumatic skid control system

BORG-WARNER PNEUMATIC SKID-CONTROL SYSTEM

BORG-WARNER PNEUMATIC LOAD-SENSING VALVE

Fig. 22 - Typical schematics of combination skid control and proportioning valve systems

1. The sensor, which is normally installed on the driving axle.

2. The air control valve.

3. The modulator.

It should be noted that in spite of the difference in names, the three basic skid control elements are present. Whenever the controlled wheels reach incipient skid, the sensor actuates the air control valve. This allows the air pressure in the brakes to vent through the exhaust line to the modulator.

The atmospheric valve in the modulator, which is controlled by the pendulum and the load-biasing spring G, adjusts the pressure to which the brakes are released. In this fashion, the amount of pressure drop is determined both by the coefficient of friction between the tire and road (which is determined by the pendulum, which measures vehicle deceleration) and the load on the bias spring G, which measures the load on the axle.

The improvement in performance of this system is shown in Fig. 21. Again, it should be noted that the performance gains are greatest at the lower coefficients because of the higher ratio of peak-to-sliding coefficients.

## COMBINATION SYSTEMS FOR ARTICULATED VEHICLES

To provide maximum cost effectiveness, a combination of skid control on the driving axle of the tractor coupled with load-sensitive proportioning valves on the trailer axles can be utilized. A schematic of this system is shown in Fig. 22. This provides a system with excellent cost effectiveness and minimum complexity.

An advantage of this approach to the problem is that jackknifing can be minimized significantly, even on interchange trailers which do not have skid control or proportioning valves.

## SUMMARY

In summary, by approaching the problems of brake controls with a broad viewpoint, this family of brake control systems allows the controls to be selected on the basis of the specific vehicle requirements. It is apparent that no single device can provide the answer for all vehicles under all conditions while maintaining maximum cost effectiveness.

## REFERENCES

1. H. A. Wilkins, "Jackknifing." Motor Transport Magazine, March 29, 1968.
2. T. A. Byrdsong, "Investigation of the Effect of Wheel Braking on Side-Force Capability of a Pneumatic Tire." NASA TND-4602, June, 1968.
3. J. L. Harned, L. E. Johnston, and G. Scharpf, "Measurement of Tire Brake Force Characteristics as Related to Wheel Slip (Antilock) Control System Design." Paper 690214 presented at SAE International Automotive Engineering Congress, Detroit January, 1969.
4. "The Jackknife Terror." Precision Magazine, March/April, 1968, pp. 21-25.
5. W. C. Eaton and I. J. Schreur, "Brake Proportioning Valve." Paper 660400 presented at SAE Mid-Year Meeting, Detroit, June 1966.
6. G. Putnam, "Brake Balancing Under All Conditions." Fleet Owner Magazine, July, 1968, pp. 53-56.
7. H. J. H. Starks, "Loss of Directional Control in Accidents Involving Commercial Vehicles." Presented at the Symposium on Control of Vehicles during Braking and Cornering, London, England, June 11, 1963.

# Electronic Anti-Skid System—
# Performance and Application*

### P. Müller and A. Czinczel
Germany

*Paper 725046 presented at the XIV FISITA
Congress, London, England, June, 1972.

SYNOPSIS   In the course of many attempts to increase motor vehicle safety, the braking behavior is becoming of particular significance.  This study shows to what extent an antiskid control system can improve traffic safety by reducing the stopping distance and retaining vehicle stability and steerability during panic stops in a turn.  In particular, it deals with the braking behavior in a turn showing that even in case of high lateral acceleration full brakings are possible whereby vehicle stability and steerability are maintained with a stopping distance shorter than that obtained with locked wheels at the same initial speed.  The behavior of various antiskid configurations in turns is described and compared to each other with regard to vital vehicle maneuvers on the road.  The Teldix antiskid system and its operation are briefly outlined.  Its performance characteristics for brakings in turns are illustrated in the form of test results.

## INTRODUCTION

A great many drivers think it is the engine that drives the motor vehicle and it is the brake that brings it to a halt.  It is true that the engine drives the wheels and that the brakes decelerate wheel rotation, but the vehicle itself is accelerated or decelerated by the friction force between tire and road.  Every driver having encountered an icy road has noticed this physical phenomenon.

Many traffic accidents are caused by people unaware of the extreme friction differences between tire and road under certain conditions. Quite often, the required stopping distance is underrated even with good tire and road surface conditions.  Therefore, all possibilities must be employed to reduce the stopping distance.  At the same time, vehicle stability and steerability are of vital importance during a panic stop.  The loss of stability leads to swerving and that of steerability to driving off the lane or running into an obstacle on the road.  This study describes the increase in traffic safety by reducing the stopping distance and retaining vehicle stability and steerability during a panic stop.

## 1.   Braking Behavior of a Vehicle under Various Road and Vehicle Conditions

The shape of friction coefficients between tire and road has been known for some time for brakings on a straight and homogeneous road (Fig. 1).  With most of the actual tire-road friction coefficients, these curves reach a maximum within a 10 % to 20 % slip range.  The friction values decrease with an increase in slip.  At wheel lockup (100 % slip), they may decrease by 40 % compared to the friction maximum.  These phenomena are employed for antiskid system control in that the automatic brake pressure control keeps the slip value within a range of maximum friction.  It is relatively simple to design a control system that reduces consider-

ably the stopping distance during braking on a straight homogeneous road.  In practice, however, frictional behavior between tire and road becomes extremely complex.  Hereby, some of the significant parameters are:

a.  Axle load distribution, height of center of gravity, front or rear axle drive, and automatic or mechanical transmission.

b.  Inhomogeneity of the road:  On an inhomogeneous road and with sudden changes in tire-road friction (e.g. ice spots, puddles, dirt on a dry road, etc.) or with different tire-road friction coefficients on the left and right side, the braking behavior changes considerably as compared to that on a homogeneous road.

## 2.   Dynamic Behavior of Vehicle Braking in a Turn

The behavior of a braking vehicle in a turn is essentially affected by the lateral force coefficient, that is, by the tire-road friction coefficient vertical to the wheel plane.  This coefficient depends on the tire slip and the tire sideslip angle $\alpha$ ($\alpha$ is the angle between the wheel plane and the velocity of the wheel's center of gravity, see Fig. 2).  This figure shows that there is a strong reciprocal action between the lateral force coefficient $\mu_s$ and the brake force coefficient $\mu_B$. The two coefficients together with the dynamic load of the individual wheels determine the stability and brake behavior of the vehicle.

This study mainly deals with the features of controlled braking in a turn.  Based on investigations by Desoyer and Slibar (Ref. 1) and assuming certain vehicle parameters [1], the brake and

1) Weight: 1000 kgf
   Height of center of gravity: 0.6 m
   Track width: 1.25 m
   Wheel base:  2.50 m
   Center of gravity central to wheel base
   Max. friction coefficient:     = 0.9

lateral forces when braking from a steady turn as a function of lateral acceleration up to the limit speed (Fig. 3) were calculated for that purpose. Starting from the state-of-the-art, it was assumed that an antiskid system continuously retains the tire slip value within a range of optimal brake force coefficients. With zero lateral acceleration, the brake force on the inner and outer wheels are identical in a turn; however, due to the dynamic axle load distribution, it is much higher on the front wheels than on the rear wheels. An increase in lateral acceleration results in a decrease of the brake force on the inner wheels. With a lateral acceleration of 0.6 g in a turn, almost no brake force is available on the inner rear wheel. Yet, the brake force of the outer rear wheel decreases only slightly with an increase in lateral acceleration and even increases on the outer front wheel so that a high resulting brake force is available in the case of a 0.6 g lateral acceleration.

Figure 3b shows the lateral forces of the individual wheels as a function of lateral acceleration. While the lateral force of the inner rear wheel – due to the high dynamic unloading – only slightly increases with lateral acceleration, the lateral forces of the outer wheels sharply increase up to the limit speed in a turn (b =0.6 g). This limit speed is characterized in that the lateral forces are just capable to equalize the centrifugal force resulting from this speed. If the limit speed is exceeded in a steady turn, the sideslip of the front wheels increases (vehicle understeer) and the vehicle drives off the curve lane; yet, vehicle stability is retained. However, in case of wheel lockup, the lateral force of the locked wheel drops to almost zero in a minimum of time (Fig. 2). That means that a sudden additional lateral force load occurs on the other wheels which cannot generate an additional lateral force but after a certain delay. At once, the vehicle becomes unstable. Thus, with severe braking in a turn, the lockup of one single wheel leads to an immediate swerving of the vehicle.

Figure 3 illustrates very clearly that an antiskid system controlling wheel speeds in an optimal range is very well capable of attaining maximum vehicle deceleration (Fig. 3a) and, at the same time assuring sufficient lateral force (Fig. 3b).

The significance of the above shall be demonstrated by means of three typical vehicle maneuvers essential for accident prevention.

a. Evasive Maneuvers

In case of a suddenly occuring obstacle, the average driver will fully apply the brake pedal and thus lock the wheels; that is, his vehicle will no longer be steerable. A control system maintaining the friction coefficients in an optimal range between brake and lateral force allows for steering past the obstacle with full braking effect as on a straight course.

b. Curves with Decreasing Radius

Particularly in mountainous areas but also on highway exits there are curves with a decreasing radius. A driver entering such a curve at a high speed generally causes a hazardous situation in that he cannot sufficiently brake his vehicle to pass that part of the curve with the minimum radius. In such a curve, a good antiskid system maintains the lateral force at a maximum value and assures a high brake force coefficient $\mu_B$ to decelerate the vehicle as shown in Fig. 2. Because of the various road conditions, only approximate values are available for the shortest stopping distances possible in a curve with a decreasing radius. It can be stated, however, that in such curves, too, stopping distances not longer than those achieved with locked wheels on a straight road are possible.

c. Different Friction Coefficients on the Left and Right Vehicle Wheels (split coefficients)

Particularly under snow and ice and even under wet conditions, the side of the road is icy or slippery while the middle of the road has already dried up offering a higher friction coefficient. If all four wheels lock up during a panic stop under the above conditions, the vehicle will start swerving. Even with a four-wheel antiskid system there is a torque about the vertical axis, that is, the vehicle tries to break off toward the high friction coefficient side (generally onto the oncoming traffic lane). Since full steerability is maintained, the driver is able to prevent any breakaway of the vehicle by countersteering during a panic stop. Hereby, the individual wheel control assures the shortest possible stopping distance resulting from the sum of the optimum controlled brake forces on all four wheels.

3. Comparison between Individual Four-wheel Control System and Simplified Control Systems

At present, a number of systems are under discussion that are to improve brake behavior with a minimum of technical input. In the following, three of such system configurations are discussed showing their advantages and shortcomings.

a. Front and Rear Axle Control

For cost savings reasons, less sophisticated systems have been proposed for brake torque control per axle. Hereby, two design approaches were adopted. One can be defined "select low" which means the lowest friction force coefficient of one of the two wheels controls the brake force. The second approach "select high" applies just the opposite. It should be mentioned that the performance of a "select low" or "select high" system is almost as good as an individual four-wheel control system when driving on a straight homogeneous road. However, great differences in performance occur when driving in a turn, during evasive maneuvers and with different friction coefficients on the left and right wheels.

The braking behavior of a "select low" system in a turn is characterized in that the brake forces of the outer wheels have to be adapted to the very low brake forces of the inner wheels (Fig. 3). Near the limit speed in a turn, only about one third of brake force is available as compared to that of an individual four-wheel control system. This leads to a considerable increase in stopping distance. It goes without saying that such a vehicle can no longer be mastered when driving in a curve with decreasing radius. The stopping distance of a vehicle with "select low" control, for instance, is about as twice as long as that of a vehicle with individual wheel control for brakings in a turn at high speed.

When braking in a turn with a "select high" system, the brake torques are controlled per axle according to the friction forces of the outer wheels (Fig. 3a). This leads to an immediate lockup of the inner wheels and to a loss of lateral force. As a result, the total lateral force is reduced by about one third (Fig. 3b). The sudden loss of lateral force is followed by vehicle instability and hazardous swerving. Because of this hazard encountered with brakings in a turn, this system is rarely taken into consideration today.

b. "Select Low" Rear Axle Control

Antiskid systems employing rear axle control only for cost reasons are already on the market today. Certainly, such systems prevent vehicle swerving under various conditions, but their basic shortcoming is the absence of steerability at a panic stop due to the uncontrolled front wheels. That means full braking in a turn or during evasive maneuvers is impossible. This system configuration does by no means meet the present traffic safety requirements.

c. Antiskid Systems with Individual Front Wheel Control and Select Low Rear Axle Control

Another control configuration - also designed under the cost savings aspect - comprises individual front wheel control and select low rear axle control. Similar to a system with individual four-wheel control, this system assures full vehicle steerability and stability with brakings in a turn or evasive maneuver. Due to the dynamic wheel load distribution during braking in a turn, the higher brake forces available on the front wheels are fully utilized. Only the lower brake force of the outer rear wheel is adjusted to that of the inner rear wheel (Fig. 3a). This system shows good results for braking in a turn: On dry roads, the stopping distance is not longer than that with locked wheels, and it is considerably shorter on a wet road. Vehicle behavior is still satisfactory on a road with different friction coefficients on the left and right hand side.

Although the above system does not offer all advantages of an individual four-wheel control system, its performance can be considered satisfactory.

4. Results Achieved with the Teldix Antiskid System

4.1 System Design

The Teldix antiskid system is an individual four-wheel control system (Fig. 4). Pulse sensors mounted on the wheels measure wheel speed, and two accelerometers (damped spring-mass systems) on the front wheels furnish deceleration and acceleration signals. The signals of all sensors are fed to an electronic control unit where they are logically processed and transmitted to the electromagnetic valves of the hydraulic unit. Depending on the control signals, the magnetic

valves effect a fast or slow (pulsed) pressure rise, a constant pressure or a fast or slow (pulsed) pressure drop. The pressure fluid becoming available during the pressure drop is led to a storage chamber and fed back to the primary brake circuit. Some of the advantages of the hydraulic unit are the extremely fast or slow (pulsed) pressure modulation, the compact design and the low weight. In addition, this unit does not come into conflict with the problems vacuum pressure modulators encounter in meeting the vehicle exhaust requirements. Figure 5 shows the components of the Teldix antiskid system with the sensors in the front and the hydraulic and electronic unit in the back.

4.2 Performance Characteristics of the Teldix Antiskid System

Figure 6 shows vehicle and wheel velocity as a function of time demonstrated during a typical braking with the Teldix antiskid system on a straight road. The design objective of any antiskid system is the control of all wheels within a very narrow slip range of maximum brake force. For that purpose, a reference voltage $V_{REF}$ is generated in the electronic control unit being similar to vehicle velocity and corresponding to the velocity with a slip of maximum brake force. This reference voltage is used to define three tire slip ranges in each of which the pressure is modulated in a different manner. If the wheel is in slip range I or II, that is, near the speed with maximum brake force only slight changes in brake torque are necessary; they are achieved by a slow pulsed pressure modulation. If a sudden decrease of friction coefficient ($\mu$-jump) results in high wheel deceleration, a deceleration signal from the accelerometer initiates at once a rapid pressure drop which continues to remain in slip range III until the wheel is reaccelerated (Fig.6, point E). Thus, even in the case of $\mu$-jumps only a relatively small tire slip occurs and the lateral force is retained.

The quality of antiskid control depends largely on the fact that the reference voltage always represents the approximate speed with a slip of maximum brake force. On one hand, this is achieved by the accelerometers that accurately sense wheel acceleration and furnish reliable data to the electronic control unit when a wheel exceeds the slip value of maximum brake force. On the other hand, it is accomplished by the extremely rapid pressure modulation in case of $\mu$-jumps and the slow pulsed pressure modulation on homogeneous roads to keep wheel velocity in a narrow slip range.

4.3 Performance Test Results

The essential results obtained with the Teldix antiskid system under actual service conditions and on test facilities can be summarized as follows: After years of testing in cooperation with Mercedes-Benz, the antiskid system has met the design requirements under all conditions. In the following, results obtained on test facilities are discussed. Figure 7 shows the layout of the Teldix skid pad which includes positive and negative $\mu$-jumps between the friction coefficients of 0.9 and 0.1. The sides of the skid pad allow for trials with split coefficients.

27

Furthermore, it is possible to drive across a so-called chess-board pattern under full braking conditions with a rapid change in friction coefficient. Hereby, the following results were obtained:

a. The antiskid system fulfilled all performance requirements in a $\mu$-range between 0.1 and 0.9. In this respect, the low friction coefficient range is of particular significance since the range between 0.2 and 0.1 is a very critical factor under low temperature and aquaplaning conditions. An antiskid system that does not perfectly operate below a tire-road friction coefficient of 0.2 would fail in a great many hazardous situations and thus offer the driver less assistance than a brake system without antiskid control.

b. It is possible to pass $\mu$-jumps from 0.9 to 0.1 without a wheel lockup of more than 200 milliseconds in the low speed range. There is absolutely no lockup at high speeds.

c. Full vehicle stability was maintained during test trials with different friction coefficients on the left and on the right. The stopping distance achieved hereby corresponded to the theoretical stopping distance based on the mean value of a pavement including high and low friction coefficients.

d. Full vehicle stability was achieved on the chess-board surface under full braking conditions. This was true for low and high vehicle speeds.

Another test facility was used to test vehicle braking in a turn. These tests confirmed the calculated values of the diagrams represented in Fig. 8a and 8b. They show the brake and lateral forces as well as vehicle velocity as a function of stopping distance with full brakings from a steady turn (radius r = 100 m) on wet asphalt at a speed of 90 km/h. At first, the antiskid system initiates a fast brake force build-up (on the outer wheels) until the maximum tire-road friction coefficient has been achieved (Fig. 8a). Then follows a slow rise in brake force up to the possible maximum so that the required lateral force is maintained at all times.

Full braking without an antiskid system leads to a very fast wheel lockup; hereby, the lateral force decreases to almost zero, and the brake force is determined by the slip coefficient of the locked wheel. As a result, the vehicle generally drives off the circular path in a swerving manner.

Figure 8b shows the reduction in speed with a controlled braking in a turn and with locked-wheel braking. As to the latter, the slip coefficient of the locked wheel determines the stopping distance. This coefficient is lower than that of the maximum tire-road friction coefficient during controlled braking in a turn. Thus, controlled braking in a turn leads to a shorter stopping distance than locked braking on a straight road. At the point where the controlled vehicle has already come to a halt, the uncontrolled vehicle is still travelling at approximately 45 % of its initial speed. In this

connection, it should be mentioned that the uncontrolled vehicle drove off the road while the controlled one has come to a halt in the turn without deviating from its course.

The stopping distances obtained during testing correspond to the calculated values. Some typical results are stated in Table 1.

5. Conclusion

The results summarized above are to show that the state-of-the-art of automatic antiskid systems assures the retention of vehicle stability and steerability with a simultaneous reduction in stopping distance even during brakings in a turn. This study shall also point out that the low-cost design solutions may become more hazardous in a number of instances than conventional brake systems. The technical complexity can be reduced, assuring the present performance characteristics, by matching suspension, brake system and antiskid system. Thus, sophisticated antiskid systems will be offered at a price acceptable for lower-price car categories. This means more safety for all.

REFERENCES

1. K. Desoyer und A. Slibar: Kraftschlußbeanspruchung und Schräglauf der Räder eines Kraftfahrzeuges bei stationärer Kurvenfahrt, Automobiltechnische Zeitung 72, Nr. 6, Stuttgart, p. 206-212.

**Table 1:** Theoretical and Experimental Characteristics for Brakings in Turns with Teldix Antiskid System ( Radius of Curvature r = 100 m )

| pavement | brake system | stopping distance theoretical | stopping distance experimental | decrease in stopping distance theoretical | decrease in stopping distance experimental | theoretical impact speed |
|---|---|---|---|---|---|---|
| dry asphalt initial speed $V_0$=100 km/h | controlled (in a turn ) | 41,5 m | 42,4 m | 10 % | 9% | 30 % $V_0$ = 30 km/h |
| | uncontrolled (straight-line skidding ) | 46,2 m | 46,5 m | | | |
| wet asphalt initial speed $V_0$= 90 km/h | controlled (in a turn ) | 44,0 m | 42,0 m | 21% | 22% | 45% $V_0$ = 42 km/h |
| | uncontrolled (straight-line skidding ) | 55,7 m | 54,0 m | | | |

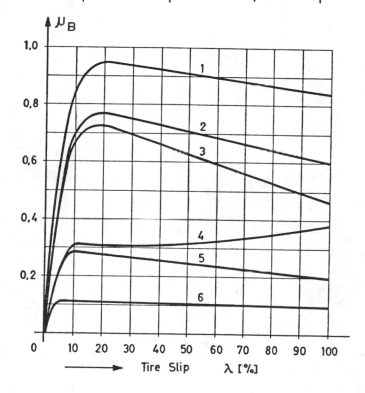

1: Dry Asphalt

2: Wet Asphalt, Thin Water Film

3: Wet Asphalt, Thick Water Film

4: Fresh Snow

5: Packed Snow

6: Glare Ice

Figure 1    Brake Force Coefficients $\mu_B$ as a Function of Tire Slip for Braking on a Straight Road with Radial-ply Tires (own tests)

29

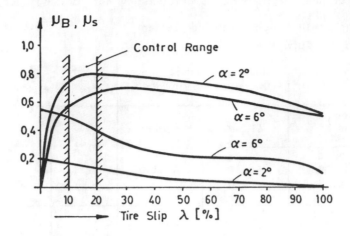

Figure 2 Brake and Lateral Force Coefficients $\mu_B$ and $\mu_s$ as a Function of Tire Slip with Two Sideslip Angles (own tests)

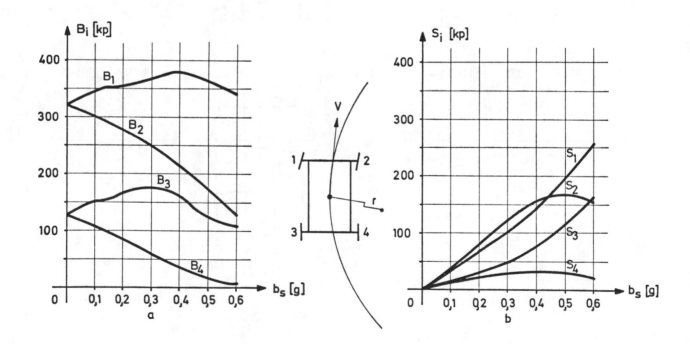

Figure 3 Maximum Possible Brake Forces and Lateral Forces as a Function of Lateral Acceleration on a Rear-wheel Driven Passenger Car

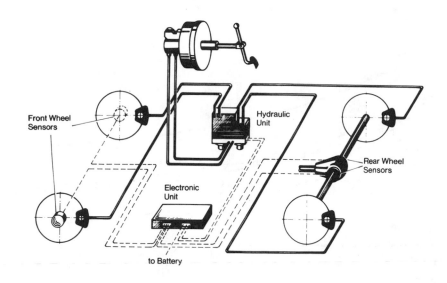

Front Wheel
Sensors

Hydraulic
Unit

Rear Wheel
Sensors

Electronic
Unit

to Battery

Figure 4    Schematic Diagramm of Teldix Antiskid System

Figure 5    Components of Teldix Antiskid System

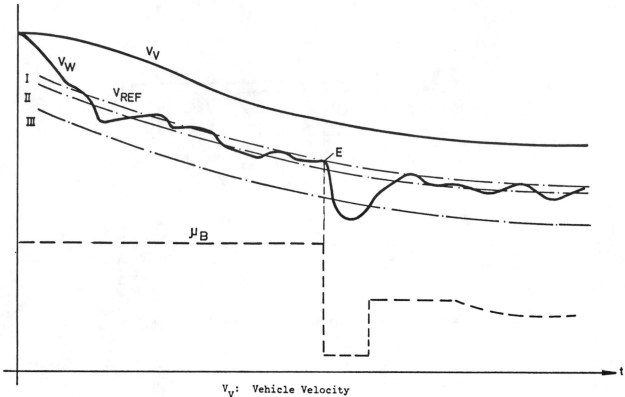

$V_V$: Vehicle Velocity

$V_W$: Wheel Velocity

$V_{REF}$ : Reference Voltage

Figure 6     Schematic Diagram of a Typical Braking on a Straight
Road with Teldix Antiskid System

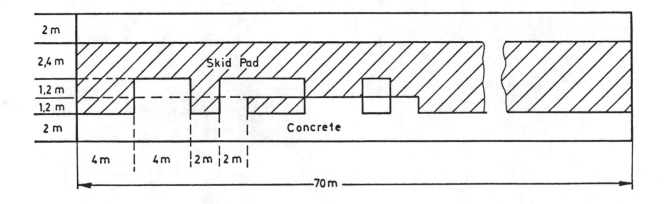

Figure 7     Layout of Teldix Skid Pad

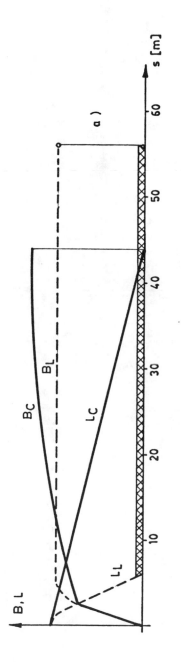

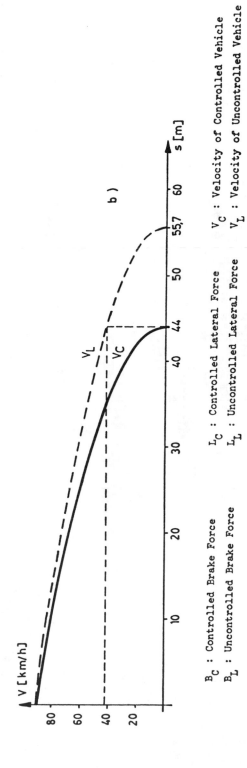

$B_C$ : Controlled Brake Force     $L_C$ : Controlled Lateral Force     $V_C$ : Velocity of Controlled Vehicle

$B_L$ : Uncontrolled Brake Force     $L_L$ : Uncontrolled Lateral Force     $V_L$ : Velocity of Uncontrolled Vehicle

Figure 8     Shape of Brake and Lateral Force Curves and Decrease of Vehicle Speed Along the Stopping Line for Brakings in a Turn with and without the Teldix Antiskid System

33

# Design Considerations of Adaptive Brake Control Systems**

**R. R. Guntur**
Vehicle Research Laboratory
University of Technology, Delft

**Paper 741082 presented at the International
Automobile Engineering and Manufacturing
Meeting, Toronto, Canada, October, 1974.

IN A PREVIOUS PAPER (1)* an attempt was made to analyze several methods of prediction and reselection. In this paper this analysis is extended to take into account the characteristics of the brake pressure modulator and the signal processing unit.

In general, an adaptive brake control system is programmed to operate in a certain fashion in order to perform a desired function. If it is agreed to call the programming part of the adaptive brake control system the software and the components of the system the hardware, the previous study (1) is evidently concerned with the software, with only occasional reference to the hardware.

In this paper, it is proposed to study the interaction of the software with the hardware. When the hardware is changed—for example, when a modulator that has two solenoid valves is employed—the software may be suitably changed to improve the performance of the wheel and the vehicle. However, the scope of the present paper is limited to rear wheel adaptive brake control systems.

## ANALYSIS THROUGH NUMERICAL EXAMPLE

The mathematical model used here is the same as the one used in Ref. 1. It is therefore convenient to study the prob-

*Numbers in parentheses designate References at end of paper.

lem through a numerical example; for the purpose of digital simulation, the following data of the vehicle will be used.

Weight of vehicle, $W = 1404$ kgf

Front axle distance, $a = 1.386$ m

Rear axle distance, $b = 1.134$ m

Height of center of gravity, $h = 0.519$ m

Mass moment of inertia of parts connected to front wheel,

$$I_f = 0.13 \text{ kgf m s}^2$$

Mass moment of inertia of parts connected to rear wheel,

$$I_r = 0.12 \text{ kgf m s}^2$$

Effective rolling radius of front wheel, $R_f = 0.273$ m

Effective rolling radius of rear wheel, $R_r = 0.273$ m

The following assumptions have been made while making the calculations:

1. A two-wheel model is assumed to be representative of the actual vehicle.

2. The air resistance on the vehicle is assumed to be small compared to the braking force on it.

3. The effective rolling radius of a wheel is equal to the height of the corresponding axle.

4. The brake force coefficient increases linearly with slip as long as the slip is below the critical slip; if the slip is above the critical slip, the brake force coefficient decreases linearly as the slip increases.

─────────────── ABSTRACT ───────────────

In this paper, some of the design aspects of adaptive brake control systems are studied, especially the interaction of the software with the hardware of the system. Two modes of operation of the brake pressure modulator have been considered; the software changes are effected to modify further the mode of operation of the system. The effect of the rate of rise of wheel cylinder pressure and the effect of rate of decay of pressure on the effectiveness and the maximum wheel slip in the first cycle have been studied. The hardware and the software are so modified as to give satisfactory performance of the wheel and the vehicle for four different forward speeds and for three different road conditions.

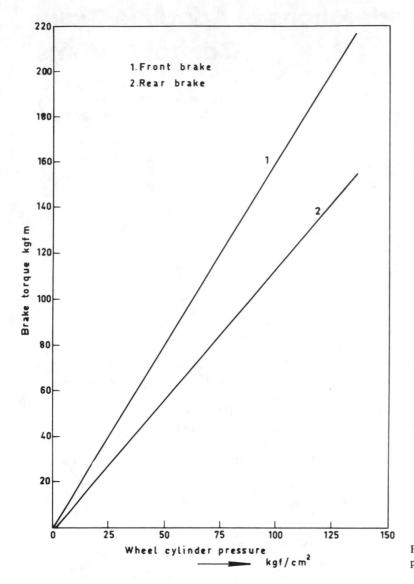

Fig. 1 - Variation of brake torque with wheel cylinder pressure

5. The pressure of the fluid in the wheel cylinder is assumed to vary linearly w.r.t time if it is not limited.

The front and rear brake torques vary with the respective wheel cylinder pressures, as shown in Fig. 1. Calculations have been made for three different road conditions and for four different initial speeds of the vehicle; the $\mu$-slip curves for the 12 different cases are represented in Fig. 2.

## HARDWARE OF ADAPTIVE BRAKE CONTROL SYSTEM

A brief description of the general layout of an antiskid system will be useful in appreciating the present study. A schematic layout of a modern antiskid system is given in Fig. 3 (2). The antiskid system consists of a sensor, a control unit, and a brake pressure modulator. A sensor measures the angular velocity of the wheel. The signal from the sensor is fed to

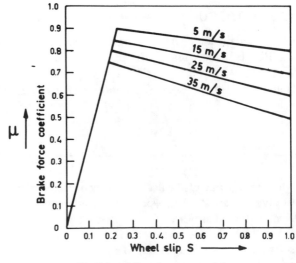

Fig. 2A - $\mu$-slip curves on road A

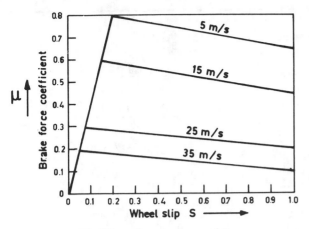

Fig. 2B - $\mu$-slip curves on road B

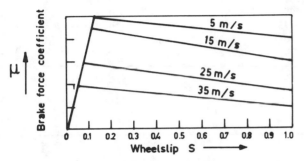

Fig. 2C - $\mu$-slip curves on road C

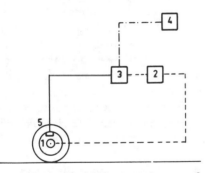

Fig. 3 - Schematic layout of antiskid system. 1. sensor; 2. control unit; 3. modulator; 4. master cylinder; 5. wheel cylinder

the signal processing unit of the control unit. The control unit monitors a solenoid valve; in some cases it monitors two solenoid valves—in the brake pressure modulator through an electronic switch.

Most of the sensors have the same basic features, and as the signal from the sensor has to be processed by the signal processing unit, only the interaction of the signal processing unit with the software has to be studied. However, it should be noted that if locking of the wheel at low speeds is allowed, the design of the sensor will be simplified (1).

Two of the most important parts of the adaptive brake control system whose characteristics interact with the software are the signal processing unit and the brake pressure modulator. Brake pressure modulators have been classified according to the sources of power used, and the relative advantages and disadvantages of them are given elsewhere (2). In this paper, concern will be focused on the characteristics of the modula-

tors; accordingly, the modulators are reclassified as follows: on-off modulators and four-phase modulators.

## OPERATION OF ON-OFF MODULATOR

The on-off modulator has only one solenoid valve. The hydraulic circuitry of the braking system incorporating the on-off modulator is arranged in such a way that whenever the valve is opened, the wheel cylinder pressure begins to decrease; if the valve is kept open, the pressure continues to decrease until it reaches the minimum value that it can attain. When the valve is closed for sufficient time, the pressure reaches the maximum value and thereafter it remains constant until the valve is opened again. Evidently, in the case of the on-off modulator, the wheel cylinder pressure remains constant only when it has attained one of the two extreme values.

In some of the on-off modulators, a provision is made to change the rate of decay of pressure and the rate of rise of pressure. The modulators that are used in the Chrysler adaptive brake control system and in the Ford Sure-Track system are on-off modulators (4, 5).

## OPERATION OF FOUR-PHASE MODULATOR

The four-phase modulator has two solenoid valves; one of the two valves is an inlet valve and the other is an outlet valve. The hydraulic circuitry of the braking system is arranged in such a way that normally the inlet valve is opened and the outlet valve is closed. If the pressure in the master cylinder is allowed to build up and if the inlet valve is opened long enough, the pressure in the wheel cylinder will reach the maximum value; it then remains constant until the outlet valve is opened. On the other hand, if the inlet valve is closed before the pressure in the wheel cylinder has reached its maximum value, the pressure in the wheel cylinder will remain constant at a high value until the outlet valve is opened. Then, if the inlet valve is kept closed and the outlet valve is opened, the pressure begins to decrease. The pressure will continue to decrease as long as the outlet valve is opened, unless it has reached the minimum value. However, if the outlet valve is closed before the pressure has reached the minimum value, the pressure will remain constant at a low value until the inlet valve is again opened. The difference between the high constant pressure and the low constant pressure depends on the time for which the inlet valve is kept open before the high constant pressure phase and also on the time for which the outlet valve is kept open before the low constant pressure phase. Thus, it is possible to keep the pressure at a constant value that is different from the two extreme values. In other words, when a four-phase modulator is used, the fluctuation of pressure can be made small, if necessary (6). The four-phase modulator is employed in the ABS system of the Daimler Benz and Teldix (7).

## SOFTWARE

The software includes the logic for prediction and reselection. However, when a four-phase modulator is used instead

Table 1 - Prediction Point Slip $S_p$ when Pressure is Limited to 110 kgf/cm$^2$

| Initial Velocity, m/s | Rate of Rise of Pressure kgf/cm$^2$/s | Road A | | Road B | | Road C | |
|---|---|---|---|---|---|---|---|
| | | $S_p$ without Time Delay | $S_p$ with Time Delay | $S_p$ without Time Delay | $S_p$ with Time Delay | $S_p$ without Time Delay | $S_p$ with Time Delay |
| 5 | 220 | – | – | – | – | 0.151 | 0.497 |
| 5 | 366.67 | – | – | – | – | 0.139 | 0.530 |
| 5 | 733.33 | – | – | – | – | 0.131 | 0.592 |
| 5 | 1466.67 | 0.069* | 0.179* | 0.069* | 0.179* | 0.069 | 0.307 |
| 15 | 220 | – | – | 0.229 | 0.310 | 0.153 | 0.287 |
| 15 | 366.67 | – | – | 0.200 | 0.322 | 0.139 | 0.293 |
| 15 | 733.33 | – | – | 0.180 | 0.329 | 0.122 | 0.308 |
| 15 | 1466.67 | 0.060* | 0.165* | 0.060 | 0.159 | 0.060 | 0.199 |
| 25 | 220 | – | – | 0.159 | 0.272 | 0.159 | 0.272 |
| 25 | 366.67 | – | – | 0.128 | 0.259 | 0.129 | 0.262 |
| 25 | 733.33 | – | – | 0.099 | 0.267 | 0.098 | 0.269 |
| 25 | 1466.67 | 0.075* | 0.162* | 0.0773 | 0.274 | 0.073 | 0.275 |
| 35 | 220 | – | – | 0.166 | 0.276 | 0.166 | 0.276 |
| 35 | 366.67 | – | – | 0.128 | 0.252 | 0.128 | 0.252 |
| 35 | 733.33 | – | – | 0.088 | 0.239 | 0.088 | 0.239 |
| 35 | 1466.67 | 0.086* | 0.157* | 0.060 | 0.242 | 0.060 | 0.242 |

*False prediction point. – shows no false prediction point. $k_2 = 1.5$ g; $k_4/k_3 = 2.5$ s$^{-1}$.

of an on-off modulator, an additional independent condition for closing the inlet valve and an additional independent condition for closing the outlet valve have to be used. Details about the software are given in the Appendix.

## PRELIMINARY DESIGN CONSIDERATIONS

In the signal processing unit of the control unit, a differentiating circuit, which incorporates a filter, has to be employed. Therefore, it is necessary to determine the desired cutoff frequency before the effect of the signal processing unit on the software is studied. Accordingly, Ouwerkerk and Rop (8) determined the angular acceleration of the wheel when the vehicle is moving at a constant speed; the results have been recorded on a tape and analyzed. The analysis of the results has indicated that there is a small peak in the spectral density of the angular acceleration at a frequency of 12 Hz; another peak has been observed at a frequency of 24 Hz. Evidently, if the noise has to be effectively filtered out, the cutoff frequency has to be 12 Hz. So a fourth-order filter having a cutoff frequency of 12 Hz has been designed; the time delay introduced by the filter is about 44 ms. If a cutoff frequency above 12 Hz is specified in designing the filtering unit, the shift in the prediction point can be made smaller than 44 ms, but, owing to the inaccuracy introduced in determining the relevant parameter, there will be additional shift in the prediction point; this additional shift cannot be found accurately, and hence a cutoff frequency of 12 Hz has been specified.

Naturally, the time delay affects the software that has to be used. To show the effect of the time delay on the prediction point slip, the calculations were made at first without taking into account the time delay and then the calculations were repeated taking into account the time delay. The results are shown in Table 1. From the results in Table 1, it is evident that, as a result of the time delay, the prediction point for a given threshold value is shifted. So the threshold values should be smaller than those without the time delay. Moreover, if the time of application is smaller than the time delay, the application of the brake will be complete even before the control unit predicts the locking of the wheel, and the resulting maximum slip will be high for a given rate of decay of pressure. Therefore, the application time should be longer than the time delay.

The time delay of 44 ms renders the low-speed reselection condition $F_r$, which was proposed in Ref. 1, unsuitable, because when this condition is used, very late reselections will take place at low speeds.

Furthermore, it should be noted that when the brake pressure is limited to 110 kgf/cm$^2$, the false predictions are not entirely eliminated unless the threshold values are 2.25 g and 3.00 s$^{-1}$. If the threshold values are 2.25 g and 3.00 s$^{-1}$, the prediction point slip will be very high under some circumstances.

To ensure the locking of the rear wheel on road A, the maximum pressure is increased to 135 kgf/cm$^2$. The prediction point slip for various rates of rise of pressure, when the time delay of 44 ms is taken into account, is given in Table 2. By comparing the results in Table 2 with those in Table 1, it is inferred that on road B when the maximum brake line pressure is increased to 135 kgf/cm$^2$, the prediction point slip for a given threshold value, under the same circumstances, is also increased.

Table 2 - Prediction Point Slip $S_p$ When Pressure Is Limited to 135 kgf/cm$^2$

| Initial Velocity, m/s | Rate of Rise of Pressure, kgf/cm$^2$/s | On Road A, $S_p$ with Time Delay | On Road B, $S_p$ with Time Delay | On Road C, $S_p$ with Time Delay |
|---|---|---|---|---|
| 5 | 220 | 1.000 | 0.863 | 0.497 |
| 5 | 366.67 | 0.523 | 0.612 | 0.530 |
| 5 | 733.33 | 0.413 | 0.562 | 0.694 |
| 5 | 1466.67 | 0.200 | 0.196 | 0.347 |
| 15 | 220 | 0.326 | 0.313 | 0.287 |
| 15 | 366.67 | 0.315 | 0.316 | 0.293 |
| 15 | 733.33 | 0.250 | 0.353 | 0.308 |
| 15 | 1466.67 | 0.176 | 0.174 | 0.214 |
| 25 | 220 | 0.327 | 0.272 | 0.272 |
| 25 | 366.67 | 0.337 | 0.259 | 0.262 |
| 25 | 733.33 | 0.348 | 0.298 | 0.269 |
| 25 | 1466.67 | 0.189 | 0.314 | 0.316 |
| 35 | 220 | 0.427 | 0.276 | 0.276 |
| 35 | 366.67 | 0.435 | 0.252 | 0.252 |
| 35 | 733.33 | 0.437 | 0.239 | 0.239 |
| 35 | 1466.67 | 0.184 | 0.277 | 0.277 |

To find the effect of the rate of rise of pressure on the average deceleration of the vehicle, computations were made assuming that the adaptive brake control system was not operative.

Thus, the average deceleration of the vehicle until the time all the wheels are locked has been determined. In Figs. 4-6, the average deceleration is plotted against the rate of rise of pressure. It is observed that in general the average deceleration of the vehicle increased with increase in the rate of rise of pressure. The increase in the average deceleration of the vehicle with the rate of rise of pressure is very pronounced at low speeds, particularly when the vehicle is moving on a high friction surface, whereas the increase in the average deceleration of the vehicle with the rate of rise of pressure is small at high speeds and on low-friction surfaces.

Thus, it is evident that the rate of rise of pressure should be as high as possible, but it has already been noted that the time of application should be larger than 44 ms. In view of the above findings, the rate of rise of pressure is fixed at 1466.67 kgf/cm$^2$/s. If it is possible to change the rate of rise of pressure, at high speeds and on low-friction surfaces it may be made smaller than 1466.67 kgf/cm$^2$/s to keep the maximum slip in a brake cycle low. However, it is suggested by the latter results that such a change in the rate of rise of pressure is not necessary.

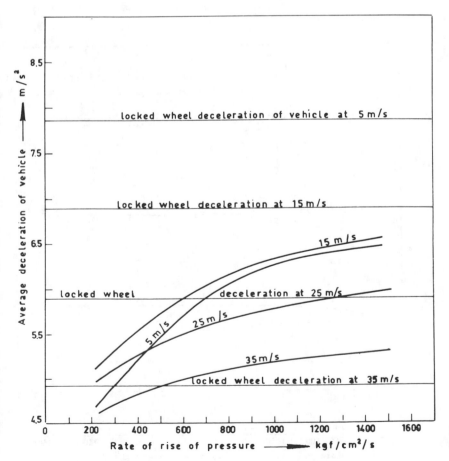

Fig. 4 - Effect of rate of rise of pressure on average deceleration of vehicle

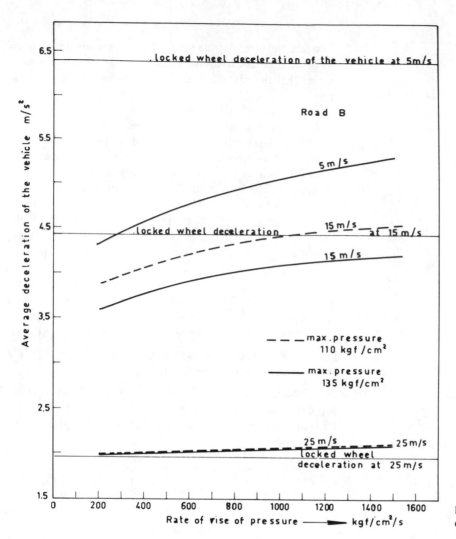

Fig. 5 - Effect of rate of rise of pressure on average deceleration of vehicle

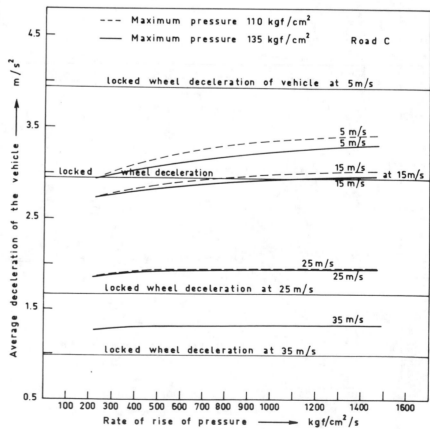

Fig. 6 - Effect of rate of rise of pressure on average deceleration of vehicle

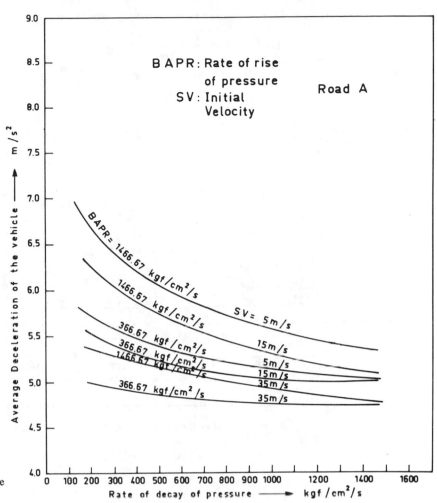

Fig. 7 - Effect of rate of decay of pressure on average deceleration of vehicle

## DESIGN CONSIDERATIONS OF ADAPTIVE BRAKE CONTROL SYSTEM USING ON-OFF MODULATOR

At first it is assumed that a modulator having a single solenoid valve is available. It is assumed that the time delay in either closing or opening the valve is about 10 ms. It is also assumed that the values of $k_2$ and $k_4/k_3$ are 1.5 $g$ and 2.50 $s^{-1}$, respectively. For evaluating the performance of the vehicle and that of the wheel, it is assumed that the second cycle commences at a slip value of 0.75 $S_c$.

OPERATION ON ROAD A - The average deceleration of the vehicle in the first brake cycle is determined for various speeds and various rates of decay of pressure on road A, and the results are shown in Fig. 7, confirming the earlier finding that the rate of rise of pressure should be as high as possible. It is also observed that when the rate of decay of pressure is low, the average deceleration of the vehicle in the first cycle is high. The increase in the average deceleration that is obtained by decreasing the rate of decay of pressure is markedly high at low speeds.

The maximum slip in the first brake cycle is determined for the various rates of decay of pressure and for three different speeds, and the results are given in Fig. 8. From this figure, it is evident that the maximum slip in the first brake cycle is higher at low rates of decay of pressure than the maximum

slip at high rates of decay of pressure. The rate of rise of pressure has an effect on the maximum slip in the first cycle. When the rate of rise of pressure is high, if the other conditions are not changed, the maximum slip is low. The change in the value of maximum slip with the rate of decay of pressure is more with a low rate of rise of pressure than it is with a high rate of rise of pressure.

In Fig. 9, the time interval between the effective release point and the beginning of the second cycle, which is called the desired recovery time, is plotted against the rate of decay of pressure. As expected, a low rate of decay of pressure has increased the desired recovery time significantly. The rate of rise of pressure has also an effect on the desired recovery time; when the rate of rise of pressure is low, the desired recovery time is longer than it is when the rate of rise of pressure is high.

In short, a low rate of decay of pressure on road A increases the average deceleration, and it increases only slightly the maximum wheel slip; it also increases the desired recovery time. Evidently, a low rate of decay of pressure is highly desirable on high-friction surfaces such as road A.

OPERATION ON ROADS B AND C - To find the effect of the rate of decay of pressure on the performance of the vehicle on road B, calculations have been made for two rates of rise of pressure and for two rates of decay of pressure. As be-

41

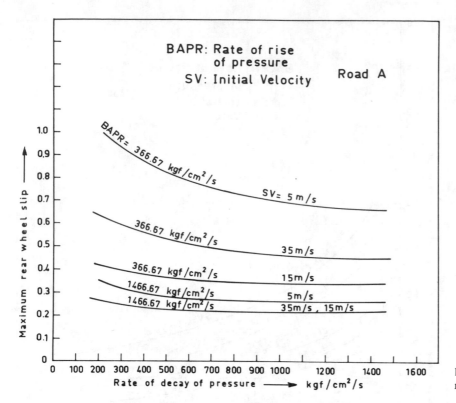

Fig. 8 - Variation of maximum rear wheel slip w.r.t rate of decay of pressure

Table 3 - Effect of Rate of Decay of Pressure on Performance of Wheel and Vehicle

| Initial Velocity, m/s | Rate of Rise of Pressure, kgf/cm$^2$/s | Rate of Decay of Pressure, kgf/cm$^2$/s | Road B | | | | Road C | | | |
|---|---|---|---|---|---|---|---|---|---|---|
| | | | Average Deceleration of Vehicle, m/s$^2$ | Maximum Wheel Slip | Desired Recovery Time, ms | Time Wheel is Locked, ms | Average Deceleration of Vehicle, m/s$^2$ | Maximum Wheel Slip | Desired Recovery Time, ms | Time Wheel is Locked, ms |
| 5 | 366.67 | 366.67 | 5.00 | 1.000 | 40 | 92 | 3.373 | 1.000 | 54 | 124 |
| 5 | 366.67 | 1466.67 | 4.714 | 1.000 | 31 | 11 | 3.176 | 1.000 | 37 | 15 |
| 5 | 1466.67 | 366.67 | 5.628 | 0.458 | 56 | 0 | 3.753 | 1.000 | 70 | 163 |
| 5 | 1466.67 | 1466.67 | 4.861 | 0.306 | 20 | 0 | 3.572 | 1.000 | 45 | 30 |
| 5 | 1466.67 | 183.33 | 5.954 | 1.000 | 74 | 79 | – | – | – | – |
| 15 | 366.67 | 366.67 | 3.972 | 0.911 | 133 | 0 | 2.919 | 0.946 | 146 | 0 |
| 15 | 366.67 | 1466.67 | 3.779 | 0.456 | 50 | 0 | 2.846 | 0.453 | 50 | 0 |
| 15 | 1466.67 | 366.67 | 4.445 | 0.614 | 115 | 0 | 3.048 | 1.000 | 136 | 141 |
| 15 | 1466.67 | 1466.67 | 4.074 | 0.307 | 40 | 0 | 3.132 | 0.519 | 58 | 0 |
| 15 | 1466.67 | 183.33 | 4.227 | 0.432 | 50 | 0 | – | – | – | – |
| 25 | 366.67 | 366.67 | 2.092 | 0.646 | 65 | 0 | 1.952 | 0.691 | 186 | 0 |
| 25 | 366.67 | 1466.67 | 2.043 | 0.374 | 68 | 0 | 1.969 | 0.381 | 80 | 0 |
| 25 | 1466.67 | 366.67 | 2.095 | 1.000 | 236 | 178 | 1.863 | 1.000 | 254 | 198 |
| 25 | 1466.67 | 1466.67 | 2.177 | 0.637 | 124 | 0 | 2.038 | 0.657 | 140 | 0 |
| 35 | 366.67 | 366.67 | 1.301 | 0.622 | 297 | 0 | 1.301 | 0.622 | 297 | 0 |
| 35 | 366.67 | 1466.67 | 1.325 | 0.368 | 146 | 0 | 1.325 | 0.368 | 146 | 0 |
| 35 | 1466.67 | 366.67 | 1.305 | 1.000 | 285 | 200 | 1.305 | 1.000 | 285 | 200 |
| 35 | 1466.67 | 1466.67 | 1.345 | 0.579 | 248 | 0 | 1.345 | 0.579 | 248 | 0 |

fore, the average deceleration of the vehicle, the maximum wheel slip, and the desired recovery time have been determined and are given in Table 3.

On road B, if the initial velocity is about 15 m/s, the average deceleration of the vehicle is high when the rate of rise of pressure is high and the rate of decay of pressure is low. If the

initial velocity is above 15 m/s, the average deceleration is high when both the rate of rise of pressure and the rate of decay of pressure are high.

On road C, if the initial velocity is 5 m/s, the average deceleration is high when the rate of rise of pressure is high and the rate of decay of pressure is low. If the initial velocity is above

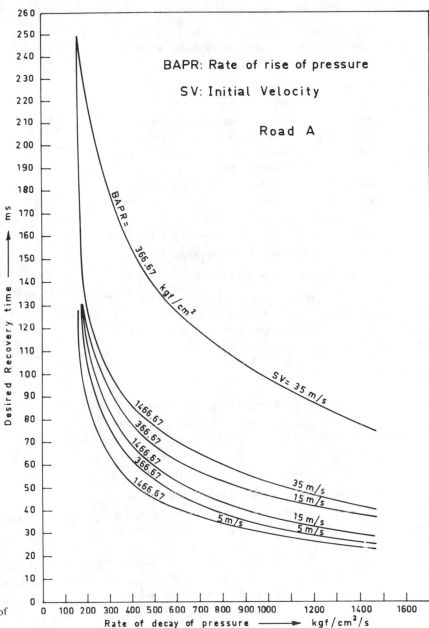

Fig. 9 - Variation of desired recovery time w.r.t rate of decay of pressure

5 m/s, the average deceleration is high when both the rate of rise of pressure and the rate of decay of pressure are high.

The maximum wheel slip in the first cycle is low when the rate of rise of pressure is low and the rate of decay of pressure is high. The maximum wheel slip in the first cycle is high when the rate of rise of pressure is high and the rate of decay of pressure is low. On road C, when the rate of rise of pressure is high and the rate of decay of pressure is low, intermittent locking of the wheel occurs even at an initial speed of 35 m/s.

On road B and at low speeds, the time for which the wheel is locked is long when the rate of rise of pressure and the rate of decay of pressure are low. When the rate of rise of pressure is high and the rate of decay of pressure is low, the wheel is locked even if the initial speed is above 5 m/s; when the initial speed is 35 m/s, the time for which the rear wheel is locked is 200 ms.

The differences in the desired recovery times under the dif-ferent road conditions may be decreased by using a low rate of decay of pressure on the high-friction surfaces and a high rate of decay of pressure on the low-friction surfaces. Thus, the results suggest that the rate of decay of pressure should be changed in accordance with the road conditions. Therefore, the control system may be designed in such a way that if the deceleration of the vehicle is higher than 0.5 $g$, the rate of decay of pressure is 183.33 kgf/cm$^2$/s; otherwise, it is 1466.67 kgf/cm$^2$/s.

DETERMINATION OF THRESHOLD VALUES IN CONDITIONS FOR RESELECTION - The noise on the angular deceleration rarely exceeds 0.25 $g$; therefore, it is assumed that a value of 0.7 $g$ for $k_8$ will render the system insensitive to the noise on the angular acceleration signal.

An examination of the desired recovery times under the various conditions of operation will show that a time delay

$D_t$ of 500 ms will be adequate to allow the wheel to recover from impending danger of locking.

Based on the assumption that the reselection has to take place when the wheel slip is 0.75 $S_c$, the desired angular velocity factor $k_6$ has been determined for different road conditions and for different speeds.

From the results, which are given in Table 4, it is evident that the range of speed of the vehicle should be divided into at least three parts. The speed range, including all the speeds from 20-35 m/s may be denoted as the high-speed range; the range of speed from 0-10 m/s may be called the low-speed range. The speed range including all the speeds between 10-20 m/s may be known as the medium-speed range.

The value of the factor $m_1$ should be about 0.9; the value of $m_2$ should be about 0.85; the value of $m_3$ should be about 0.75; the value of $m_4$ should be about 0.45 (see Appendix).

The on-off system suffers the following disadvantages: first, the maximum slip on road C is relatively high; second, false predictions cannot be avoided unless very high threshold values are selected.

## MODIFIED ON-OFF SYSTEM

To decrease the value of the maximum slip in a brake cycle, the software is modified. The threshold values in both the conditions for prediction are changed as follows: if the deceleration of the vehicle is above 0.5 $g$, the threshold value $k_2$ is 2.25 $g$ and the value of $k_4/k_3$ is 3.00 $s^{-1}$; otherwise, these values are 1.5 $g$ and 2.00 $s^{-1}$, respectively. By effecting this change in the software, false predictions are avoided and the maximum slip on road C is decreased. Further decrease in the maximum slip on road C is brought about by decreasing the delay in opening the valve from 10 to 5 ms. The resulting system is referred to as the modified on-off system.

For determining the performance of the vehicle and that of the wheel, the rate of rise of pressure is assumed to be 1466.67 kgf/cm$^2$/s. In addition, the following adaptive feature is used: if the deceleration of the vehicle is above 0.5 $g$, the rate of decay of pressure is 183.33 kgf/cm$^2$/s; otherwise, it is 1466.67 kgf/cm$^2$/s.

The results for the four different speeds on different roads are presented in Table 5. By comparing the results in Table 5 with the results in Table 3, the following conclusions are drawn: if the modified on-off system is used, the performance of the wheel and that of the vehicle is improved under most operating conditions. On low-friction surfaces such as road C, the maximum wheel slip is decreased; on high-friction surfaces such as road A (see also Fig. 8), the maximum slip is increased. The average deceleration of the vehicle under all the conditions is improved, except on roads B and C, when the initial speed is 15 m/s. When the modified on-off system is used, on roads B and C the rear wheel will lock intermittently if the velocity of the vehicle is not higher than 5 m/s. However, the time the wheel is locked on road C is less when the modified on-off

#### Table 4 - Values of $k_6$ for On-Off System

|  | Initial Velocity of 5 m/s | Initial Velocity of 15 m/s | Initial Velocity of 25 m/s | Initial Velocity of 35 m/s |
|---|---|---|---|---|
| Road A | 0.536 | 0.740 | 0.820 | 0.886 |
| Road B | 0.440 | 0.730 | 0.850 | 0.930 |
| Road C | 0.620 | 0.780 | 0.900 | 0.930 |

#### Table 5 - Performance of Wheel and Vehicle in First Cycle when Modified On-Off System Is Used

| Road | Initial Velocity, m/s | Average Deceleration of Vehicle, m/s$^2$ | Maximum Wheel Slip | Desired Recovery Time, ms | Time Wheel is Locked, ms |
|---|---|---|---|---|---|
| A | 5 | 6.905 | 0.530 | 118 | 0 |
| A | 15 | 6.338 | 0.290 | 116 | 0 |
| A | 25 | 5.904 | 0.281 | 136 | 0 |
| A | 35 | 5.417 | 0.285 | 156 | 0 |
| B | 5 | 5.979 | 1.000 | 76 | 146 |
| B | 15 | 4.243 | 0.449 | 54 | 0 |
| B | 25 | 2.178 | 0.407 | 85 | 0 |
| B | 35 | 1.371 | 0.498 | 208 | 0 |
| C | 5 | 3.616 | 1.000 | 43 | 16 |
| C | 15 | 2.969 | 0.205 | 32 | 0 |
| C | 25 | 2.061 | 0.415 | 88 | 0 |
| C | 35 | 1.371 | 0.498 | 208 | 0 |

system is used than when the on-off system is used. On road B, the wheel is locked at low speeds for about 146 ms, whereas when the on-off system is used, the wheel is locked for only 79 ms; this longer time for which the wheel is locked on road B cannot be regarded as a disadvantage of the modified on-off system because locking of the wheel on road B at low speeds does not constitute a danger to the stability of the vehicle.

CONDITION FOR RESELECTION IN MODIFIED ON-OFF SYSTEM - Because the largest desired recovery time with the modified on-off system is the same as the maximum desired recovery time with the on-off system, the threshold value in the time delay condition need not be changed. For determining the desired angular velocity factor $k_6$, the procedure outlined earlier is followed and the corresponding results are presented in Table 6.

The operating range of speeds may be divided into three parts as before. If the vehicle speed is above 20 m/s, it is in the high-speed range; if it is not above 10 m/s, it is in the low-speed range. If the speed of the vehicle is between 10-20 m/s it is in the medium-speed range. The results in Table 6 suggest that $m_1$ should be about 0.90, $m_2$ should be about 0.85, $m_3$ should be about 0.75, and $m_4$ should be about 0.40.

MERITS AND DEMERITS OF MODIFIED ON-OFF SYSTEM - When the modified on-off system is used instead of the on-off system, the stability of the vehicle is improved on low-

## Table 6 - Values of $k_6$ for Modified On-Off System

|  | Initial Velocity of 5 m/s | Initial Velocity of 15 m/s | Initial Velocity of 25 m/s | Initial Velocity of 35 m/s |
|---|---|---|---|---|
| Road A | 0.500 | 0.740 | 0.850 | 0.800 |
| Road B | 0.400 | 0.790 | 0.900 | 0.940 |
| Road C | 0.630 | 0.850 | 0.900 | 0.940 |

friction surfaces; the average deceleration of the vehicle under most of the operating conditions is also improved. False predictions could be completely eliminated.

The modified on-off system suffers from one disadvantage—that is, the maximum wheel slip on road C is relatively high, even though it is not as high as when the on-off system is employed. Further reduction in maximum wheel slip may be obtained by limiting the wheel cylinder pressure. However, it is not possible to limit the pressure with an on-off modulator unless an additional valve is incorporated. The following sections are devoted to the study of the systems incorporating a modulator with two solenoid valves.

## DESIGN CONSIDERATIONS OF ADAPTIVE BRAKE CONTROL SYSTEM USING FOUR-PHASE MODULATOR

So far, it is presumed that a modulator having only one valve is available. As explained above, in a four-phase modulator it is possible to modulate the pressure in such a way that its fluctuation is limited to a small range. Thus, in the Teldix modulator, which has three solenoid valves, it is possible to operate the valves independently to maintain either a high constant pressure or a low constant pressure in the wheel cylinder. Furthermore, when two inlet valves and an outlet valve are present, the effective rate of rise of pressure may be altered by using a pulsating signal for monitoring the inlet valves (7). Similarly, the effective rate of decay of pressure may be changed by using a pulsating signal for monitoring the outlet valve.

The idea of using a pressure-limiting or brake-proportioning device is not new (9). However, pressure proportioning is more difficult to arrange than pressure limiting. Moreover, it is also observed while making the calculations that pressure proportioning adversely affects the effectiveness of the braking system on high-friction surfaces because it reduces the effective rate of rise of pressure. Therefore, only a pressure-limiting feature is used in the four-phase system.

To evaluate the performance of the wheel and that of the vehicle, it is assumed that the valve time delays are 5 ms. The compound condition for prediction is the same as the one used in the on-off system, except that the threshold value $k_4/k_3$ is changed from 2.50 to 3.00 $s^{-1}$; hence, at high speeds, false predictions are completely eliminated. Maximum wheel cylinder pressure is still 135 kgf/cm$^2$, but the pressure is

limited to 85 kgf/cm$^2$ if the deceleration of the vehicle is not above 0.5 $g$.

The rate of rise of pressure is not changed. As the pressure is limited to 85 kgf/cm$^2$ on the low-friction surfaces, the optimum rate of decay of pressure on the low-friction surfaces has to be determined. For this purpose, the performance of the vehicle and that of the wheel for different rates of decay of pressure has been found. The results, which are represented in Figs. 10 and 11, indicate that when the vehicle speed is low (about 5 m/s), the rate of decay of pressure should be as low as 183.33 kgf/cm$^2$/s, because with this rate of decay of pressure the average deceleration of the vehicle is high and the maximum wheel slip is only 20%. When the speed is above 5 m/s, the rate of decay of pressure should be increased in order to obtain an improvement in the performance of the wheel. When the speed of the vehicle is above 15 m/s, a high rate of decay of pressure not only decreases the maximum slip, but also increases the average deceleration of the vehicle.

Accordingly, the following adaptive feature is used for controlling the rate of decay of pressure. If the deceleration of the vehicle is above 0.5 $g$ or if the speed of the vehicle is lower than 15 m/s, the rate of decay of pressure has to be adjusted to a value of 183.33 kgf/cm$^2$/s; otherwise, it has to be 1466.67 kgf/cm$^2$/s. The results, which relate the performance of the four-phase system on the three roads (roads A, B, and C), are given in Table 7.

By comparing the results in Table 7 with those in Table 5, the following conclusions are drawn: On road C, for the initial speeds 5, 25, and 35 m/s, both the vehicle and the wheel performance are improved when the four-phase system is used instead of the modified on-off system. When the initial speed is 15 m/s, the vehicle performance improves, but the wheel performance deteriorates, in that the maximum wheel slip increases from 0.205 to 0.350.

On road B, both the vehicle and wheel performance are improved when the four-phase system is used for all conditions, except when the initial speed is 5 m/s. If the initial speed is 5 m/s on road A or on road B, the average deceleration of the vehicle is low when the four-phase system is used instead of the modified on-off system.

CONDITION FOR RESELECTION IN FOUR-PHASE SYSTEM - A review of the desired recovery times in Table 7 reveals that a time delay $D_t$ of 500 ms will be adequate to allow the wheel to recover from impending danger of locking. Then the reselection is assumed to take place at a slip value of 0.75 $S_c$ and the desired angular velocity factor $k_6$ is determined for different roads and different speeds. From the results (Table 8), it is evident that the speed range may be divided into three parts, as has been done before.

In the present case, the condition for reselection has to be altered in the following way: the desired angular velocity factor $k_6$ should be made equal to $m_2$ only if the vehicle is moving on a low-friction surface at a medium speed. Or if $M_s$ F = 1 (see Appendix); then $k_6 = m_2$.

If the vehicle is moving on a high-friction surface and if the

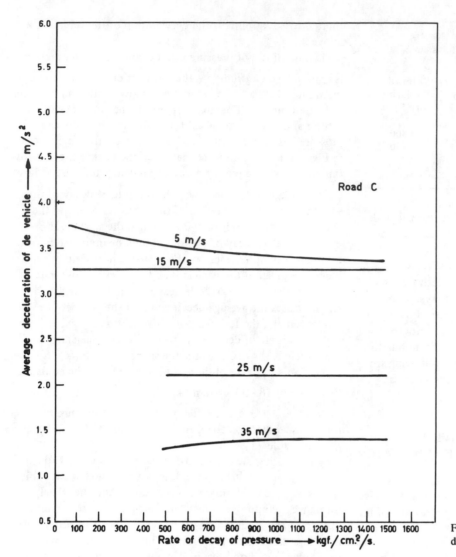

Fig. 10 - Effect of rate of decay of pressure on average deceleration of vehicle

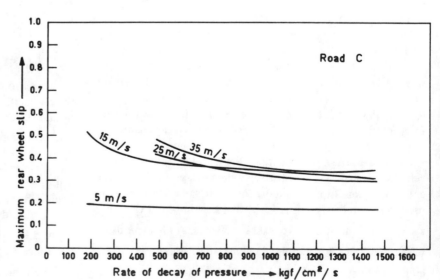

Fig. 11 - Effect of rate of decay of pressure on maximum rear wheel slip

46

Table 7 - Performance of Wheel and Vehicle with Four-Phase System

| Road | Average Deceleration of Vehicle, $m/s^2$ | Maximum Wheel Slip | Desired Recovery Time, ms | Time Wheel is Locked, ms | Initial Velocity, m/s |
|---|---|---|---|---|---|
| A | 6.730 | 0.292 | 140 | 0 | 5 |
| A | 6.437 | 0.290 | 180 | 0 | 15 |
| A | 5.978 | 0.291 | 232 | 0 | 25 |
| A | 5.420 | 0.720 | 374 | 0 | 35 |
| B | 5.971 | 0.688 | 144 | 0 | 5 |
| B | 4.264 | 0.449 | 53 | 0 | 15 |
| B | 2.212 | 0.301 | 62 | 0 | 25 |
| B | 1.418 | 0.316 | 124 | 0 | 35 |
| C | 3.750 | 0.200 | 120 | 0 | 5 |
| C | 3.213 | 0.350 | 45 | 0 | 15 |
| C | 2.130 | 0.306 | 74 | 0 | 25 |
| C | 1.418 | 0.316 | 124 | 0 | 35 |

Table 8 - Values of $k_6$ for Four-Phase System

| | Initial Velocity of 5 m/s | Initial Velocity of 15 m/s | Initial Velocity of 25 m/s | Initial Velocity of 35 m/s |
|---|---|---|---|---|
| Road A | 0.540 | 0.725 | 0.760 | 0.740 |
| Road B | 0.517 | 0.783 | 0.900 | 0.943 |
| Road C | 0.750 | 0.823 | 0.880 | 0.943 |

vehicle speed is not low, or if it is moving on a low-friction surface at a low speed, $k_6$ should be equal to $m_3$.

If

$$(\overline{L}_s \cdot \overline{F} + L_s \cdot F) = 1$$

$$k_6 = m_3 \qquad (1)$$

The value of $m_1$ should be about 0.90; the value of $m_2$ should be about 0.8; the value of $m_3$ should be about 0.75; the value of $m_4$ should be about 0.50.

ADVANTAGES AND DISADVANTAGES OF FOUR-PHASE SYSTEM - In general, the four-phase system gives improved performance of the wheel and the vehicle under most operating conditions. However, the four-phase system still suffers from one disadvantage: false predictions at low speeds are not avoided.

MODIFIED FOUR-PHASE SYSTEM

To eliminate false predictions and to lower the maximum slip on low-friction surfaces, the software of the four-phase system may be modified. If the deceleration of the vehicle is above 0.5 $g$, the value of $k_2$ is made equal to 2.25 $g$ and the

Table 9 - Performance of Wheel and Vehicle with Modified Four-Phase System

| Road | Initial Velocity, m/s | Average Deceleration of Vehicle, $m/s^2$ | Maximum Wheel Slip | Desired Recovery Time, ms | Time Wheel is Locked, ms |
|---|---|---|---|---|---|
| A | 5 | 6.994 | 0.530 | 160 | 0 |
| A | 15 | 6.437 | 0.290 | 180 | 0 |
| A | 25 | 5.978 | 0.292 | 232 | 0 |
| A | 35 | 5.420 | 0.720 | 906 | 0 |
| B | 5 | 5.986 | 1.000 | 67 | 146 |
| B | 15 | 4.264 | 0.449 | 53 | 0 |
| B | 25 | 2.134 | 0.227 | 52 | 0 |
| B | 35 | 1.411 | 0.284 | 116 | 0 |
| C | 5 | 3.750 | 0.200 | 120 | 0 |
| C | 15 | 2.870 | 0.142 | 24 | 0 |
| C | 25 | 2.073 | 0.229 | 52 | 0 |
| C | 35 | 1.411 | 0.284 | 116 | 0 |

Table 10 - Values of $k_6$ for Modified Four-Phase System

| | Initial Velocity of 5 m/s | Initial Velocity of 15 m/s | Initial Velocity of 25 m/s | Initial Velocity of 35 m/s |
|---|---|---|---|---|
| Road A | 0.445 | 0.725 | 0.760 | 0.740 |
| Road B | 0.465 | 0.780 | 0.908 | 0.950 |
| Road C | 0.750 | 0.880 | 0.905 | 0.950 |

value of $k_4/k_3$ is made equal to 3.00 $s^{-1}$; otherwise, these threshold values are 1.5 $g$ and 2.00 $s^{-1}$, respectively.

The calculations have been made taking into account the above changes in the software, and the results are presented in Table 9.

Evidently, on the high-friction surfaces, if the speed is low (about 5 m/s), the performance of the modified four-phase system is better than that of the four-phase system. On road A, if the speed is not below 15 m/s, the performance of the modified four-phase system is the same as that of the four-phase system.

On road C, if the speed is low, the performance of the modified four-phase system is the same as that of the four-phase system. On road C, if the speed of the vehicle is not below 5 m/s, the average deceleration of the vehicle is decreased when the modified four-phase system is used instead of the four-phase system. However, on low-friction surfaces, if the speed is higher than 5 m/s, the proposed modification to the software brings a reduction in the value of the maximum slip.

CONDITION FOR RESELECTION - The threshold value in condition $A_r$ need not be changed. The desired angular velocity factor $k_6$ for the different operating conditions is determined; the results are presented in Table 10. From the results, it is gathered that the condition $D_r$ should be similar to the one used in the four-phase system. The value of $m_1$ should be 0.90; the value of $m_2$ should be 0.85; the value of $m_3$ should be about 0.75; the value of $m_4$ should be about 0.45.

ADVANTAGES AND DISADVANTAGES OF MODIFIED FOUR-PHASE SYSTEM - The modified four-phase system offers one advantage over the four-phase system; that is, it eliminates false predictions. However, on low-friction surfaces, the average deceleration of the vehicle is decreased. It is difficult to say with conviction whether or not the modification to the software is necessary unless additional information about the allowable average and maximum slip is available. However, as the differences in the maximum slip on road C with and without the modification are small, it is concluded that the change in the value of $k_4/k_3$ need not be effected.

The adaptive feature used to change the threshold value $k_2$ is useful, because when it is used the effectiveness on high-friction surfaces is improved.

## FINAL CHANGES TO FOUR-PHASE SOFTWARE

The modified four-phase system does not improve performance under all conditions. To improve the performance of the four-phase system on high-friction surfaces at low speeds, the value of $k_2$ is to be changed in a stepwise manner. On high-friction surfaces (if the deceleration of the vehicle is above 0.5 $g$), the value of $k_2$ should be 2.25 $g$; on low-friction surfaces it should be 1.5 $g$.

Operation of the four-phase system at high speeds and on high-friction surfaces is improved if a high rate of decay of pressure is used. Thus, on road A, if the initial speed is 35 m/s, a change in the rate of decay of pressure from 183.33 to 1466.67 kgf/cm$^2$/s will decrease the maximum slip from 0.72 to 0.53.

The results for the four-phase system may be summarized as follows: $k_4/k_3 = 3.00$ s$^{-1}$, $k_8 = 0.7$ $g$, $D_t = 500$ ms, $k_9 = 0.5$ $g$, $k_{10} = 20$ m/s, $k_{11} = 10$ m/s, and $k_{12} = 85$ kgf/cm$^2$. On high-friction surfaces, $k_2 = 2.25$ $g$ and on low-friction surfaces, $k_2 = 1.5$ $g$. The rate of rise of pressure is 1466.67 kgf/cm$^2$/s. The values of the desired angular velocity factor $k_6$ are given in Table 11.

## DISCUSSION OF RESULTS

It is gathered that the effectiveness is increased by increasing the rate of rise of pressure. It is also noted that the maximum slip may be decreased by decreasing the valve time delays, by decreasing the threshold values in the conditions for prediction, by limiting the maximum pressure in the wheel cylinder, or by increasing the rate of decay of pressure. On high-friction surfaces, the recovery of the wheel from a condition of incipient locking may be slowed down by decreasing the rate of decay of the pressure, whereas, on low-friction surfaces, in order to reduce the maximum slip the rate of de-

Table 11 - Values of Factors $k_2$ and $k_6$

| | High-Friction Surface | | | Low-Friction Surface | | |
|---|---|---|---|---|---|---|
| | High Speed | Medium Speed | Low Speed | High Speed | Medium Speed | Low Speed |
| $k_2$ m/s$^2$, $g$ | 2.25 | 2.25 | 2.25 | 1.5 | 1.5 | 1.5 |
| Rate of decay of pressure kgf/cm$^2$/s | 1466.67 | 183.33 | 183.33 | 1466.67 | 1466.67 | 183.33 |
| $k_6$ | 0.75 | 0.75 | 0.45 | 0.90 | 0.80 | 0.75 |

cay of pressure should be high. However, once the wheel has started to accelerate, there is no danger of increasing the wheel slip in the same cycle; therefore, further recovery of the wheel may be slowed down by closing the outlet valve. Thus, the effectiveness of the braking system may be improved by allowing the wheel slip to decrease slowly.

In short, it is advantageous to operate the control system in such a way that the wheel quickly reaches critical slip during the brake-on period. Then the control system should limit maximum slip and allow the wheel to recover from an impending danger of locking. Once the wheel has started to accelerate, the control system should allow the wheel to recover slowly.

If the approach to the problem of slip control that has been outlined above is compared with the current practice, a striking difference will be noted. In existing antilock systems, a low rate of rise of pressure is used on low-friction surfaces even at low speeds; in addition, a high rate of decay of pressure is preferred under all conditions of operation (7).

## CONCLUSIONS

Even if the time delay in the signal processing unit is taken into account, general conclusions regarding the methods of prediction and reselection (1) will hold good. When the time delay is present, the threshold values have to be decreased; owing to the time delay, the low-speed reselection condition $F_r$, mentioned (1), cannot be used. However, after effecting suitable changes in the threshold value, the desired angular velocity condition can be used instead of the low-speed reselection condition $F_r$. The four-phase system gives improved performance only if the software is changed suitably.

## ACKNOWLEDGMENTS

The author wishes to thank Professor ir. H.C.A. van Eldik Thieme, director of the Vehicle Research Laboratory (Delft), for permission to publish this paper.

## REFERENCES

1. R. R. Guntur and H. Ouwerkerk, "Adaptive Brake Control System." Proceedings IME, Automobile Division, Vol. 186.

2. H. Ouwerkerk and Ramachandra Rao Guntur, "Skid Prediction." Vehicle System Dynamics, Vol. 1, No. 2, 1972.

3. James B. Pond, "Goodrich Goes On-Highway With Anti-Skid." Automotive Industries, October 1972.

4. J. W. Douglas and J. C. Schafer, "The Chrysler Sure-Brake—The First Production Four-Wheel Anti-Skid System." Paper 710248 presented at SAE Automotive Engineering Congress, Detroit, January 1971.

5. R. H. Madison and H. E. Riordan, "Evolution Of Sure-Track Brake System." SAE Transactions, Vol. 78 (1969), paper 690213.

6. R. R. Guntur and H. Ouwerkerk, "Onderzoek En Ontwikkeling Betreffende Antiblokkeerinrichtingen Voor Per-sonenauto-Remsystemen." Laboratorium Voor Voertuigtechniek, Technische Hogeschool, Delft.

7. H. Leiber and A. Czinczel, "Der Elektronische Bremsregler Und Seine Problematik." A.T.Z., Vol. 74, No. 7, 1972.

8. H. Ouwerkerk and K. Rop, "Meting En Analyse Wielhoekversnellingen Simca 1100 Bij Constante Rijsnelheden." Rapport No. A 134, Laboratorium Voor Voertuigtechniek, Technische Hogeschool, Delft.

9. H. Strien, "Trends In Development Of Vehicle Brakes and Anti-Skid Devices In Europe." Paper 3040 presented at SAE Automotive Engineering Congress, Detroit, January 1961.

# APPENDIX

The following notations are used in describing the software:

| | |
|---|---|
| $\omega$ | = angular velocity of wheel |
| $\dot{\omega}$ | = angular acceleration of wheel |
| $\dot{V}$ | = acceleration of vehicle |
| $\omega_b$ | = angular velocity of wheel at point of braking in each cycle |
| $P_{wr}$ | = pressure of fluid in rear wheel cylinder |
| $k_2, k_3, k_4, k_6, k_8, k_9, k_{10}, k_{11}, k_{12}$ | = constants |
| $m_1, m_2, m_3, m_4$ | = constants |
| $A_p, B_p, C_p$ | = conditions for prediction and also corresponding binary logic variables |
| $A_r, D_r$ | = conditions for reselection and also corresponding binary logic variables |
| $C_R, L_s, H_s, M_s, F, BPR$ | = various conditions and corresponding binary variables |
| RPHI, RPHIC, RPHID | = command signals (binary logic variables) |
| $D_t$ | = time delay in fixed time delay condition |

## SOFTWARE TO BE USED IN CONJUNCTION WITH ON-OFF MODULATOR

As there is only one solenoid valve to be monitored in an on-off system, only one command signal is required. The command signal is a digital signal with one value when the valve has to be opened and the other value when the valve has to be closed. As the valve has to be opened whenever there is a danger of locking the wheel, a digital signal should be produced to indicate the performance of the wheel. The process of finding whether or not the wheel is on the point of locking is known as prediction and the method employed for this purpose is known as the method of prediction. Once the valve is opened it is necessary to find out whether or not danger of locking is averted; the process of finding whether or not danger of locking the wheel is averted is called reselection.

METHOD OF PREDICTION - The relative advantages and disadvantages of several methods of prediction have already been discussed elsewhere (1). The most satisfactory method makes use of a compound condition, which is a combination of the fixed threshold deceleration condition and the variable threshold deceleration condition. While making the calculations, it is assumed that this method of prediction is employed.

The fixed threshold deceleration condition is satisfied whenever the deceleration of the wheel exceeds a predetermined value; it is satisfied whenever the following inequality is satisfied:

$$\dot{\omega} R_r \geq k_2 \qquad (A-1)$$

The above condition will be referred to as $A_p$.

The variable threshold deceleration condition $B_p$ is satisfied whenever the deceleration of the wheel has exceeded a threshold value that is proportional to the angular velocity of the wheel. In other words, the variable threshold deceleration condition is satisfied whenever the following inequality is satisfied:

$$- k_3 \dot\omega - k_4 \omega \geqslant 0 \qquad (A\text{-}2)$$

or

$$- \dot\omega / \omega \geqslant k_4/k_3 \qquad (A\text{-}3)$$

The compound condition for prediction named $C_p$ is given by the following relation:

$$C_p = A_p \cdot B_p \qquad (A\text{-}4)$$

Therefore, in making all the relevant calculations, the braked wheel is predicted to be on the point of locking whenever both conditions $A_p$ and $B_p$ are satisfied.

METHOD OF RESELECTION - In our previous study (1), the relative advantages of several methods of reselection were enumerated. In order to give satisfactory performance under all circumstances, a compound condition has to be used in the method of reselection. A necessary but not a sufficient condition is $\overline{C}_p$, which is satisfied only when the compound condition for prediction is no longer satisfied. The additional compound condition for reselection should consist of three conditions: a fixed time delay condition, a desired angular velocity condition, and a low-speed condition. However, for reasons that are given in this paper, the low-speed condition proposed in Ref. 1 cannot be used. Presently, it is proposed to alter the threshold value of the desired angular velocity conditions at low speeds, and use this condition as the low-speed condition. Thus, in reality, the compound condition consists of only two conditions: the fixed time delay condition and the desired angular velocity condition. The fixed time delay condition, which is denoted by $A_r$, requires for

its fulfillment that a fixed time has elapsed during the brake release period. The time for this condition should be taken from some point during the brake-off period, but in order to ensure reapplication only after the wheel has actually started to accelerate, this time has to be taken from the point at which the acceleration of the wheel has exceeded a predetermined value. Ideally, the time has to be taken from the zero acceleration point, as mentioned in Ref. 1, but, owing to the noise on the acceleration signal, a positive threshold value has to be used. Thus, the time for the fixed time delay condition is taken from the point at which the following inequality is satisfied:

$$\dot\omega \, R_r > k_8 \qquad (A\text{-}5)$$

The above condition is referred to as $C_R$.

If the time $\tau$ is taken from the point at which the condition

$C_R$ is satisfied, the fixed time delay condition is satisfied only when the following inequality is satisfied:

$$\tau \geqslant D_t \qquad (A\text{-}6)$$

In formulating the desired angular velocity condition denoted by $D_r$, the desired angular velocity is computed in the following way. At first the signal proportional to the angular velocity of the wheel at the point of application of the brake in each cycle is stored in a simple track and hold circuit. The desired angular velocity of the wheel is assumed to be proportional to the already held value. If the held signal is equal to $\omega_b$, the desired angular velocity is equal to $k_6 \omega_b$, where the value of $k_6$ depends on the acceleration of the vehicle and also on the stored value $\omega_b$. The desired angular velocity condition is satisfied whenever the following inequality is satisfied:

$$\omega \geqslant k_6 \omega_b \qquad (A\text{-}7)$$

As the desired angular velocity should depend on the condition of the road and also on the speed of the vehicle, the following additional conditions are also used to alter the value of $k_6$. To determine whether the vehicle is moving on a high-friction surface or a low-friction surface, the vehicle deceleration condition is used. This condition, which is denoted by F, is satisfied whenever the following inequality is satisfied:

$$- \dot V \leqq k_9 \qquad (A\text{-}8)$$

The operating range of vehicle speed is divided into three parts. For this purpose, $R_r \omega_b$ is assumed to indicate the vehicle speed. If $R_r \omega_b$ is higher than a predetermined value, then the vehicle speed is in the high-speed range; if it is lower than another predetermined value, the vehicle speed is in the low-speed range. The condition to determine whether or not the vehicle speed is in the high-speed range, which is denoted by $H_s$, is satisfied whenever the following inequality is satisfied:

$$R_r \omega_b \geqslant k_{10} \qquad (A\text{-}9)$$

The low-speed condition $L_s$ is satisfied whenever the following inequality condition is satisfied:

$$R_r \omega_b \leqq k_{11} \qquad (A\text{-}10)$$

If the two conditions $H_s$ and $L_s$ are not satisfied, then the vehicle speed is in the medium-speed range. Thus, the medium-speed condition $M_s$ is given by the following relation:

$$M_s = \overline{(H_s + L_s)} \qquad (A\text{-}11)$$

If the vehicle is moving on a low-friction surface at a high speed, $k_6$ is made equal to $m_1$. Or if $H_s \cdot F = 1$, then $k_6 = m_1$.

If the vehicle is moving on a high-friction surface at high speed or if it is moving on a low-friction surface at medium speed, $k_6$ is made equal to $m_2$. Or if $(H_s \cdot \bar{F} + M_s \cdot F) = 1$, then $k_6 = m_2$.

If the vehicle is moving on a high-friction surface at medium speed or if it is moving on a low-friction surface at low speed, $k_6$ is made equal to $m_3$. Or if $(M_s \cdot \bar{F} + L_s \cdot F) = 1$, then $k_6 = m_3$.

If the vehicle is moving on a high-friction surface at low speed, $k_6$ is made equal to $m_4$. Or if $L_s \cdot \bar{F} = 1$, then $k_6 = m_4$.

Suppose the digital signal required for operating the single solenoid valve of the on-off modulator is RPHI, then, as indicated elsewhere (1), the signal RPHI should satisfy the following relation:

$$RPHI = A_p \cdot B_p + RPHI \cdot (\overline{A_r + D_r}) \qquad (A\text{-}12)$$

Thus, as long as the digital signal RPHI is one, the wheel is on the point of locking. The danger of locking is averted only when RPHI is zero.

The condition F is also used for the purpose of changing the rate of decay of pressure.

## SOFTWARE TO BE USED IN CONJUNCTION WITH FOUR-PHASE MODULATOR

The digital signal RPHI used in conjunction with the on-off system is also used in the four-phase system. In the case of the four-phase system, when the signal RPHI attains one value the inlet valve is closed and the outlet valve is opened. Thus, under some circumstances, operation of the four-phase system is similar to that of the on-off system. However, under other circumstances, the four-phase system operates differently as its software includes two additional conditions.

The digital signal for monitoring the inlet valve is formulated in the following way. The digital signals RPHI and F are obtained, as has been mentioned above. In addition, a pressure-limiting condition BPR is used; the condition BPR is satisfied whenever the following inequality is satisfied:

$$P_{wr} \geqslant k_{12} \qquad (A\text{-}13)$$

The digital signal RPHIC given by the following relation is used for monitoring the inlet valve:

$$RPHIC = RPHI + F \cdot BPR \qquad (A\text{-}14)$$

Thus, on a high-friction surface, the inlet valve is closed only when the digital signal RPHI is equal to one. On low-friction surfaces, the inlet valve is closed either when the digital signal RPHI is equal to one or when the pressure has attained a predetermined value.

The digital signal RPHID given by the following relation is used for monitoring the outlet valve:

$$RPHID = RPHI \cdot \bar{C}_R \qquad (A\text{-}15)$$

Thus, the dump valve is opened as long as both conditions are satisfied. When RPHI is zero and $C_R$ is not satisfied, the outlet valve is closed and the inlet valve is opened; in this case, the operation of the four-phase system is similar to that of the on-off system. When RPHI is one and condition $C_R$ is satisfied, the outlet valve will be closed, but the inlet valve will not be opened; in this case, the operation of the four-phase system will be different from that of the on-off system.

# Wheel Lock Control Braking System*

**Robert A. Grimm**
AC Spark Plug Div.
General Motors Corp.

*Paper 741083 presented at the International
Automobile Engineering and Manufacturing
Meeting, Toronto, Canada, October, 1974.

THE WHEEL LOCK CONTROL BRAKE SYSTEM to be discussed in this paper has been in production for several years. This system (rear wheel control only) allows the driver to achieve the following braking advantages during maximum braking stops:

1. Better lateral stability control may be achieved by automatically pumping the rear brake. This prevents continuous rear wheel lock-up which is one cause of rear-end skidding.

2. Shorter stopping distances may be generally achieved by automatically providing the average rear brake pressure necessary for maximum stopping force.

These advantages are generally achieved over a wide variety of road surfaces, weather conditions, and driving situations. However, the extent of the wheel lock control brake system advantages is determined by many factors: road surfaces, weather conditions, driver proficiency, vehicle speed, tire tread wear, tire inflation pressure, and the condition of other vehicle systems, including brakes and suspension components.

During less than maximum braking stops, for example, when the rear wheel brake pressure is insufficient to cause lock-up, the brake system operates in the normal manner. During maximum braking stops, when the rear brake pressure would be sufficient to cause lock-up the wheel lock control brake system senses the impending lock-up and automatically releases and applies the rear brakes as required to control wheel speed. This cycling rate is approximately four cps and continues throughout the maximum braking stop until the car is slowed to approximately five mph, or until the brakes are released by the driver.

After a review of each system component and its respective function, there follows a discussion of the requirements of the system as a whole, and of the electronic control unit in particular.

## SYSTEM COMPONENTS

The wheel lock control brake system is composed of the following component parts (Fig. 1): speed sensors, controller, solenoid valve, modulator, brake system warning light, and electrical harness.

SPEED SENSORS - The front wheel drive vehicle sensor consists of a magnet, pole piece, and a coil (Fig. 2). This device senses the presence of a magnetic material. When the magnetic material is in close proximity to the pole piece, the lines of flux from the magnet are collected as shown by the solid lines ($B_1$). When the magnetic material is replaced by air, the lines of flux collapse to position B as shown. The voltage induced into the coil is proportional to the frequency at which the magnetic material moves by the pole piece. On the front wheel drive vehicle, there are slots in the hub which turns with each rear wheel (Fig. 3). The probe is mounted to each brake backing plate with its tip in close proximity to the hub. Since there are 90 teeth on the hub, and the tire's rolling

---

ABSTRACT

Automobile and truck manufacturers have given increasing attention to electronic wheel lock control brake systems during the last few years.

These systems prevent continuous wheel lock-up during maximum braking stops, thus aiding the driver in retaining lateral stability and generally improving stopping distances.

This presentation discusses a system for preventing continuous rear wheel lock-up of an automobile during maximum braking stops. Included is a description of the control system components, tire and road characteristics, brake and vehicle dynamics, and an analysis leading to the requirements for optimum control.

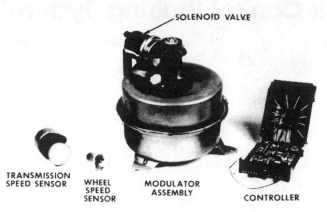

Fig. 1 - Wheel lock control system components

Fig. 4 - Transmission speed sensor

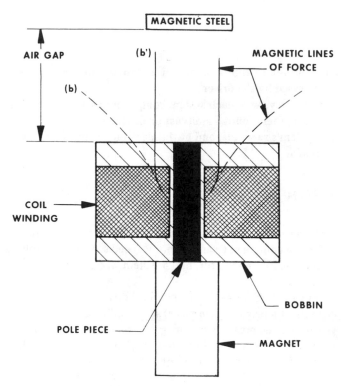

Fig. 2 - Wheel speed sensor

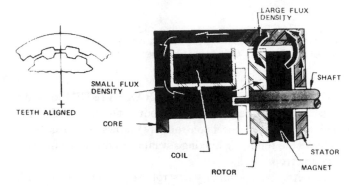

Fig. 5 - Cross section transmission speed sensor

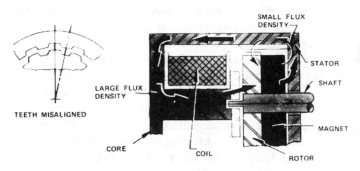

Fig. 6 - Cross section transmission speed sensor

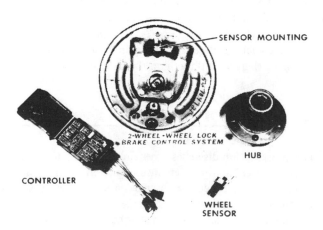

Fig. 3 - 2-wheel-wheel lock brake control system

radius is approximately one ft, the frequency of the voltage induced into the coil is 18.75 Hz/mph.

The transmission speed sensor (Fig. 4) used on rear wheel drive vehicles is a mechanically driven electromagnetic device which produces an a-c voltage with frequency proportional to the input shaft speed. Basic parts of the sensor include a coil, permanent magnet, stator, and rotor (Fig. 5).

Both the stator and rotor have a number of teeth or grooves equally spaced around their circumferences. The magnetic circuit has two parallel paths. One is through the stator and rotor teeth, and the other directly links the coil. When the stator and rotor teeth are aligned (Fig. 5), the flux linking the coil is small. When the stator and rotor teeth are not aligned (Fig. 6), the flux linking the coil is larger. The change in flux linking the coil induces an a-c voltage in the coil. The frequency of this voltage is proportional to the number of teeth and speed of the rotor.

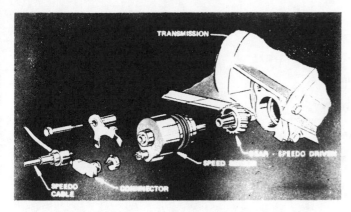

Fig. 7 - Typical mounting transmission speed sensor

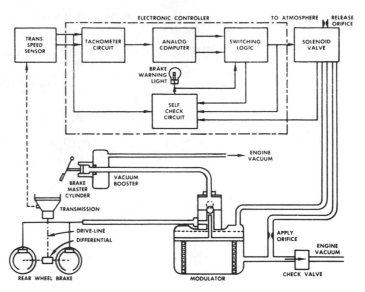

Fig. 8 - Pictorial diagram wheel-lock control system

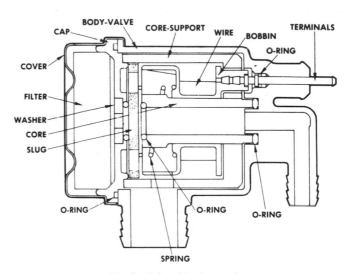

Fig. 9 - Solenoid release valve

A typical transmission speed sensor installation is illustrated (Fig. 7). The sensor mounts in the transmission in place of the speedometer driven gear assembly. Both the driven gear and flexible shaft coupling are integral parts of the sensor. Since the rear wheels of a rear-wheel drive vehicle are connected through the differential and transmission, the one transmission speed sensor produces an average speed signal. With 90 teeth on the stator, 15-in wheels, and a gear ratio of 1.33:1, the resulting transmission sensor output frequency is 25 Hz/mph of vehicle speed.

CONTROLLER - The controller receives and continuously monitors the input speed signals from the two wheel speed sensors in the case of a front-wheel drive vehicle, or from the transmission speed sensor in the case of a rear-wheel drive vehicle.

Its basic functions are to determine from the input speed signal when rear wheel lock-up is impending, to generate an electrical output which automatically releases the rear brakes, to determine when the impending rear wheel lock-up situation ceases to exist, and to remove the output signal which again applied the rear brakes. It interconnects with the other system components. (Fig. 8)

Also illustrated is the fact that the controller contains additional failure warning circuitry which is used in conjunction with the brake system warning light.

SOLENOID VALVE - The solenoid valve is a three-way pneumatic valve which is operated by the electromagnetic action of the coil. It is mounted in a grommet as an integral part of the modulator. Sensing that rear-wheel lock-up is imminent, the controller sends an electrical signal to the solenoid valve. Its basic function is to convert this electrical signal into a pneumatic signal which is used by the modulator.

This section view (Fig. 9) illustrates the internal parts, construction, magnetic operation, and pneumatic operation. There are two pneumatic inputs to the valve. Atmospheric air is supplied through a filter, and engine vacuum is supplied through a hose connection. These two pneumatic inputs may be alternately connected to the modulator through a large output port by the action of the valve.

With no electrical signal supplied to the solenoid, there is no magnetic action. The spring pushes the slug against the apply mode O-ring seal. This blocks the flow of atmospheric air through the filter and allows vacuum to draw air through

the valve from the modulator. The resulting action of the modulator is to apply the rear brakes.

When the valve receives an electrical signal from the controller, the magnetic action of the solenoid opposes the spring and pulls the slug against the release mode O-ring seal. This blocks the vacuum supply and allows the modulator to draw atmospheric air through the filter and valve. The resulting action of the modulator is to release the rear brakes. During a maximum stop, the valve will continue to cycle, alternately supplying air or vacuum to the modulator.

MODULATOR - The modulator, as pictured with the integral solenoid valve, (Fig. 10) is mounted in the engine compartment. In response to the pneumatic input from the solenoid valve, its basic function is to control or cycle the hydraulic pressure to the rear brakes during maximum braking stops. This figure illustrates the component parts and construction of the modulator which is basically a spring-loaded, pneumatically operated, diaphragm actuator. Operated by the diaphragm is a piston which controls a hydraulic

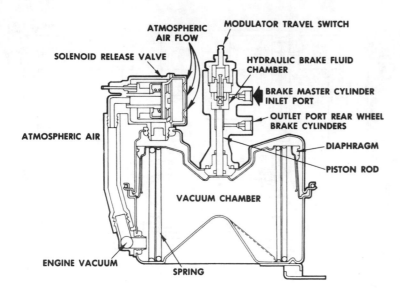

Fig. 10 - Modulator apply position

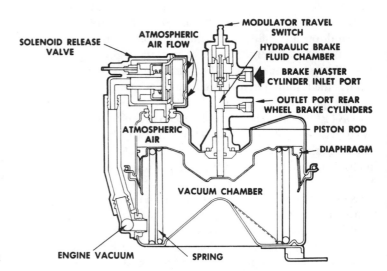

Fig. 11 - Modulator release position

check valve and travel switch. Also included as part of the overall modulator assembly are a pneumatic check valve and the solenoid valve discussed above.

In the apply mode, engine vacuum is connected through the pneumatic check valve to one side of the diaphragm and through the solenoid valve to the other side of the diaphragm. With the same pressure on both sides of the diaphragm, the spring pushes the diaphragm and piston to the top of the modulator. The piston opens the hydraulic check valve and connects the hydraulic pressure from the master cylinder, or proportioner, to the rear brake wheel cylinders. In the release mode, (Fig. 11), the vacuum connection through the solenoid valve is blocked and atmospheric air is connected through the solenoid valve to one side of the diaphragm. The resulting differential pressure across the diaphragm creates a force opposing the spring force and forces the diaphragm and piston toward the bottom of the modulator. When the piston moves down, the hydraulic check valve closes and disconnects the master cylinder from the rear brake wheel cylinders. Further movement of the piston increases the volume of the rear brake lines, reducing the rear brake pressure. During maximum brak-

ing stops, the modulator continues to cycle and pump the rear brake until the vehicle speed is reduced to 5 mph or until the brakes are released by driver.

The pneumatic check valve provides the extra vacuum capacity in case of loss of engine vacuum. The extra capacity is stored in the modulator and normally would be sufficient to provide several cycles of operation during one maximum braking stop.

## HYDRAULIC, PNEUMATIC, AND ELECTRICAL INTERCONNECTIONS

The total wheel lock control brake system for a front wheel drive vehicle is shown in Fig. 12. It shows the position of each component in the vehicle, approximate routing of the hydraulic, pneumatic, and electrical lines, and pictorial enlargement illustrating details of the hydraulic, pneumatic, and electrical connections. The wheel lock control brake system contains three separate functional control circuits; hydraulic, pneumatic, and electrical.

HYDRAULIC CONTROL CIRCUIT - This circuit is shown

56

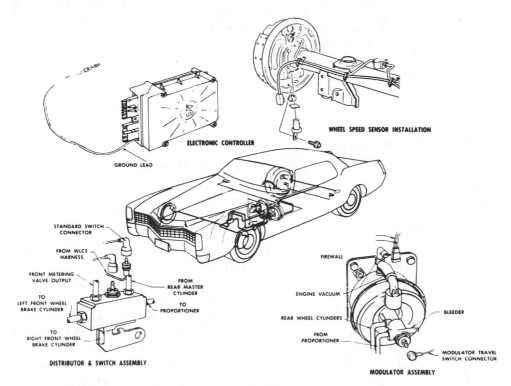

Fig. 12 - Installation wheel lock control system

in dashed lines (Fig. 13). In addition to the brake system components (dual master cylinder, front metering valve, distributor and switch assembly, front wheel cylinders, rear wheel cylinders, connecting brake lines, and proportioner), the hydraulic control circuit also includes the hydraulic check valve. This valve, which is an integral part of the modulator assembly, is connected between the master cylinder and rear brake wheel cylinders. During maximum braking stops, the hydraulic check valve controls the pressure applied to the rear brakes.

PNEUMATIC CONTROL CIRCUIT - This circuit, shown in double solid lines (Fig. 13) includes vacuum hose connection from the engine manifold, pneumatic check valve, valve section of the solenoid valve, and actuator section of the modulator assembly. The pneumatic control circuit provides the mechanical power necessary to operate the hydraulic check valve.

ELECTRICAL CONTROL CIRCUIT - This circuit is shown in solid lines (Fig. 13). For all cars, the electrical control circuit includes the controller, solenoid section of the solenoid valve, speed sensor, differential pressure switch, modulator travel switch, brake system warning light, and harness. Electrical power is supplied to the wheel lock control brake system when the ignition switch is turned to all positions except "Off" and "Accessory."

Having discussed the function of each of the components and their interface with the vehicle, we are now in a position to discuss the overall control philosophy.

## CONTROL PHILOSOPHY

A wheel under the influence of braking has two major tractive torques acting on it, brake torque and tire torque (Fig.

14). Brake torque arises from the application of brake pressure through the brake mechanism, and tire torque is generated by the friction of the tire-road interface as wheel slip occurs.

Brake torque, $T_B$, can be approximated as being proportional to brake pressure, $P_B$, with brake gain, $K_B$, the constant of proportionality:

$$T_B = P_B \cdot K_B$$

Tire torque, $T_T$, is related to the tire-to-road friction coefficient, $\mu$, as follows:

$$T_T = \mu N R$$

where N is the normal load on the tire, and R is the rolling radius of the tire.

The coefficient, $\mu$, is a nonlinear function of wheel slip, the latter being defined as:

$$S = \frac{(V - R \cdot W)}{V}$$

where V is vehicle velocity, and W is wheel angular velocity. Slip equals unity for a locked wheel, and zero for a freely rolling wheel.

For the free body consisting of the brake, wheel, and tire, the torque equation is

$$J\alpha + T_B - T_T = 0$$

57

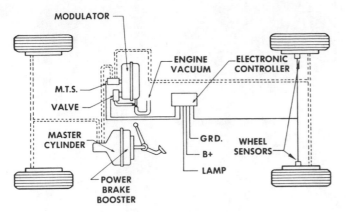

Fig. 13 - Typical vehicle system

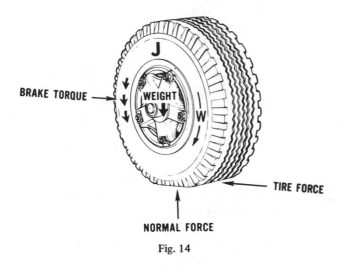

Fig. 14

where:

$J$ = wheel polar moment of inertia, lb-Ft-sec.$^2$,

$\alpha$ = wheel angular acceleration, Rad/Sec.$^2$.

Rearranging gives

$$\alpha = (1/J)(T_T - T_B)$$

when the difference between the tire torque and brake torque is positive, the wheel accelerates; and when negative, the wheel decelerates.

Once brake force coefficient is established, vehicle deceleration A is obtained as follows:

$$A = \frac{G}{W}(N_{LR}\mu_{LR} + N_{RR}\mu_{RR} + N_{RF}\mu_{RF} + N_{LF}\mu_{LF})$$

where

$G$ = acceleration of gravity (32.2 ft/s$^2$)
$W$ = vehicle weight,
$N$ = normal load on left rear wheel

Vehicle velocity is then obtained by integrating vehicle de-

celeration:

$$V = V_O \int_0^T A\text{-}dt$$

$V_O$ = initial vehicle velocity

$T$ = time

Vehicle stopping distance follows:

$$X = \int_0^T V\text{-}dt$$

Fig. 15 illustrates a variety of $\mu$-slip curves for various road surfaces. Curve "1" is typical of some pavements at highway speeds. Maximum braking is achieved at 100% slip. Curve is typical of pavement at lower speeds below 50 mph. Maximum braking is achieved at low wheel slip. Curve "3" is typical of wet pavements where maximum braking is significantly greater than 100% slip braking. Curve "4" is for wet ice. Braking force is very low, and there is no significant peak. Curve "5" is for loose gravel. 100% slip braking is significantly greater than braking at low wheel slip.

The initial portion of the $\mu$-slip curve is a linear relationship related to the tire type and condition. Smooth or bald tires tend to exhibit a steep rise in the $\mu$-slip curve initially, while snow tires show a lesser slope. The shape of the curve past the peak as a function of tire condition is not well documented.

Vehicle stability (rotationally) depends upon maintaining a significant lateral force at the rear wheels. (Fig. 16) This is an important part of the wheel lock control function.

Fig. 17 shows a typical $\mu$-slip curve with a lateral force curve superimposed. It is apparent from this data that one cannot expect to obtain maximum retarding force and maximum lateral stability simultaneously, since maximum lateral stability occurs at 0% slip and maximum braking force occurs in the range of 10-100% slip.

Optimum braking will be a compromise between maximum lateral stability and maximum braking force, generally requiring control of wheel slip between 0-25% (Fig. 18).

Referring again to Fig. 15, notice that some of the curves have a definite peak or extremal point. Wheel lock control systems have often been called extremal control systems because they generally cycle the wheels about this peak. Even though some surfaces such as gravel contain no true mathematical extremal point, (a point where slope of curve goes through zero) these exhibit a sudden reduction in slope in the neighborhood of 12% slip, which is sufficient to allow the systems to perform their basic functions satisfactorily. Thus, it is still meaningful to refer to these as extremal control systems. This terminology is better appreciated from the following: The brake force characteristic can also be interpreted as tire torque coefficient versus wheel slip (Fig. 19). Brake torque coefficient versus wheel slip can be plotted on these same coordinate axis as shown here. Wheel angular accelera-

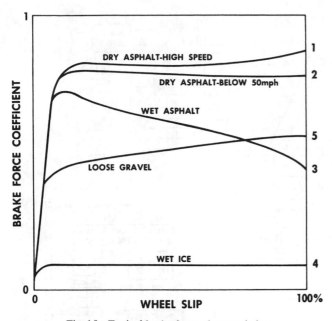

Fig. 15 - Typical brake force characteristics

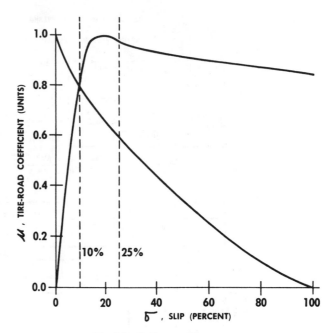

Fig. 18 - Optimum slip range

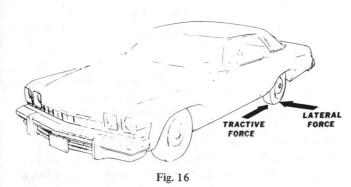

Fig. 16

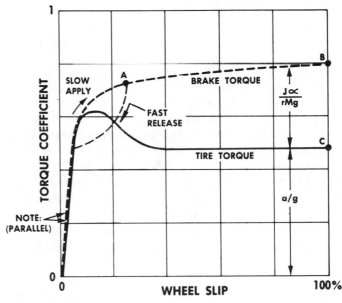

Fig. 19 - Dynamic trajectory

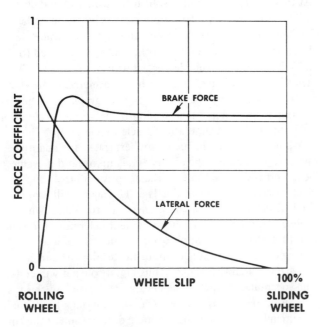

Fig. 17 - Typical tire-to-road force characteristics

tion is directly proportional to the difference between tire torque and brake torque. As the driver applies brake pressure, brake torque will eventually exceed tire torque, causing the wheel to decelerate and wheel slip to increase.

Fig. 19 shows brake torque increasing along the trajectory from 0 to A. If brake pressure continues to increase as with a standard braking system, the wheels will lock up at Point B. Brake torque will then decrease to Point C where it is equal to tire torque, since wheel acceleration goes to zero.

On the other hand, if a WLC system is installed, wheel lock will be averted. When wheel deceleration reaches a pre-selected value (Point A), the controller energizes the solenoid which ports atmosphere into the top of the modulator, (Fig. 11).

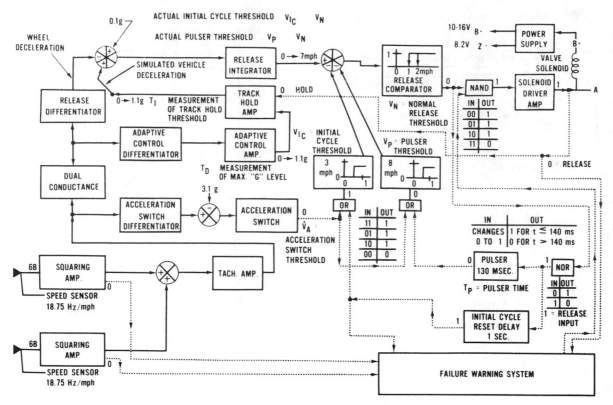

Fig. 20 - Failure warning system

The modulator diaphragm moves down, compressing the spring. The piston riding on the diaphragm moves down, allowing the hydraulic check valve to seal, preventing further increases in pressure. The piston continues to travel downward creating a chamber for the brake fluid in the rear system to occupy. Since the volume in the rear system has been increased, and the master cylinder checked off, the pressure applied to the brakes is reduced, thus releasing the brakes. As brake torque falls below tire torque, the wheel accelerates to Point D where the controller de-energizes the solenoid and brakes are reapplied. The system will continue to cycle in this fashion from Point D to A to D, etc., above and below the so-called extremal point.

Thus, one can see the intent of the wheel lock control function and it should be apparent why the extremal control terminology is sometimes used. Controlling the wheels to a predetermined slip level would be a simple task if one had continuous access to both vehicle speed and wheel speed information. However, true vehicle speed information is not available without employing an additional system component such as a fifth wheel or an accelerometer. Such a system would be completely impracticable, so it is an important objective to design the electronic controller in such a way that it is able to perform its necessary function with no information other than the wheel speeds. That which follows is intended to show how this is accomplished.

## ELECTRONIC CONTROLLER LOGIC

Examining the block diagram (Fig. 20), we see the speed sensors feed squaring-amplifiers to convert a varying amplitude and frequency signal from the wheel or transmission senders to a square wave of fixed amplitude and changing frequency. The output of the squaring-amplifiers feeds a tachometer-amplifier which converts the input frequency to a DC signal proportional to speed, approximately .090 volts per mph. The signal is then fed through a dual conductance filter to allow some filtering of the tachometer ripple without loss of sensitivity to a sudden decrease in voltage due to a sharp deceleration.

At this point, the circuit is split into two paths and changed from speed-sensitive to deceleration-sensitive through the use of differentiating circuitry. One path will retain wheel deceleration information and the other path will be used to establish a simulated vehicle deceleration.

The circuits used to simulate vehicle deceleration are the adaptive control amplifier and track-hold amplifier. The adaptive control amplifier filters and limits the wheel deceleration signal, establishing a simulated vehicle deceleration. The track-hold circuit, as the name implies, tracks the adaptive control signal and has the ability to clamp or hold its value. The circuit is designed so that the adaptive control follows wheel deceleration with a time lag. This signal will represent vehicle deceleration after the wheels start to brake. Wheel deceleration is coincident with vehicle deceleration for a short period following brake application, while tire torque is increasing on the linear synchronous portion of the curve.

Due to the adaptive control time lag, its output will continue to simulate vehicle deceleration, and based on this information, the wheels can be released. The track-hold amplifier remembers the simulated vehicle deceleration at the time of release. This hold capability allows continued reference to

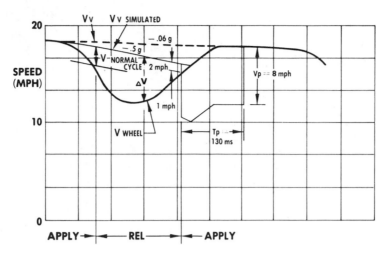

Fig. 21 - Second and later cycles low coefficient

simulated vehicle speed for us in computing reapply times. Based on this simulated vehicle deceleration signal, the brakes will be reapplied and the track-hold circuit released from the hold mode to again track the vehicle deceleration signal from the adaptive control amplifier. This can be done since the wheels are now back to synchronous speed and again represent vehicle speed.

The wheel deceleration and simulated vehicle deceleration information paths are brought back together for the release integrator input. Here the difference in the two signals is fed into the release integrator to convert the deceleration information back to speed information which will be the computed difference between vehicle speed and wheel speed. As long as the wheel speed is equal to the simulated vehicle speed, the output of the release integrator is 0 mph. To compensate for the small electrical differences between the wheel deceleration and simulated vehicle deceleration inputs to the release integrator, a 0.1 g reverse bias is introduced. This prevents the release integrator from giving a false output.

When the wheel deceleration becomes greater than vehicle deceleration, as in a hard brake application, the release integrator calculates the difference between wheel speed and estimated vehicle speed. The speed difference is fed into a release comparator which produces a brake release signal when the wheel speed drops approximately 2 mph below the vehicle speed, and produces the re-apply signal when the wheels have recovered to within approximately one mph of vehicle speed. The release comparator signals are then amplified to drive the solenoid valve.

Fig. 21 illustrates a typical cycle on a low coefficient surface. Note the true vehicle speed $V_v$, the simulated vehicle speed $V_v$, simulated wheel speed V wheel, and $\Delta V$ the controller's estimate of the difference between vehicle and wheel speed. V wheel appears at the tachometer amplifier output, and $\Delta V$ appears at the release integrator output (Fig. 20).

This covers the basic performance of the controller. However, testing revealed driving conditions during which the basic performance had to be modified.

INITIAL CYCLE CIRCUIT - To assure that the wheels are going toward lock-up before the WLC system is required to start cycling the brakes, the first cycle of operation must require a greater difference between vehicle and wheel speeds, approximately 5 mph instead of 2 mph (Figs. 22-23). After this initial cycle of operation, the system then returns to the basic performance as described earlier.

The inital cycle circuit works by sensing the output of the release comparator (Fig. 20). When this output is a non-releasing state 0, the initial cycle circuit injects in front of the release comparator a 3 mph speed signal. The release integrator then has to accumulate a 3 mph signal to overcome the initial cycle input and then the 2 mph required to turn on the release comparator or a 5 mph difference between vehicle and wheel speeds to obtain the first brake release in a skid stop. However, the initial cycle circuit is one-shot in nature and requires approximately one second to reset after its input changes back to the non-releasing 0 state.

Due to the short cycles during normal maximum effort stops, the initial cycle circuit never resets during a WLC stop, and the system continues to cycle based on a 2 mph release comparator threshold until the stop is completed.

PULSER CIRCUIT - During WLCS cycling on high coefficient surfaces, as the wheels recover and approach vehicle speed, the suspension dynamics can allow the apparent wheel speed to overshoot and go through damped oscillation as they recover to synchronous speed (Figs. 23-24). This wheel oscillation can cause premature releasing of the brakes and an increase in stopping distance. To prevent this, the pulser circuit (Fig. 20) senses when the release comparator asks for reapply of the brakes following which the oscillation would normally occur.

For a set period of time, approximately 130 ms, the pulser injects an 8 mph speed signal into the release comparator. Thus, the release integrator has to accumulate an 8 mph signal to overcome the pulser and a 2 mph signal to get a release signal from the release comparator (total 10 mph). This is greater than the normal speed oscillation. After the 130 ms,

61

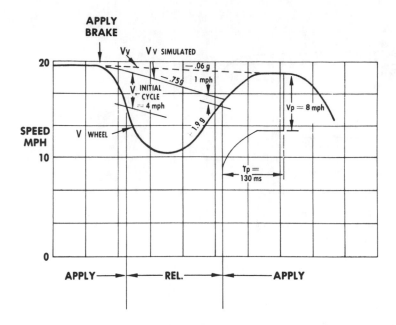

Fig. 22 - First cycle low coefficient

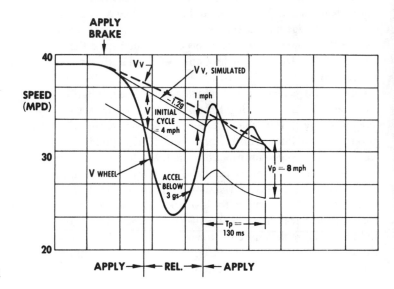

Fig. 23 - First cycle high coefficient road

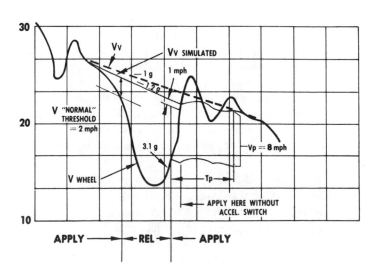

Fig. 24 - Second and later cycles high coefficient road

the 8 mph signal drops out and the system returns to the normal 2 mph release threshold. The shaded areas of Figs. 21-24 depict the time duration and amplitude of the pulser threshold contribution.

ACCELERATION SWITCH CIRCUIT - To improve stopping distance on dry roads, it was desired that the system be returned to an apply condition before the release comparator circuit would normally allow this to happen. To accomplish this, the acceleration switch is provided (Fig. 20). It differentiates the tachometer output to obtain wheel acceleration information. When the wheel acceleration exceeds 3 g's, a fast wheel recovery that cannot occur on low coefficient roads, the acceleration switch feeds in both the pulser and initial cycle threshold speed levels (Fig. 20). This combination of speeds (11 mph) can always overcome the release integrator output causing re-application of brakes. Fig. 24 illustrates the fact that on high coefficient surfaces, the acceleration switch is indeed capable or providing an earlier reapply signal than would otherwise result from the release comparator logic by itself (wheel speed recovered to within 1 mph of the simulated vehicle speed).

## SUMMARY

The Wheel Lock Brake System employing the extremal control techniques have been in production for several years and have performed very satisfactorily.

## ACKNOWLEDGMENTS

This system was developed for application on General Motors Vehicles under the guidance of AC Spark Plug Div., Buick Motor Div., Cadillac Motor Car Div., Delco-Moraine, and Oldsmobile Div.

The author wishes to express his gratitude for the patience and assistance of the engineering personnel of these Divisions of General Motors.

## REFERENCES

1. John L. Harned and Laird E. Johnston, "Anti-Lock Brakes." GM Engineering Staff Publication 3760, July 1968.

2. David Van Ostrom, "Adaptive Control Anti-Lock Brake System." U.S. Patent No. 3,709,567.

3. J. H. Moran and R. A. Grimm, "Max Trac—Wheel Spin Control by Computer." Paper 710612 presented at SAE International Mid-Year Meeting, Montreal, June 1971.

# Influence of Antiskid Systems on Vehicle Directional Dynamics**

**E. Bisimis**
Alfred Teves GmbH
Frankfurt, Germany

**Paper 790455 presented at the Congress and Exposition, Detroit, Michigan, February, 1979.

THE PURPOSE OF THIS PAPER is to present some results concerning the changes of steering behavior of cars, due to the influence of braking forces, particularly when an anti-skid system is used to prevent wheel locking.

Modern antiskid systems have reached a high technological standard; sophisticated sensing devices as well as hydraulic and electronic control units enable these systems to fulfil their main functional requirement, namely adequate control of brake fluid pressure for lock-free braking under a wide variety of external conditions. However, steering behavior of a car during combined braking and turning maneuvers will, for a number of reasons, always deviate from the steering behavior at constant velocity, to which the driver is mainly accustomed. The principal reason is the manner in which tires behave, when longitudinal and lateral tire forces interact. It is the intention of the simplified analysis presented in this paper, to derive a quantitative approximation of the changes of steering response due to the mentioned tire properties. This will be used to determine the influence of an antiskid system, the configuration of which will also be briefly described.

## 1. THEORETICAL BACKGROUND AND METHOD USED

For a passenger car moving straight ahead at constant speed, a useful representation of the directional dynamics can be done in terms of the yaw rate response (1-3)*. This is governed by the transfer function

$$\left(\frac{\dot{\psi}}{\delta}\right) = \left(\frac{\dot{\psi}}{\delta}\right)_{ss} \frac{1 + T_r s}{1 + \frac{2D}{V}s + \frac{1}{V^2}s^2} \qquad (1)$$

where

| | |
|---|---|
| $\dot{\psi}, \delta$ | = yaw rate, steering wheel angle |
| $\left(\dfrac{\dot{\psi}}{\delta}\right)_{ss}$ | = gain factor |
| $T_r$ | = numerator time constant |
| $D$ | = damping ratio |
| $V$ | = undamped natural frequency |

*Numbers in parentheses designate References at end of paper.

───────── ABSTRACT ─────────

The results presented demonstrate the influence of longitudinal tire slip and load transfer during braking on steering behavior of cars in terms of parameters which have been found to be important for the function of the driver/vehicle control loop. This is followed by a description of an antiskid system and a comparison of directional properties during braking with conventional brake system and with antiskid. The importance of appropriate selection of slip levels during adaptation of antiskid systems to a given vehicle is pointed out.

Interpretation of the transfer function Eq. (1) is as follows: A rapid steer input of $\delta_0$ (at constant forward velocity) will cause the vehicle to follow a circular path, after a certain transition time which depends on the parameters $T_r$, $D$, $V$. The yaw velocity at steady state will be

$$\dot{\psi}_0 = \delta_0 \ (\frac{\dot{\psi}}{\delta})_{ss}. \qquad (2)$$

That is, the gain factor determines the necessary magnitude of steer control for a desired steady state motion, whereas numerator time constant, damping and natural frequency govern the transition effect, the key factors of which are peak response time, overshoot, transition time.

The magnitude of the parameters of the yaw transfer function depend on vehicle data, forward velocity, tire and road surface data. The corresponding functional relationships are listed in the Appendix. The cornering coefficients of the tires (slope of the lateral force-slip angle curve in the linear range) are of particular importance for the purposes of this study, since they are strongly dependent on longitudinal tire slip which is the result of braking forces acting at the tire-road interface.

A change of cornering coefficients due to longitudinal tire slip will cause changes of all parameters of the yaw transfer function. It is important to note, that these parameters are not independent since they are all functions of vehicle and tire data as shown by the Equations listed in the Appendix. Hence, it will be, in general, sufficient to investigate the behavior of two of these parameters (1), to demonstrate the influence of braking forces on steering dynamics. The most suitable parameters to do this, are yaw rate gain factor and numerator time constant. A relationship between these parameters and subjective driver opinion has been found (1), (4). According to this, acceptable steering dynamics at constant speed can be expected, if $(\dot{\psi}/\delta)_{ss}$ and $T_r$ are within a specific optimum region.

It is, of course, not possible to use these results without any restriction to describe braking maneuvers, since they are based on linear vehicle behavior at constant speed. The necessary assumptions are: (1) step input of braking torque which results in constant remaining longitudinal tire slip after a short transition effect; (2) the influence of variable speed on the parameters of the yaw transfer function can be neglected for a sufficient short time period during braking. This can be interpreted in practical terms as follows: At initial speed $v_0$, the brakes of the straight ahead moving car are rapidly actuated. After a certain time period, at speed $v_1$, the driver recognizes the necessity of a steering maneuver, to avoid an obstacle. He will then have to handle a car whose steering behavior, as described by the initial change of the parameters $T_r$ and $(\dot{\psi}/\delta)_{ss}$ deviates (more or less unexpectedly, depending on driver skill) from that at the same speed, but without braking.

To determine the cornering coefficients as functions of longitudinal slip, an analytical tire model (5), (6) will be used. Input for this model are vehicle speed, longitudinal and lateral tire slip, wheel load. The model uses a set of parameters describing tire-road conditions and yields longitudinal as well as lateral tire forces.

## 2. STEERING BEHAVIOR OF CARS DURING BRAKING WITH CONVENTIONAL BRAKING SYSTEM

To determine the values of $T_r$ and $(\dot{\psi}/\delta)_{ss}$ for various braking conditions, the following procedure was used:

2.1) Select vehicle data, compute brake pressure distribution.

2.2) For given vehicle deceleration, compute necessary braking forces $F_x$ on front and rear axle, compute longitudinal load transfer, determine actual wheel loads.

2.3) Select a small value of slip angle $\alpha$, select a value for longitudinal slip $s_x$.

2.4) Run the tire model program, compute trial braking force $\overline{F}_x$ and lateral force $F_y$.

2.5) Adjust value of $s_x$ and repeat step 2.4 until deviation between $F_x$ and $\overline{F}_x$ is small enough. Compute cornering coefficient $\gamma = F_y/\alpha$.

2.6) Enter cornering coefficients and vehicle data into the equations given in the Appendix, to determine initial values of $T_r$ and $(\dot{\psi}/\delta)_{ss}$.

Some results of computation examples are shown in Fig. 1 and 2 (vehicle used and tire data are listed in the Appendix). In these diagrams the cornering coefficients for front and rear tires are plotted versus normalized brake line pressure. Corresponding values of vehicle deceleration are also shown in the plots.

These data reflect the effects of longitudinal slip and load transfer. It can be seen that the cornering coefficients remain almost constant up to decelerations of $\sim 6$ m/s$^2$. In this region, severe decreases of tire cornering ability became apparent. At wheel lock the cornering ability is almost zero.

The effects of reduced tire cornering coefficients on the key steering control parameters $(\dot{\psi}/\delta)_{ss}$ and $T_r$ are shown in the examples of Fig. 3 and Fig. 4. It can be observed from the plot of Fig. 3 that, at a deceleration of 8 m/s², the decrease of yaw gain factor is 40%. The gain factor contains the inverse value of steering ratio. Thus, for illustration, a 40% decrease of gain factor is comparable to an unexpected 67% increase of steering ratio. At wheel lock, the gain factor becomes negative, indicating unstable vehicle behavior. The yaw time constant $T_r$ (Fig. 4) increases rapidly in the region between $\ddot{x} \approx 8$ m/s² and wheel lock. Increasing $T_r$ makes steering control increasingly difficult according to experimental findings (1), (2).

## 3. BASIC CONFIGURATION OF ANTISKID SYSTEM

Modern antiskid systems are designed to prevent wheel lock on all wheels of a car by sensing wheel angular speed, processing the sensor signals in suitable digital electronic control circuits and command electrohydraulic actuators to control brake fluid pressure.

Fig. 5 shows a simplified block diagram in which the main antiskid system components are indicated. The controlled elements (braked wheels) are subject to a variety of external disturbances such as variable inertia (engaged or disengaged clutch), additional engine braking torques, friction coefficient variations in the brake as well as in the tire-road interface, dynamic wheel load fluctuations, influence of side forces. Regulating against all these unpredictable disturbances requires a considerable amount of complex nonlinear control loops which must

operate fast enough to give the system the required high response speed. In addition, the system must have the capability of detecting automatically a wide variety of possible mechanical, hydraulic and electric sub-system failures. Upon detection of faulty sub-system behavior, the system must react in a fail/safe manner, that is, partial or total turn-off, generation of warning

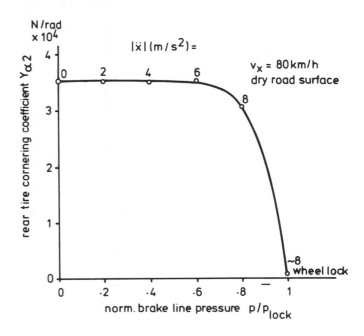

Fig. 2 - Rear tire cornering ability during braking

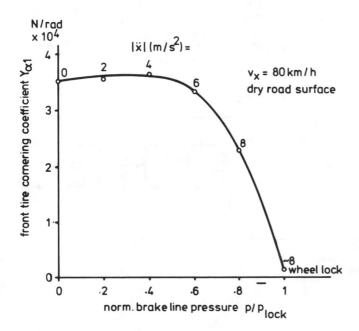

Fig. 1 - Front tire cornering ability during braking

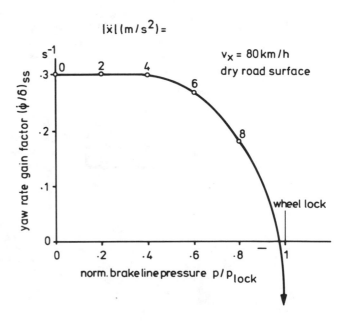

Fig. 3 - Influence of braking on yaw rate sensitivity

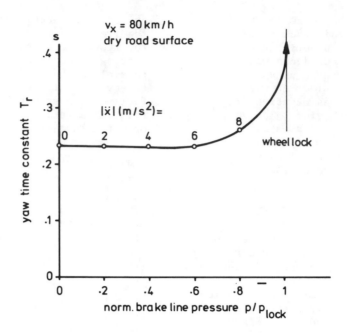

Fig. 4 – Influence of braking on yaw rate
time constant

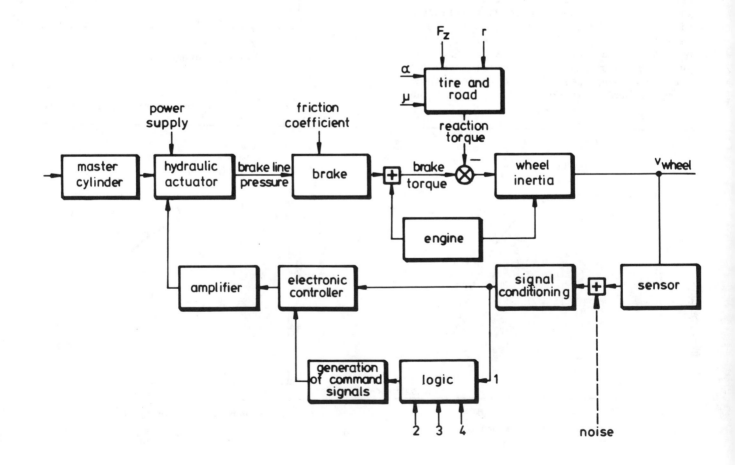

Fig. 5 – Block diagram of antiskid system

signal for the driver and return to conventional brake operation which must be guaranteed under any circumstances.

Extensive experimental testing of the antiskid system shown in the diagram of Fig. 6, under a variety of severe conditions and simulated component failures, has been carried out. A number of redesign and repeated testing steps of the various components was necessary before the present fully satisfactory standard could be reached.

A subject of special interest for system development and assessment is the adaptation of an antiskid system to a given vehicle. Initially, the major requirement was to achieve shortest possible brake distances. For an antiskid system, this means that longitudinal slip must be controlled in the region of maximum longitudinal force value. However, because of the growing interest in the braking-in-a-turn maneuver as a measure of motor vehicle performance, further require-

ments concerning controllability and stability receive increasing attention.

For the antiskid system shown in Fig. 6, it is possible to adjust brake line pressure slopes to maintain mean longitudinal slip in different regions for front and rear wheels. The effects of this on steering behavior will be described in the following paragraph.

4. STEERING BEHAVIOR DURING BRAKING WITH ANTISKID SYSTEM

In Fig. 7, computed cornering coefficients are plotted versus longitudinal tire slip, for two different vertical wheel loads, representing the front and rear wheel loads, including longitudinal load transfer at a high deceleration level on dry road surface. It can be observed from this diagram that the amount of cornering coefficient decrease due to tire slip is considerably influenced by the actual wheel load. If the antiskid

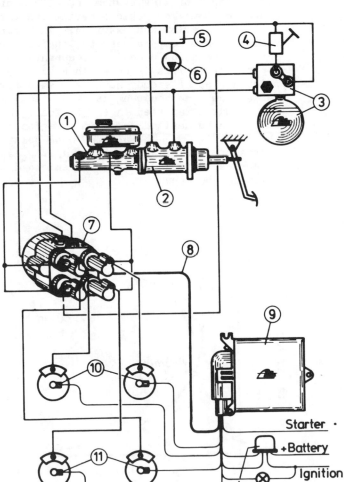

1) Tandem Master Cylinder

2) Hydraulic Booster H 31

3) Accumulator with loading valve

4) Power steering

5) Reservoir

6) Energy source (power steering pump)

7) Hydraulic control unit (4 control channel for brake-system with diagonal brake circuit split)

8) Cable harness with connectors

9) Electronic controller

10) Sensors (front wheels)

11) Sensors (rear wheels)

12) Relay

Starter

+ Battery

Ignition

Fig. 6 - Components of antiskid system combined with hydraulic booster system is adjusted to maintain the same mean slip at front and at rear wheels (setting 1, Fig. 7) then the loss of cornering ability at the rear wheels will be significantly higher than at the front wheels. This would lead to unacceptable steering behavior, as reflected by the values listed in Fig. 7 for the parameters $(\psi/\delta)_{ss}$ and $T_r$. If, on the other hand, the antiskid system is adjusted to maintain lower slip values at the rear tires (setting 2, Fig. 7) the ratio of cornering coefficients for front and rear axle remains almost constant. The corresponding values for gain factor and yaw time constant also listed in Fig. 7 indicate a

considerable improved steering behavior in comparison to the previous discussed setting.

Corresponding results for wet road surface are presented in Fig. 8. The wheel load difference is lower here than in Fig. 7 due to the lower achievable maximal deceleration.

Again, two different possible settings are shown. The optimum slip value for maximum normalized longitudinal force is considerably lower than for dry road. With appropriate setting of the antiskid system (setting 2, Fig. 8) the values of the cornering coefficients are almost the same as for dry road. Hence, the same quality of steering behavior can be expected.

## 5. COMPARISON OF RESULTS

Thus far, the influence of braking with conventional braking system and with antiskid system on some key vehicle parameters for steering behavior was examined. In order to summarize and compare results, the kind of data presentation used in (4) appears to be appropriate. This is shown in Fig. 9 in which a plot of steady state yaw velocity gain versus yaw time constant is presented. Tentative results reported in (1) indicate that there can be defined an optimum region as shown in Fig. 9. The plotted data points represent initial values for $(\psi/\delta)_{ss}$ and $T_r$ at a speed of 80 km/h for the vehicle data given in the Appendix.

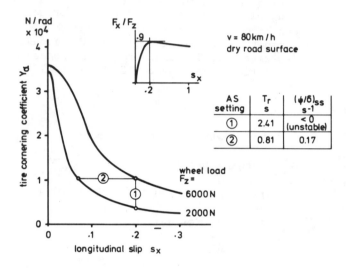

Fig. 7 - Cornering ability during braking with antiskid

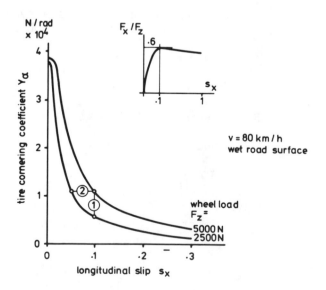

Fig. 8 - Cornering ability during braking with antiskid

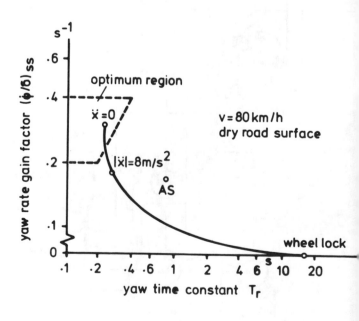

Fig. 9 - Influence of braking with conventional braking system and antiskid on vehicle response

For zero deceleration the vehicle is represented in Fig. 9 by a point within the optimum region. Braking on dry road surface with sufficiently high coefficient of friction leads to decreased gain factor and increased yaw time constant. If wheel lock occurs, the vehicle becomes uncontrollable, as shown in Fig. 9 by a point lying far away from the optimal region. Braking with antiskid raises the yaw gain factor again to an acceptable level. However, yaw time constant is significantly higher than the value corresponding to the boundary of the optimum region.

## 6. CONCLUDING REMARKS

Longitudinal tire slip and load transfer during braking are key factors having vital importance for cornering capability and steering behavior of cars. Modern antiskid systems can reliably prevent wheel lock. However, this major functional requirement is not sufficient for safe car behavior during combined steering and braking maneuvers. The aim of this paper has been to give an approximate quantification of these relationships. It should be further pointed out that correct adaptation of an antiskid system to a given car is of great importance.

## REFERENCES

1. D.T. McRuer, et.al., "Automobile Controllability - Driver/Vehicle Response for Steering Control." NHTSA Repts. DOT HS-801 407 and DOT HS-801 406, Feb. 1975.

2. E. Bisimis, "The Use of System Identification Methods for the Determination of Vehicle Parameters from Measured Time Histories", 5th VSD - 2nd IUTAM (Symposium on Dynamics of Vehicles on Roads and Trucks, Wien, Sept. 1977.)

3. E. Bisimis, H.-D. Beckmann, R. Rönitz, A. Zomotor, "Lenkwinkel - Sprung and Übergangsverhalten von Kraftfahrzeugen ", ATZ 79 (1977), Nr. 12.

4. D.H. Weir, D.T. McRuer, "Review and Correlation of Driver/Vehicle Data." STI Report, Oct. 1976.

5. H. Dugoff, P.S. Fancher, L. Segel, "An Analysis of Tire Traction Properties and Their Influence on Vehicle Dynamics Performance." SAE Paper 700 377 (1970).

6. P. Wiegner, "Über den Einfluß von Blockierverhinderern auf das Fahrverhalten von Personenkraftwagen bei Panikbremsungen." Dissertation, TU Braunscheig, 1974.

## APPENDIX

### I. PARAMETERS OF YAW TRANSFER FUNCTION

Basic vehicle parameters are:

1. Total vehicle mass    $m$
2. Wheelbase    $l$
3. Distance mass center-front axle    $a$
4. Distance mass center-rear axle    $b$
5. Height of mass center    $h$
6. Effective cornering coefficient of front tire    $Y_{\alpha 1}$
7. Effective cornering coefficient of rear tire    $Y_{\alpha 2}$
8. Steering ratio    $i_L$
9. Forward speed    $v_x$

For yaw inertia, the approximation $I_{zz} = mab$ is used (1).

With $p = a\, Y_{\alpha 1}/b\, Y_{\alpha 2}$, the parameters of the yaw transfer function are as follows:

1. Steady state yaw rate gain

$$(\dot{\psi}/\delta)_{ss} = \frac{p}{i_L l}\, v_x \; \frac{1}{p + T_r \frac{v_x}{l}(1-p)}$$

2. Numerator yaw time constant

$$T_r = \frac{m}{2}\frac{v_x}{Y_{\alpha 2}}\frac{a}{l}$$

3. Damping ratio

$$D^2 = \frac{(1+p)^2}{4\left(p \pm T_r \frac{v_x}{l}(1-p)\right)}$$

4. Undamped natural frequency

$$\mathcal{V}^2 = \frac{1}{T_r^2}\left(p + T_r \frac{v_x}{l}(1-p)\right)$$

### II. VEHICLE AND TIRE DATA

The vehicle and tire/road data used for the calculated examples are

$$m = 1600 \text{ kg}$$
$$l = 2.60 \text{ m}$$
$$a = 1.17 \text{ m}$$

$$h = 0.55 \text{ m}$$
$$i_L = 20$$

Brake torque distribution rear/front = 0.316

The following tire/road data are taken from (6)

  1. Longitudinal stiffness

$$C_S = \phantom{0}96937 \text{ N} \quad \text{(dry)}$$
$$C_S = 118800 \text{ N} \quad \text{(wet)}$$

  2. Cornering stiffness

$$C_\alpha = 34368 \text{ N/rad} \quad \text{(dry)}$$
$$C_\alpha = 39243 \text{ N/rad} \quad \text{(wet)}$$

  3. Friction coefficient

$$\mu = f_o (1 - k v_g)$$
$$\text{with } v_g = v_x \sqrt{s_x^2 + \tan^2 \alpha}$$

and

$$f_o = 1.067 \quad \text{(dry)}$$
$$f_o = 0.74 \quad \text{(wet)}$$
$$k = 0.0146 \text{ s/m} \quad \text{(dry)}$$
$$k = 0.0235 \text{ s/m} \quad \text{(wet)}$$

# An Analytical Approach to Antilock Brake System Design**

**John W. Zellner**
Dynamic Research, Inc.

**Paper 840249 presented at the International Congress and Exposition, Detroit, Michigan, February, 1984.

## ABSTRACT

An analytical method applicable to design and development of antilock brake systems is described. Dynamic components of antilock systems---including vehicle, sensor, and modulator--are examined using nonlinear feedback control techniques. An overall design approach is illustrated via an example involving a motorcycle front brake and typical pneumatic modulator. A computer simulation is used to generate time and frequency responses of system components. These data are used to identify the preferred feedback structure. Results show that a stable antilock limit cycle can exist for wheel angular acceleration feedback, among other possibilities. Overall the method and results can provide additional insight into detailed requirements for antilock components and systems, and may hold potential for reducing development time and costs.

HISTORICALLY, ANTILOCK BRAKE SYSTEMS have usually been developed empirically, through iterative design and test techniques. This technical process is made especially challenging by the complexities of tire/roadway friction and dynamic behavior of vehicles and brake systems. To date, little insight has been provided by theoretical approaches, and the development process remains mostly empirical, with accompanying costs and risks.

Recently, other technologies are being brought to bear in the antilock area. These include, to mention a few: digital and microprocessor based systems; all mechanical antilock; and fluidic sensing and control. As these new technologies join the several dozen antilock systems in various stages of development and use--as reviewed in (1)* and elsewhere--the need for analytical input continues.

As part of such antilock development (2), a preliminary effort to analytically treat antilocks was undertaken and is reported here, in part.

The objective of this portion of the development study was to analyze antilock dynamics in a preliminary way; and to determine desirable and feasible feedback control structures (i.e., which variable to use as a feedback). A complementary goal was to identify antilock component response requirements. The approach used was largely analytical, based on nonlinear control systems analyses.

This paper focuses on the analytical method used, beginning with antilock and control system requirements known from past research. An example preliminary design for a motorcycle antilock is then used to illustrate the analytical approach.

## REVIEW OF ANTILOCK REQUIREMENTS

Basically, an antilock brake system is a device installed in the brake line between the upper brake system (manipulator lever, master cylinder) and lower brake system (caliper, pads, disc) which acts so as to prevent the braked wheel from locking up while the vehicle is in motion. The antilock system does this by automatically modulating the applied brake pressure so that the braked wheel maintains a tangential speed not much less than the vehicle's forward speed. By doing this the tires are able to retain most of their lateral force capability, allowing the vehicle to remain steerable, and the operator/ vehicle

---

* Numbers in parenthesis designate references at end of paper.

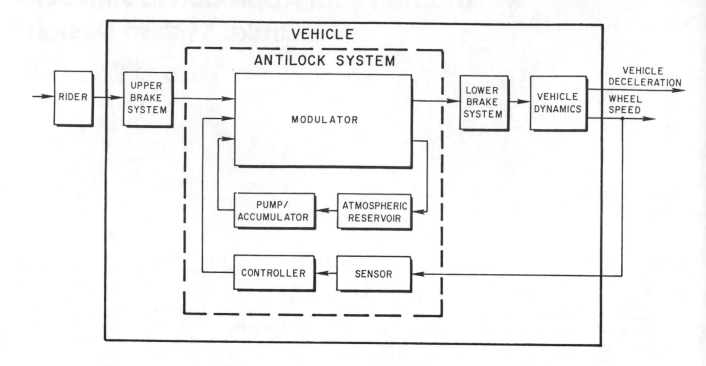

Fig. 1 - Schematic of Driver/Vehicle/Antilock System

system to remain stable. Also, in most cases the stopping distance is shortened in comparison to a locked wheel stop.

As shown schematically in Fig. 1, the antilock system consists of several key components. These are the

- Modulator
- Controller
- Sensor
- Pump (or energy source), and
- Reservoir (usually atmospheric pressure)

Together these elements, which can be implemented and integrated in a variety of ways, provide the antilock system function. Referring to Fig. 1, the operator provides the force command to the system. This, acting through the brake system, causes the vehicle and braked wheel to decelerate. Some sensed variable -- usually wheelspeed --is fed back to the antilock system "controller." When a certain threshold (of some function of wheelspeed) is exceeded, the controller unit generates a signal which causes hydraulic pressure to be reduced. As the wheel begins to recover speed, a second threshold is crossed and pump begins to reapply the brake effort. The system continues cycling until the vehicle comes to (nearly a) rest, or the brake command is reduced below the first threshold.

Antilock system requirements fall under literally dozens of different areas, for example response, performance, operational, reliability, etc. These have been reviewed elsewhere in some detail, e.g., (1). In this paper, the main focus is on response centered requirements, and some examples of these are excerpted from (1), below.

RESPONSE REQUIREMENTS - Antilock response requirements are determined by the:
- Open loop response of the vehicle, and
- Closed loop (vehicle/antilock system) response features.

These are briefly reviewed.

Open loop brake response - At the limit of performance, the braking response is nonlinear. Past analytical and experimental work has shown that, in the worst case, the response for motorcycle brake systems has:
- Wheel lockup times as rapid as 25 msec, from peak friction to full lock.
- Vehicle capsize divergence times (time to triple roll angle amplitude) of 0.2 to 0.3 sec for front wheel lockup.
- Nonlinear sensitivity to: shape of the input; initial conditions; and operating variables.

Each of these places demands on the antilock system design. The nonlinear sensitivity is especially significant, and many antilock systems have adaptive control features (i.e., controller parameters change, depending sensed conditions) to help account for this.

Closed loop brake response - The above considerations lead to requirements for closed loop antilock system response. Basically, the antilock system should:
- Eliminate the divergence in wheel speed, and reduce the error between

wheel speed and vehicle
speed to an acceptable
level.

In specific and practical terms this means
maintaining peak wheel slip ratio near the
break in the mu-slip curve (e.g., in the range
of 0 to 20 percent slip). To do this, all
successful antilock systems have had to:

- Establish the potential
  for a stable limit cycle.

The concept of antilock braking as a type of
limit cycle has been noted in the automotive
literature, yet the concept may deserve more
emphasis. Antilock systems can be said to fall
into a specific category of limit cycling
termed hard self excitation (3). This denotes
a system which is forced into sustained
oscillation by a specific set of circumstances
(in this case, a sufficiently large input and
nonzero forward velocity).

   Virtually all realizable antilock systems
do this by:

- Dumping the line pressure to
  atmosphere as rapidly as possible,
  upon detection of incipient lockup;
  then after a threshold delay,
- Reapplying the pressure in an open
  loop, ramp-like way, attempting to
  reach a compromise between
  excessive wheel slip and excessive
  resultant stopping distance.
- Establishing a stable limit cycle
  with a frequency of at least
  (usually) 2 Hz. (No known antilock
  cycles slower than about 1.5 Hz.)

In addition, some antilock systems use:

- Emulation, i.e., comparison of sensed
  wheel velocity to an "ideal" wheel
  velocity (e.g., assuming the initial,
  prelockup deceleration is maintained).
- Adaptive control, to account for
  differences in operating conditions.
  Usually this means allowing the wheel
  to recover more speed on slippery
  surfaces (i.e., when wheel deceleration
  is large) or when there is transmission
  or brake drag, before reapplying
  pressure.

   This type of on-off control system can be
analogous to "bang-bang" type controllers in
the presence of nonlinear controlled element
dynamics. These can be analyzed using either
classical methods or modern optimal techniques.
For reasons of simplicity and insight, the
former approach was selected.
   Note that to date, no published work using
either of these approaches, has been
encountered.
   Next, linear control design methods are
briefly reviewed.

## REVIEW OF CONTROL SYSTEM DESIGN PRINCIPLES

   Feedback control theory suggests that the
controller and controlled element of a system
be tailored to one another in order to result
in a system which has a balance among such
things as

- Stability,
- Rapid, well damped response to
  commands,
- Insensitivity to disturbances, and
- Insensitivity to small changes in
  controlled element or controller
  parameters.

In particular, a crossover model of a feedback
control system requires that:

- The series, open loop response of
  controller and controlled element look
  like an integrator times a gain, in the
  frequency range of interest, and

- The "frequency range of interest"
  is taken to be around the crossover
  frequency, which is the frequency at
  which adequate phase and gain margins
  exist, and which also represents the
  required bandwidth of the closed loop
  system (in the antilock case, 2 Hz or
  greater).

So, basically, the design of the feedback
controller involves: selecting the proper
feedback signal; and selecting the proper
controller "shape" or frequency response, to
meet the above criteria. The latter process is
referred to as "equalization", and can involve
proportional, integral, differential (PID),
discrete, and other types of controller
elements.

## SUMMARY OF APPROACH

   Based on the above considerations, the
preliminary analysis and design of an antilock
system can involve the following steps:

- Developing a mathematical model and
  simulation of the vehicle/brake system.
- Determining representative values of
  vehicle/brake system parameters, to run
  the simulation.
- Mathematically modeling the modulator.
- Analyzing the response of the various
  subsystems to determine desired
  feedback and controller shape.
- Verifying the closed loop functioning
  of the resultant preliminary design.

## DESIGN EXAMPLE

   The above design approach can be

illustrated by means of an example. The case selected is a motorcycle front wheel and a typical pneumatic/hydraulic modulator.

In this example, the simplest feedback system was sought, that would result in a feasible, workable control system.

The example proceeds with descriptions of the responses of the
- Vehicle
- Modulator, and
- Controller,

and concludes with feedback loop structuring and an example of resulting closed loop antilock response.

## VEHICLE DYNAMIC RESPONSE

A mathematical model of a motorcycle with independent front and rear brake systems was derived. A description of the vehicle model is presented in the Appendix. As an example, the model's response to sinusoidal inputs of brake lever force is shown in Fig. 2. The sinusoidal modulation occurs at a series of discrete frequencies (i.e., 1,2,...10Hz), with a dwell time of 1 second at each frequency. Responses are given in terms of vehicle velocity (UP), tire slip ratio (SR1), front wheel angular velocity (OM1), angular acceleration (OM1D), angular jerk (OM1DD) and caliper pressure (PSI1). The basic level of the input (FB1) is sufficient to cause lockup, in the absence of the sinusoidal modulation.

These responses can be used to determine the frequency responses of the several motion variables to hydraulic pressure variations. The technology for computing these frequency responses is well established, namely that of discrete (fast) Fourier transforms. For example, Okada (4) has developed and applied FFT techniques for handling transient response data, and Barter and Little (5) have applied such techniques to swept sinusoidal data as well.

Since the vehicle dynamics involve important nonlinearities, the resulting Bode diagrams correspond to sinusoidal "describing functions" of the vehicle/brake system. These can serve as fundamental building blocks of the antilock feedback control system design. Note that in general -- because of the presence of amplitude nonlinearites in the vehicle dynamics -- a family of describing functions is needed at a series of input amplitudes to more completely describe the response.

Resulting example Bode plots for hydraulic pressure inputs are shown in Figs. 3 to 6. These show the motion response amplitude ratio in decibels and phase angle in degrees as a function of frequency in radians per second.

The results show important trends. Wheel speed, acceleration and jerk differ by 20 db per decade slope and 90 degrees phase angle, as would be expected for differentiation. Slip ratio shows a loss of coherence in the mid-frequency region, probably due to the additional nonlinearities present in it. At

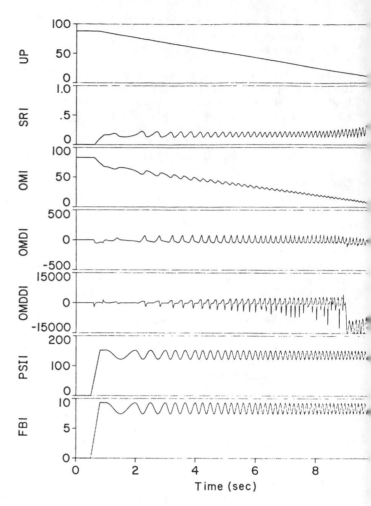

Fig. 2 – Simulated Vehicle Braking Response for Swept Sine Input

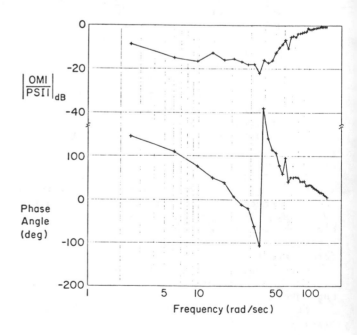

Fig. 3 – Frequency Response of Front Wheel Speed to Pressure Variation

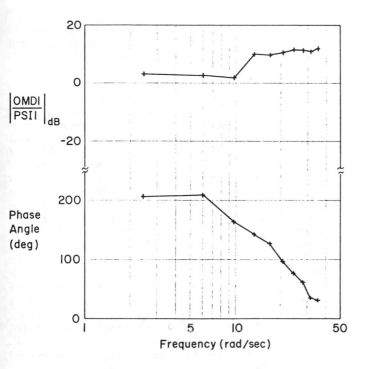

Fig. 4 - Frequency Response of Wheel Angular Acceleration to Pressure Variation

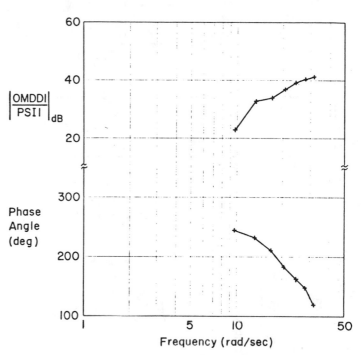

Fig. 5 - Frequency Response of Wheel Angular Jerk to Pressure Variation

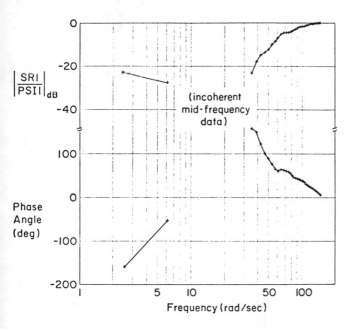

Fig. 6 - Frequency Response of Tire Slip Ratio to Pressure Variation

low and high frequencies, it appears generally similar to wheel speed, which is its main component.

The specific shapes and trends these Bode plots -- along with those of the modulator/controller -- determine the feasibility and quality of the various candidate feedbacks. The modulator and controller responses will be discussed subsequently.

## MODULATOR RESPONSE

So far, we have illustrated the response of the vehicle brake system portion of the overall system. Next it is appropriate to discuss the modulator contribution to system dynamics.

Figure 7 shows a schematic diagram of a typical brake modulator in its relation to the closed loop antilock brake system.

Virtually all low cost, automotive type antilock brake systems developed to date have used a discrete type modulator to vary the brake line pressure in response to controller commands. This is because a rapid cycle, dump/reapply device is the only known low cost solution to the requirement for rapid wheel lockup dynamics (lockups occurring as rapidly as 25 msec). Note that, in general, a linear type servoactuator with equivalent response characteristics could be an order of magnitude greater in cost and complexity.

In order to examine the response of a discrete type modulator, it was necessary to select an existing representative antilock modulator that would be appropriate for front wheel motorcycle application. A Mitsubishi pneumatic modulator recently introduced into the US market on the Starion sedan (6), is sized for rear axle use on a light vehicle, and could be suitable for touring motorcycle front wheel application. This modulator was selected to be modeled.

The Appendix describes the derived math model of this modulator. Figure 8 shows an example time response of vehicle motion to modulator command inputs. In other words the

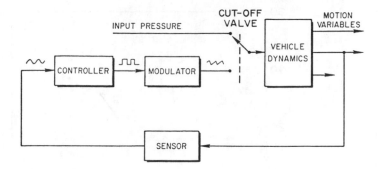

Fig. 7 – Schematic of Antilock System
Showing Relation of Controller
and Modulator

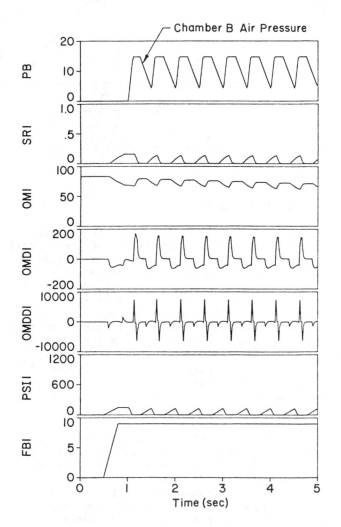

Fig. 8 – Time Response of Modulator and
Vehicle to Square Wave Command

modulator release valve is switched on and off at a 2 Hz switching frequency and the vehicle response is computed. The hydraulic pressure trace shows the same cyclic behavior as the test data for the Starion in (6) (the mechanical parameters of the modulator, including piston size, spring rate, plenum area, etc., have been selected to match the response in (6).

Figure 9 shows the frequency response of the modelled modulator. At low frequency, the amplitude ratio is flat, and at about 6 r/s (1 Hz), it transitions to a broad –20 db/decade slope, up to about 6 Hz. This therefore, gives the desired integrator-like behavior in the forward loop. This is a new result and shows why this particular type of ramping (rate limited) modulator is feasible and universally used in automotive antilock. Also shown in Fig. 9 is the subsequently measured response of the modulator. This shows the actual modulator has more gain, more phase lag, and is of generally the same shape as compared with the math model. The detailed differences are no doubt due to the use of assumed rather than measured modulator dimensions, constants, etc. in the simulation example. Nonetheless the overall shapes of the frequency responses are remarkably similar. Given the integrator like behavior of the modulator, it is apparent that

• The series open loop combination of vehicle dynamics and controller should be close to a pure gain.

This is a keystone in the subsequent design of an antilock controller.

CONTROLLER ANALYSIS

Now, various discrete output (bi-state) controllers are available, whose outputs are compatible with solenoid valve, on/off type modulators. These include (1):

• Contactors (i.e., comparators)
• Toggles (i.e., negative deficiencies),

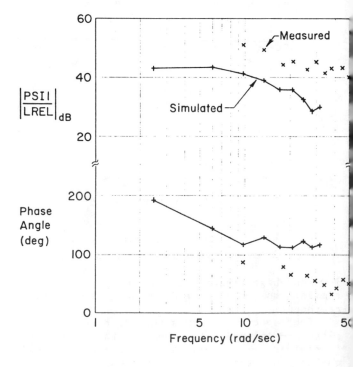

Fig. 9 – Frequency Response of Modulator to
Square Wave Command

the former being a special case of the latter. In fact combinations of these -- along with logical gates -- are among the universal building blocks of all known antilock systems.

Both contactors and toggles have the notable characteristic that their amplitude ratio and phase are frequency independent (3). To the system, they appear as pure gain with constant phase lag, where the gain and lag are functions of input amplitude and toggle "width". Figure 10 illustrates the effective amplitude ratio and phase charaterisics of a toggle.

Likewise, logical gates might be thought of as frequency invariant elements, and in any case, the simplest antilock systems do not require such gates.

A generic toggle controller is shown in Fig. 11. Of course, the input is the selected motion feedback signal and the output is the command to the modulator. The response is determined by the width parameters a and b. For example, in a typical antilock case the input (feedback) variable would start at the origin, with the toggle in the "off"

position. The input variable might then increase, and when it reached "a" it would trigger the output of the toggle to "on". The variable might further increase (overshoot) and then -- if the feedback closure is suitable -- it can decay toward and through zero, triggering the output to "off" at "b". After a further overshoot, the variable may be returned again, pass through the origin, and -- if conditions for a stable limit cycle exist -- repeat the cycle.

To verify the transfer characteristics of a toggle, an open loop response test was made with the Fig. 11 toggle in series with the modulator. The results as shown in Fig. 12 for toggle width a = b = 0, for a sinusoidal input of unity amplitude. Comparison with Fig. 9 verifies the pure gain character.

FEEDBACK SYNTHESIS

Summarizing, it has been found that

- To construct a workable control system, the open loop series combination of controller/modulator/vehicle dynamics should look like an integrator (times a gain) in the 2 to 6 Hz region. Also, the series combination should have sufficient open loop phase margin to allow crossover frequencies in the 2 to 6 Hz region.
- Practical antilock modulators by themselves tend to look like an integrator in this frequency region (due to their on/off and rate limiting features).
- Single input/output controllers compatible with these modulators look like a pure gain/constant phase lag (where gain and phase lag are functions of feedback signal amplitude and toggle width).
- Therefore, the necessary and sufficient motion feedback variable should look like a pure gain across this frequency region.

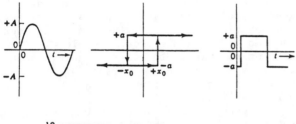

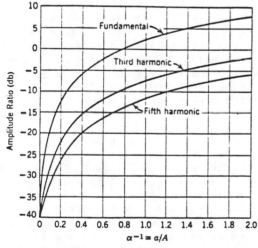

Fig. 10 - Sinusoidal Describing Function for the Toggle Characteristic (3)

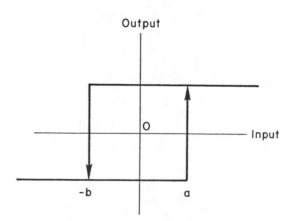

Fig. 11 - Toggal Input/Output Relationship

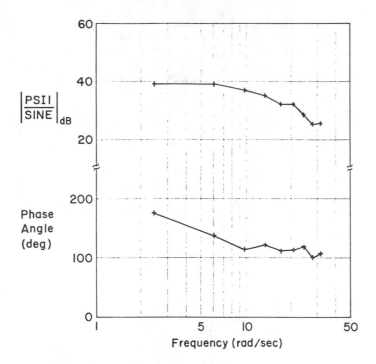

Fig. 12 - Frequency Response of Toggle (a=b=0)
in Series with Modulator

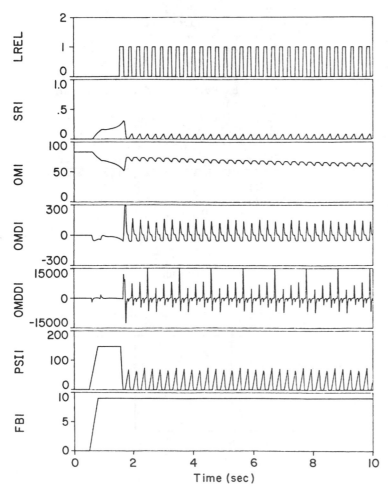

Fig. 13 - Example Closed Loop Antilock
Response

Examining the Bode plots (Figs. 3 to 6) for the four candidate feedbacks, it is clear that, without further signal processing,

- Only angular acceleration appears to have the necessary pure gain-like characteristics and needed phase margin in the frequency region of interest.

In contrast, angular speed shows a -20 db/decade slope, while angular jerk shows a +20 db/decade slope in the 1 to 6 Hz region. Their amplitude ratios are therefore too lag and lead-like, respectively. Also, angular speed has 90 degrees more phase lag than angular acceleration. When combined with the phase lags in the controller and modulator, this would not allow crossovers (i.e., limit cycles) at frequencies greater than about 2 Hz, which is -- as discussed -- marginal. Slip ratio, on the other hand, even if it weren't difficult to sense, shows some incoherence in the mid-frequency region. Further analysis might clarify this; however, slip ratio at best might be similar to angular speed (its main component) and therefore suffer the same shortcomings.

A conclusion which could be drawn from these preliminary analyses is that angular acceleration can provide a fundamental, innermost antilock feedback loop. Another conclusion might be that angular speed (or possibly slip ratio) -- because of its response qualities -- may provide an outer loop for "trimming" control.

These results are not surprising in retrospect, in light of the fact that several prototype and production antilocks (e.g., the prototype Girling all mechanical system described in (1)) use angular acceleration as their single feedback loop. Also, many other production systems seem to use slip ratio as a supplementary (outer loop) feedback, for either adaptive toggles or as second inputs to logical gates.

EXAMPLE CLOSED LOOP RESPONSE

To demonstrate the feasibility of the above feedback structure, an example loop closure was performed by feeding the angular acceleration signal back into a toggle (see Fig. 7). The main task then was to select suitable values for the toggle width parameters a and b. Basic gain-phase plot analyses, not shown here, indicated that, for the modeled vehicle, brake system, and modulator, a and b values around 36 r/sec should result in a stable limit cycle condition at about 3.5 Hz. Note also that the sense of the toggle is reversed, to account for the sign reversal (180 phase shift) between angular acceleration and pressure.

Using these toggle values, the closed loop response of Fig. 13 was generated. This indeed shows the predicted stable limit cycle behavior at about 3.5 Hz (as seen in the release valve signal and other variables), demonstrating antilock feasibility.

Recognizing that Fig. 13 is only a preliminary example, it also shows that the limit cycle, though stable, occurs about a very small value of slip ratio (about .04 mean) rather than the desired .15 value. Continuing analyses are aimed at trimming the operating point of the limit cycle. This can be accomplished with a combination of different a and b toggle width values, pressure release and build rates (as could be readily adjusted by pneumatic orifice sizes) and possible filtering of the feedback signal. All this can be done by trial and error techniques; it can also be accomplished -- efficiently and with some insight -- by analytical gain-phase plot techniques (e.g., (3)).

## CONCLUSIONS AND RECOMMENDATIONS

As shown, analytical approaches to antilock design -- based on feedback control techniques -- can be a useful complement to full scale activities. They may be used to help reduce development time and costs.

Based on preliminary analyses and design of an example antilock brake system for a motorcycle front brake, several conclusions can be drawn, as follows:

- To carry out an analysis and design of an antilock system, as a starting point a modulating means can be selected. For example, in the current analysis, a typical, discrete type (pneumatic) modulator was selected.
- To establish a stable antilock limit cycle, and meet other fundamental control requirements, the openloop, series combination of controller, modulator and vehicle responses needs to have certain known frequency response characteristics.
- After modelling and analyzing these responses, it was found that -- of several potential motion feedbacks -- wheel angular acceleration (or a facsimile thereof) was a preferred and sufficient feedback variable for simple, single loop antilock control.
- Bandpass requirements for sensor, modulator, and controller emphasize the 2 to 6 Hz region.
- A time domain simulation verified the feasibility of an example angular acceleration feedback design, which produces a stable limit cycle when a toggle and a discrete pneumatic modulator are used.

Further application of these techniques can be used to examine various potential antilock development areas,for example:

- Adjusting and optimizing antilock design in consideration of measured component response characteristics.
- Fine tuning component response qualities, for example, toggle biases.
- Assessing the sensitivity to

- Noise,
- Parameter variations (eg., due to temperature)
- Road surface characteristics,
- Vehicle load.
- Evaluating the feasibility and requirements for various candidate sub-systems, e.g.,
    - Digital, hybrid, or analog electronics,
    - Mechanical or fluidic sensors, controllers or modulators, or
    - Alternate power sources.

## ACKNOWLEDGEMENT

The results described in this paper were obtained in part under subcontract to the Jet Propulsion Laboratory under contract RE 182-0230 to the United States Army Harry Diamond Laboratories. The views expressed herein are those of the author, and not necessarily those of JPL or the US Army/HDL.

## REFERENCES

1. Zellner, J. and White, M., Advanced Motorcycle Brake Systems, Final Report, DOT -HS-806-095, November 1981.
2. "Development of Fluidic Antilock Brake System for Motorcycles", Contract RE 182-0230, between US Army Harry Diamond Laboratories and the Jet Propulsion Laboratory, 1982.
3. Graham, D., and McRuer, D., Analysis of Nonlinear Control Systems, Dover Publications, Inc., New York, N.Y., 1961.
4. Okada, T., "A Practical Method of Obtaining the Frequency Response from the Transient Response", Japan Journal of Applied Physics, Vol. 7, 1968.
5. Barter, N.F., and Little, J., "The Handling and Stability of Motor Vehicles," Part 7, Frequency Response Measurements and Their Analysis", Motor Industry Research Association Report 1970/10.
6. Yoneda, S., et.al., "Rear Brake Lock-Up Control System of Mitsubishi Starion", SAE Paper 830482, presented at SAE Automotive Engineering Congress and Exposition, Detroit, February, 1983.

## APPENDIX - SUMMARY OF VEHICLE AND MODULATOR EQUATIONS OF MOTION

This appendix summarizes the motorcycle longitudinal equations of motion and modulator response equations.

## SUMMARY OF VEHICLE EQUATIONS

The assumptions included in the mathematical model of the motorcycle include:

- Equations written in body fixed (stability) axis system
- Five vehicle dynamic degrees of freedom (sprung mass pitching and vertical velocities, front and rear wheel angular velocities and total mass longitudinal velocity)
- Two kinematic degrees of freedom (sprung mass pitch angle and vertical position)
- Linearized suspension forces proportional to sprung body pitch angle and vertical position and rates thereof,
- Both wheels in contact with ground at all times,
- Rigid tires,
- Rear caliper brake forces borne completely by swing arm, so as to give anti-lift reaction forces,
- Tire tractive forces are the product of: the non-linear function relating coefficient of friction and slip ratio (Fig. 14); and tire vertical force (which is dynamically varying).
- Braking torques are linear functions of time varying hand or foot lever input forces, and include an effective time delay (assumed to be lumped into calipers and lower part of brake systems).

Figure 15 shows vehicle dimensions and coordinate systems.

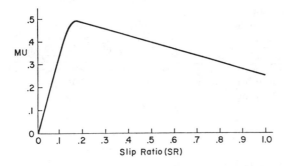

Fig. 14 - Relation of Tire/Road Coefficient of Friction to Slip Ratio

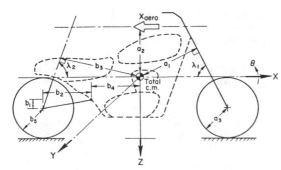

*All quantities positive as drawn*

Fig. 15 - Vehicle Coordinate System and Dimensions

**EXTERNAL MOMENT EQNS:**

$$\dot{Q} = \frac{1}{I_{ys}} \left( a_1 F_{s_1} + Z(F_{xt_1} + F_{xt_2}) - b_3 F_{s_2} \right.$$

$$\left. + b_4 F_{\tau_2} + a_2 F_{x_{aero}} \right)$$

$$\dot{\omega}_1 = \frac{-1}{I_{y_1}} \left( a_3 F_{x_{t_1}} - \tau_1 \right)$$

$$\dot{\omega}_2 = \frac{1}{I_{y_2}} \left( b_5 F_{x_{t_2}} - \tau_2 + \tau_{motor} \right)$$

**EXTERNAL FORCE EQNS:**

$$\dot{U}' = \frac{1}{m_s + m_1 + m_2} \left( F_{x_{ext}} \right)$$

$$\dot{W}' = \frac{1}{m_s} \left( -\sin \lambda_1 F_{s_1} - \sin \lambda_2 F_{s_2} \right.$$

$$\left. + F_{\tau_2} + m_s g \right) + Q U'$$

**KINEMATIC RELATIONS:**

$$\dot{z} = W' - U' \theta$$

$$\dot{\theta} = Q$$

**SUSPENSION FORCES:**

$$s_1 = -d_1 + d_2 \theta - d_3 z$$

$$s_2 = -d_4 - d_5 \theta - d_6 z$$

$$\dot{s}_1 = d_2 \dot{\theta} - d_3 \dot{z}$$

$$\dot{s}_2 = -d_5 \dot{\theta} - d_6 \dot{z}$$

$$F_{s1} = k_1 (s_{01} - s_1) - c_1 \dot{s}_1$$

$$F_{s2} = k_2 (s_{02} - s_2) - c_2 \dot{s}_2$$

$$F_{\tau_2} = \tau_2 / b_2$$

**TIRE FORCE:**

Given $\mu peak$, $\mu final$, SR peak

$$SR_1 = 1 + \frac{\omega_1 r_1}{U'}$$

82

$$SR_2 + 1 + \frac{\omega_2 \, r_2}{U'}$$

$$SR_{t_1} = (.8) \, SRpeak$$

If $-SR_{t_1} < SR < SR_{t_2}$

$$\mu_1 = M_1 SR_1$$

$$\mu_2 = M_1 SR_2$$

where $\mu = \mu peak$ at $SR = SR$ peak

$$M_1 = \frac{\mu peak}{SR \ peak}$$

If $SR_{t_1} \leq |SR| < SR_{t_2}$

$$\mu_1 = ASR_1^2 + C \, |SR_1| + D$$

$$\mu_2 = A \, SR_2^2 + C \, [SR_2] + D$$

Where A is given by input

$$C = M_1 - 2A \, \left| SR_{t_1} \right|$$

$$D = A \, SR_{t_1}^2$$

If $\left| SR_1 \right| \leq SR_{t_2}$

$$\mu_1 = M_2 \, SR_2 + B$$

$$\mu_2 = M_2 \, SR_2 + B$$

Where $B' = \dfrac{\mu peak - \mu final \ SRpeak}{1 - SRpeak}$

$$M_2 = \mu final - B'$$

$$SR_{t_2} = \frac{M_2 - C}{2A}$$

$$B = ASR \, t_2^2 + C \, |SRt_2| + D - M_2 \, |SRt_2|$$

Therefore:

$$N_1 = \sin\lambda_1 \, F_{s1} + m_1 g$$

$$N_2 = \sin\lambda_2 \, F_{s2} + m_2 g + \tau_2 \frac{b_2 - b_1}{b_1 \, b_2}$$

$$F_{xt_1} = \mu_1 \, N_1$$

$$F_{xt_2} = \mu_2 \, N_2$$

$$F_{xaero} = x_{aero} \, (U')^2$$

$$F_{x_{ext}} = -F_{xaero} - (F_{x_{t_1}} + F_{x_{t_2}}) - m_s g \, \theta$$

## SUMMARY OF MODULATOR EQUATIONS

The mathematical model of the Mitsubishi Starion type modulator is described by a series of equations. The modelling of the modulator included a number of assumptions, and is loosely based on the cutaway drawing and narrative in (6). Figure 16 shows the resultant interpretation of the functional properties of the modulator. This comprises a simplified model with solenoid valves, orifices, spring loaded plunger, and check valve. An important feature of this and most antilock brake modulators is the effective compliance of the brake line between the modulator and caliper; and this is modelled as a linear spring/piston combination. All elements are considered to be massless, so that static equilibrium is assumed for each subelement.

MODULATOR EQUATIONS -- The equations for the modulator are summarized as follows:

Plunger displacement:

Given $F_B$ (= resultant plunger force due to chamber B and brakeline pressures):

If $F_B < F_{so}$;

$$X_p = 0$$

If $F_B \leq F_{so}$;

$$X_p = \frac{F_B - F_{so}}{K_p}$$

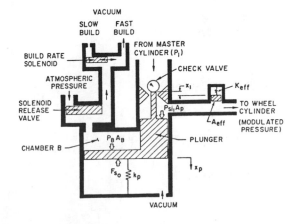

Fig. 16 - Simplified Schematic of Modulator Mechanism

Expansion chamber volume change versus hydraulic pressure change:

$$\Delta_{vol} = X_p A_p \quad \text{(Volume change due to plunger motion)}$$

Assuming the line contains some effective spring rate ($K_{eff}$):

$$X_p = \frac{\Delta_{vol}}{A_p}$$

$$F = K_{eff} X_p$$

$$PSI_1 = PSI_0 - \frac{F}{A_{eff}} ;$$

$$PSI_0 = \text{Initial pressure as system becomes isolated}$$

$$PSI_1 = PSI_0 - \frac{K_{eff}\Delta_{vol}}{A_p A_{eff}}$$

Pressure in chamber B:

$$P_B = P_B + \dot{P}_i\, t$$

where:

$$\Delta t = \text{Step size of simulation}$$

$$\dot{P}_1 = \text{Pressure drop or increase rate}$$

and:

$$\dot{P}_1 = \text{Dump (positive) pressure rate}$$

$$\dot{P}_2 = \text{Slow build (negative) pressure rate}$$

$$\dot{P}_3 = \text{Fast build (negative) pressure rate}$$

$$\dot{P}_3 > \dot{P}_2$$

Boundary conditions:

$$0 \leq P_B \leq 14.7 \text{ psi}$$

$$\text{Max. } \Delta_{vol} = 0.11 \text{ in}^3 \quad \text{(limiting plunger displacement)}$$

$$\text{Modulated } PSI_1 \leq P_1$$

NOMENCLATURE

| | |
|---|---|
| $a_1$, $a_2$, $a_3$ | — Vehicle dimensions See Fig. 15 |
| $b_1$, $b_2$, $b_3$, $b_4$, $b_5$ | |
| $c_1$, $c_2$ | — Front, rear suspension linear damping coefficient |
| $d_1$, $d_4$ | — Front, rear, spring preloaded length |
| $d_2$, $d_5$ | — Front, rear spring extension per unit pitch angle |
| $d_3$, $d_6$ | — Front, rear spring extension per unit vertical position |
| $F_{s_1}$, $F_{s_2}$ | — Front, rear suspension force |
| $F_{xt_1}$, $F_{xt_2}$ | — Front rear tire/roadway tangential force |
| $F_{x_{aero}}$ | — Longitudinal aerodynamic force |
| $F_{\tau_2}$ | — Rear antidive vertical force |
| $g$ | — Acceleration due to gravity |
| $I_{ys}$ | — Sprung body pitch moment of inertia |
| $I_{y_1}$, $I_{y_2}$ | — Front, rear wheel spin moment of inertia |
| $k_1$, $k_2$ | — Front, rear spring rate |
| $m_s$ | — Sprung body mass |
| $m_1$, $m_2$ | — Front, rear unsprung mass |
| $N_1$, $N_2$ | — Front, rear tire normal force |
| $Q$ | — Sprung body pitch velocity |
| $r_1$, $r_2$ | — Front, rear tire rolling radius |

$\dot{s}_1$, $\dot{s}_2$ — Front, rear spring length

$s_{o_1}$, $s_{o_2}$ — Front, rear spring free length

$s_1$, $s_2$ — Front, rear damper extension rate

SN — Pavement skid number

$SR_1$, $SR_2$ — Front, rear tire slip ratio

U' — Sprung body forward velocity

W' — Sprung body vertical velocity

$X_{aero}$ — Aerodynamic coefficient ($\frac{1}{2}\rho C_D Area$)

z — Sprung body vertical position relative to roadway

$\lambda_1$, $\lambda_2$ — Front, rear suspension spring inclination angles

$\mu_1$, $\mu_2$ — Front, rear tire roadway coefficient of friction

$\omega_1$, $\omega_2$ — Front, rear wheel angular velocity

$\theta$ — Sprung body pitch angle

$\tau_1$, $\tau_2$ — Front, rear brake torque

$\tau_{motor}$ — Rear axle drive torque

# Antiskid Systems and Vehicle Suspension*

## C. Tanguy
### Bendix France - Technical Division

*Paper 865134 presented at the XXI FISITA
Congress, Belgrade, Yugoslavia, June, 1986.

## ABSTRACT

Only measurements available for actuating
antiskid systems are the measurements of the
wheel speeds sensed at the wheel hub. The angular
wheel speed gives a signal which integrates several parasitic phenomena such as :
- horizontal displacement and
- vertical movement of the wheels relating to
the vehicle chassis.
By analysis of the behaviour of the connections
between the wheels and the chassis (suspension),
we will define the nature and amplitude of these
phenomena as well as a logic model for processing
the signals to avoid a reduction in braking efficiency.

AN ANTISKID SYSTEM modulates brake torque at the
wheels of the vehicle at a frequency and with an
amplitude depending essentially on the charater-
istics of the road surface which the vehicle is
braking on. These modulations of brake torque
induce reactions in the wheel suspension which
appear, schematically, as horizontal and vertical
displacements of the wheel axis relative to the
chassis. Given that the data utilised by the
electronic calculator which controls the antiskid
function is essentially the speeds of rotation
of the wheels measured at their axis, it is necessary to analyse the manner in which suspension
reactions affect this data.

## MODELLING

In order to simplify the presentation, we
will consider only the study of the behaviour of
one front wheel of the vehicle, this wheel being
connected to the vehicle by a Mc Pherson type
suspension. During the braking sequences, only
this front wheel will subjected to torque modulations avoiding wheel lock ; the other three
wheels will be braked by a constant torque which
is slightly less than the torque corresponding
to the limit of adhesion.

## GENERAL CHARACTERISTICS OF THE VEHICLE
- Wheel base : 2.42 m
- Centre of gravity : 0.42 m
- Pitch inertia : 1000 m2kg
- Static loads :
  . front axle  : 6000 N
  . rear axle   : 4100 N
- Suspension
  Flexibility at the wheel for 1000 N :
  . front      : 55 mm
  . rear       : 47 mm
- Vertical damping : See FIG 1
- Wheels and tyres :
  . unsprung mass : 36.5 kg perfront wheel
  . apparent rolling radius : 0.28 m
  . tyre vertical stiffness : $1.5 \times 10^5$ N/m
  . tyre damping : 150 N/m/s

N.B. Tyre stiffness and damping are very approximate figures.

SUSPENSION MODEL FOR ONE FRONT WHEEL is
shown diagrammatically in FIG 2. The model consists
of :
- an arm AB comprising ball-joints at points A
and B and supporting the shock absorber and suspension spring
- a triangle BCD hinged on the chassis along an
axis CD
- a wheel rotation axis OY, which is rigidly supported by the arm AB.
The positions of the points A, B, C and D are defined relative to a set of orthogonal axes OXYZ,
OY being the wheel axis, OZ being the upwards vertical axis and OX the rearwards horizontal axis.
The table in FIG 3 gives the position of the different points according to the following code :
ZA, ZB projection of points A, B onto OX
XA, XB projection of points A, B onto OY
YA, YB projection of points A, B onto OY
All dimensions are in mm

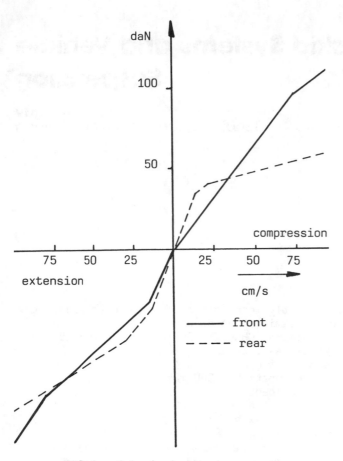

FIG 1 - Schock absorbers

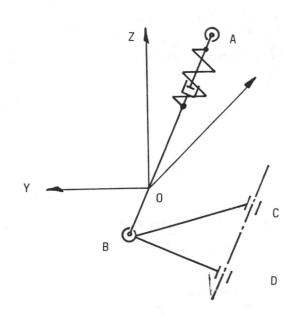

FIG 2 - Suspension diagram

| XA | 26 | YA | -156 | ZA | 495 |
|----|----|----|------|----|-----|
| XB | -5 | YB | 51 | ZB | -93 |
| XC | 205 | YC | -377 | ZC | -103 |
| XD | -13 | YD | -362 | ZD | -103 |

FIG 3 - Suspension : position of the main points.

As an illustration, the axial stiffness of the ball-joint support at point A is estimated at $6 \times 10^5$ N/m. In fact, this value will be neglected and we will define the overall deformation of this suspension under load by writing :
    horizontal displacement of the wheel axis :
= K x drag of the braked wheel and
    vertical displacement of the wheel axis :
= N % x horizontal displacement
    ROAD SURFACE MODEL. Whether the surface is flat or uneven, the tyre-to-road connection will be represented by the characteristics (cf FIG 4) in which MU, the adhesion utilisation is given by:
MU = T/P
where T : drag of the braked wheel, P : load applied by the tyre to the road and G : the slip is given by : G = 1 - VA x R/VX
VA : angular speed of the wheel measured at the hub
R = apparent radius of the wheel
VX : horizontal speed of movement of the projection onto the ground of the wheel axis.

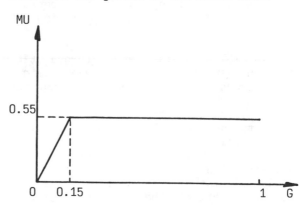

FIG 4 - Road surface characteristic

ANTISKID OPERATING LOGIC MODEL. The model used is intentionally simplified compared to the real models used in the calculator but remains representative of the operation of the antiskid system.
The feature of the model are :
- a rapid reduction in brake torque is triggered as soon as the wheel speed drops below the reference speed (defined later) by more than a predetermined threshold value

- a slower reduction in torque once the wheel starts to accelerate again
- an increase in torque, which is rapid at first, during a pre-determined time, then slower until impending wheel lock reappears.
The point A at which the torque increase begins is defined by a threshold value of acceleration which is a function of the difference between reference speed and wheel speed
- each wheel speed signal of the vehicle is associated with a limited slope signal (for example : 10 m/s/s). If VF (i) is the limited slope signal associated with wheel (i), the reference speed is defined as being :
MAX (VF (i) ) i = 1 to 4
- a delay is introduced between the moment when the change of rate of variation of brake torque is defined and the moment when it is applied (approx. 20 ms including the delays of the calculator, of the torque variation control units and the brakes.)

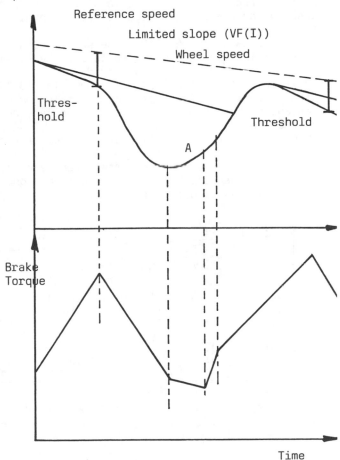

FIG 5 - Antiskid operation logic

PRESENTATION OF THE RESULTS. The results will be presented below in the form of recordings of the variations with time of the different parameters.
The identification codes for the parameters and the corresponding scales are as follows :

| Parameter | Plot code | Units |
|---|---|---|
| vehicle speed | ——— | km/h |
| reference speed | ——×—— | km/h |
| wheel speed | ——○—— | km/h |
| wheel axis displacement along OX | ——△—— | cm |
| variation of tyre-to-road load | ——Z—— | N. |
| brake torque | ——□—— | N.m |
| time | | sec |

BRAKING WITHOUT ANTISKID ACTION

This first part of the work was used to adjust the values characteristics of suspension deformation defined by the parameters K and N referred to at § "Suspension model for one front wheel" to obtain good correlation between simulation results and vehicle measurements.
The results of the adjustments are :
- adoption of a variation of the value of K as a function of the horizontal displacement X of the wheel relative to the chassis
$K = 1/a/X^2$ (a = 1.5  $10^9$ N/m3)
- estimation of N as 20 % approx
- use of horizontal wheel movement damping of 500 N/m/s.
BRAKING ON A FLAT SURFACE. Two brake applications were made, one as shown in FIG 6 for a slow application of brake torque (3000 Nm/s) and the other as shown in FIG 7 for a rapid application (15000 Nm/s)

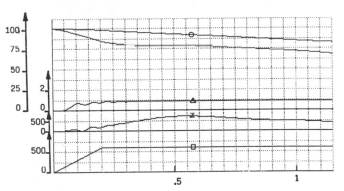

FIG 6 - Slow application of brake torque

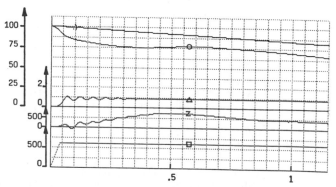

FIG 7 - Rapid application of brake torque

The only points to be noted in the two plots are :
- a slight oscillation in the tyre-to-road load
- a corresponding oscillation in the position of the wheel relative to the chassis along the OX axis.
These two oscillations are more marked the higher the rate of increase in pressure.

BRAKING ON SINE-WAVE SURFACE. In order to avoid too many interactions between the parameters to be analysed, this braking was simulated for a value of N = 0.
The road surface used presented a sinusoidal irregularity of ± 0.01 m amplitude and variable wave-length so as to maintain an excitation period of 0.1 sec at the wheel considered.
Only the wheel studied is subjected to vertical excitation produced to the road.
FIG 8 shows the response of the model to this excitation and gives very interesting data in particular as to the perturbations of the wheel speed signal :
- modulation frequency of the wheel speed signal 10 Hz (excitation frequency)
- modulation amplitude ± 6 km/h for a modulation of tyre load of ± 3250 N and a lag of the wheel relative to the chassis of about 0.015 m maximum.
It is concluded from this simulation that, contrary to expectations, horizontal modulation of the position of the wheel axis relative to the vehicle has a negligeable influence on the form of the speed signal sensed at the wheel hub but that variations in vertical load can perturb the signal very significantly.

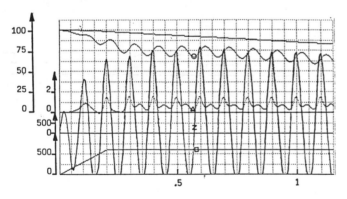

FIG 8 - Braking on sine-wave surface

### ASSOCIATION WITH AN ANTISKID SYSTEM

FLAT ROAD SURFACE. Returning to the suspension deformation model characterised by the values of K and N defined at § "Braking without antiskid action" and applying the action of the antiskid system whose logic model was defined at § "antiskid operation logic", we find the results shown in FIG 9, using the following parameters :
* rapid rate of increase of torque : 15000 Nm/sec
* slow rate of increase of torque : 3000 Nm/sec
* rapid rate of decrease of torque : 20000 Nm/sec

* slow rate of decrease of torque : 3000 Nm/sec
* "difference" for triggering torque reduction : 3 m/s

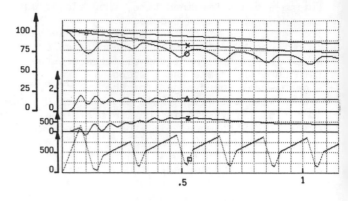

FIG 9 - Antiskid action on flat road surface

The variations of brake torque are somewhat too big compared to real systems. This arises solely from the fact that the logic model used for processing the wheel speed signals was extremely simplified.

SINE-WAVE ROAD SURFACE. The same road surface simulation model was used as for the simulation of operation without antiskid.
The result of the simulation is shown in FIG 10 from which it is clear that very quickly the modulations of brake torque occur with the same period as the ground excitation.
This demonstrates a coupling effect and, if certain precautions are not taken in the antiskid control logic, the coupling effect can lead to progressive reduction in brake torque and consequantly a loss of brake efficiency.

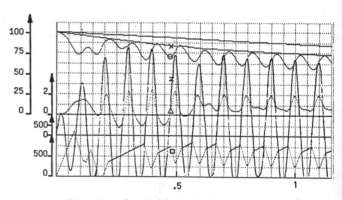

FIG 10 - Antiskid on sine-wave road surface

INFLUENCE OF DAMPING OF HORIZONTAL WHEEL DISPLACEMENTS. FIG 11 shows the form of the speed signals after about one second of antiskid operation for the case where damping of horizontal wheel displacements is zero.
The wheel speed signal is again heavily perturbed and precautions must be taken to avoid reduction in brake efficiency.

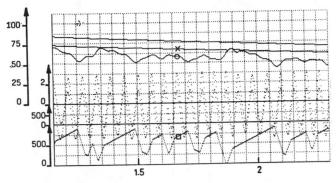

FIG 11 - Influence of lack of horizontal
         damping


CONCLUSIONS

This study demonstrates that wheel speed
signals detected at the wheel hub can be heavily
perturbed by variations in the variations in
vertical wheel load generated :
- by road surface features
- by lack of damping of horizontal wheel displacements.
The influence of poor operation of hydraulic
suspension shock absorbers was not simulated
but leads to results of the same type as the
lack of horizontal damping.
In pratice, faced with the problems posed by
the perturbed wheel speed signals, the antiskid
control logic has to be adjusted by means such
as :
-permanent adaptation of the value of the trigger
threshold
- permanent adaptation of the parameters which
are characteristic of the variations in brake
torque.
These adjustments are defined by continuous
analysis of the speed variations.

Bibliography : in this study, we have used
- a vehicle model close to that presented by
Mr. T. Kapitaniah in his paper "The influence
of vehicle suspension displacement on the working of antilock braking systems" I. Mech. E.
conference publications
- numerical data kindly communicated by the
company P.S.A. and the REGIE NATIONALE DES USINES
RENAULT
- an antiskid model using several characteristics of the system developed by BENDIX FRANCE.

*Systems*

# Hydraulic Brake Actuation Systems under Consideration of Antilock Systems and Disc Brakes*

**Otto Depenheuer and Hans Strien**
Alfred Teves GmbH

*Paper 730535 presented at the National
Automobile Engineering Meeting, Detroit,
Michigan, May, 1973.

THE DEVELOPMENT of hydraulic power supply systems for passenger car and light truck brake actuation appears necessary and important for three main reasons:

1. The increasing application of fuel injection and the exhaust emission regulations have led to an inlet manifold vacuum of only 40-55% instead of the previous 75-85%. In order to maintain the same degree of servo assistance to the driver's pedal effort while retaining the vacuum booster (and the trend is to even greater assistance), the booster diameter must be increased 15-30% versus today's units. Thus, the already existing packaging problems will become more acute, and can be alleviated only by boosters operating at a higher pressure level.

2. Investigations by the H.S.R.I. of the University of Michigan and the industry have shown that the application time for brakes (that is, the time elapsed between the driver pressing the pedal and a buildup of brake torque) increases with decreasing pressure differential. This means unwanted and even dangerous lengthening of the stopping distance.

3. Antilock systems require in all cases an energy source at least equivalent to that of a servo brake system, and as a rule even greater, because the assisted pedal effort of the driver is not available during the antilock operation. Logically, the same energy source can be used for service brake and antilock system. The installation space for an antilock system with hydraulic control, including the energy source (pump, reservoir, accumulator), is only a fraction of that for a vacuum-actuated antilock system. When establishing parameters for a new braking system, a hydraulic energy source should at least be planned as an alternative, with the possibility of easily installing an antilock system at a later date.

## SYSTEMS

CONTINUOUS-FLOW AND ACCUMULATOR SYSTEMS - For the actuation of a braking system, continuous-flow systems as well as accumulator layouts can be considered. The basic designs of both systems are diagrammatically shown in

_____ ABSTRACT

Hydraulic power braking systems for use in passenger cars and light trucks are attracting considerable automotive design attention. This is due to their compactness, smaller space requirements, and better operating "feel," as well as their more direct control over the braking function, which has extremely short application and release times. Moreover, they are readily adaptable to the energy source and controls of an antilock system, and they contribute to (or even form the basis of) a central power-supply system that would provide servo assistance to other vehicle systems. This paper describes and explains ways of creating these brake systems and gives design calculations of brake layouts based on standard values and comparative judgment criteria.

Figs. 1 and 2. The continuous-flow system may at first seem to be more attractive, considering the smaller number of its elements. Instead of the pressure-controlled suction valve in the accumulator system, a suction orifice is all that is needed, and the accumulator and nonreturn valve are eliminated completely. Also, the pressure regulator valve is simpler in the continuous-flow system compared to other systems. On closer examination, two factors speak against the continuous-flow system.

*Acceleration* - With currently available friction brakes, the volume requirements for maximum deceleration with 0.10 to 0.15s application time is 30-40 cm$^3$/s/1000 kg vehicle weight (0.83-1.13 in$^3$/s/1000 lb vehicle weight). This would have to be delivered at idling engine speeds (600-800 rpm) by a full-flow pump. In other words, for a car weighing 2000 kg (4400 lb) the pump would have to deliver 6-7 cm$^3$/rev. The energy absorption of such a constantly running pump at higher engine speeds is not inconsiderable, in spite of the suction orifice.

For accumulator systems, the same vehicle would use a pump with only 0.9-1.1 cm$^3$/rev, which operates only when charging the accumulator.

*Stall or Pump Failure* - Should the engine stall or the pump be damaged, the continuous-flow system would fail without warning. Even if this system were used only to augment the driver's pedal effort, its failure would be critical when the driver had to apply more than two or three times the normally assisted effort to obtain the same retardation. Experience

shows that most drivers are at such a time psychologically and physiologically overstressed. Therefore, dependent on the vehicle weight, continuous-flow and power-assisted systems (next section), might require an extra pump, either electrically or transmission driven, which in the event of a damaged main pump or stalled engine would be automatically actuated.

*Brake Efficiency Deterioration* - The sudden failure of an accumulator is less probable, for with a correct design volume a predetermined number of brake applications with a gradual diminishing efficiency can be made, should the pump be damaged or the engine stall.

With these points in mind, the accumulator system and the development of its associated hardware was given preference.

## POWER-ASSISTED AND FULL-POWER SYSTEMS

POWER-ASSISTED SYSTEM - A system that permits the driver's pedal effort to be assisted by servopower can (according to the present European regulations) be applied to a vehicle whose gross weight is such that, despite loss of power assistance, 30% deceleration with max 50 kp* (110 lb) pedal effort is still available.

*Efficiency Factor* - We shall now introduce a dimensionless value for the comparison between various brake systems, called the "efficiency factor" ($W^X$). This shows the relationship between the total brake force of a vehicle and the pedal effort to generate this force, and is expressed as

$$W^X = \frac{\text{total brake force}}{\text{pedal effort}} = \frac{aG}{P_F}$$

It is readily established that for current vehicles without boosters (index 0), the efficiency factor $W_0^X$ = 12-15. Therefore, without power assistance and in fulfillment of the given legal requirements, vehicles of 2000-2500 kg (4400-5500 lb)** can use a power-assisted brake system.

With all brakes intact, decelerations of 85-90% with pedal efforts of 25-30 kp (55-66 lb) can be achieved. This means that for vehicles in the 2000-2500 kg (4400-5500 lb) range,

$$W^X = 60\text{-}90$$

and there is no doubt that even higher values can and will be reached.

If we consider the fully working system as a basis (100%), then the "effectiveness coefficient"

$$w = \frac{W_0^X}{W^X} 100 \ (\%)$$

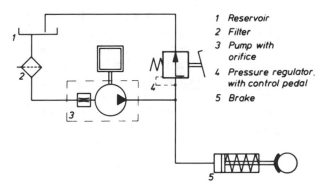

1 Reservoir
2 Filter
3 Pump with orifice
4 Pressure regulator, with control pedal
5 Brake

Fig. 1 - Hydraulic circuit plan of continuous-flow system

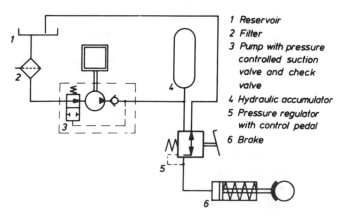

1 Reservoir
2 Filter
3 Pump with pressure controlled suction valve and check valve
4 Hydraulic accumulator
5 Pressure regulator with control pedal
6 Brake

Fig. 2 - Hydraulic circuit plan of accumulator system

*According to FMVSS 105a, pedal effort of 68 kp (150 lb) is specified.

**In the United States this weight would be 2700-3400 kg (6000-7500 lb).

shows by what percentage the system efficiency suffers for the same pedal effort compared with intact brakes. For the aforementioned vehicle weights the usual value for the effectiveness coefficient lies in the range 18-25%. Although such figures have been legally acceptable up to the present, they are not adequate. If the assistance to the driver's pedal effort suddenly fails, the value for "w" should not be less than 35-40%.

*Backup System* - If the 30% deceleration at maximum 50 kp (110 lb) pedal effort with failed power assistance cannot be achieved, European legislation calls for a reservoir of energy that must be large enough to enable 6-10 full brake applications at 30% deceleration to be made without exhausting the accumulators. (The number of stops varies from country to country.) Heavier vehicles, requiring in any case a larger energy supply, will probably be equipped with full-power braking systems, with efficiency factors $W^x$ of 200 or more.

As already pointed out, there is a clear trend toward further reduction of the driver's effort to effect a stop. This will have the effect of pulling down the vehicle weight class for which full-power braking can be applied. Therefore, such systems for passenger cars must be granted the same degree of importance as power-assisted systems.

## HYDRAULIC POWER-ASSISTED BRAKE SYSTEMS

### APPLICATION WITH INTERPOSED BOOSTER (ZHS 2.0)

- Fig. 3 shows a system where the vacuum booster has been simply replaced by its hydraulic counterpart, the original tandem master cylinder being retained. The energy supply is covered by the "hydraulic module" (pump and electro motor), which consists of the following units:

1. Electric motor with 3800 rpm nominal.
2. Radial piston pump, with a delivery of only 0.1 $cm^3$/rev.
3. Accumulator, gas loaded, whose design volume can be 250, 500 or 750 $cm^3$ (15, 32, or 48 $in^3$) depending on the brake system. Fig. 4 shows the relationship of fluid reserve, $V_1$, to nominal volume, $V_N$, available for braking at: maximum accumulator pressure (170 bar = 2400 psi); charging pressure (150 bar = 2150 psi); and the pressure at which the driver is warned in case no recharging occurs (130 bar = 1850 psi).

4. Pressure switch, by means of which the electric motor is switched off at 170 bar (2400 psi) accumulator pressure and switched on at a pressure of 150 bar (2150 psi).
5. Warning switch, giving an optical or acoustic signal if the pressure falls below 130 bar (1850 psi).
6. Pressure relief valve, adjusted to open at approximately 190 bar (2700 psi).

The pump is connected to a special reservoir, item 2 in Fig. 3. The pump could also be engine driven (Figs. 2 and 10), in which case the reservoir for the pump could then supply the brake actuation as well (Fig. 7); this will be discussed later.

Fig. 5 shows a cutaway drawing of a tandem master cylinder with an interposed pressure regulator valve. As such a valve arrangement always exhibits a certain leakage rate when modulating high pressures, a check valve is installed between accumulator feed and the regulator valve housing. This valve opens only when the brake pedal is pressed, and thus prevents the accumulator emptying itself when the vehicle is parked or when the brakes are not operated for long periods.

As the pedal is depressed, the control piston advances to

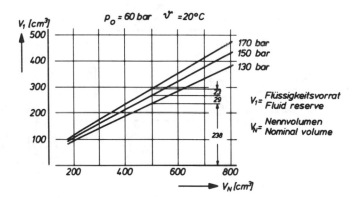

Fig. 4 - Fluid reserve in hydraulic accumulator

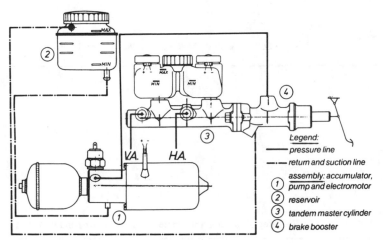

Fig. 3 - Power-assisted brake system with interposed booster

97

contact the booster piston (area $F_1$) and accumulator pressure (Sp) is simultaneously admitted. Once the control piston porting overlaps the feed ports in the booster piston, pressure is applied to the full area ($F_1$) as well as to the cross-sectional area of the control piston ($d_s$). Thus, the value of the regulated pressure is directly proportional to the pedal effort and can be exactly controlled with a high degree of response.

With such an actuation, the boosting need be only single-circuit, if 50 kg (110 lb) pedal effort can achieve 30% deceleration with a failed booster (so-called auxiliary brake effect); that is, if the efficiency factor $W_0^x$ is equal to or greater than six times the maximum vehicle weight (in 1000 kg). Because the chance of an accumulator failure is extremely remote, the efficiency factor $W^x$ may be large, and the effectiveness coefficient "w" can be made small because the driver will be warned that the accumulator is not being recharged and he

can gradually adapt to a reduced performance of the brake system.

APPLICATION WITH INTERGRAL BOOSTER (ZHS 2.1) - Means of improving the cost and installation factors of the interposed booster are shown in Figs. 6 and 7.

As in the interposed booster, the energy supply is provided by a "hydraulic module." item 1 in Fig. 6. A brakefluid reservoir, item 2, with a dividing wall supplies the dynamic brake circuit, consisting of pump, accumulator, and pressure regulator valve, as well as the hydrostatic brake circuit comprising master cylinder and wheel brakes. According to regulations, a fluid-level warning device must also be provided.

The ZHS 2.1 booster, item 3 in Fig. 6, combines the spool-valve controlled booster and a single-circuit master cylinder in one unit; Fig. 7 shows its operation.

When depressing the brake pedal, the control piston advances to contact the master cylinder piston so that the primary seal lip passes the recuperation port (at B in Fig. 7)

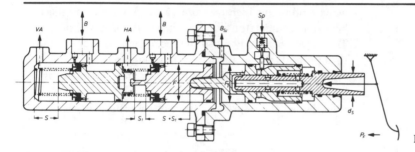

Fig. 5 - Interposed booster with tandem master cylinder (ZHS 2.0

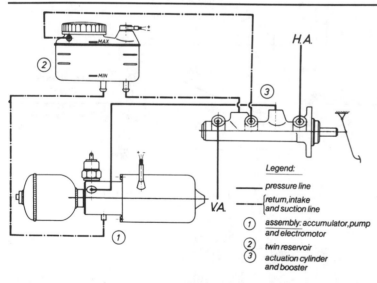

Legend:

_____ pressure line

{ return, intake
{ and suction line

(1) assembly: accumulator, pump and electromotor

(2) twin reservoir

(3) actuation cylinder and booster

Fig. 6 - Power-assisted brake system with integral booster

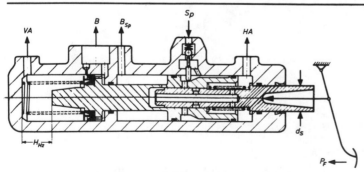

Fig. 7 - Integral booster and single-circuit master cylinder (ZHS 2.1)

and the check valve is lifted open by the ramp on the piston. Accumulator pressure then enters the annular control valve chamber and, passing through the overlapping porting, is applied to the full cross-sectional area at the rear of the master cylinder piston, passes to the control piston cross section (diameter $d_s$), and via the outlet marked HA enters the brake circuit. The pressure generated in the hydrostatic circuit passes via outlet VA to the other brake circuit. Due to seal drag and differences in the two spring loads, this pressure is slightly lower than that in the boost chamber, the latter value being governed by the input load $P_F$ and the area $d_s$.

On failure of the accumulator pressure, or a leak in the HA brake circuit, the master cylinder and the VA circuit are mechanically operated. Should the VA circuit fail, the master cylinder piston advances to the bottom of the bore, at which stage the accumulator pressure is directed to the HA circuit.

With the ZHS 2.1 booster, the cross-sectional area of the master cylinder can be considerably less than that of the ZHS 2.0 because it feeds only one brake circuit; in the case of the common front-rear split, this circuit would be for the front brakes, and by a diagonal split, for one rear and one front brake. With the same pedal ratio and pedal effort a higher pressure can thus be generated in the intact hydrostatic circuit than could be with the ZHS 2.0 layout, achieving the same brake efficiency as with a lower pressure in a fully intact system.

Contrary to ZHS 2.0, a sudden failure of the HA circuit would rapidly drain the accumulator. Therefore, when using ZHS 2.1, the value of "w" should (where possible) not fall below 35-40%.

APPLICATION WITH INTEGRAL BOOSTER (ZHS 2.2) - On failure of the power assistance or of a brake circuit, the regulations call for 30% deceleration; that is, only a third of the maximum available with an intact system. Consequently, the fluid volume, even with the "auxiliary system," can be reduced and the unboosted mechanical master cylinder stroke need not be so large as with ZHS 2.0 or ZHS 2.1 systems.

The following example will serve to illustrate the case for a standard front-rear split: With all brakes intact, the brake force for the front axle is

$$B_{VA} = (1 - \phi)a_{max} G = P_{max}K_{VA} \qquad (1)$$

where:

$\phi$ = brake force proportion of the rear axle
  $= B_{HA}/aG = B_{HA}/(B_{VA} + B_{HA})$
$1 - \phi$ = brake force proportion of the front axle
  $= B_{VA}/aG = B_{VA}/(B_{VA} + B_{HA})$
$a_{max}$ = maximum deceleration
$G$ = vehicle weight, kp
$P_{max}$ = max. line pressure at $a_{max}$, bar

$K_{VA}$ = a constant, incorporating wheel cylinder diameter, brake factor, etc., kp/bar
(See Nomenclature for definitions of symbols.)

After failure of the boosting and the rear brakes, the front brakes alone must create a brake force of

$$B'_{VA} = a'G = p'K_{VA} \qquad (2)$$

where:

$a'$ = deceleration with failed circuit
$p'$ = line pressure at $a'$

Combining Eqs. 1 and 2, the required line pressure must be

$$p' = p_{max} \frac{a'}{a_{max} (1 - \phi)} \qquad (3)$$

If a maximum deceleration of $a_{max} = 0.75$-$0.85$, requiring a line pressure $p_{max} = 100$-$120$ bar (1420-1700 psi), then the pressure $p'$ needed to achieve the deceleration $a' = 0.3$ is only $[40/(1 - \phi)]$ to $[4.25/(1 - \phi)]$ bar. If $1 - \phi = 0.6$ to $0.7$, then $p'$ is only 56-71 bar (800-1000 psi) or 55-60% of the maximum pressure.

Fig. 8 shows the essential relationship between volume requirement and line pressure for disc brakes (curve "a") and drum brakes (curve "b"). If for the maximum theoretical line pressure a volume of 100% is required for both brake types, it can be seen that at 55-65% of maximum pressure, the fluid volume required for disc brakes is 26-33% below the maximum ($\Delta V_a$), whereas the volume for drum brakes is only 8-11% below maximum ($\Delta V_b$). Thus, in the case of booster failure and mechanical operation of the master cylinder, the stroke needs to be approximately 70% of the maximum available for disc brakes and a full 90% when drum brakes are used.

The other well-known features, such as insensitivity to friction changes and fade resistance, also make disc brakes highly desirable whenever they can be possibly used.

On pressing the brake pedal, with accumulator intact, the preloaded outer simulator spring (which is captive) initially

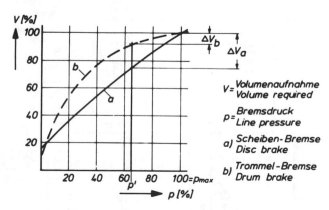

Fig. 8 - Fluid consumption characteristics

moves the spool control value and effects three actions; see Fig. 9:

1. Isolates the booster chamber from the fluid reservoir, $B_{Sp}$.

2. By means of the two sliding pins in the control valve bushing, moves the master cylinder primary seal lip past the recuperation port.

3. Lifts the ball check-valve in the accumulator feed port ($S_p$).

Further travel causes the rear edge of the control valve to overlap the rearmost annular recess in the control valve bushing. This causes the accumulator pressure to build up in the booster chamber and (via the lower transfer passage) in the chamber at the rear of the master cylinder piston, thus feeding the brake circuit marked HA. Pressure level in the latter is controlled by the relationship between the driver's pedal effort and the cross-sectional area of the control valve piston $F_1$.

The stroke of the control valve piston is governed by the travel/load characteristics of the simulator and can be a maximum $s_1$. On the other hand, the stroke of the master cylinder piston is governed by the volume/pressure characteristic of the VA brake circuit, and can be $s_3$, independent of the control valve stroke. (If, for reasons of poor bleeding, etc., in the VA circuit, 80% of the maximum available stroke is attained, the driver is given a warning signal).

If the accumulator fails, the master cylinder piston can be mechanically operated to a maximum stroke $s_2$ via the two

pins; at the same time, play "s" is overcome and the central valve bushing is urged forward without compressing the strong inner simulator spring. By a correct design of the simulator springs, the control valve piston and thus the pedal travel remain virtually the same for intact brakes or for failure of one circuit. This shows a clear improvement of ZHS 2.2 over ZHS 2.0 and ZHS 2.1.

Dependent upon application or vehicle design, the ZHS 2.2 offers a choice of advantages: application to higher vehicle weights and lower pedal efforts, or shorter pedal travel, or an improvement of the effectiveness coefficient "w" (see formulas in Appendix A).

## HYDRAULIC FULL-POWER BRAKE SYSTEM (ZHS 3)

The conditions described under "Backup System" can be satisfied by the system shown in Fig. 10. The engine driven pump 2 is fed by the reservoir 1 and charges both accumulators 3. The pressure level is either controlled from the charging valve 4 or from a pressure-controlled suction valve integral with the pump (see also Fig. 2). Generally, the lower cut-in pressure is 150 bar (2150 psi) and the upper, 170 bar (2400 psi).

The accumulator unit also incorporates a nonreturn valve that prevents a transfer of fluid from one accumulator to the other when the pressure falls below 135 bar (1900 psi). Each accumulator has a warning switch set at 130 bar (1850 psi) and a pressure relief valve set at 190 bar (2700 psi). When depressing the brake pedal, the dual-circuit pressure control

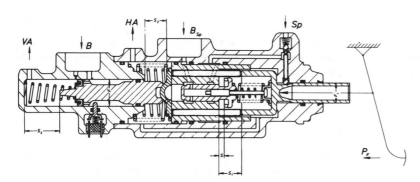

Fig. 9 - Modified integral booster and single circuit master cylinder (ZHS 2.2)

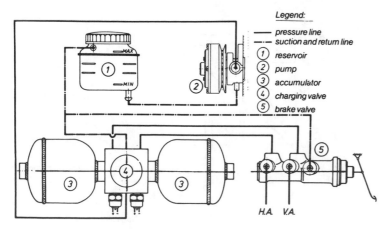

Legend:

— pressure line
–·– suction and return line
① reservoir
② pump
③ accumulator
④ charging valve
⑤ brake valve

Fig. 10 - Hydraulic full-power brake system

valve 5 allows pressure to flow to VA and HA brake circuits, each circuit being fed by its own accumulator.

As in ZHS.2 range, check valves are fitted between accumulators and the control valve, and a load-travel simulator is provided.

On failure of the pump, a number of brake applications, dependent on accumulator volume, charge pressure, and gas pressure, are possible. If the charge pressure falls to that of the gas pressure in the accumulator, no more assistance is possible. Therefore, the gas-filling pressure is chosen sufficiently high to retain—even in the final brake operation—an auxiliary braking effect of a = 30%.

For such a brake system it is important to know how the accumulator pressure behaves when its capacity is repeatedly tapped of volume, and how many brake applications can be made in a certain pressure range. Fig. 11 presents these factors in a diagrammatic form, applied to one brake circuit (approximately 50%) of a 4000 kg (9000 lb) vehicle.

The heavy curve shows the volume usage, and the left coordinate shows the pressure change "p" in a 500 cm$^3$ (32 in$^3$) accumulator: Line 1 shows the number of braking operations, item Z, for a deceleration of 20-25%; line 2 shows the number of braking operations, item Z, for a deceleration of 82-87%.

When the warning lamp operates at 130 bar (1850 psi), the volume for 12 applications has already been used (or 6.5 brake applications when considering line 2). After 20 applications more, the pressure remaining is still 92 bar (1300 psi) for line 1 and 70 bar (1000 psi) for line 2. This system thus has plenty in reserve to cope with the severest legal requirements.

The basic design of the full-power control valve is shown in Fig. 12. In the "brake-off" position, the check valves located between each of the two pressure control valves and the accumulators are closed, and both brake circuits VA and HA are in contact with the fluid reservoir, $B_{sp}$. On operating the brakes, both circuits are initially isolated from the fluid return; then both check valves open and, with a fractional advance, pressure is built up in the VA circuit. This pressure also actuates the control valve of the HA circuit, thus equalizing the pressure buildup in both circuits.

The stepped-diameter input piston is influenced by the reaction force from the VA-circuit control valve, $F_2$, transmitted rearward by the simulator springs, and by the HA-circuit line pressure acting on the ring area $F_1$. Thus, despite a control valve diameter of only 0.85 cm (0.355 in), the reaction load from the driver is such that this ZHS 3 control valve can be used with existing pedal layouts.

Should the HA circuit fail after opening the check valve, the forward control valve moves to its stop; if the VA circuit fails, the HA control valve is mechanically (not hydraulically) brought into play.

## HYDRAULIC CONTROL FOR USE IN ANTILOCK SYSTEMS

For antilock control, the system of actuation adopted for the service brake plays a decisive role. Some solutions are considered below.

CONTROL WITH A PLUNGER UNIT - If the brake actuation relies on a master cylinder with limited volume

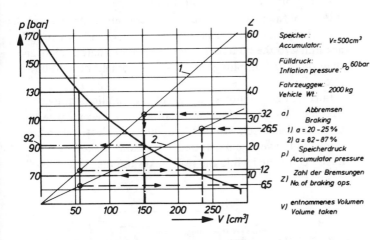

Speicher:
Accumulator: V = 500 cm³

Fülldruck:
Inflation pressure: $p_0$ 60 bar

Fahrzeuggew.:
Vehicle Wt.: 2000 kg

a) Abbremsen
   Braking
   1) a = 20-25%
   2) a = 82-87%

p) Speicherdruck
   Accumulator pressure

z) Zahl der Bremsungen
   No. of braking ops.

v) entnommenes Volumen
   Volume taken

Fig. 11 - Accumulator pressure related to number of brake applications and volume used

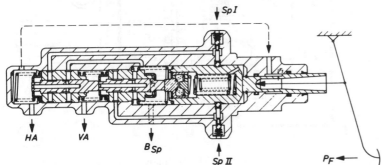

Fig. 12 - Dual-circuit full-power control valve (ZHS 3)

output, the fluid can be retained in the brake circuit and the antilock operation carried out by depressurizing and repressurizing the fluid already in the brake's wheel cylinders.

Fig. 13 shows that for each brake supplied by the master cylinder, a plunger unit with a reliable sliding seal is provided. When the antilock function is triggered, the inlet valve 3a opens and outlet valve 3b closes, the spring-loaded stepped piston is moved to the left, and the cutoff valve 4 closes so that fluid from the wheel cylinders flows into the space vacated by the plunger (pressure drop). When the "repressurize" signal is given, valve 3a closes and the spring pressure forces the plunger to the right, reapplying the brakes and opening the cutoff valve to complete the normal brake circuit again.

CONTROL WITH A VALVE BLOCK - If the service brake is directly supplied from an accumulator, then the fluid from the wheel cylinder can be led back to the reservoir, and when the "repressurize" signal is given, the brakes are reapplied, using the energy source or accumulator.

Fig. 14 shows a simplified layout of a valve block without plunger. When the antilock signal is given, the inlet valve 3b closes and the outlet valve 3a opens, releasing fluid from the brakes. At the end of the antilock control, the valve functions return to that shown and the wheel cylinders are repressurized.

THE COMPACT-BREL - For a vehicle equipped with power-assisted brakes (Fig. 6) using ZHS 2.1 or ZHS 2.2 boosters (Figs. 7 and 9) and an antilock system, it is practical to use a plunger unit (Fig. 13) for the one circuit and a valve block for the other (Fig. 14). Thereby the spring loading of the plunger

can be replaced by accumulator pressure, which is reduced or increased during the antiskid operation.

Furthermore, to save space, hydraulic connections, and pipework, the energy source and hydraulic actuation can be conveniently integrated into one BREL unit, diagrammatically shown in Fig. 15. ("BREL" is short for brake-force regulator-electronic.) Apart from this unit, only the ZHS 2 type booster 14 with integral reservoir 9 and a 500 cm$^3$ (32 in$^3$) diaphragm accumulator 12 are required to complete the brake system.

The electric motor driven pump "M" charges the membrane accumulator 12 as well as the piston accumulator 4, which is preloaded by the powerful spring 3. With this latter pressure flowing via the open solenoid valves 5b and passage 6, the plungers 1 and 13 are held against their stops, whereby the cutoff valves 7 remain open. In this position the service brake functions quite normally.

When an antilock function is triggered by the computer, valve 5b closes and 5a opens, cutting off the pressurized chamber 8 from the piston accumulator 4 and allowing the fluid to return to the reservoir 9. The line pressure in the upper wheel cylinders urges the plunger 1 to the left and the cutoff valve 7 closes off the return to the ZHS booster.

Thus, the space vacated by the plunger "s" is occupied by the wheel cylinder volume "s" and the brake force in the one or more regulated brakes reduced.

Repressurization occurs after the solenoid valves have returned to their original position and accumulator pressure is reintroduced to chamber 8. If the accumulator pressure fails, one circuit remains intact. The vehicle can be braked with the hydrostatic circuit because the spring 3 and its cup 10 urge the tappets 11 against the plunger 1 and mechanically keep the cutoff valve open all the time.

Because the plungers 1 are located concentrically around the central piston accumulator 2, their number can be increased according to the number of wheels to be regulated. With more than three or four plungers, the spring 3 might have to be strengthened.

Finally, Fig. 16 shows the convincing simplicity and compactness of a total vehicle installation with power-assisted brakes and antilock with the hydraulic and electrical connections. In the vehicle illustrated, both front brakes are individually controlled while the rear brakes use the "select-low"

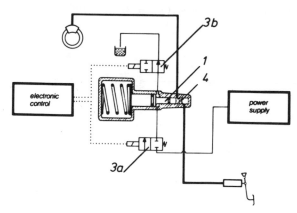

Fig. 13 - Plunger type hydraulic antilock control

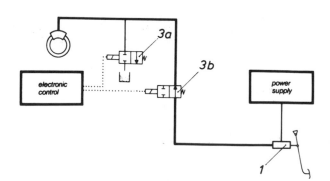

Fig. 14 - Valve block hydraulic antilock control

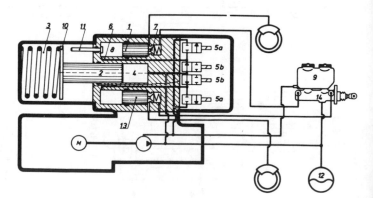

Fig. 15 - Compact BREL with power-assisted brake system

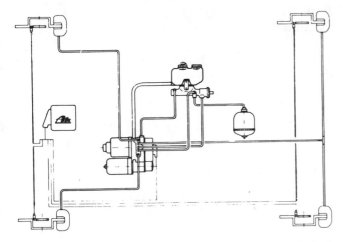

Fig. 16 - Vehicle installation of hydraulic service brakes and antilock system

principle. Dependent on the space available, the separately mounted accumulator can be integrated with the BREL or with the hydraulic booster.

## SUMMARY

Three variations of hydraulic boosters for power-assisted brake systems and a dual-circuit control valve for hydraulic full-power brake systems are described. The units all make use of the accumulator, for this offers the driver more security than the continuous-flow system.

It has been suggested that brake systems be classified by means of an "efficiency factor" (total brake force/driver's pedal effort) and that an "effectiveness coefficient" be introduced as a comparison value to correlate the efficiency factor of a brake system having failed boosting with that of an intact system. Applying these values to brake systems using the described boosters helps judge the advantages and disadvantages, and emphasizes the importance of a new ZHS 2.2 booster, especially for heavier vehicles with disc brakes.

A hydraulic actuation system is especially attractive when a vehicle is also to be equipped with antilock. By means of the compact BREL unit, hydraulic actuation of the antilock function and the energy source for service and antilock braking can be integrated into one unit so that a relatively simple installation with a minimum of connections and piping is now possible.

## NOMENCLATURE

| | | |
|---|---|---|
| $a$ | = | % deceleration |
| $B$ | = | brake force; kg; wheel cylinder |
| $C^*$ | = | brake factor |
| $d$ | = | diameter, cm |
| $F$ | = | cross-sectional area, $cm^2$ |
| $G$ | = | weight, kg |
| $HA$ | = | rear axle |
| $i_p$ | = | pedal ratio |
| $K$ | = | constant |
| $M$ | = | master cylinder |
| $p$ | = | hydraulic pressure, bar |
| $P_F$ | = | pedal effort, kp |
| $r$ | = | effective radius, cm |
| $R$ | = | rolling radius, cm |
| $s$ | = | stroke, cm |
| $S$ | = | control piston |
| $S_F$ | = | pedal travel, cm |
| $VA$ | = | front axle |
| $w$ | = | effectiveness coefficient |
| $W^x$ | = | efficiency factor |
| $W_0^x$ | = | efficiency factor, unboosted |
| $\eta$ | = | efficiency |
| $\vartheta$ | = | temperature, °C |
| $\lambda$ | = | max. wheel cylinder piston travel, cm |
| $\psi$ | = | rear-axle brake force proportion |

# APPENDIX A
## CALCULATION OF EFFICIENCY FACTOR AND
## EFFECTIVENESS COEFFICIENT

Brake force on a vehicle with four hydraulically operated disc brakes is expressed as

$$B = p(F_{VA} + F_{HA}) \eta_B \left(C^* \frac{r}{R}\right) 2 \quad kp \quad (A-1)$$

if $p$, $C^*$, $r/R$ on the front and rear axle are equal.

Given

$$K = \eta_B C^* \left(\frac{r}{R}\right) 2 \quad (A-1a)$$

and

$$p = \frac{P_F i_p}{F_M} \eta_M \quad bar \quad (A-1b)$$

and substituting Eqs. A-1a and A-1b into Eq. A-1,

$$B = aG = P_F i_p \eta_M K \frac{F_{VA} + F_{HA}}{F_M} \quad kp \quad (A-2)$$

Balancing volumes,

$$S_M F_M = \left(\frac{s_F}{i_p}\right) F_M = 4\lambda(F_{VA} + F_{HA}) \quad cm^3 \quad (A-3)$$

The efficiency factor of the unboosted brake system is

$$W_0^x = \frac{aG}{P_F} \quad (A-4)$$

Combining Eqs. A-2, A-3, A-4 gives

$$W_0^x = i_p K \eta_{M'} \frac{F_{VA} + F_{HA}}{F_M} = \frac{s_F}{4\lambda} K \eta_M \quad (A-5)$$

With an interposed, intact booster ZHS 2.0, as in Fig. 5,

$$p = \frac{P_F i_p}{F_s} \eta_s \quad bar \quad (A-6)$$

and since Eq. A-3 still applies, the efficiency factor with booster is

$$W^x = i_p K \eta_s \frac{F_{VA} + F_{HA}}{F_s} = \left(\frac{s_F}{4\lambda}\right) K \eta_M \left(\frac{F_M}{F_s}\right) \quad (A-7)$$

The percentage effectiveness of the brake system with failed

booster compared with that for an intact system is given by the effectiveness coefficient

$$w = \frac{W_0^x}{W^x} (100) \quad \% \quad (A-8)$$

For a system with ZHS 2.0,

$$w = \frac{F_s}{F_M} (100) \quad \% \quad (A-9)$$

For ZHS 2.1 (Fig. 7), Eq. A-6 still applies, but instead of Eq. A-3, since the VA circuit only is supplied from the master cylinder,

$$S_M F_M = \left(\frac{s_F}{i_p}\right) F_M = 4\lambda F_{VA} \quad cm^3 \quad (A-10)$$

and since

$$F_{VA} = (1 - \phi)(F_{VA} + F_{HA}) \quad cm^2 \quad (A-11)$$

the efficiency factor of the brake system with intact ZHS 2.1 booster is

$$W^x = i_p K \eta_s \frac{F_{VA} + F_{HA}}{F_s} = \frac{s_F}{4\lambda} K \eta_s \frac{F_M}{F_s(1 - \phi)} \quad (A-12)$$

If the accumulator pressure fails, only the VA circuit will be operated; thus

$$B = pF_{VA} K = aG \quad kp \quad (A-13)$$

and combining Eqs. 1b, 10, 11,

$$W_0^x = i_p K \eta_M (1 - \phi) \frac{F_{VA} + F_{HA}}{F_M} = \left(\frac{s_F}{4\lambda}\right) K \eta_M \quad (A-14)$$

Now, dividing Eq. A-14 by Eq. A-12 gives the effectiveness coefficient

$$w = \frac{F_s (1 - \phi)}{F_M} (100) \quad \% \quad (A-15)$$

For ZHS 2.2 (Fig. 9), the pedal travel causes a shorter travel $\lambda'$ at the wheel cylinders. Thus, the area $F'_M$ can be smaller than $F_M$; in this case, instead of Eq. A-10, the following applies:

$$\left(\frac{{}^sF}{i_p}\right) F'_M = 4\lambda' F_{VA} = 4\lambda' (1 - \phi)(F_{VA} + F_{HA}) \quad cm^3$$

$$(A-16)$$

and the efficiency factor for brake systems with a ZHS 2.2 booster will be

$$W^x = i_p K \eta_s \frac{F_{VA} + F_{HA}}{F_s} = \frac{{}^sF}{4\lambda'} K \eta_s \frac{F'_M}{F_s(1-\phi)}$$

$$(A-17)$$

If the accumulator fails, the efficiency factor is similar to Eq. A-14:

$$W_0{}^x = i_p K \eta_M (1 - \phi)\frac{F_{VA} + F_{HA}}{F'_M} = \frac{{}^sF}{4\lambda'} K \eta_M$$

$$(A-18)$$

and thus the effectiveness coefficient is

$$w = F_s \frac{(1 - \phi)}{F'_M} (100) \quad \% \qquad (A-19)$$

Table A-1 - Summary of Most Important Equations

|       | ZHS 2.0 | ZHS 2.1 | ZHS 2.2 |
|-------|---------|---------|---------|
| $W_0{}^x$ | A-5 | A-14 | A-18 |
| $W^x$ | A-7 | A-12 | A-17 |
| $W$   | A-9 | A-15 | A-19 |

NOTES:
If $F_M$ for ZHS 2.1 $= (1 - \phi)$ ($F_M$ for ZHS 2.0), then equation applications are as follows:

$$A-5 = A-14; \quad A-7 = A-12; \quad A-9 = A-15$$

If $\lambda' < \lambda$ and $F'_M/\lambda' = F_M/\lambda$, then

$$A-18 > A-14; \quad A-17 = A-12; \quad A-19 > A-15$$

## APPENDIX

It will no doubt be of interest, that the development work on the hydraulic systems has been persued with success since the original paper was written in February 1973. Particularly, the ZHS.2.2 system has been improved (see Fig. A-1 in comparison with that illustrated in Fig. 9).

Fig. A-2 shows the new "hydraulic booster unit" with brakes off. The cut-off valve, 2, is closed and prevents accumulator pressure leaking away over the spool valve, 3, when the brakes are not operated for long periods.

On pressing the pedal, the reaction piston, 4, advances to the left and the cut-off valve is lifted, the return to the reservoir, 5, being blocked as the spool valve, 3, advances. The modulated pressure then acts on:

1. The rear face of the master cylinder piston, 6, which is hydraulically moved to the left, thereby operating brake circuit I (VA).

2. The sleeve, 8, surrounding the travel-simulator, 7, which is held against the step in the bore (left to right).

3. The cross-sectional area of the reaction piston, 4, which transmits to the driver a "feel" corresponding to the modulated pressure level.

The load-travel characteristic at the brake pedal is influenced only by the rate of the simulator spring-pack.

If the accumulator pressure fails, the sleeve, 8, is no longer held on its stop, and a load applied to piston 4 is transmitted without compressing the simulator spring-pack mechanically to piston 6. In so doing the comparatively light coil spring in chamberR1 is compressed, and the intact master cylinder circuit brought into operation. In this case the load-travel

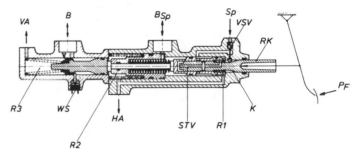

Fig. A-1 - Modified integral booster and single-circuit master cylinder (ZHS 2.2)

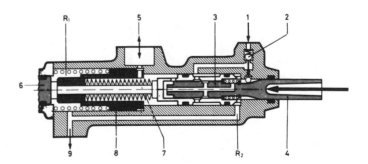

Fig. A-2 - Hydraulic booster unit

characteristic at the brake pedal is governed by the pressure-volume characteristic of the brake circuit supplied from the master cylinder. Chamber R2 can be connected at port 9 to operate brake circuit II (HA) directly.

# Introduction of Antilock Braking Systems for Cars*

**Hans Christof Klein and Werner Fink**
ITT-Teves Germany

*Paper 741084 presented at the International
Automobile Engineering and Manufacturing
Meeting, Toronto, Canada, October, 1974.

IN THE EARLY 1970S, various manufacturers introduced their antilock systems to the public. Only a small number of systems entered series production or were actually sold to customers as prototypes. There are two main reasons for this situation:

The concept was always that of an add-on unit to the existing brake system, resulting in a complicated and expensive total package.

Second, before introducing a system, comprehensive durability testing was carried out, which proved to be not always sufficient to allow for mass production methods and did not provide the level of reliability expected in everyday service life. This was particularly the case for 4-wheel antilock, and these factors were instrumental in postponing the introduction date of such systems.

Both these facts must be considered during the further development of antilock systems.

## BASIC DESIGN OF PRESENT SYSTEMS

This description will only cover systems whose basic design concept is the same, although they may be produced by various manufacturers. The advantages and drawbacks of the different designs will not be dealt with in detail, but the most suitable systems will be highlighted.

SENSOR - All sensors work on the induction principle: Self-induced current generation is used, whereby the magnetic field is created by permanent magnets. Fig. 1 shows the basic design of a single pole inductive sensor. Fig. 2 shows the sensor output in relation to frequency. The approximately sinus-

wave form of current generated at the sensor can be converted to a right angled form either in the sensor body itself, or in the electronic control unit.

Fig. 3 shows a ring sensor design; the field changes occur over the whole of the ring's circumference.

ELECTRONIC CONTROL - This enables the so-called antilock philosophy to be achieved, whereby locking of the vehicle's wheels is prevented. The word philosophy is used in a true context, as no uniform consideration of this complex problem has been evolved. It is of decisive importance for the function and reliability of the system how this control philosophy is reflected in the electronic circuitry. Here too, almost every possible application offered by electronics is utilized, encompassing pure analog, pure digital and mixed circuits, whereby the number of individual items is kept to a minimum by partial integration.

A check- and safety-circuit monitors and controls the electronic part of the antilock system. As the complexity would be too great to control every theoretically possible fault, only those are covered which might in any way result in a reduction of braking effect.

BRAKE LINE-PRESSURE MODULATION - For this part of the system, the comments made initially on complexity of design are particularly applicable. The reason for this lies in the small displacement volume available in the master cylinder. Current brake systems consist of one or two hydrostatic brake circuits. The word "static" means that brake fluid is compressed by the driver exerting effort on the brake pedal. The possible energy expended is limited by the available pedal travel and the pedal load acceptable to the driver. Brake

──────── ABSTRACT ────────

The overall question of brake actuation, antilock control and other vehicle functions requiring servo assistance will be discussed. Too much effort has been devoted in treating each of these units on an individual basis. This paper describes ways in which this complex subject can be dealt with in a unified manner, to provide a more compact and reliable solution.

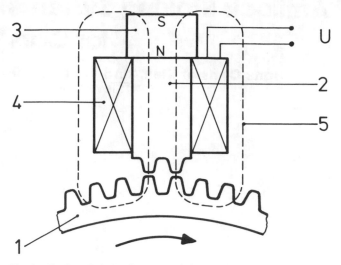

Fig. 1 - Single pole inductive sensor. The temporary change of the magnetic field $\varphi$ is proportional to the $\frac{d\varphi}{dt} \sim U$ voltage U. 1 = rotating toothed ring; 2 = pole; 3 = permanent magnet; 4 = induction coil; 5 = magnetic field.

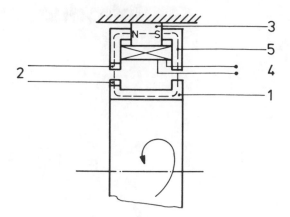

Fig. 3 - Inductive sensor in ring form. 1 = rotating part; 2 = equal number of teeth opposite each other on fixed and moving parts; 3 = ring formed permanent magnet; 4 = induction coil; 5 = magnetic field.

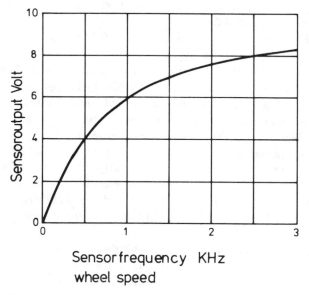

Fig. 2 - Sensor output as function of frequency or wheel speed.

boosters do not alter this picture to any great degree, as the minimum performance set by the legislators on the vehicle deceleration, is that when the booster has failed.

Because of the limited volume in the master cylinder, a sophisticated design is needed for line pressure modulation in an antilock system to occur independent of the driver. The operating principle used is shown in Fig. 4. The energy expended by the driver at the brake pedal (F × S) is fed into a spring accumulator $A_S$ in the event of a "depressurize" signal. For this to happen, the plunger P must be depressed against the spring (see Fig. 4), the energy for which is usually provided hydraulically, or by using the vacuum available in the engine inlet manifold. By activating the solenoid valve $V_E$, hydraulic pressure is admitted to the ring area of the plunger, moving the latter downwards, and closing cutoff valve $V_M$ onto its seat

and interrupting the connection from master cylinder to brake cylinder. The line pressure in the brake can then be released into the annular plunger chamber. This procedure can also be described in that the energy in the wheel brakes is stored in the spring accumulator during pressure release, and that the energy F × S, created by the driver, can be stored for a short time in the very reliable spring of the accumulator.

## DESIGN OF FUTURE SYSTEMS

These vary from current designs principally in the way line pressure modulation is achieved. Therefore, the following description will first be considered.

BRAKE ACTUATION WITH INTEGRATED ANTILOCK - The power source discussed in Fig. 4 was either a hydraulic pump or vacuum. The latter is the most economical source, assuming a sufficient quantity is available. However, exhaust emission regulations and fuel injection have much reduced the degree of vacuum. Future designs will require a hydraulic energy source, consisting of a pump and a gas-loaded accumulator. This type of accumulator replaces the spring unit $A_S$ in Fig. 4. The energy storage capacity of a gas-filled accumulator in relation to its weight is very much higher than for a spring accumulator. If by way of comparison, the gas accumulator contains ten times more energy, then this energy can be utilized to operate the wheel brakes directly, so that the limited ability of the driver can be overcome. Thus, a much simpler design can be introduced. Here, by means of a pressure regulating valve, the driver modulates a line pressure in the brakes proportional to the pedal effort. Such a system is described as "dynamic," as, contrary to the "static" system, a piston is not employed to compress the brake fluid. A further difference is that the fluid volume required for each brake application is not pumped to and fro, but returns to the reservoir R via return pipe $R_P$.

For antilock modulation, the brakes are depressurized by actuating solenoid valve $V_E$, the fluid returning to the reservoir R. The brakes are reapplied by switching the valve $V_E$

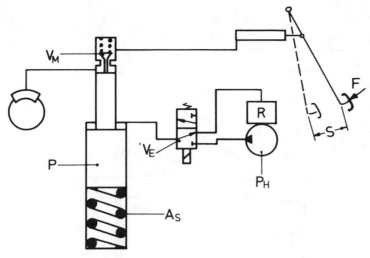

Fig. 4 - Displacement plunger principle for antilock line pressure modulation with "static" brake circuit. F = pedal effort; S = pedal travel; $A_S$ = spring accumulator; P = plunger; $V_E$ = electromagnetic valve; $V_M$ = mechanical valve; $P_H$ = hydraulic pump; R = reservoir.

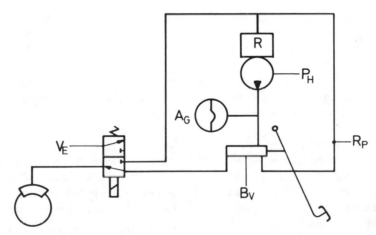

Fig. 5 - Valve layout for antilock line pressure modulation with "dynamic" brake circuit. $V_E$ = electromagnetic valve; $B_V$ = brake valve; $A_G$ = gas accumulator; $P_H$ = hydraulic pump; R = reservoir; $R_P$ = return pipe.

back to its original position, whereby the brake actuating valve $B_V$ allows pressure from the accumulator $A_G$ to reapply the brakes.

When comparing the "static" with the "dynamic" system, including antilock control, the reduction in the number of necessary components is immediately apparent. The plunger P and valve $V_M$, with spring $A_S$ are replaced by the accumulator $A_G$. This simplification is specially relevant when several wheels are controlled independently, because a "static" system requires one plunger unit for each controlled wheel.

A further advantage of this concept, shown in Fig. 5, is that the accumulator $A_S$ comes into use with every brake application, whereas the plunger system only works when antilock is needed. The use of an accumulator is therefore not tied directly to the introduction of antilock, and considerable advantages can be gained in conventional brake systems if a "dy-

namic" actuation is chosen, as described in the SAE Paper 730535 (1)*.

SENSOR - The self-energizing inductive signal generator is not ideal for this application. The toothed ring rotating with the wheel is located relative to the sensor by the wheel bearing, so that radial runout and bearing axial clearance induce additional voltage noise, which is superimposed on the actual signal required. As the sensor voltage output is very small at low road speeds (Fig. 2), these signal noise simulate a false road speed. This basic drawback can be avoided if remote energized signal generation is resorted to, so that even at low wheel speeds, a high voltage amplitude in possible.

In spite of this disadvantage, the inductive principle will continue to be popular, as it is of simple design, and, compared

---

*Numbers in parentheses designate References at end of paper.

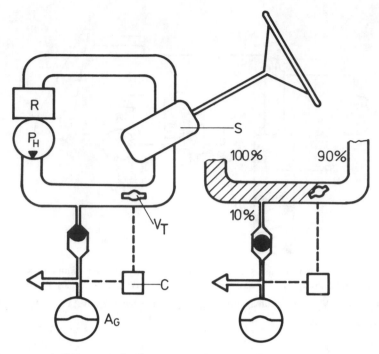

Fig. 6 - Combining power steering pump "open-center" flow with accumulator charging unit. $P_H$ = hydraulic pump; R = reservoir; S = power steering; $A_G$ = gas accumulator; $V_T$ = throttle valve; C = charging unit.

with electronic recording devices (photo-cells, Hall generators, etc), can be subjected to operating temperatures of 160°C. Such temperatures are met with on non-driven wheels in the vicinity of the brake disc.

ELECTRONIC CONTROL - The target for further development is a major reduction in the number of components and connections, in order to improve the reliability. To render the electronic circuitry insensitive to manufacturing variations and temperature changes, digital techniques will be adapted. As a clear command signal for every switching function is achieved by utilizing the total voltage span of several volts, the interim conditions which met with analog circuitry cannot occur. The subject of application of electronic circuitry in road vehicles will not be further dealt with in this paper.

INTRODUCTION OF FUTURE SYSTEMS

SINGLE CIRCUIT ANTILOCK IN CONJUNCTION WITH POWER STEERING - Single circuit antilock means that the antilock control only applies to one brake circuit of a normal two circuit brake system.

As described in Fig. 5, an accumulator charged by a pump is used to provide a "dynamic" brake system with integral line pressure modulation.

The energy source can be elegantly provided by utilizing the already available power steering pump (Fig. 6). The pressure level of the accumulator $A_G$ is held between upper and lower limits by the automatic operation of the throttling valve $V_T$.

As only 10% of the circulating oil is directed to the accumulator during the charging phase, the driver, should he at that moment be using the power steering, hardly detects anything, as 90% of the normal flow of oil is still available for steering purposes.

Two operating mediums are used (automatic transmission fluid in the steering circuit and brake fluid in the brake system), and as the steering pump output pressure alone is insufficient to provide a high enough line pressure level in the brakes, the basic design is adapted accordingly and is shown in Fig. 7. Pistons $P_S$ separate the two fluids and simultaneously provide the required pressure ratio. Brake circuit $C_R$ is actuated purely hydraulically by modulated accumulator pressure from the proportional valve $P_V$. Similarly the piston $P_S$ of the circuit not provided with antilock is normally hydraulically actuated, but the latter can be mechanically operated by the driver should the energy supply fail. This system therefore retains a "static" master cylinder for the front brakes and uses "dynamic" operation for the rear brakes.

Apart from the usual front/rear circuit split, there are several variations in Europe, of which the most logical one appears to be the diagonal circuit split (2). Here too, the control of one circuit is also feasible. The one controlled rear wheel ensures the straight-ahead stability, while the other controlled front wheel provides some degree of steerability.

THREE CIRCUIT ANTILOCK - Three circuit antilock means that there is independent antilock regulation of each front wheel and common regulation of both rear wheels.

In this case, the brake line pressure must be modulated in both brake circuits so that the one "static" circuit of the previous system must be converted into a "dynamic" one to provide a more compact and reliable unit. The vehicle then has a full-power brake system.

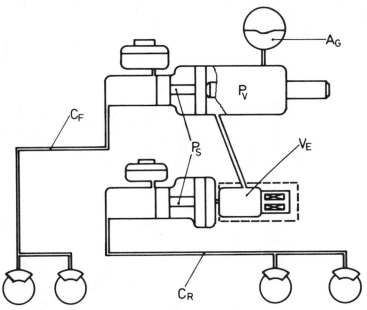

Fig. 7 - Integration of brake actuation and valve system for single circuit antilock. $A_G$ = gas accumulator; $P_V$ = proportional valve; $P_S$ = separating piston; $C_F$ = front brake circuit; $C_R$ = rear brake circuit; $V_E$ = electromagnetic valve.

The continued use of the steering pump as an energy source is in principle possible, but results in a complex design due to different pressure levels in both systems. It is thus desirable that such power-consuming units as brakes, steering and perhaps levelling suspension work at only one pressure level with one fluid medium.

This demand could mean in certain conditions a change of concept in power steering design. Today they work on the "open-center" principle, whereby a continuous circulation of oil is throttled, in other words pressure is raised when hydraulic energy is needed. In order to achieve the pressure level required by the brakes, the overall pressure must be raised, for which the current vane pumps are not suitable. Therefore a change to piston pumps must be considered. A further possibility would be to use the "closed-centre" principle for power steering, which means that the energy requirements for steering would be provided by an accumulator. Such a system is already known in Europe, however the steering "feel" is open to judgment.

SINGLE AND TWO CIRCUIT ANTILOCK FOR VEHICLES WITHOUT POWER STEERING - Where no steering pump is specified, a hydraulic pump must be fitted in any case. The pressure level and fluid medium can be tailored to suit the brake system of a particular vehicle. The additional cost of an extra pump installation will only be accepted by the public when the necessity for antilock is recognized, or when it is demanded by legislation. The possible concepts for such cases will be discussed at a later date.

SUMMARY

The basic design of current antilock systems is described. They are all characterized by being back-up units to the existing brake systems with "static" brake circuits being retained. A reduction in complexity is possible with a new design, whereby the brake actuation consists of one "dynamic" and one "static" circuit with single circuit antilock, or two "dynamic" circuits for three circuit antilock. In the development of the sensors and electronic circuitry, similar degrees of change are not so likely to take place.

REFERENCES

1. Otto Depenheuer, Hans Strien, and Alfred Teves Gmbh, "Hydraulic Brake Actuation System Under Consideration of Antilock Systems and Disk Brakes." Paper 730535 presented at SAE National Automobile Engineering Meeting, Detroit, May 1972.

2. D. Banholzer, "Negative (or Outboard) Scrub Radius and Diagonal Circuit Split." ATZ Report 11, 1972.

# Electronic Control Systems for Ground Vehicles*

**Edward J. Hayes and George W. Megginson**
Kelsey-Hayes Co.
Romulus, MI

*Paper 790457 presented at the Congress and Exposition, Detroit, Michigan, February, 1979.

ANTI-WHEEL LOCK ELECTRONIC CONTROL SYSTEMS have been in wide spread use on air craft since 1960. The transfer of this technology to ground vehicles required over ten years of research and development because of the high system cost and technical problems associated with the wide range of operating conditions for both automobiles and trucks. In our opinion, the wide spread use of antilock on ground vehicles was recently interrupted in the United States due to consumer attitudes, lack of trained maintenance personnel, priority use of technical resources to meet gas economy and emission standards, and opposition to government controls. To assist the continuity of product development following this interruption, we believe it worthwhile to review some of the decisions that have evolved to date.

## EVOLUTION OF AIR BRAKE SKID CONTROL

An electro-mechanical antilock system using a flywheel sensor was being used on trailers in 1960. This was replaced with an electronic system used on both tractors and trailers by 1970. These early units were wheel control systems with an independent system for each wheel. To reduce the complexity and to eliminate side to side torque unbalance, axle systems were first introduced in 1972. A relatively small specialty market was developing until 1975 when FMVSS-121 became effective requiring most vehicles with air brakes to be equipped with an antilock system to meet the no-wheel lock requirement of the standard.

Air brake skid control has been produced in relatively large quantities for three years. Until a few months ago, it appeared to be developing into a stable product with long-term potential. The Ninth Circuit Court of Appeals has ruled that since the National Highway Traffic Safety Administration (NHTSA) did not adequately test antilock systems to assure their reliability, their use must be discontinued. It is understood that this oversimplification is our opinion of a complex legal decision.

INFORMATION RATE - One of the first debates that developed was the information rate. The basic input required by the skid control performance circuit is wheel speed. The information used by the electronic input circuit to determine wheel speed is the number of electrical pulses per wheel revolution. The most popular information rates were 60, 90, or 120 pulses per revolution. With a 10:00 x 20 tire, the 60 pulses gives about 8.5 Hertz per mile per hour, and the 120 pulses gives about 17 Hertz per mile per hour.

After reviewing the problem, control engineers requested 240 pulses per revolution or more. However, using conventional manufactur-

## ABSTRACT

The wide spread installation of anti-wheel lock systems in the United States has been interrupted. To assist the continuity of development after this interruption, we believe it worthwhile to review some of the designs that have evolved to date. Potential for antilock to improve vehicle stability during heavy braking has been demonstrated. However, with the current situation in the United States, Europe and Japan will probably lead the way in the popular use of electronic antilock systems.

ing techniques, and restricting the exciter to the brake drum area, it soon became evident that 120 pulses per revolution was the maximum that could be economically and reliably provided, and 60 pulses per revolution would produce an exciter that was easier to make and mechanically stronger at a lower cost. All major OEM's now use 60 pulses per revolution or have plans to convert to 60 pulses per revolution. The practical result of this lack of information is usually a degree of over control -- a short period of synchronous running following a control cycle. If a required stopping distance of less than 245 feet from 60 M.P.H. is needed, this reduced information rate may make it difficult for certain vehicle configurations to comply. However, if the requirement is 275 feet or more, there should be very few vehicles that will experience difficulty in complying.

TYPE OF SENSOR - Annular and point-type sensors are both used. However, due to oil seal problems and the difficulty of fitting annular sensors to drive axles, the annular sensor is more popular on trailers, and the point sensor is more popular on tractors.

The annular sensor produces the most consistent noise-free signal. The self-contained annular sensor which can be located completely within the trailer axle provides the best signal combined with ease of installation and maintenance.

The single pole variable reluctance sensor provides the lowest cost sensor and the smallest package. However, it places a heavy demand on the electronics to eliminate the noise generated by mechanical vibration. A proximity pick-up that reduces the magnitude of the signal produced by mechanical vibration reduces the complexity of the required electronic input circuit.

An exciter attached to the brake drum with sixty (60) equally spaced ripples is the most popular exciter.

ELECTRONIC INPUT CIRCUIT - All input circuits have a clipping network with a low pass filter. However, much of the noise, due to mechanical vibration, is in the 400 to 600 Hertz range. Normal operation for a sixty (60) pulse per revolution system is zero (0) to 800 Hertz. Most circuits now include a dead band or hysteresis in addition to a clipping circuit and a low pass filter to improve the noise rejection capability. At least one system uses a switch filter to greatly attenuate any signal above 200 Hertz if the vehicle speed is below 5 M.P.H. As the vehicle accelerates through 5 M.P.H., the filter is switched off to allow normal operation.

PERFORMANCE CIRCUITS - Performance circuits require both a wheel deceleration rate threshold and a percent slip threshold to be exceeded before a signal is provided to reduce brake pressure. One or both of these thresh-

olds are modulated to improve the adaptability. To directly determine slip vehicle speed in addition to wheel speed must be known accurately. Radar or an unbraked wheel has been used to measure vehicle speed. However, there are no commercially available units for air brake vehicles using this method. Slip is calculated by integrating wheel speed or by comparing it to a predicted vehicle reference speed.

Methods used to generate a vehicle reference and to modulate the rate and slip thresholds are the subject of many patents, resulting in a wide variety of performance circuits.

MANUAL BACK-UP - During normal operation, the driver can manually modulate brake pressure below the pressure required to lock the wheel. At pressures above this, the antiskid automatically modulates the pressure to avoid wheel lock. Electronic circuits are so designed that the failure of any single component will not cause loss of brakes on the controlled axle. All system designs provide for immediate return to full manual brake operation if system power is lost or the valve shorts or opens. Many of the circuits will revert to full manual brakes within .1 second if the input transistor shorts or opens. The tractor system also monitors in the driving mode and reverts to manual operation if abnormal operation is noted. All systems have a monitor that reverts the system to full manual operation if the valve signal remains on for more than two or three seconds.

VALVES - On-off solenoid control valves and pneumatic modulated solenoid valves are the most popular air control valves. However, digital and hold-mode valves are still provided with some systems. The on-off valve response time is controlled by its size and the size of other air system components. At least one company uses a dual solenoid on-off valve with two different rates of pressure release and apply. Dual performance circuits select the appropriate solenoid to energize.

The pneumatic modulated valve reduces pressure similar to the on-off valve. It then reapplies rapidly to a pressure just below that required for the last skid cycle and then slowly applies more pressure until the skid cycle is repeated. This valve reduces the number of skid cycles and in general provides smoother operation with less air consumption.

RADIO FREQUENCY INTERFERENCE - It is difficult to talk about radio frequency interference susceptibility in specific numbers. However, for the purpose of this discussion, far field equivalents will be used. Initially, units could withstand about 20-V/M. This was compatible with tube type mobile transmitters and early solid state units. Today's units must tolerate 100-V/M to be compatible with new higher wattage mobile transmitters. Direct radiation is not the major source of radio frequency interference. The interconnecting wiring picks up the radiated energy and conducts it into the electronic package.

SYSTEM CONFIGURATION - Initial units were all axle control. A sensor was used on each wheel, and a single module and valve were used for each axle. This included the front tractor as well as both axles of a tandem trailer.

In an effort to reduce the system complexity, a bogie system replaced the axle system on most trailers. A sensor was used on each wheel of the leading axle of the tandem, and a single module and valve controlled all four brakes on the tandem. Tandem control was evaluated for tractor use. However, most users preferred the individual axle control.

When the 60 M.P.H. stopping distance requirement of FMVSS-121 was increased to 293 feet, the front brake torque on most tractors was reduced allowing these vehicles to comply with the "no-wheel lock-up" requirement with no front axle skid control.

## EVOLUTION OF AUTOMOTIVE SKID CONTROL

Automotive skid control was not the result of legislation. It was perceived as a need by the automotive industry. From 1969 to 1977, skid control was standard on the Lincoln Mark II to Mark IV. It was optional for at least one year during this period on the Ford LTD, Mercury Marquis, Thunderbird, Buick Riviera, Cadillac Eldorado, Oldsmobile Toronado, Chrysler, and IHC Travelall. Initial units were expensive. However, in 1974, the future of rear wheel skid control with a drive-line sensor looked very bright. The system was standard on the Lincoln, optional on the Granada, Monarch, Mercury Marquis, and Thunderbird, and for the first time, sales volumes were projected to justify a cost to the vehicle manufacturer of less than $50.00. However, with the economic recession of 1975, government mandated fuel economy and emission requirements, there were too many demands upon technical resources to justify continuing development of a safety-related product that was being vigorously opposed by the trucking industry and not well understood by the motoring public. No vehicle manufacturer in the United States has plans to market an antilock system after 1980.

SENSORS - The first production sensors were variable reluctance type mounted on the rear wheel axle shaft and provided 65 pulses per revolution. The current production units are variable reluctance mounted on the drive line and provide 48 pulses per revolution. However, the information rate with the drive sensor is actually increased by the rear end ratio (2.7:1). This results in an information rate of 27 Hertz per mile per hour when a 13 inch tire is used. New systems will probably use drive line or transmission mounted sensors for the drive wheels and individual wheel mounted sensors for the non-drive wheels.

ELECTRONIC CIRCUITS - The automotive electronic circuits have followed the same development pattern discussed for the air brake system.

ACTUATOR - The antilock actuator used with the automotive master cylinder powered brake system must have a source of power to restore wheel cylinder pressure following an antilock release. Either the engine vacuum or the power steering hydraulic pump is used for this power.

The vacuum powered actuator tends to be larger, lighter, and a bit slower than a hydraulic powered actuator. Both actuators provide acceptable performance.

SYSTEM CONFIGURATION - The most popular automotive system consists of a single driveline sensor, electronic module, and actuator controlling the rear wheels only. This configuration was chosen to introduce skid control because of its relative simplicity and the fact no new driver skills are necessary. The rear wheel system assists the automobile to stop with a greatly reduced tendency to spin out. The four-wheel skid control system provides lateral stability and allows the driver to steer during heavy braking. This increased control will be new for most drivers but will give them maximum control of the vehicle during a panic situation.

## CONCLUSIONS

Antilock has proven to be a valuable aid to safe vehicle operation during heavy braking. The public is becoming more aware of its potential benefits. The advent of wide spread use of electronics to solve gasoline economy and emission problems must result in maintenance personnel with the skills required to maintain a complex electro-mechanical system. The vehicle manufacturers and the government recognize the potential for antilock to improve vehicle stability during heavy braking. It will only be a matter of time until the two can negotiate mutually agreeable conditions to govern its use.

Europe and Japan have recently accelerated their interest in antilock. Several European countries currently have a brake regulation requiring front wheels to lock before the rear wheels for all operating conditions. The common market is drafting a control document to allow antilock as an option to meet this requirement. With the current situation in the United States, Europe and Japan will probably lead the way in the wide spread use of electronic antilock systems.

## REFERENCES

1. R. H. Madison and H. E. Riordan, "Evolution of Sure-Track Brake System." SAE Paper 690213.

2. E. J. Hayes and G. W. Megginson, "Electronic Braking System." SAE Paper 780856.

4

3. T. C. Schafer, D. W. Howard, and R. W. Carp, "Design and Performance Considerations for a Passenger Car Anti-Skid System." SAE Paper 680458.

4. U. S. Department of Transportation, "Technical Assessment of FMVSS-121, Air Brake System." Document No. 75-16-GR-038, February 24, 1978.

# REGULATION AND TESTING

# The Design of the MIRA Straightline Wet Grip Testing Facility*

## C. Ashley, H. C. Allsopp, V. E. Davis, F. Fielden, and K. S. MacKellar
### Motor Industry Research Association, Nuneaton, Warwickshire

*Paper C194/85 is published as part of **Anti-lock Braking Systems for Road Vehicles** and is reproduced by permission of the Council of the Institution of Mechanical Engineers© Institution of Mechanical Engineers 1985.

SYNOPSIS  This paper reviews the design and application of the new Low Friction Wet Grip Testing Facility at MIRA, which will allow testing of tyres and vehicle braking systems at speeds up to 160 km/h over a range of surfaces including basalt, polished concrete and Bridport pebble. The disposition of the surfaces together with extensive wetting arrangements will allow a large range of test work to be performed including legaslative tests to Annex 13 of Regulation 13 concerning anti-lock.

## 1   INTRODUCTION

The braking performance of vehicles in wet conditions has been the basis of much research in previous years, particularly with respect to the effect of tyres. The improvement in car tyre grip characteristics has been quite marked over a 20 year period. The improvement in truck tyre wet grip characteristics has not been as apparent, this has been due in the main to constraints put on tyre designers in terms of operating conditions and severities of trucks tyre loads. These prevent the design of intricate tread patterns and the use of high u tread compounds which are common in most car tyres.

In the 1950s and 60s the UK vehicle industry was a world leader in wet grip research and developments. In more recent times progress has been inhibited by the lack of modern, comprehensive outdoor test facilities which companies in other countries have developed. This has particularly affected those firms involved in the development and manufacture of tyres and anti-lock braking systems.

Braking systems of vehicles have also been the subject of research and development and the recognized advantages of preventing wheel lock-up, not only to improve vehicle stability under heavy braking but to optimize the frictional characteristics of tyres with slip are now being studied and is the topic of this conference.

Tyre manufacturers need to assess the performance of their patterns and tread compounds over a range of speeds and on a variety of surfaces, the characteristics of which correlate with the road surfaces in general use where tyre/road adhesion is critical. In carrying out these tests it is also essential to study the effects of varying water depth up to and including the full aquaplaning condition. In order to gain a fuller understanding of tyre/road interaction it is also essential to be able to study, visually, the tyre/ground contact patch by means of a glass plate and underground chamber.

When developing anti-lock braking systems it is necessary to check the efficiency of the systems over a wide range of surface friction levels and vehicle speeds and, in particular, to check their effectiveness under 'split u' conditions and during transitions from high to low and low to high friction surfaces. In addition to the requirements of development there is impending European legislation which will require manufacturers to meet certain performance criteria under specified test conditions. Legislation already exists in the U.S.A. on the minimum wet grip performance of tyres and may be introduced in Europe at some future date.

Certain limited facilities do exist in the UK but none are sufficiently comprehensive or sophisticated to allow the precise control of surface and watering conditions which modern development testing demands. The facility described will meet all present and foreseeable future needs for wet grip, brake, tyre and whole vehicle testing in a straight ahead mode (Fig 1).

Under the present financial conditions the cost to independant organizations of designing and constructing their own purpose built facility is prohibitive.

It makes sense therefore that any such facility should be financed by a consortium of interested parties and be sited at M.I.R.A. where it would be available to all member companies for either independant or joint research projects.

On this basis it would be possible to provide a comprehensive facility to cover the requirements of all interested parties in terms of surface specification, watering systems, test speeds and test methods.

## 2  GENERAL DESIGN PRINCIPLES

A detailed ground investigation confirmed that the site was underlain by Glacial Boulder Clays resting upon the Keuper Marl Formation. Laboratory California Bearing Ratio Tests, (C.B.R.), were carried out on samples of the Boulder Clay, recompacted to its insitu density, and gave CBR values ranging from 5 per cent to 18 per cent. For design purposes a value of 5% was adopted. As weathering could rapidly soften the clay formation it was also recommended that the sub-base to the road should be laid as soon after exposure as possible. Soluble sulphate content determinations of the near surface soils and ground water showed that Ordinary Portland Cement would be suitable for use in buried concrete.

Apart from the test surfaces the circuit has been designed as flexible construction in accordance with Road Note 29 (Ref 1). The test areas themselves however require a variety of surfaces incorporating concrete, bitumen, and basalt blocks. In order to provide continuity of construction over the area of the test surfaces, where levels are extremely critical due to the watering requirements and any differential settlement between tracks could have disastrous effects, they have all been designed as rigid construction with the bituminous surfaces and blockwork forming an overlay to the main construction. Again the design of the rigid pavement has been carried out in accordance with Road Note 29.

It is normal practice to design pavements to have a structural life of 40 years in rigid construction and 20 years in flexible construction. The shorter design life of the flexible construction reflects the relative ease of extending the life of a flexible pavement by means of overlaying at a later date. However, with the comparatively low traffic volumes included and the difficulty that would be experienced in overlaying due to the critical levels on the site, both pavement forms have been designed for a structural life of 40 years. The estimated volume of traffic carried over this 40 years' period is equivalent to $6.39 \times 10^6$ standard axles (a standard axle being 8,200 kg) for the approach roads and slightly less for the test surfaces. A figure of $7.0 \times 10^6$ standard axles has been used for design purposes.

## 3  CHOICE OF TEST SURFACES

A range of six different test surfaces were chosen in order to meet the various friction levels and surface textures required for the many aspects of wet braking tests performed. They take into account the requirements of Annex 13 to Regulation 13 for anti-lock braking as well as those for tyre testing and vehicle handling. They are as follows, together with their anticipated Skid Resistance Values when measured with the Pendulum:-

|  | | SIZE, m | SRV |
|---|---|---|---|
| 1) | Basalt tile | 200 X 7.5 | 20 |
| 2) | Bridport pebble | 100 x 4 | 30 |
| 3) | Polished concrete | 200 x 4 | 50 |
| 4) | Open textured bitmac | 100 x 4 | 68 |
| 5) | Sand asphalt | 100 x 4 | 72 |
| 6) | Delugrip | 100 x 4 | 77 |

Note that 100 metres of the polished concrete can be converted into an aquaplaning trough whenever required.

A short description of each surface is given below.

### 3.1  Basalt tile

This is a new test surface and is fully described in Section 5.

### 3.2  Bridport pebble

Bridport pebble has been used extensively by the tyre industry in the past but when incorporated into a bituminous binder has exhibited poor durability. This problem has been overcome by using the pebble in a concrete matrix suitably treated to leave the aggregate exposed at the surfaces.

A 10% loading by volume of 6-8mm quartzite chippings have been included with the 10-12mm Dorset Pebble in order to increase the micro texture. This has the effect of increasing the friction level by about 10% over 100% Pebble aggregate. Tread compound differentiation is greatly enhanced without undue tyre abrasion.

### 3.3  Polished concrete

This is a very durable smooth surface whose frictional characteristics remain fairly constant over long periods of time.

If however over a period of years the surface became polished to an unacceptable level it would be possible to grind and regenerate the surface texture.

### 3.4  Open textured bituminous macadam

This surface was included at the behest of the Michelin company as being representative of a type of road surface in general use in France.

### 3.5  Sand asphalt

This has similar texture characteristics to smooth concrete but slightly higher friction levels.

### 3.6  Delugrip

This is a modern high friction road surfacing material which gives good adhesion under wet conditions. As shown in Fig 1 it is interspersed between the low friction surfaces.

## 3.7 Aquaplaning trough

This is formed, when required, by inserting a flexible rubber wall into prepared slots in the concrete. Adjustable height dams are then used to control the depth of water in the trough up to a maximum of 25mm.

## 4 BASALT TILES

This is a new and innovative low friction test surface designed and developed by MIRA at the instigation of the Department of Transport who were looking for a suitable area on which both development and legislative testing of anti-lock braking systems could be carried out.

### 4.1 Requirement

The detailed characteristic required of the surface is that the coefficient of adhesion shall be about 0.3. Also, the surface must be controllable in the sense that if the friction level varies due to wear, contamination or other factors, the surface can be brought back to the specified value fairly readily. It may also be presumed that its speed sensitivity should be comparatively low.

### 4.2 Description

The 'Contidrom' at Jeversen, near Hanover, is a comprehensive tyre testing facility owned by Continental Tyres. One of the features is a large - 500m x 7.5m - area of natural basalt sets (Fig 2). These sets were originally laid some considerable time ago as the wearing surface of a street in Hamburg. They were removed from Hamburg to the 'Contidrom' in 1967 and were re-laid in exactly the same order as they had been in Hamburg. As can be seen the sets vary in size but on average they are an 80mm cube. Although this surface is claimed by the Germans to meet all the requirements of a low friction surface for anti-lock testing, one could be forgiven for equating it to pave - a rough durability road surface - and wondering how the two entirely dissimilar type of tests can possibly be reconciled.

MIRA has endeavoured to produce a surface somewhat similar to that at the 'Contidrom' but without the pave input, and one that could readily be made available to any interested party. The availability of large quantities of worn basalt sets from roads in this country or Europe was found to be virtually nil. It was, therefore, decided to investigate the manufacture of new basalt sets to a design that would to some extent resemble those of the 'Contidrom'.

Besides the material, there were two major aspects of the 'Contidrom' sets that it was considered vital to reproduce. The first of these was the surface finish and contour, a product of countless years of wear from horseshoes and metal clad cart wheels, resulting in the elimination of all sharp corners. Secondly, the joint or groove formation, thought to be a continual source of lubrication to the tyres during braking. Fig 3 shows the resulting design that was considered as meeting the above requirements and which was used for the subsequent manufacture of the sets.

Although a slightly domed surface might have been preferable to the flat area indicated, this was rejected since any subsequent grinding of such a shape would immediately introduce flats and, also, sharp edges.

After exhaustive enquiries a firm was located that imported reconstituted basalt i.e., mined basalt rock melted down and recast into tiles or blocks, from Czechoslovakia. In this form it is widely used in this country by the N.C.B. and the Electricity Boards to line coal slurry shutes where its wearing properties are claimed to be fifteen times that of mild steel. After due discussion with their suppliers, this firm undertook to supply the required number of basalt sets to our design and to subsequently grind and polish them as necessary. A sample quantity of six sets were produced initially and these were assessed as being satisfactory viz, a fairly smooth top surface that would require little in the way of heavy grinding prior to the polishing operation.

### 4.3 Trial Area

A trial area put down on the Proving Ground in early 1979 (Ref 2) has proved all the parameters that were being sought. Although not large enough for full scale testing to be carried out, it has been in use, on and off, for the past six years. During this time we have experienced extremes of weather not likely to be exceeded to any great extent in this country, and the surface shows no signs of deterioration.

At the proposed legislative test speed of 50 km/h the average peak u figures were 0.35 for cars and 0.20 for commercial vehicles (Fig 4).

Any attempt to polish the surface further in order to lower the value for cars must also lower it for commercial vehicles as well. As it is considered that a lower value for commercial vehicles would be inadvisable, making it virtually impossible to carry out the prescribed anti-lock test, it would seem that a test surface having peak u readings of 0.35 for cars and 0.2 for commercial vehicles could prove to be very satisfactory.

The problem of properly defining the friction level of a surface still remains, and if the ability to control its level is to be utilised then some standard procedure using standard tools should be agreed. (Ref 3).

The tile and joint grooves performed their function in holding water and there was no evidence of dry tracks being left behind by sliding front wheels.

It was decided to lay the tiles with the

grooves in-line as that was most approaching the lay-out of the 'Contidrom', (Fig 5), despite its varying size of set. However, subsequent experimentation has indicated that an offset pattern is preferable when testing motorcycles, (Fig 6), and this has been adopted in the final design, (Ref 4).

### 4.4 Wear Properties

It can be confidently predicted that no wear problems will be encountered with Basalt.

Cast Basalt is Basalt rock melted, cast and annealed to form a finely crystallised 'glass ceramic' of extreme hardness (8.5 on the Mohs scale). It is highly resistant to wear, corrosion and extremes of temperature, but is brittle. This latter property is of little importance as one would not normally expect heavy metallic objects to be dropped onto a low friction test surface.

Should any accidental damage to the surface take place then pre-ground and polished tiles could readily be substituted for the damaged ones.

Whilst it is envisaged that the surface will only be used when wet, no damage as such will result if it is used in a dry condition. However, a change in u value due to tyre rubber shedding is a possibility.

### 4.5 General Comments

The evolved surface would seem to equate reasonably well with the friction levels of the 'Contidrom' but it is superior to the 'Contidrom' in that the surface is smoother and flatter.

The trial area of basalt surface has given consistent low friction levels which it is considered will fulfil the requirements for type approval testing of anti-lock braking systems.

The friction level with commercial vehicle tyres is lower than with car tyres and for this reason it is necessary to accept differences between surface friction levels in the testing of such vehicles on a single test surface. The levels obtained with the Basalt surface at 50 km/h viz, 0.35 for cars and 0.2 for commercial vehicles would seem to be a satisfactory compromise.

It is a relatively quick and easy process to return to a set friction level if it should change for any reason.

Finally, it is a surface that could be duplicated at any time, anywhere, and finely tuned to any chosen u value within the range of say 0.5 to 0.2.

## 5 TEST SURFACE WETTING

In order to produce the required levels of surface friction it is necessary to wet the surfaces during testing. This is normally accomplished by means of spray bars or sprinkler systems designed to give predetermined rates of precipitation up to about 80mm/hr. Because of their very nature however, spray systems have certain disadvantages primarily due to the effect of wind on the uniformity of water distribution and the reduced visibility of test drivers. While for the majority of test work these factors are more annoying than important it was considered desirable by some prospective users of the facility that for certain test work, particularly that associated with tyre testing, an alternative watering system be devised which gave greater uniformity and controllability than could be obtained from spray systems.

The best alternative watering system found in use was that on the 55 metre Turning Circle at the 'Contidrom', Hanover, West Germany shown diagrammatically in Fig 7. A very uniform covering of water is achieved on this circuit with good control of the water depth produced. With the present proposal consisting of several tracks adjacent to each other the upstand formed by the cover to the water supply main was unacceptable with the possibility of vehicles crossing, either accidentally or deliberately from one track to the next. To overcome this problem the system has been modified, basically by inverting it to provide what is in essence an overflow weir.

With no information available in the literature to relate quantities of water to crossfall and water depth a series of tests were undertaken at Birmingham University using a perspex model and a free standing road slab capable of being titled both longitudinally and transversely. As a result of this limited series of tests and applying a strict discipline to the number of tracks that could be watered at any one time it was estimated that the quantity of water required for such an overflow system was 53,000 gals/hour. In addition to this the basalt tile strip which could only be effectively watered by a spray system, because of its geometry, required 8,000 gals/hour based upon a rate of precipitation of 20mm/hour. Thus there is a total water requirement of 61,000 gals/hour for the straight line braking facility.

While these watering requirements do not appear to be that great this circuit has been designed in conjunction with another proposed circuit also requiring water and located in the same area. The quantity required by this proposal at 202,000 gallons/hour make it necessary to make provision for a total of 265,000 gallons/hour.

It was also decided that in order to provide uniformity of test conditions from MIRA to other proving grounds, if required by users, the strips with the overflow watering would also be provided with a back up spray system. All spray heads are positioned below track level to remove the risk of damage by vehicles and the sprays have a low trajectory to minimise wind effects.

Water supply is by means of rainfall collection and a holding/supply reservoir with full recirculation of water used. Analysis of the rainfall data for the area over a 15 year period 1967 to 1981, revealed there would be a shortfall between supply and demand of between 44.2 million and 66.1 million gallons per year. This analysis indicated an optimum reservoir size of 2 million gallons. To make up this shortfall a borewell has been sunk to a depth of 210 metres into the lower Keuper Sandstone which underlies the site. Anticipated yields from this pumped 300mm borewell is 2,000 gallons/hour. Suitable phasing of pumping from the borewell in relation to the level of water in the reservoir will ensure sufficient water for all year round testing.

## 6 TEST METHODS

It is envisaged that the facility will be used in the main by one vehicle at a time, and certainly by only one member company at any one time. Most car testing can be contained within the facility, the length of run-up, some 850m, being sufficient to reach test speeds up to 160 km/h. Commercial vehicles with a lower maximum test speed around 130 km/h and high speed car tests will require a much longer run-up and this is catered for by a slip road coming off the adjacent part of the High Speed Circuit. At the end of the test strips there is a length of 170m for braking and turning, terminating in a Lytag vehicle arrester pit in case of brake failure.

The following are some of the uses for which the facility will be available:-

1) Anti-lock brake testing and development
2) Type approval testing of anti-lock brake systems
3) Tyre testing and tyre design evaluation
4) Steering system feel and feedback on varying u surfaces
5) Brake force distribution tests and development
6) Vehicle stability eg, aquaplaning
7) Tyre noise
8) Vehicle noise tests and evaluation
9) Tyre footprint photography - static and dynamic

Some thought has been given to the positioning of the test surfaces so that certain specific tests can be carried out. High friction surfaces have been placed alongside low friction ones so that 'split u' tests can be carried out. It will be possible to carry out locked wheel tests on a low friction surface whilst the wheels on the other side of the test vehicle are free running on a high friction surface in order to give directional stability. Vehicles fitted with a special test wheel in the centre of the vehicle will have no problems. It will be possible to wet a length of high friction surface both before and after the low friction Basalt sets so that braking can be carried out in a high-low and low-high mode.

## 7 ANCILLIARIES

### 7.1 Noise Surfaces

It was decided to use the long straight approach road to the braking straights for the provision of noise generating surfaces.

Six different surfaces will be constructed to give a range of noise generating surfaces from nominally "quiet" to "very noisy".

They are as follows:-

(a) Open textured Bitumen Macadam with 12mm nominal size aggregate.

(b) Laterally grooved concrete.

(c) Hot rolled asphalt to BS594, 20mm pre-coated chippings 1.5mm texture depth (motorway specification).

(d) Guss asphalt - a surface used extensively in Germany.

(e) I.S.O. specification - smooth.

(f) I.S.O. specification - coarse.

⎫ Ref 5
⎭

Each surface has a length of 150 metres to allow a 5 second sound recording at 100 km/h to be made.

The final surfaces are positioned 150 metres from the start of the braking straights so that any disturbance on the vehicle by the noise surfaces has decayed before any braking tests are started.

### 7.2 Glass Plate

To enable the study of tyre tread contact patch areas a glass plate has been sited immediately before the braking straights. An underground laboratory will permit high speed photographs to be taken of the glass/tread interface. The maximum wheel loading of 6.5 tonnes will allow trucks up to 38 tonnes GVW to be used.

## 8 ACKNOWLEDGEMENT

This is made to all the sponsoring members not only for their financial support but also for their technical input towards the formulation and design of the facility. They are as follows:-

SP Tyres UK Ltd
Department of Transport
Michelin Tyre Company
Department of Industry

MIRA, besides being deeply involved in the technical input, has supplied the required land area and, also, the project supervision.

## REFERENCES

(1) Road Note 29. A guide to the structural design of pavements for new roads. Road Research Laboratory. Department of the Environment.

(2) ALLSOPP, H.C. The assessment of Basalt Tiling as a low friction area. MIRA Report No. K33541, 1980.

(3) ALLSOPP, H.C. Road surface friction measurement methods and machines. MIRA Report No. 1983/1.

(4)  ALLSOPP, H.C., CART, J.  The assessment
     of  Basalt tile alignment on  the  MIRA
     and Goodyear test  areas.   MIRA Report
     No. K545001, 1983.

(5)  ECE  GRRF  ad-hoc group on methods  for
     measurement    of    tyre/road    noise.
     Revised 1984-09-21 according to meeting
     1984-01-17/18.    Annex   5   Reference
     surfaces, page 50.

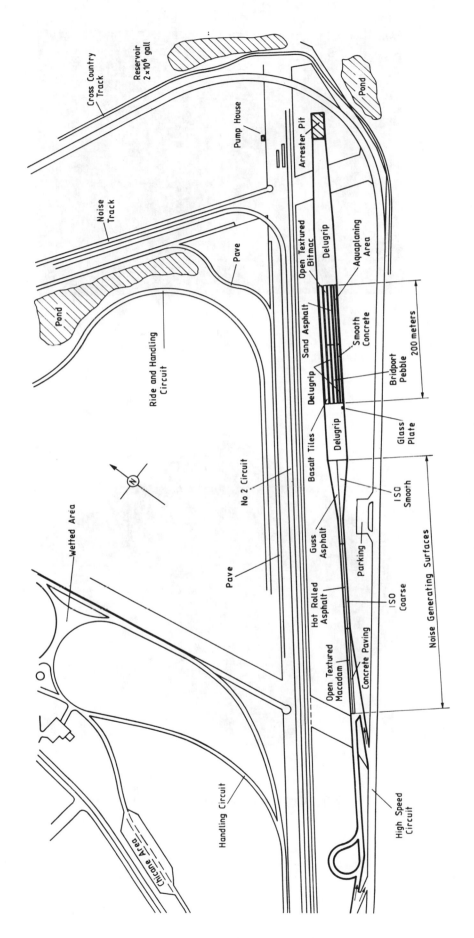

Fig 1 Plan view of test facility

125

Fig 2    The 'Contidrom' at Hanover

Fig 3    A basalt tile showing the pendulum test area

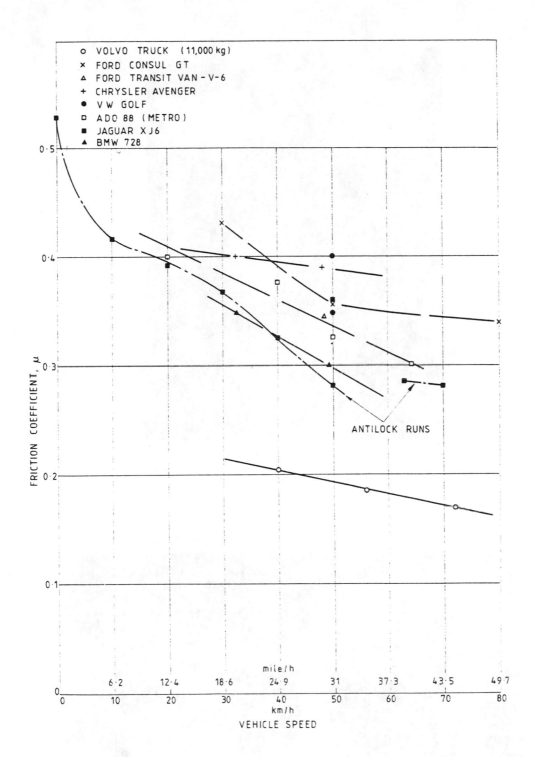

Fig 4   Peak friction levels — wet

Fig 5    General view of the MIRA experimental test area

Fig 6    The Goodyear offset basalt tile area

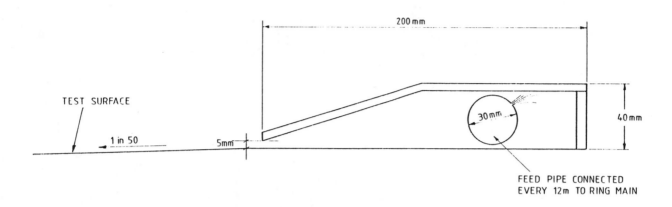

Fig 7    A 'Contidrom' watering system

# Electronic Braking System*

## Edward J. Hayes and George W. Megginson
### Kelsey-Hayes Company
### Romulus, Michigan, U.S.A.

*Paper 780856 presented at the International Conference on Automotive Electronics, Dearborn, Michigan, September, 1978.

ABSTRACT

Electronic braking systems for air-brake vehicles have been required by the Federal Motor Vehicle Safety Standard 121 since 1975. The need for the standard and the difficulty in documenting its benefits are discussed. Public criticism, court and legislative action threaten to cause it to follow the seat belt inter-lock. Major opposition derives from misunderstanding and opposition to government regulation.

IN 1975, ELECTRONIC BRAKING SYSTEMS became standard equipment for most air-brake trucks, tractors, and trailers. This major innovation to vehicle braking systems was required to meet the "no-wheel lock" requirement of a Department of Transportation Safety Standard known as FMVSS-121.

Prior to the implementation of the standard, heavy trucks required 330 to 450 feet to stop from 60 M.P.H. on a dry surface. A passenger car can make the same stop within 200 feet. To reduce this difference and avoid lateral instability, Standard 121 requires air-brake vehicles to stop within 293 feet and remain within a 12-foot lane without sustained wheel lock. Most vehicles require an electronic braking system known as antilock or antiskid to meet the no-wheel lock requirement.

The antilock system strives to maintain wheel roll during panic braking to achieve optimum retarding force consistent with lateral stability. It is relatively easy to demonstrate stopping distance and lateral stability. Sixty (60) M.P.H. panic stops with antilock can be made within 293 feet with confidence. Any panic stop above 30 M.P.H. without antilock is extremely hazardous. Dramatic lateral stability differences can be observed during lane changing and split coefficient tests. It is more difficult to relate improved braking to a reduction in accidents.

First of all, a 121 system can have no effect on non-braking related accidents, and it cannot completely eliminate all braking related accidents.

Several studies have been performed to determine or predict the actual reduction of accidents which can be expected from the implementation of FMVSS-121. One such study was made by the Road Safety Division of Transport Canada in which 1,200 accidents were investigated, and an analysis was made to determine whether a 121 type standard would have been effective in eliminating the accident. The results showed that 8 to 15% of the accidents could have been eliminated or reduced in severity. A conclusion of this study states "the results suggest that the requirement for antilock braking on the rear axles of air-brake vehicles is the most effective single feature of the standard and should produce a substantial reduction in accidents resulting from instability while braking." In addition, the author states, "It may be noted that the achievement of the more important performance improvements embodied in the FMVSS-121 depends rather critically on good maintenance practices and a high level of reliability in the antilock system."

The Highway Safety Research Institute has issued a preliminary report on a 30-month study designed to determine the reduction in accidents due to FMVSS-121. The observed data from 3,600 vehicles indicates that approximately 1,800 121-equipped vehicles had 19% less accidents than a similar number of pre-121 vehicles. However, to extrapolate this to the total 121 heavy vehicle population raises many questions. Is the sample truly representative? Is the reported data consistent? Other studies have indicated braking related accidents account for about 20% of the total vehicle accidents. Is it possible for the system to eliminate nearly all braking related accidents? A final report was due in July, 1978. However, I understand the study has been extended.

Burlington Fleet Services reported that 508 121-equipped vehicles had an accident rate of 2.2 accidents per million vehicle miles. 536 pre-121 vehicles had 3.4 accidents per million vehicle miles.

Criticism of antilock systems apparently stems from (1) maintenance cost, (2) misunderstanding of operation, (3) the cost/benefit, (4) system failures, and (5) resistance to government regulations.

Maintenance costs can be reduced by taking

advantage of the manufacturer's warranty system which will give feedback to the vehicle manufacturer and antilock manufacturer to expedite improvements and by training drivers and mechanics to more quickly identify the sources of problems. Most component manufacturers warranty the systems for one year or 100,000 miles. However, systems are designed to exceed the ten-year life of the average truck. Comparative tests have indicated relay valves used with 121 brake systems have four times the expected life of the relay valves used with the pre-121 systems. Sensors and computers have a high degree of reliability. With proper maintenance of interconnections and protection from physical abuse, ten years of trouble-free service is practical. Vehicle and component manufacturers have provided maintenance information, schools, and field service assistance.

There is a great deal of misunderstanding among drivers as to what the 121 braking system should or should not do. Drivers complain of the brakes releasing under wet conditions. These are normal cycles of the antilock, and it is, in fact, performing as it should. However, these cycles are considered to be malfunctions by the drivers due to their lack of understanding of how the system is supposed to work. Heavy Duty Trucking published an excellent pamphlet, "Driving With The New Brakes." Nearly 200,000 copies have been distributed. The American Trucking Association and the International Brotherhood of Teamsters prepared a truck drivers' manual on 121 brakes. The National Highway Traffic Safety Administration (NHTSA) is in the process of publishing and distributing this manual.

The requirements of FMVSS-121 do constitute approximately a 3% to 4% increase in the cost of heavy-duty vehicles. However, these costs are necessary to achieve the level of overall performance required by 121. The antilock portion of that increase in vehicle cost is approximately 30%. Some fleets have indicated that it is not the initial purchase cost that concerns them as much as it is the anticipated maintenance costs and downtime.

A major issue of the controversy about FMVSS-121 is that some users feel that antilock is unreliable, does not failsafe, and in fact, is the cause of accidents. During a recent DOT survey of 500 drivers, 100 stated they believed "the new brakes" had helped avoid accidents while two drivers stated they believed "the new brakes" had contributed to an accident. A teamsters' survey showed 13% of those polled felt antilock had prevented an accident. Warranty return data submitted by seven antilock component manufacturers indicates current generation components range from 94 to 99.99% probability of operating for one year without replacement.

A NHTSA task force has examined available data and investigated all serious accidents with alleged braking involvement. To date, there have been no serious or fatal accidents which have factual supporting evidence indicating FMVSS-121 involvement. The few accidents attributable to 121 factors have typically been low speed or "loading dock" type accidents. NHTSA and the manufacturers are vigorously analyzing each situation to effect product improvements. These improvements result in a product with a greater tolerance for the harsh environment experienced by heavy vehicles.

There can be no doubt that some owner-operators, fleet owners, vehicle manufacturers, and an occasional engineer have not utilized their full resources and capabilities to properly effectuate the standard. There may be a number of reasons why this is true. The FMVSS-121 package did increase vehicle cost and people naturally resist changes that affect their pocket book. The employers' approach to the 121 brake system has a very direct bearing on drivers acceptance of the standard, particularly with their perception of the owners' reluctance to maintain and use the system. As part of human nature, people resist things they do not understand and government dictates.

The Ninth Circuit Court of Appeals has ruled that since NHTSA did not adequately test antilock systems to assure their reliability, its use should be discontinued. It is understood, I am sure, that this oversimplification is my opinion of a complex nineteen page legal document. The Justice Department has appealed this ruling to the Supreme Court. They have asked for a ruling on two questions:

1. Whether the National Highway Traffic Safety Administration (NHTSA) acted arbitrarily or capriciously in determining that its performance standard for air-braked trucks is "practicable" at the time of its promulgation or at the time of its subsequent implementation.

2. Whether NHTSA must conduct independent tests to "assure" the reliability of equipment designed by manufacturers to meet safety performance standards.

Bills are pending in both the House and Senate to delay the use of antilock. DOT has published proposals to impose a moratorium on antilock for trailers and to enforce antilock maintenance on tractor drive axles. Currently, 121 applies to the vehicle manufacturer. This proposal would require the vehicle owner to maintain the system. Further action on this rule making and legislation has apparently been delayed until the result of the court case is determined.

Antilock is a safety device. When in use, a portion of the "mayhem" on our nation's highways can be avoided. We have paid the price of three years of "in-production development." No product can reach dependable maturity without the agony of early production. We are on the threshold of public acceptance of a dependable driver aid to safe vehicle operation. It will be unfortunate indeed if it is lost to society because of misunderstanding or the opposition to government regulation.

Air-brake vehicles are increasing in size and weight. Passenger cars are becoming smaller. If they are to continue to share all of the nation's highways, the heavy-truck industry must share the cost of increased safety. A commitment by government, manufacturers, and vehicle operators is required to allow this milestone in highway safety to be utilized to its maximum and preserve our resources of lives and property on our nation's highways.

REFERENCES

1. Transcript hearing on the Federal Brake Standard, December 6 and 7, 1977, before the Senate Committee on Government Affairs, Sub-Committee on Government Efficiency and the District of Columbia. Senator Thomas Eagleton, Chairman. Court Reporting Services, Inc., 1911 Jefferson Davis Highway, Suite 1006, Arlington, Virginia 22202. (703) 922-6155.

2. Transcript public meeting, December 15, 1977, on Standard No. 121, Air Brake Systems, U. S. Department of Transportation, NHTSA. Ace-Federal Reporters, Inc., 444 North Capital Street, Washington, D. C. 20001. (202) 347-3700.

3. Transcript hearing on Federal Motor Vehicle Safety Standard 121, Air Brake Systems, February 27, 1978, House of Representatives Committee on Interstate and Foreign Commerce, Sub-Committee on Consumer Protection and Finance. Honorable Bob Eckhardt, Chairman. Alderson Reporting Company, Inc., 300 Seventh Street, S.W., Washington, D. C. 20001. (202) 554-2345.

4. Technical Assessment of FMVSS-121 Air Brake Systems, February 27, 1978, U. S. Department of Transportation, National Highway Traffic Safety Administration, Document No. 75-16-GR-038.

# Antilock Braking Regulations**

**Paul Oppenheimer**
Lucas Girling Ltd.
England

**Paper 860507 presented at the International Congress and Exposition, Detroit, Michigan, February, 1986.

## ABSTRACT

The increasing adoption of antilock braking systems has focussed attention on the corresponding standards and regulations. The most notable antilock specifications are contained in the European ECE and EEC regulations, within Annex 13 and Annex X, respectively.

This paper reviews the construction and performance requirements within these two important antilock Annexes, with particular emphasis on the new requirements adopted during 1985, viz. the introduction of antilock 'categories', additional definitions and new test procedures, including a 'split-friction' test.

ANTILOCK BRAKING REGULATIONS may be 'voluntary' or 'mandatory'. For example, the pan-European braking regulations described in this paper include certain antilock Annexes which "Do not make it compulsory to fit vehicles with antilock devices, but if such devices are fitted to a road vehicle, they must meet the requirements of this Annex". In other words, the vehicle manufacturers retain the choice of fitting antilock braking systems to any vehicle models, and either as an option or as standard equipment.

On the other hand, there are proposals in Europe which would make it mandatory to fit antilock braking systems on every new vehicle of certain categories, such as buses and coaches, heavy trucks and trailers, articulated vehicle combinations and vehicles carrying dangerous or hazardous goods. The antilock braking systems on these vehicles would have to satisfy the above mentioned Annexes.

Whilst these antilock Annexes only apply in Europe, and not in America, these Annexes are establishing certain minimum standards for antilock braking systems which may become generally accepted. Furthermore, the eventual 'harmonization' of the European and American braking regulations for passenger cars may lead to the adoption of at least some of these European antilock requirements within the US Federal Standards. At the same time, there is a worldwide upsurge of actual antilock applications on almost every type of road vehicle, and new types of antilock braking systems are being developed towards even greater market penetration.

For all of these reasons, an explanation of the latest European antilock braking regulations may be of interest to automobile engineers in America, and worldwide, at this time.

## BACKGROUND

The most comprehensive antilock braking regulations have been promulgated by the United Nations' Economic Commission for Europe (ECE) in Geneva and the Common Market European Economic Community (EEC) in Brussels. The basic road vehicle braking regulations from these two important organizations have been explained and summarized on previous occasions. (1) (2)*

Briefly, ECE Regulations and EEC Directives establish uniform conditions for the type-approval and reciprocal recognition of road vehicle equipment, thus removing technical barriers to trade within Europe. For example, any vehicle or equipment approved to an ECE Regulation or EEC Directive may be sold and registered in any other European country which

* Numbers in parentheses designate references at end of paper.

accepts this same ECE Regulation or EEC Directive.

The relevant regulations for road vehicle braking systems are -:

a) ECE Regulation No. 13 (3) which is accepted in Belgium, Czechoslovakia, Federal Repulic of Germany, France, German Democratic Republic, Hungary, Italy, Luxemburg, Netherlands, Romania, Sweden, United Kingdom (England) and Yugoslavia.

The latest publication of ECE Regulation No. 13 includes the 03 series of amendments; subsequently, the 04 and 05 series of amendments have been published in 1981 and 1985, respectively.

The "Requirements applicable to tests for braking systems equipped with anti-lock devices" are contained in Annex 13 to ECE Regulation No. 13, which was included in the 03 series of amendments published in 1979.

A new version of Annex 13, bringing it into line with the corresponding EEC Annex X, is to be published as a complement to the 05 series of amendments during 1986.

b) EEC Directive 71/320 (4), which is accepted in Belgium, Denmark, Federal Republic of Germany, France, Greece, Ireland, Italy, Luxemburg, Netherlands, Portugal, Spain and the United Kingdom.

There have been two 'adaptations to technical progress' in subsequent Directives 75/524 and 79/489; a third adaptation has been agreed for publication early in 1986 and this document will include the official text of the new Annex X: "Requirements applicable to tests for vehicles equipped with anti-lock devices."

The historical development of ECE Annex 13, and the subsequent updating revisions within EEC Annex X, have recently been reported to the Institution of Mechanical Engineers' conference on "Anti-Lock Braking Systems" in London (5), in a paper which describes the work of two successive ad-hoc groups within the ECE forum between 1973/75 and 1983/85, respectively. Therefore, this present paper will concentrate only on the technical aspects of ECE Annex 13 and particularly on the latest revisions introduced within EEC Annex X.

Other previous regulatory developments and standards iniatives on antilock braking systems may be summarized as follows -:

The earliest antilock braking regulations were proposed in Sweden in about 1968 (6), but the requirement that "brake systems on vehicles shall be so constructed that all available friction between wheels and the road surface can be utilized for braking without the risk of any wheel locking" has never been enforced. An international meeting of antilock/braking experts in Stockholm in 1971 concluded that such requirements were premature at that time.

The next antilock braking regulations were introduced in the United States in 1975, when a new Federal Motor Vehicle Safety Standard (FMVSS) 121 became effective for new air-braked trucks, buses and trailers. Strictly speaking, FMVSS 121 did not demand antilock braking systems: however, the stringent stopping distance requirements on high and low friction road surfaces were so difficult to guarantee, that every American vehicle manufacturer decided to fit antilock braking systems. And although the FMVSS 121 stopping distances were subsequently relaxed, the universal application of antilock braking systems on trucks, buses and trailers continued until the famous court ruling in 1978 forced the National Highway Traffic Safety Administration (NHTSA) to withdraw the stopping distance and 'no-lockup' requirements from FMVSS 121. After this, the antilock market in the US collapsed as quickly as it had appeared only 3 years earlier.

The SAE also initiated action on antilock braking systems during the 1970's, principally in relation to FMVSS 121, and developed a "Wheel slip brake control system road test code" (SAE J.46) and subsequently "Minimum requirements for wheel slip brake control system malfunction signals" (SAE J.1230). These SAE Recommended Practices addressed some of the basic features of antilock braking systems and attempted to establish uniform procedures in accordance with industry practice -:

- SAE J.46 includes stopping distance measurements with and without the antilock braking systems in operation, on a variety or uniform low, medium and high friction road surfaces and on 'split-friction' surfaces; also, additional tests on 'changing-friction' surfaces (low-to-high and vice versa) and a 'lane-change' test on the low friction surface. There are no (minimum) performance requirements.

- SAE J.1230 describes the functional areas of antilock braking systems which should be monitored and distinguishes between malfunction and failure warning signals to be provided.

Neither of these standards have been adopted within legislation, although the following European regulations include some very similar tests.

ECE ANNEX 13

Annex 13 is based on a series of investigations by Krugel and Hoffmann from the German Technical Services, who proposed certain "Guidelines for the testing of anti-lock devices" (7). This proposal was debated and amended within an ECE ad-hoc group of government and industry antilock braking experts between 1973 and 1975, eventually being approved for publication within the 03 series of amendments to ECE Regulation No. 13 (3). Thus, Annex 13 "entered into force" on 4 January 1979 and an increasing number of European vehicles have since been 'type-approved' to Annex 13. In consequence, both the manufacturers and the Technical Services, especially in England, France and Germany, have gained experience in the application of Annex 13 and the various associated new test procedures. For example, a recent paper by E. P. Williams (8) describes the procedures adopted by the UK Department of Transport in testing antilock braking systems to Annex 13, particularly in relation to energy consumption, utilization of adhesion, semi-trailer axles and electro-magnetic interference.

The following review of the principal requirements and tests in Annex 13 also highlights some of the ambiguities, controversies and novelties.

Voluntary Antilock - Paragraph 1 (of Annex 13) explains that "this Annex does not make it compulsory to fit vehicles with antilock devices, but if such devices are fitted to a road vehicle they must meet the requirements of this Annex." This statement clearly leaves the choice of fitting antilock braking systems with the vehicle manufacturer.

Compatibility - Paragraph 1 also refers to the compatibility between towing vehicles and trailers equipped with compressed-air braking systems. Although such vehicles may be exempt from the requirements of Annex 10 in ECE Regulation No. 13, which covers the distribution of braking between the axles of individual vehicles and the compatibility between towing vehicles and trailers, they must still meet the compatibility bands in the fully laden condition.

Exemption From Annex 10 - Paragraph 3 is one of the most important features of Annex 13; it also includes some of the most mis-understood statements.

Paragraph 3.1 must be read in conjunction with paragraph 1 of Annex 10: "Vehicles which are not equipped with an antilock device as defined in Annex 13 shall meet all the requirements of this Annex." In other words, a vehicle equipped with an antilock device which satisfies Annex 13 is exempt from the Annex 10 requirements relating to distribution of braking and compatibility (unladen). In practice, this exemption means that a vehicle with antilock does not have to fit apportioning valves, such as rear-axle load-sensing valves. Whilst some vehicle manufacturers might choose to fit both antilock and load-sensing valves, there are obvious commercial advantages in omitting the load-sensing valve when fitting antilock. In the case of heavy trailers, for example, Lucas Girling introduced a simple antilock braking system which effectively 'competed' against the load-sensing

valve installation (either antilock or load-sensing valves must be fitted under the ECE and EEC regulations) such that more than 80% of all new UK trailers built since 1982 are equipped with antilock braking systems only.

Paragraph 3.1 defines which antilock installations are exempt from Annex 10 and paragraph 3.1.1. explains that in the case of a motor vehicle, if there are only two antilock controlled wheels, they must be placed diagonally or both on the rear axle; therefore, in the case of a passenger car with antilock control only on the front wheels, such a vehicle would not be exempt from Annex 10. In other words, such a vehicle would have to meet both Annex 10 and Annex 13 --- but, and this is vitally important --- such an antilock installation is not prohibited by this Regulation. This situation is further clarified in paragraph 3.2.

Failure Indication - Paragraph 3.3 prescribes the minimum warning requirements, i.e. only electrical supply and wiring failures must be indicated to the driver. Most manufacturers monitor other potential malfunctions in antilock braking systems (SAE J.1230), but it has proved very difficult to agree internationally on a specific list of failures (or malfunctions) that should be indicated to the driver. Of course, any hydraulic leakage failures from an antilock device must be indicated under the basic Regulation No. 13 requirements.

Annex 13 does not specify the color of the antilock warning signal, nor the words or symbols to be affixed. A yellow warning light with the letters "ABS" (Antilock Braking System) appears to be the preferred European solution and this proposal has been submitted for worldwide standardization to the relevant technical sub-committee of the International Standards Organization (ISO). Daimler Benz AG have signified their agreement to permit the general use of their ABS trademark, when the ISO accepts the ABS symbol for standardization.

Post-failure performance - Paragraph 3.4 does not differentiate between various types of antilock failures, i.e. after any electrical, hydraulic or mechanical failure in the antilock device, the normal secondary and residual braking performance levels must be assured. These terms have been explained previously (2): briefly, the lower residual performance must be available from the service brake control, whilst the higher secondary performance may be obtained via another control (e.g. handbrake). Although it was recognized that the majority of antilock failures may be caused by electrical faults, including diagnostic checks inhibiting the antilock function, it was not considered appropriate to demand the full service brake performance level after such failures --- because otherwise apportioning valves might need to be fitted and this would negate the previous explicit exemption from Annex 10.

Electro-magnetic Interference - Paragraph 3.5 suggests that "The operation of the (antilock) device must not be affected adversely by magnetic fields." This succinct statement hides years of experimentation and lengthy committee debates, without agreement on an appropriate practical test procedure. Several attempts have been made to ascertain and measure the external interference which might be encountered by road vehicles, but without much success. Various test procedures developed by the SAE, and others used by vehicle manufacturers, the aerospace industry and for military applications have been investigated. As an interim solution, the UK Department of Transport has introduced a simple component test, whereby the electronic parts of the antilock braking system, together with their electrical wiring, are installed in a parallel plate strip line and subjected to electro-magnetic fields. However, the degree of correlation with the on-vehicle situation has yet to be established; meanwhile, whole vehicle testing is under consideration, possibly using a 'bulk current injection' method. The importance of this subject in relation to road safety, and the increasing application of electronic systems within road vehicles, is expected to ensure the necessary impetus to secure an eventual solution.

Energy Consumption - Paragraph 4 describes the test procedure which aims to ensure that antilock braking systems will maintain their performance during extended periods of operation. This is considered important when descending long slippery gradients and when encountering sudden patches of ice on motorways. Annex 13 therefore prescribes a period of antilock cycling related to the maximum speed capability of the vehicle, followed by four full static brake applications; afterwards, at least the secondary braking performance level should remain available.

Many experts might consider this test excessive, arguing that few drivers would drive at the maximum speed (up to 160 km/h) under icy conditions. Nevertheless, the test has been introduced and applied, particularly for air-braked vehicles. The test (for motor vehicles) must be carried out on a low friction road surface with a peak tire-to-road adhesion (K) of less than 0.3. The method of determining K is explained in the Appendix to Annex 13 (later in this paper). But because of the difficulties in preparing suitable test surfaces, Annex 13 permits testing on low friction surfaces with K values up to 0.4. However, even these surfaces are rarely long enough to permit the energy consumption test to be carried out in one single stop; Annex 13 therefore allows the calculated period of

antilock operation (up to 23 seconds) to be achieved in up to four phases. It should also be noted that this calculation formula is based on a full stop on a surface of K = 0.2.

The energy consumption test has caused certain difficulties for air-braked vehicles and this has led to system improvements. Passenger cars with hydraulic braking systems may not be unduly affected by this test, although any accumulators should be designed to maintain antilock operation for up to 23 seconds with the engine idling.

Utilization of Adhesion - Annex 13 recognizes that even the best contemporary antilock braking systems cannot achieve the 'ideal' braking performance, corresponding to braking with the peak prevailing adhesion K at all wheels and a utilization of adhesion of 100%. Paragraph 5 prescribes a minimum adhesion utilization (E) of 75%. This is defined as the braking performance with the antilock device in operation (Zmax), compared to the theoretical ideal braking performance, measured on both the low friction and high friction road surfaces. The method of measuring and calculating E is defined in the Appendix to Annex 13, and explained later.

This test procedure provoked the greatest arguments during the ad-hoc meetings. Many experts pointed to the usual method of comparing the antilock braking performance against braking with locked wheels; this is the difference that most drivers would experience when making an emergency stop, it is also the usual procedure in antilock performance demonstrations ("with and without antilock") and the locked-wheels comparison has also been adopted in SAE J.46. Finally, the method of measuring the peak tire-to-road adhesion appeared both difficult and controversial. But despite all these arguments, certain government delegates considered braking with locked wheels to be dangerous and not in line with the other ECE Regulation No. 13 test procedures; even more importantly, the shape of the brake force/tire slip curves varies considerably on different road surfaces and makes it difficult to prescribe an adequate adhesion utilization performance.

The eventual minimum value of 75% was considered excessive, especially on low friction road surfaces, in relation to the requirements for non-antilock vehicles in Annex 10, e.g. a minimum value of 50% adhesion utilization on a peak friction K of 0.2. Furthermore, very high values of braking utilization tend to compromise the tire lateral force coefficient and the steering/handling performance. However, the original German proposal prevailed and most manufacturers have now accepted this procedure and the various refinements which have been introduced.

Additional Checks - Paragraph 6 lists several additional tests to check the general vehicle behavior with the antilock braking system in operation on low and high friction road surfaces, from low and high initial vehicle speeds. Although brief periods of wheel-locking are allowed (up to ½ second in the UK interpretations) --- provided the vehicle does not deviate from its initial course --- the prohibition of wheel-locking of controlled wheels appears to penalize select-high control systems unduly. The test from 120 km/h on the low friction road surface has also been criticized as potentially dangerous. Finally, there is a test from the high friction to the low friction road surface, to check that the antilock braking system adapts quickly enough to prevent wheel-locking; this test is also carried out at 40 and up to 120 km/h.

Trailers - Paragraphs 7, 8 and 9 describe the energy consumption, utilization of adhesion and additional checks for trailers. These tests are very similar to the motor vehicle tests, except that all trailer tests are confined to the high friction road surface and carried out with the trailer unladen. The principal reason is that the low achievable deceleration, when braking only the trailer in a vehicle combination, would require very long low friction road surfaces. It also simplifies and shortens the trailer tests considerably.

The energy consumption test for trailers is based on a fixed antilock cyling period of 15 seconds, followed by the usual four full static applications and a subsequent minimum braking performance. The utilization of adhesion must again be not less than 75% of the optimum braking performance. And there are general vehicle behavior checks from 40 and 80 km/h.

Coefficient of Adhesion (K) - The Appendix to Annex 13 prescribes the method of determining the coefficient of adhesion (K) and the utilization of adhesion (E). Briefly, K is measured with the vehicle under test, by disconnecting the antilock device and braking with only one axle. From the maximum deceleration which can be achieved without wheel-lock, and the corresponding dynamic axle weight, the peak tire-to-road adehesion (K) can be calculated. The same procedure is repeated for other axles equipped with antilock devices.

In practice, most vehicle tests show different K values at the front and rear axles (although the tires are identical) on the same vehicle. A higher rear axle K value could be explained in the case of wetted low friction road surfaces, because the front wheels might have dispersed some of the water; however, despite extensive investigations, this theory has not been confirmed and occasionally the front axle K values have been higher than the rear axle K values on the

same vehicle. The axle loading may also influence the value of K. When two different K values have been measured on the same vehicle, most authorities use an average K value to calculate the utilization of adhesion (E) --- unfortunately, the UK Technical Services prefer to use the higher K value. This tends to avoid values of E exceeding 100%, which has also happened.

Another problem with this procedure relates to the method of measuring the deceleration, both for the determination of K and again for measuring the braking performance with the antilock device in operation (Zmax). It is well known that the deceleration varies considerably during a brake application, even with a constant input force on the brake pedal; in the case of antilock, there may be additional cyclic fluctuations. It has therefore been agreed to base all deceleration measurements on the time to reduce the vehicle speed from 40 to 20 km/h; this avoids initial variations (stops are usually made from 50 km/h) and wheel-locking effects towards the end of the stop.

Further difficulties have arisen in the case of semitrailers, because of the problem of measuring or calculating the dynamic axle load on one axle of a 2- or 3-axle bogie, when braking only one axle to determine the ideal braking performance (maximum braking without wheel-lock). Eventually, it was agreed to remove the wheels from the other axles during the test. Both this simplification, and the above method of measuring the deceleration, have been incorporated within the new version of Annex 13, as described under EEC Annex X.

EEC ANNEX X

Annex X is an update of ECE Annex 13, based on discussions and agreements within a second ECE ad-hoc group of antilock experts' meetings between 1983 and 1985. The reasons for updating Annex 13 may be summarized as follows -:
- some European government representatives wanted to introduce additional, more stringent requirements and test procedures;
- others were prepared to maintain the existing Annex 13, but wanted to propose amendments to permit simpler antilock braking systems;
- experience with Annex 13 had shown the need for various 'interpretations' to be clarified and agreed.

The diverging government opinions about various sophisticated/simple, 'full'/'partial' types of antilock braking systems eventually resulted in the acceptance of antilock 'categories', which recognized all known antilock variations and differentiated between them purely in performance terms.

For administrative reasons, the final

text of the new Annex 13 was approved in Brussels (EEC Annex X) before it was approved in Geneva (new ECE Annex 13); neither text has yet been published officially and therefore the following explanations and comments are based on the latest version of Annex X in the draft of the new EEC Commission Directive (9). A consolidated version of the new ECE Annex 13 is also being prepared for publication as a complement to the 05 series of amendments to ECE Regulation No. 13.

Annex X includes all the requirements of Annex 13, described previously, together with some important additions, described below:

DEFINITIONS - Paragraph 2 (of Annex X) introduces definitions for sensors, controllers and modulators, as illustrated in a typical antilock braking system in Figure 1. Paragraph 2 also introduces the concept of "directly" and "indirectly" controlled wheels, examples of which are shown in Figure 2:

(A) represents a 2-axle bogie of a truck, bus or trailer, where only 2 wheels on opposite sides of the vehicle are sensed (direct control DC), but a single modulator also acts on the other 2 wheels of the bogie (indirect control IDC);

(B) depicts a front-wheel-drive passenger car with diagonally-split hydraulic braking circuits in which sensors at the 2 front wheels (DC) also control and modulate the diagonally opposite rear wheels (IDC); the Lucas Girling 'Stop Control System' employs this antilock control technique.

## DIRECTLY & INDIRECTLY CONTROLLED WHEELS

### (DC & IDC)

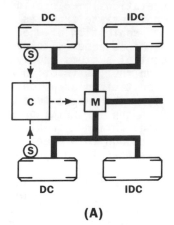

**(A)**

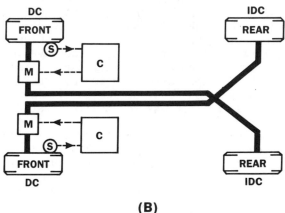

**(B)**

Figure 2

## TYPICAL ANTILOCK SYSTEM

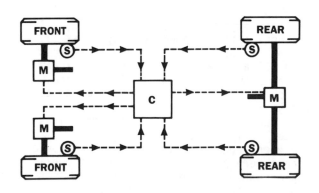

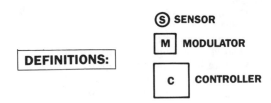

Figure 1

The reasons for defining directly and indirectly controlled wheels are twofold -:

- firstly, Annex X includes a new philosophy that "any individual axle without at least one directly controlled wheel may fulfill the conditions of adhesion utilization and the wheel-locking sequence prescribed in the EEC Directive 75/524 (the EEC version of ECE Annex 10) --- instead of the adhesion utilization requirements prescribed in paragraph 5.2 of Annex X." For example, in the case of the Stop Control System in Figure 2(B), the front wheels must satisfy the adhesion utilization prescribed in Annex X, but the rear wheels may instead satisfy the corresponding conditions of 75/524;

- secondly, directly controlled wheels usually possess their own sensors, whilst indirectly controlled wheels are modulated via information from sensors on other wheels; thus, in the case of axle control by the "select-low" principle, as shown on the rear axle of Figure 3, both rear wheels are considered to be directly controlled; but, in the case of axle control by the "select-high" principle, as shown on the front axle of Figure 3, one wheel is considered to be indirectly controlled (even though it has its own sensor) --- thereby permitting the extended locking of one "select-high" wheel during various tests, even on nominally uniform friction surfaces.

## SELECT–LOW & SELECT–HIGH WHEEL CONTROL

**SELECT–LOW = DC/DC**
**SELECT–HIGH = DC/IDC**

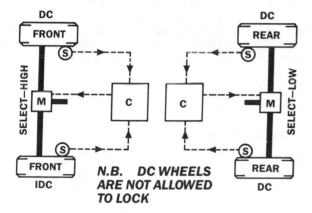

N.B. DC WHEELS ARE NOT ALLOWED TO LOCK

Figure 3

CATEGORIES - Paragraph 3 lists the categories of antilock devices for motor vehicles and the performance requirements which must be fulfilled. There are two principal reasons for the introduction of these controversial antilock categories:
(1) it permits simpler antilock braking systems to be approved, without apparently lowering the standards laid down in Annex 13;
(2) it opens the door for eventual mandatory antilock application (on certain vehicle types), with defined minimum performance characteristics.

On the other hand, many experts have deplored the introduction of antilock categories with different performance levels: they refer to the accepted principle in safety regulations that prescribe only one minimum performance standard. Perhaps further analyses may enable these conflicting philosophies to be reconciled in the future.

Category 1 Antilock - Figure 4 shows a typical antilock braking system which may be a category 1 antilock device. Note that an examination of the antilock system layout in

terms of sensors, controllers and modulators does not necessarily define the appropriate antilock category. Several experts in the ad-hoc group would have preferred to define the antilock categories in this way, i.e. by analyzing the antilock system design, and then omitting such tests as the 'split-friction' test; this would certainly simplify both the regulation and the approval process. However, other experts argued that all regulations should be based on performance standards, and not on design criteria. Furthermore, there is no absolute guarantee from a design analysis that the antilock system in Figure 4 would be satisfactory on 'split-friction' surfaces; for example, the use of individual front wheel control in conjunction with large positive steering offsets (as used on trucks and buses) could even be dangerous. Therefore, eventually, the fundamental concept of performance-based requirements was accepted and paragraph 3 clearly states that, "A vehicle equipped with a category 1 antilock device shall meet all the relevant requirements of this Annex." Conversely, a vehicle with an antilock braking system which meets all the relevant requirements of this Annex (including the 'split-friction' stability and stopping distance demands) will be accepted as a category 1 antilock device.

'Relevant', in this context, means that a motor vehicle does not have to meet the trailer requirements, that a passenger car does not have to meet the tractor/trailer connector

## CATEGORY 1 ANTILOCK

**TYPICAL SYSTEM:**

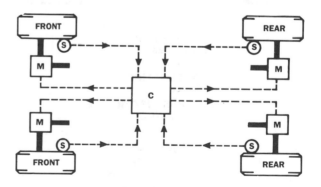

**MUST MEET ALL REQUIREMENTS IN ANNEX X**

**—INCL. SPLIT-FRICTION STABILITY**

**—INCL. SPLIT-FRICTION STOP. DISTANCE**

Figure 4

requirements and that a mechanical antilock device does not have to meet the electrical failure or electro-magnetic interference requirements.

Category 2 Antilock - Figure 5 shows a vehicle with an antilock braking system comprising "select-low" control on both the front and rear axles. This is certainly a typical category 2 antilock device; it cannot be a category 3 antilock device (because there are no axles without directly controlled wheels) and it should not be a category 1 antilock device (because it should not be able to meet the 'split-friction' stopping distance formula). In fact, the only difference between a category 1 antilock device and a category 2 antilock device is that the former must satisfy a stopping distance requirement in the 'split-friction' test, which the latter should not be able to meet. The 'split-friction' test and the associated stopping distance formula are explained later in this paper.

# CATEGORY 2 ANTILOCK

**TYPICAL SYSTEM:**

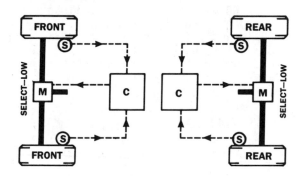

MUST MEET ALL
REQUIREMENTS IN
ANNEX X

—INCL. SPLIT-FRICTION
STABILITY

—EXCEPT SPLIT-FRICTION
STOP. DISTANCE

Figure 5

Category 3 Antilock - This category has been introduced to clarify the requirements for simpler types of antilock braking systems, such as the "rear-axle-only" layout shown in Figure 6. Such antilock installations can provide significant safety benefits: for example, in the case of an articulated vehicle combination, a simple antilock braking system on the tractor drive-axle can prevent dangerous jack-knifing accidents. There is general agreement that

such antilock applications should be permitted, perhaps even encouraged.

Paragraph 3 explains that the axle(s) with directly controlled wheels must meet the adhesion utilization requirements of Annex X (paragraph 5.2), whilst the axle(s) without directly controlled wheels (the front axle in Figure 6) must meet the adhesion utilization and wheel-locking sequence conditions prescribed in 75/524 (Annex 10). The conditions in 75/524 are usually checked by calculating and plotting the front and rear axle adhesion curves; in the case of an antilock braking system, its effects are ignored in these calculations --- but if the curves indicate that the rear axle might lock prematurely (which could well happen in the example vehicle shown in Figure 6), then a practical check is envisaged on low and high friction road surfaces to ensure that the antilock braking system prevents premature rear-wheel lock-up and provides satisfactory vehicle stability. Category 3 antilock devices are not subject to the 'split-friction' test.

# CATEGORY 3 ANTILOCK

**TYPICAL SYSTEM:**

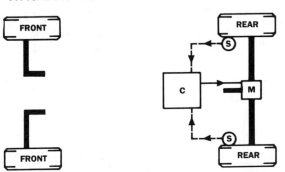

N.B. AXLES WITHOUT DC WHEELS MUST MEET WHEEL-LOCK
SEQUENCE & ADHESION UTILIZATION OF 75/524/EEC—
INSTEAD OF ANNEX X.

MUST MEET ALL
REQUIREMENTS
IN ANNEX X
—EXCEPT SPLIT-
FRICTION TESTS.

Figure 6

Trailers - No antilock categories have been introduced for trailers. Partly this was due to lack of time in the ad-hoc group, but there are also significant technical problems. For example, trailers are only tested on high-friction road surfaces (according to Annex 13), but the category principle would involve at least a 'split-friction' test for trailers.

This might be difficult to define and would certainly complicate the trailer antilock approval process. Furthermore, in practice there will probably be only category 1 and category 2 trailer antilock devices; it may be difficult to justify much additional testing to separate these two categories.

TRACTORS & TRAILERS - Annex X includes two new requirements for towing vehicles and trailers over 3500 kg maximum weight:

Warning Lights - Paragraph 4.1 clarifies that there must also be an antilock failure warning light for the trailer. Paragraph 4.2 adds that this must be a separate warning light in the driver's cab, or alternatively, there must be an 'information light' in the cab which warns the driver if an attached trailer is not equipped with an antilock braking system. In this way, the 'information light' is not obligatory and the associated sensing devices may be omitted.

Electrical connectors - Paragraph 4.3 specifies that the trailer antilock braking system must be powered via the special ISO.7638 antilock connector (10). However, an important footnote explains that until this ISO connector is in general use, one of the following electrical connector combinations must be used (Figure 7):

## TRAILER POWER SUPPLY & FAILURE INDICATION

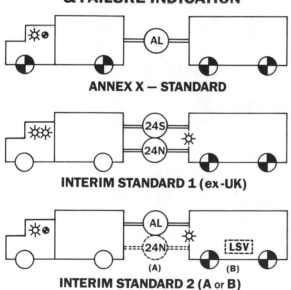

**ANNEX X — STANDARD**

**INTERIM STANDARD 1 (ex-UK)**

**INTERIM STANDARD 2 (A or B)**

Figure 7

- Interim Standard 1: the ISO.3731 24 volt supplementary connector (using pins 2 and 6 for antilock failure warning and power supply, respectively) + the ISO.1185 24 volt normal lighting connector (using pin 4 of the stop-

lamp circuit for antilock power supply); this method has been adopted in the UK, because all towing vehicles and trailers built since 1979 include the ISO.3731 connector for rear fog lights and the additional stoplamp powering ensures that every trailer antilock braking system will be operative behind every new or old towing vehicle; the diagram also shows the two separate antilock failure warning lights in the driver's cab. Interim Standard 2: the ISO.7638 special antilock connector (using 5 pins for antilock power supply/return, electronic supply/return and failure warning) + the ISO.1185 24 volt normal lighting connector (as above); some trailer antilock braking systems cannot be powered via the stoplamp circuit: in this case, the trailer must comply with 75/524 (Annex 10) by fitting load-sensing valves, for example, to assure the regulatory minimum performance when towed by a tractor without the ISO.7638 special antilock connector. The diagram also shows the alternative warning light + information light option in the driver's cab.

ADDITIONAL CHECKS - Annex X has added two new tests in this section:

Low-to-High Transition - Paragraph 5.3.3 prescribes a test from the low friction to the high friction road surface at about 50 km/h, to check that the antilock braking system will adapt sufficiently quickly and increase the vehicle deceleration to the appropriate higher value. Various tests were carried out in order to create more specific performance requirements and the deceleration, stopping distance and system pressure rise time were investigated, both on individual axles and for the whole vehicle, but no acceptable solution was found. In general terms, the ad-hoc group expected the new deceleration level to be reached between 1 and 2 seconds after crossing the surface transition.

Split-friction Test - Paragraphs 5.3.4, 5.3.5 and 5.3.7 describe the split-friction test, the stability/performance requirements and the permitted steering corrections. This controversial test was eventually included in Annex X, after exhaustive arguments in the ad-hoc group meetings, to differentiate between category 1 and category 2 antilock devices. Despite strong opposition, the ad-hoc group decided that a vehicle with "select-low" antilock control on all wheels should not be a category 1 antilock device --- although a vehicle with "select-low" antilock control on one axle (which might provide more than 50% of the total vehicle braking forces) would be accepted within category 1. These decisions were translated into performance requirements, culminating in the formula in Appendix 2 of Annex X, which relates the minimum required vehicle deceleration to the peak tire-to-road adhesion (K) of the two split-friction road surfaces.

Figures 8 and 9 show the Annex X minimum performance limit for category 1 antilock devices, on two typical high friction road surfaces of 0.6 and 0.8, respectively. The corresponding low friction road surfaces are permitted to range up to 0.3 and 0.4, respectively. The shaded areas for 'individual wheel control' and 'select-low wheel control' (on all wheels) reflect the permitted range of adhesion utilization from 75% to 100%. The figures demonstrate that "select-low" control on one axle, or some form of modified individual wheel control on the front axle of trucks and buses for steering/handling purposes, should be capable of satisfying the category 1 criteria. The figures also confirm that the "select-low" performance deteriorates by comparison with individual wheel control, as the ratio of the high friction to the low friction road surface adhesion coefficients increases. However, it was recognized that the split-friction test would be difficult to carry out with one wet and one dry road surface; therefore, it was assumed that both road surfaces would be wet. In consequence, Annex X allows the high friction adhesion coefficient K1 to be as low as 0.5, in deference to the low wet adhesion of many truck tires.

And bearing in mind the difficulties of providing suitable low friction test surfaces with an adhesion coefficient K2 below 0.2. Annex X only demands a 2 : 1 minimum ratio for K1 : K2 for the split-friction (and transition) tests.

SUMMARY

Antilock braking systems are not mandatory anywhere in the world, but if they are fitted to road vehicles in Europe, they must meet the requirements of Annex 13 to ECE Regulation No. 13 or of Annex X to the EEC Directive 71/320. Annex 13 is being updated to the Annex X level, such that both standards will be identical and include -:
- exemption from Annex 10 (75/524) for appropriate antilock installations;
- electrical antilock failure indication;
- post-failure performance requirements;
- electro-magnetic interference specifications;
- energy consumption provisions;
- utilization of adhesion tests;
- additional checks:
  - on high friction road surfaces;
  - on low friction road surfaces;
  - from high-to-low friction transition;
  - from low-to-high friction transition;

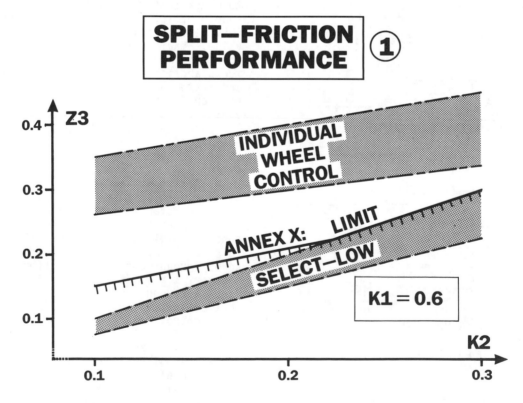

Figure 8

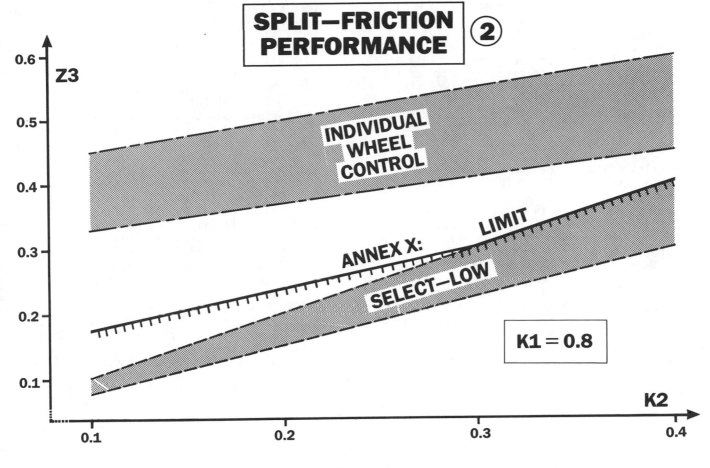

Figure 9

- split-friction stability;
- split-friction deceleration.

Antilock braking systems on motor vehicles will be classified according to their performance potential. The specific tests applicable to each category of antilock devices, and the limited requirements for trailers, are summarized in Figure 10.

CONCLUSIONS

The author of this paper has been priviledged to participate in most of the government and industry meetings leading to the development of ECE Annex 13 and EEC Annex X. During this period, the fundamental principle of creating special requirements for vehicles equipped with antilock braking systems was rarely challenged; but the introduction of numerous additional tests and more stringent performance requirements, compared with vehicles without antilock braking systems, was consistently opposed by the industry delegations.

## SUMMARY OF ANTILOCK REQUIREMENTS IN ANNEX X

| | MOTOR VEHICLES ANTILOCK CATEGORY: | | | TRAILERS |
|---|---|---|---|---|
| | 1 | 2 | 3 | |
| **ANNEX X** | | | | |
| ELECTRICAL FAILURE INDICATION | ✔ | ✔ | ✔ | ✔ |
| POST-FAILURE PERFORMANCE | ✔ | ✔ | ✔ | — |
| ELECTROMAGNETIC INTERFERENCE | ✔ | ✔ | ✔ | ✔ |
| ENERGY CONSUMPTION | ✔ | ✔ | ✔ | ✔ |
| ADHESION UTILIZATION | ✔ | ✔ | DC AXLE | ✔ |
| ADDITIONAL TESTS—: | | | | |
| —HIGH-FRICTION | ✔ | ✔ | ✔ | ✔ |
| —LOW-FRICTION | ✔ | ✔ | ✔ | — |
| —HIGH-TO-LOW FRICTION | ✔ | ✔ | ✔ | — |
| —LOW-TO-HIGH FRICTION | | ✔ | ✔ | — |
| —SPLIT-FRICTION STABILITY | ✔ | ✔ | — | — |
| —SPLIT-FRICTION STOP DISTANCE | ✔ | — | — | — |
| **75/524/EEC** | | | | |
| LADEN COMPATIBILITY | ✔ | ✔ | ✔ | ✔ |
| ADHESION UTILIZATION | — | — | NON-DC | — |
| WHEEL-LOCK SEQUENCE/ STABILITY CHECK | — | — | AXLE | — |

Figure 10

There is universal agreement that anti-lock braking systems, which effectively release the brakes when they are most wanted, should be subject to a regular performance test with the antilock device in operation --- such as the US/FMVSS 'cold effectiveness test' or the ECE/EEC 'Type-0' test on a dry surface with good adhesion, and applying the same performance requirements as for the corresponding vehicles without antilock braking systems. There is also good justification for the following additional requirements:
- partial failure (secondary/residual) performance after any failure in the antilock device;
- optical warning to the driver after certain electrical failures in the antilock device;
- the antilock device should not be adversely affected by electro-magnetic interference.

There may even be an argument to introduce some limited 'energy consumption' test, but the adhesion utilization requirements and various 'additional checks' are difficult to justify within any regulation aiming to establish minimum safety standards.  In principle, all road vehicles (within specific categories) should be subject to the same minimum requirements, irrespective of the type of equipment fitted.

## ACKNOWLEDGMENTS

The author appreciates the opportunity to publish this information and recognizes the assistance provided by Lucas Girling Ltd. and Lucas Industries plc towards publication of this paper.

## REFERENCES

(1)  P. Oppenheimer "Braking Regulations in Europe"
SAE Transactions, Vol. 83 (1974), paper 740313

(2)  P. Oppenheimer "Braking Regulations for Passenger Cars"
SAE Paper 770182 presented at International Automotive Engineering Congress, February 1977

(3)  ECE Regulation No. 13 "Uniform provisions concerning the approval of vehicles with regard to braking"
Published by the United Nations in Geneva (E/ECE/TRANS/505 - Rev. 1/Add. 12/Rev. 2 dated 5 February 1979)

(4)  EEC Directive 71/320 "On the approximation of the laws of the Member States relating to the braking devices of certain categories of motor vehicles and of their trailers"
Council Directive of 26 July 1971, published in the Official Journal of the European Communities No. 202, 6.9.71.

(5)  P. Oppenheimer "The development of international antilock braking regulations"
Paper C.190/85 presented at the I. Mech. E. conference on "Anti-lock braking systems for road vehicles" in London, September 1985.

(6)  "Regulations for braking systems on motor vehicles and trailers"
Swedish Safety Standard F.18-1971 (para.7.2)
Translated and published by Intereurope Regulations Ltd., Wokingham, England

(7)  M. Krugel and H. J. Hoffmann "Richtlinien for die Prufung von Blockierverhinderern"
Published in the VDI Journal, Vol. 237, 1974 (Federal Republic of Germany)

(8)  E. P. Williams "Testing anti-lock braking systems --- The early years"
Paper C.180/85 presented at the I. Mech. E. conference on "Anti-lock braking systems for road vehicles" in London, September 1985

(9)  EEC Draft Commission Directive III/92/85 Rev.II, dated 26 July 1985, Brussels: "Adapting to technical progress Council Directive 71/320/EEC"

(10)  ISO.7638-1985 "Road vehicles --- Brake anti-lock device connector"
Published by the ISO and available via national standards associations

# PAST ABS
# TECHNOLOGY

*Systems*

# Design and Performance Considerations for a Passenger Car Adaptive Braking System**

**Thomas C. Schafer and Donald W. Howard**
Brake and Steering Div., Bendix Corp.
**Ralph W. Carp**
Automotive Electronics Div., Bendix Corp.

**Paper 680458 presented at the Mid-Year
Meeting, Detroit, Michigan, May, 1968.

THE PURPOSE OF THIS PAPER is to discuss design and performance considerations for an adaptive braking system for passenger cars which will provide: shorter stopping distances; vehicle stability; and vehicle steerability. This system must be capable of satisfying the above requirements in an automotive environment at a price which is commensurate with customer expectations.

The control concept philosophy behind an adaptive braking system is based on the behavior of the tire in contact with the road generally depicted by the $\mu$-slip curve. The shape of the $\mu$-slip curve and its variations are influenced by: the type and condition of pavements; the type and condition of the tire; and vehicle speed. Published data, much of which is theoretical, (1-5)* indicates that

*Numbers in parentheses designate References at end of paper.

these $\mu$-slip curves are generally of four types, (Fig. 1). Curve "a" is the most common in which the coefficient peaks at some value of slip and then decreases with further increase in slip. Curve "b" illustrates a characteristic where the coefficient increases to some peak value and then tends to flatten out as slip increases further. This condition is generally associated with deep loose gravel, and/or deep wet snow. Curve "c" is an example of the type of characteristic sometimes experienced at higher speeds (above 60 mph) where the coefficient increases, flattens out for a short time, and then slowly increases again. However, as vehicle speed diminishes, the $\mu$-slip characteristic changes in a direction to approach that characteristic described by Curve "a." Curve "d" appears on most surfaces at very low speeds (below 1 mph) and can be ignored for adaptive braking system philosophy design work.

The initial portion of the $\mu$-slip curve is a linear relationship related to the tire type and condition. Smooth,

ABSTRACT

The basic philosophy for a logical approach to the design and implementation of an adaptive vehicular braking system is defined. Tentative system specifications, goals and objectives are outlined. The many technical factors, both external to the vehicle and within the vehicle, necessary to establish a general mathematical model of an adaptive braking control system are emphasized. Techniques utilized to obtain experimental data to support systems analysis and computer studies are discussed. Various control configurations are presented.

Vehicle performance is summarized for two configurations of an electronically controlled vacuum actuated adaptive braking system.

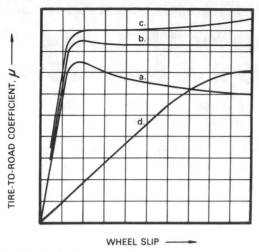

Fig. 1 - Typical tire-to-road coefficient versus wheel slip characteristics

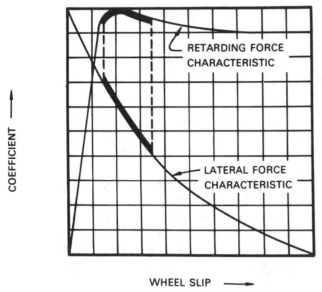

Fig. 2 - Typical retarding force and lateral force characteristics versus wheel slip

or bald tires tend to exhibit a steep rise in the $\mu$-slip curve initially, while snow tires show a lesser slope. The shape of the curve past the peak as a function of tire condition is not well documented.

Fig. 2 shows a typical $\mu$-slip curve with a lateral stability curve superimposed. It is apparent from these data that one cannot expect to obtain maximum retarding force and maximum lateral stability simultaneously, since maximum lateral stability occurs at 0% slip and maximum retarding force occurs in the range of 5-25% slip. Therefore, in order to achieve these goals, an adaptive braking system must create sufficient retarding force to reach the peak of the $\mu$-slip curve and then modulate brake pressure in such a manner as to stay at or near this peak. In this way, it is possible to achieve near maximum retarding force, while retaining a high degree of lateral stability.

PERFORMANCE SPECIFICATIONS
AND GOALS

In order to accomplish the goals set forth above, it was necessary to establish performance specifications for an adaptive braking system. These specifications have not been published but were used for the preliminary development of the Bendix Adaptive Braking System.

GENERAL SYSTEM SPECIFICATIONS

1. Vehicle stopping distance must be less than or equal to those which would result with locked wheels under the same braking conditions.

2. The system must maintain maximum vehicle stability and steerability under existing conditions consistent with obtaining maximum retarding force.

3. The system must prevent wheel lock-up of durations which would preclude meeting the preceding two requirements.

FUNCTIONAL REQUIREMENTS

The system shall be capable of:

1. Functioning over a range of tire-to-road coefficients from approximately 0.08-1.0.

2. Functioning with static wheel loads between 600 and 1500 lb.

3. Operation through a speed range from 5-85 mph.

4. Functioning with both drum and caliper type brakes as long as there is sufficient pressure supply to cause the wheels to lock.

5. Operation with brake pressure application rates up to 50,000 psi/sec.

6. Operation in situations where, without adaptive braking control, the wheel would decelerate from synchronous speed to lock-up at rates up to 50 g's.

7. Functioning in vehicles having an average suspension system bandpass of 1 Hz and approximate resonant frequency ranges in the unsprung wheel system of:

| Wheel hop | - | 10 Hz |
| Fore-and-aft | - | 15 Hz |
| Torsional | - | 30 Hz |

OTHER ADAPTIVE BRAKING SYSTEM REQUIREMENTS

In addition to those requirements specified in the preceding sections, the adaptive braking system shall conform to the following requirements:

1. Operation shall be free from roughness, oscillations or other operating characteristics objectionable to the driver and/or passengers.

2. The system shall have no influence on braking under situations where wheel lock-up is not imminent

3. The system shall require only simple initial adjustment to establish specified operation.

4. There shall be no adjustments or controls requiring attention of the driver of the vehicle.

5. The system shall revert to what is now normal braking in case of any control system failure.

6. The system shall function in a normal automotive type environment, which includes temperature variations,

altitude, vibration and contamination, and shall be immune to transients generated by other vehicular subsystems.

## ANALYSIS AND SIMULATION

A study of these performance specifications suggested simulation techniques should be used to search out a suitable solution. To simplify the simulation, it was confined to studying the behavior of one wheel. In order to represent actual vehicle conditions, certain baseline data were required.

Adaptive braking systems must function over a wide range of $\mu$-slip conditions. For the purposes of simulation, it was believed that a 0.08 locked wheel tire-to-road coefficient was a realistic low value, and 0.9 was a good high value. Fig. 3 illustrates typical $\mu$-slip curves for high and low coefficient surfaces. As shown, there is a greater potential percentage reduction in stopping distance on low coefficient surfaces than on high.

The tire-to-road coefficient changes as a function of speed. Here, again, this variable was introduced into the simulation in order to more accurately create realistic conditions. Fig. 4 illustrates typical data obtained in this area for high and low coefficient surfaces.

The tire rolling radius varies from one size to another. Table 1 shows typical rolling radius values for numerous tire sizes at 0 and 60 mph.

Weight transfer in the vehicle as a function of deceleration is another important consideration. During a stop, the loading increases on the front wheels and decreases on the rear wheels. This characteristic was introduced into the simulation using Eq. 1.

$$W = \pm W_t \left(\frac{C}{L}\right) \left(\frac{A_v}{g}\right) \qquad (1)$$

where:

W = Change in wheel loading-lb.

$W_t$ = Weight on the wheel at zero deceleration-lb.

C = Height of center of g-ft.

L = Wheelbase-ft.

$A_v$ = Vehicle deceleration, ft/sec.$^2$

g = 32.2 ft/sec.$^2$

The polar moment of inertia of the wheel and tire assembly must be known to effect simulation of the single wheel adaptive braking control system. The classical torsional pendulum technique affords a simple means to obtain these data experimentally. The period of oscillation is related to the gross inertia by Eq. 2.

$$T_t = 2\pi \sqrt{\frac{I_t}{K}} \qquad (2)$$

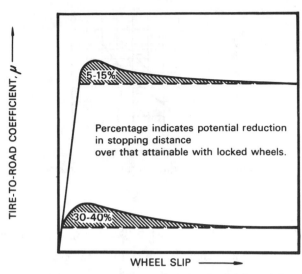

Fig. 3 - Typical $\mu$-slip characteristics for high and low coefficient surfaces

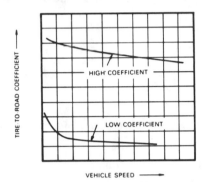

Fig. 4 - Typical tire-to-road coefficient characteristics versus vehicle speed

Table 1 - Static and Dynamic Tire Rolling Radius Data for a Leading Brand of Tires

| Tire Size | Static Rolling Radius at 24 psi, in. | Rolling Radius at 60 mph, in. |
|---|---|---|
| 6.50x13 | 11.3 | 12.1 |
| 7.00x13 | 11.5 | 12.2 |
| 6.95x14 | 11.7 | 12.4 |
| 7.35x14 | 12.1 | 12.8 |
| 7.75x14 | 12.3 | 13.1 |
| 8.25x14 | 12.6 | 13.4 |
| 8.55x14 | 12.7 | 13.5 |
| 8.85x14 | 12.9 | 13.8 |
| 7.35x15 | 12.2 | 13.0 |
| 7.75x15 | 12.4 | 13.2 |
| 8.15x15 | 12.6 | 13.4 |
| 8.45x15 | 12.9 | 13.8 |
| 8.85x15 | 13.1 | 14.1 |
| 9.00x15 | 13.4 | 14.3 |
| 9.15x15 | 13.1 | 14.0 |

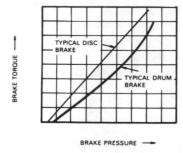

Fig. 5 - Typical brake torque versus brake pressure characteristics

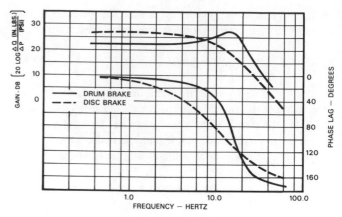

Fig. 6 - Brake frequency response for typical drum and disc brakes

where $T_t$ is the period in seconds, $K$ is the spring constant of the supporting shaft in in.-lb and $I_t$ is the gross inertia in in.-sec.$^2$ If the polar moment of the wheel is designated as $I_w$, and $T_p$ is the period of the pendulum less wheel and tire, then as shown in Eq. 3:

$$I_w = \frac{K}{4\pi^2}\left(T_t^2 - T_p^2\right) \qquad (3)$$

Typical vales of wheel inertia were found to be 9-18 in.-lbs-sec$^2$.

Fluid displacement as a function of brake pressure on a static and dynamic basis were defined to support systems analysis in establishing the mathematic model of the control system. Data was obtained for several sizes of both disc and drum brakes. Fluid displacement is a significant parameter with respect to fill time upon initial brake application, and also its relationship to wheel cylinder pressure during modulation by the control system.

A major portion of representative brake torque-to-pressure characteristics is shown on Fig. 5 for drum and disc brakes. In modeling an adaptive braking system on the computer, the incremental slope of the torque-pressure curve appears as a gain term in the forward part of the control loop. The magnitude of the gain term varies with brake pressure, vehicle speed and lining temperature.

The dynamic performance of brakes is more significant than the static characteristics in the formulation of the mathematical model. The Bendix automatically programmed multiple dynamometer (6) proved a valuable tool in obtaining the brake pressure-to-torque gain and phase lag as a function of frequency. Fig. 6 illustrates the gain and phase lag of typical disc and drum brakes.

The brake pressure as measured at the wheel cylinder was sinusoidally modulated about a fixed pressure level at several frequencies between 0.5 and 30 Hz. The pressure modulation was derived by superimposing the output of a function generator on the normal pressure command signal and using closed loop servo control to regulate brake pressure. Initial conditions, including speed and drum and lining temperatures, were maintained thoughout the test to obtain consistent data.

Experimental test data indicated that the transfer function could be described by a quadratic if the magnitude of the input forcing signal ($\triangle$brake pressure) was limited to values in the order of $\pm 10\%$ of the average or steady-state pressure. Data (Fig. 6) was obtained by matching the phase lag rather than the amplitude ratio characteristic. It should also be noted that the dynamics of the flexible hose connection between the brake and rigid tubing in the actual vehicle installation should also be considered to obtain an accurate computer model.

The lowest test frequency (0.5 Hz) was selected based upon the initial speed (60 mph) and time available to record at least two cycles before the speed had decreased to zero. The upper limit (30 Hz) was based upon the bandwidth of the dynamometer control system. However, it is evident from these data that this was adequate.

After determining the baseline data, a control concept had to be established. Systems based on the use of vehicle speed or acceleration were evaluated using analog simulation techniques. An accelerometer was required to implement these concepts. The accelerometer as a source of intelligence presented three immediate problems:

1. Component cost.
2. Grade sensitivity which introduced error on nonlevel surfaces.
3. System's inability to adapt to varying tire and road conditions.

These shortcomings led to investigations using wheel speed as the source of intelligence. Further exploration substantiated this approach as being feasible.

At this point, various approaches for obtaining power to modulate brake pressure were analyzed. Consideration was given to: hydraulics; pneumatics; and electric approaches.

VACUUM-ELECTRONIC CONCEPT

It was decided to use engine vacuum as the source of modulating power in combination with electronics for pro-

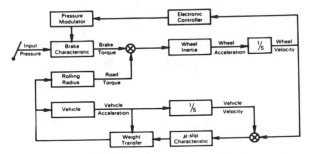

Fig. 7 - Simulation model for evaluation of control concepts

gramming the modulator. The primary factors influencing this decision were:

1. Adaptability to present brake systems.
2. Simplicity of component design.
3. Advantages in flexibility and response achievable with electronics.

The control system employs wheel speed intelligence as an input to the electronic controller. The wheel speed signal is differentiated electronically to provide a rate of change of wheel speed. The resulting signal is compared to a predetermined program and the controller then issues proper signals to the modulator valves. The operation of these valves adjusts the vacuum powered modulators to produce the correct brake pressure for maximum braking effort. During operation, the modulator isolates the controlled brake pressure to the wheel cylinder from the supply pressure out of the master cylinder. In this way, no brake pedal movement is present as the pressure is modulated to the wheel cylinder. In addition, it is necessary to provide a low-speed cutout for the system because of the dynamic range limitation of the wheel speed sensors. Also, provisions have been included which distinguish between a wheel lock-up when the vehicle speed is above the low-speed threshold and when the vehicle is below the threshold and/or parked at the curb. All subsequent discussions in this paper are based on this approach.

Using the available input data and making assumptions in the absence of rear inputs, various control cycles were investigated. The result was a set of specifications for component performance. Hardware was designed to meet these specifications, fabricated, installed on a vehicle and evaluated. Test results were used to upgrade the simulation to improve performance. This process was repeated until a suitable system was developed.

Fig. 7 is a block diagram of a single wheel adaptive braking system as simulated on the analog computer. The brake application function was simulated as a linear pressure ramp, but nonlinear applications may be simulated using function generators. One common variation of the input function was a "spike" application of the brake pedal. A diode limiter was used to limit maximum supply pressure.

The brake assembly was simulated as a second order quadratic in conjunction with a brake pressure versus displacement function, and a brake torque versus pressure func-

tion. Brake torque and road torque were summed to act on the moment of inertia of the wheel to generate wheel acceleration.

The retarding force generated at the tire-to-road interface was converted to road force acting through the wheel rolling radius and fed back to the brake torque summation point. Retardation force also acts to decelerate the vehicle through the appropriate vehicle mass and dynamic weight transfer function. Vehicle deceleration was coupled back to the weight transfer block to adjust the weight on the braked wheel for the existing conditions. Integration of vehicle deceleration provided the vehicle velocity which was summed with wheel velocity for generation of wheel slip.

Adaptive braking control was provided by controlling the brake pressure modulator. Wheel velocity signals were sensed and processed in the block labeled "Control Electronics." This block also contained the logic and signal generation required to provide proper pressure modulator control.

CONFIGURATION

When an acceptable control cycle was established on a one-wheel basis, it then became necessary to consider the control configuration on the vehicle (the manner in which the vehicle wheels are controlled). Fig. 8 shows various configurations considered for passenger car adaptive braking systems. The simplest control system involves controlling only the rear wheels with one pressure modulator. This arrangement gives some improvement in stopping distance and a significant degree of vehicle stability.

The input intelligence can originate from the propshaft or from each of the rear wheels. Experience to date indicates that individual wheel sensors provide the more satisfactory approach. This means that one wheel must be selected to control the pressure modulator at any given point in time. In order to maintain good lateral stability, it was decided to let the wheel which approached lock-up first be the controlling wheel. This can be defined as a "select low" system, which means the lowest speed wheel provides the controlling signal. The other approach when using individual wheel sensors is the opposite -- "select high." The primary disadvantage of propshaft sensing lies in the area where one wheel may inadvertently lock. This condition reduces the propshaft speed to one-half its original value. This makes it necessary to change the system gain simultaneously with the change in propshaft speed, which was considered undesirable.

The next logical configuration is individual rear wheel control. This requires two speed sensors, two control circuits and two modulators. The advantages versus cost for this approach were minor since the percentage gain in retarding force was not significant over that attainable with the rear axle system.

A rear axle-front axle configuration would give shorter stopping distances, vehicle stability and some degree of steerability. Here, again, a "select high" or "select low"

155

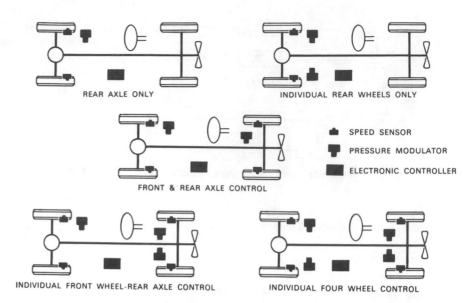

Fig. 8 - Typical adaptive
braking configurations

REAR AXLE ONLY

INDIVIDUAL REAR WHEELS ONLY

FRONT & REAR AXLE CONTROL

SPEED SENSOR
PRESSURE MODULATOR
ELECTRONIC CONTROLLER

INDIVIDUAL FRONT WHEEL-REAR AXLE CONTROL

INDIVIDUAL FOUR WHEEL CONTROL

system can be employed. Preliminary tests indicate that
stopping distance may be compromised under some conditions
with "select low" on the front axle. In addition, steerability
is not maximized with "select high" since one wheel will
lock in some cases. This results from variations in side-to-
side road coefficient and/or brake effectiveness. Individual
front wheel control and rear axle control satisfies all the
goals initially outlined with a minimum sacrifice in performance. All the subsequent 4-wheel performance data is
based on this configuration. Individual 4-wheel control is
the optimum approach, but, as discussed earlier, rather small
gains result from going to individual rear wheel control.

INSTRUMENTATION AND
TEST FACILITIES

A wide variety of instrumentation was used during the
development and performance evaluation of the various
adaptive braking control system concepts and configurations. In the initial development phases data was recorded
by light beam oscillographs. This was done to maximize
dynamic range and to provide the necessary time resolution
to analyze the interrelationships of wheel speeds, vehicle
deceleration, brake pressure and controller performance.
A vehicle installation of the recording oscillograph and signal conditioning equipment is shown in Fig. 9. Extreme
care was exercised in the selection of the transducers and
other instrumentation elements to avoid masking the control
performance by introducing bandwidth limitations and/or
saturation bounds.

Preliminary comparisons of adaptive braking control ver-
versus locked wheel performance were noted by both the
magnitude of the vehicle deceleration trace and the time
duration from a referenced initial speed to a full stop for
a range of tire-to-road coefficients. The strip chart recordings also provided considerable insight as to the smooth-
ness of control, transport lags between the various elements
of the control system, and as a means of evaluating the noise

Fig. 9 - Vehicle instrumentation

immunity between the adaptive braking control electronic
package and the other accessory systems on the vehicle.

When the adaptive braking control system performance
was considered satisfactory, the recording oscillograph instrumentation was replaced with a conventional fifth wheel
(including a vehicle speed tachometer) to measure stopping
distances for various combinations of tires, road surfaces,
and control configurations. Dye markers fired by explosive
charges were also utilized to periodically verify the instrumentation calibration.

The electronic readout device included an in-line digital
readout which measured distances up to 999 ft and two vehicles speed memory circuits. The initial speed, $V_1$, was
memorized upon brake application, while a second speed,
$V_2$ was memorized upon the release of the pedal or after
a preset time interval. This instrument is illustrated by
Fig. 10. The capability of memorizing initial and final
speeds permitted determination of locked wheel coefficient

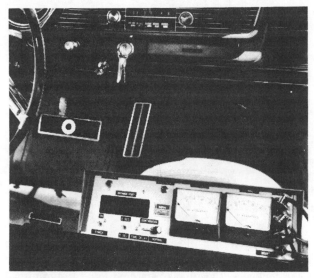

Fig. 10 - Speed memory and stopping distance counter

Fig. 12 - Aerial photograph of the low coefficient test facility

Fig. 11 - Aerial photograph of the high coefficient test facility

over a relatively narrow speed range, reducing tire wear and eliminating the lag in the vehicle speedometer. Because this instrument had "track and hold" capability, variations in the initial target speed, $V_1$, were known and data could be corrected accordingly. High-speed photography of vehicle performance was utilized to further support the evaluation. Vehicular adaptive braking control test programs were conducted at the Bendix Automotive Development Center. High coefficient tests were run on a three-mile high-speed oval. The surface coefficient ranges from 0.6-0.7 dry, down to 0.5 wet. Fig. 11 is an aerial photograph of this facility.

A skid pad, shown on Fig. 12, 17 ft wide by 500 ft long, served as the low coefficient surface. This surface is smooth asphalt coated with enamel paint. The tire-to-road coefficient was approximately 0.12 when covered with a thin film of water.

In addition, a coarse aggregate asphalt strip adjacent to the wetted paint surface affords an opportunity to evaluate adaptive braking performance for side-to-side unbalanced conditions. This combination was extremely valuable in determining vehicle stability and steerability for the 4-wheel configuration.

The high-speed track and skid pad are supplemented by gravel and unimproved roads, permitting testing on these surfaces in combination with grades and turns.

VEHICLE PERFORMANCE

With the preceding material as background, it is now appropriate to discuss adaptive braking system performance. Both rear axle and individual front wheel-rear axle control data are presented. This discussion centers around comparing stopping distances obtained with braking control with that obtained under locked wheel conditions. However, vehicle stability and steering tests under maximum braking conditions were also conducted. The results of these tests were observed visually and documented by the use of photographic techniques. Lateral stability was evaluated by braking the vehicle on an unbalanced side-to-side coefficient surface and also by braking in a turn. Steerability was demonstrated by making evasive maneuvers during maximum braking conditions on both high and low coefficient surfaces.

Three factors, initial vehicle speed, condition of the test surface, and brake application rate, were found to have a significant influence on stopping distance and, therefore, had to be controlled as closely as possible during stopping distance comparison tests.

In order to measure initial speed accurately, the "track and hold" speed indicator was used as discussed earlier. Stopping distance was then corrected for variations in initial speed, using Eq. 4.

$$S_c = S_a \left[ 1 + \left( \frac{2\,\Delta V}{V_1} \right) \right] \qquad (4)$$

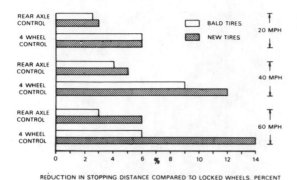

Fig. 13 - High coefficient performance data from various initial speeds

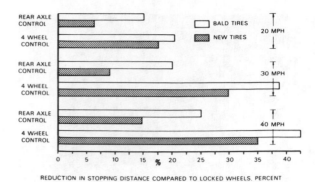

Fig. 14 - Low coefficient performance data from various initial speeds

where:

$S_c$ = Corrected stopping distance

$S_a$ = Actual stopping distance

$\Delta V$ = Difference between actual and target speed

$V_1$ = Target speed

To maintain the condition of the test surface, care was exercised in keeping the surface clean, making all stops in the same general area on the test surface, and, in the case of the skid pad, ensuring it was wet before each run. On the main track, an accumulation of skid marks was unavoidable, but periodic locked wheel stopping distance checks did not reveal a significant change in the tire-to-road coefficient.

To reduce stopping distance error because of variations in the initial brake application rate, the test drivers were instructed to make all stops under "panic" type conditions. The time variation in developing brake torque was, therefore, minimized and variations in stopping distance due to application rate were considered negligible.

To further reduce the effects of test variations, a minimum of five stops was made at each test condition and the results averaged.

Performance tests were run using new tires on all four wheels. These tires were then ground down and run several hundred miles to obtain bald tires. The bald tires were then tested under the same conditions as the new tires. It was felt that if the system functioned well with these two extremes, it would function satisfactorily for intermediate tire conditions.

Fig. 13 represents adaptive braking system performance obtained on a high coefficient surface for bald and new tires over a range of speeds for rear-axle and 4-wheel control. These data are shown as the percentage reduction in stopping distance compared to locked wheel stopping distance. As expected, the 4-wheel control configuration exhibited a higher degree of performance than rear axle control.

The data show that for new tires the percentage reduction in stopping distance increased with speed. In the case of bald tires, there is a lower percentage reduction in stopping distances for vehicle speeds above 40 mph. This in-

dicates that the locked wheel coefficient with bald tires increases at the higher speeds. Therefore, the marginal between lock wheel and peak retarding force is diminished, which results in a smaller improvement in stopping distance.

A similar group of tests were conducted on a low coefficient surface. The test procedure was identical to that used above except the maximum speed was limited to 40 mph. These data (Fig. 14) verified that significant reductions in stopping distance can be obtained on low coefficient surfaces. In all cases, the 4-wheel performance exceeded that for rear axle control. The percentage improvement in stopping distance was greater with bald tires than with new tires. This indicates again that the $\mu$-slip characteristic varies with tire condition.

The principal advantage of individual front wheel-rear axle control is the retention of steerability, which becomes a factor under the following maximum braking conditions:

1. In a curve, rear axle control results in a straight stop, while the 4-wheel control system makes it possible to keep the vehicle within the proper curved lane throughout the stop.

2. On side-to-side unbalanced coefficients, the rear-axle configuration maintains the vehicle heading, but the vehicle tends to drift laterally toward the high coefficient side. Here again, 4-wheel control allows the stop to be made within specified lane and heading.

3. On a heavily crowned road of uniform coefficient, the vehicle tends to drift to the shoulder with the rear axle configuration, while the 4-wheel concept permits steering compensation resulting in a straight stop.

4. The 4-wheel configuration makes it possible to perform evasive maveuvers, that is, lane changing, while retaining the benefits of shorter stopping distances. Rear axle control, of course, provides no steering ability.

SUMMARY AND CONCLUSIONS

The goals and objectives for an adaptive braking system were defined. The approach for seeking out a concept to accomplish these goals and objectives was outlined. Vehicle performance data was presented to substantiate the technical feasibility of an adaptive braking system.

In conclusion, it was found that:

1. The steerability, stability and stopping distance of a passenger car under maximum braking conditions can be improved by the use of an adaptive braking system.

2. A rear axle control system provides somewhat better stopping distance and greatly improved vehicle stability, where a 4-wheel system can significantly improve these parameters and provide the additional feature of steering capability under maximum braking conditions.

3. The steering capability feature (4-wheel system) improves over-all performance by making the system significantly more adaptive to changes in road geometry and/or unusual variations in tire-to-road coefficient.

4. Engine vacuum can provide a suitable power source for modulating brake pressure.

5. Electronics provide the flexibility and response capability necessary to attain the performance goals previously outlined.

REFERENCES

1. C. G. Giles, "Some Recent Developments in Work on Skidding Problems at the Road Research Laboratory." Paper presented at 1964 Highway Research Board Meeting.

2. W. B. Horne, "Skidding Accidents on Runways and Highways Can Be Reduced." Astronautics and Aeronautics, August 1967, p. 48.

3. G. Kulberg, "Method and Equipment for Continuous Measuring of the Coefficient of Friction at Incipient Skid." Highway Research Board Bulletin 348, 1962, p. 18.

4. H. W. Kummer and W. E. Meyer, "Skid or Slip Resistance?" Journal of Materials, Vol. 1, No. 3 (September 1966). p. 667.

5. G. L. Goodenow, T. R. Kilhoff and F. D. Smithson, "Tire-Road friction Measuring System -- A second Generation." SAE Transactions, Vol. 77 (1968), paper 680137.

6. D. W. Howard and J. L. Winge," Automatically Programmed Quadruple Dynamometer for Vehicle Brake Testing". SAE Transactions, Vol. 76, paper 670144.

# Evolution of Sure-Track Brake System**

**R. H. Madison**
Ford Motor Co.
**Hugh E. Riordan**
Kelsey-Hayes Co.

**Paper 690213 presented at the International Automotive Engineering Congress and Exposition, Detroit, Michigan, January, 1969.

## ABSTRACT

The history, system philosophy, design evolution, and performance of the Sure-Track anti-lock automotive braking system are presented and discussed. Considerations of performance, driver skill, reliability, and commercial acceptance resulted in the choice of a vacuum-electronic rear wheel anti-lock system that incorporates individual wheel speed sensing and control of braking as a pair.

The system provides superior directional stability under "panic" braking conditions while maintaining stopping distance equal to or shorter than those for locked wheels under most road conditions.

IN THE MORE THAN SEVEN DECADES of vehicular brake development, evolutionary improvements have been seen in, successively, the adoption of four-wheel systems, the switch from mechanical to hydraulic actuation, the addition of power assistance, the improvements in brake lining materials, and the introduction, more recently, of disc brake systems, dual master cylinders, and system failure warning lights.

A significant and dramatic advance in that evolution has been the recent introduction of an automatic anti-lock system, under the trade name Sure-Track, for mass production vehicles.

The possibility of significantly improving the performance of the vehicle-driver system by automatic assist in the braking function has been studied for many years. In particular the improvement of directional stability and reduction of stopping distance by automatically preventing locking of braked wheels was proposed at least as early as 1932. (1)*

The fact that maximum tractive force between the tire and road is obtained at values of slip greater than zero was apparently first clearly recognized around 1955. (2, 3)

The physical groundwork thus was established which showed that improvements in both stopping distance and vehicle stability should be realizable with an automatically controlled braking system.

Meanwhile, speed and weight of motor vehicles were increasing and density of traffic was growing. These trends, combined with increasing public acceptance of a higher level of sophistication in automotive auxiliary equipment, provided the motivation and the confidence in commercial feasibility which have led to the development of the first commercially feasible automatically controlled braking system for passenger cars.

## HISTORY

Some 15 years ago, the aircraft industry succeeded in developing experimental systems to prevent wheel lock during braking. Their interest centered mainly around the problems of tire flattening and tire explosion which took place when a wheel locked at high landing speeds. Up to this time, the problem had been coped with by an overly cautious use of brakes at touchdown, resulting in longer landing runs.

Since the experimental aircraft system exhibited improved stability performance on the runway, engineers of the Ford Product Research organization became interested. They obtained an anti-lock unit from a French aircraft and installed it on a 1954 Lincoln sedan with a full power brake system. Although testing indicated improvements in stopping distance and stability, the installation created serious vibration problems in the suspension and caused the front wheels to vibrate about the king pin axis. This was typical of the many developmental problems that had to be overcome before a commercially acceptable design could be introduced.

In succeeding years, Ford research engineers experimented with internal designs, evaluated designs presented by domestic and foreign component manufacturers, and jointly set about establishing anti-lock system requirements.

Kelsey-Hayes first became interested in automatic braking systems in 1957 and started an exploratory development program at that time. Various approaches to the problem of optimum braking control were investigated. As a result of this work it was soon concluded that the single most use-

*Numbers in parentheses indicate References at end of paper.

ful contribution to brake control would be a means of preventing loss of control and reducing stopping distance under panic braking conditions.

Sure-Track anti-locking system represents the optimum workable system presently available as a result of these Ford and Kelsey-Hayes investigations. The technical team which developed it also included the Hydro-Aire Div. of Crane Co., Texas Instruments, Inc., Brown Engineering Div. of Teledyne, Cornell Aeronautical Laboratory, Battelle Memorial Institute, Rensselaer Polytechnic Institute, Booz Allen & Hamilton, and numerous other consultants and specialized manufacturing firms.

## FUNDAMENTAL TECHNICAL CONSIDERATIONS

Some improvement in vehicle stability can be achieved with very simple equipment by automatically "pumping" or pulsing the brakes in an arbitrary preprogrammed way.(5) However, basic tire friction data (6) showed that, under certain conditions, controlling the brakes so that wheel slip remained in the vicinity of an optimum value could increase brake effectiveness by a theoretical factor of as much as 2.3. This made it seem worthwhile to use a feedback approach which used data sensed on the vehicle during deceleration.

The variety of physically feasible methods of producing the necessary information for closed-loop braking control and of using this information is enormous. In order to arrive at a rational basis of choice it was necessary to establish standards of performance properly related to the practical realities of economics, driver skill, system reliability, and acceptability to drivers.

Based on literature surveys, mathematical analyses, computer simulations, and vehicle tests, we arrived at the following conclusions, which served as a guide in all development work:

1. It is possible (with a relatively simple braking control system) to provide a significant improvement in vehicle stability during panic braking with road friction from 1.0 to as low as 0.05 and for a complete range of driving speeds.

2. In general, there need be no significant increase in stopping distance with a simple control system, except under certain specific -- and uncommon -- road conditions (for example, loose gravel, loose dry snow), and there can be major reductions in stopping distances under many common driving conditions (for example, wet pavement, ice).

3. Simple means can be incorporated in the brake control system to preclude loss of normal braking capability for any probable mode of failure of the control system.

4. It is preferable to incorporate additional functions in a control computer rather than employ complex sensing systems.

5. Until a higher level of public driving skill can be achieved than now exists, an automatic braking system should require no changes in driver habits or reflexes for maximum safety.

6. The cost to the user can be sufficiently low so that the system will come into reasonably widespread use.

7. Service and repair should require a level of skill no higher than that required for present vehicle systems.

8. Assembly on the vehicle should be compatible with present methods.

These requirements also were confirmed in a study conducted by Kelsey-Hayes in 1967 for the General Services Administration. (7)

Although it is not within the purpose of this paper to present the pros and cons of two-wheel and four-wheel brake application control systems, the authors are well aware of the many comparisons of complexity, performance, cost, and state-of-the-art that can be made.

One overriding argument for developing the two-wheel system was that its inherently simpler design promised earlier development success than did the four-wheel system.

A second, important consideration favoring the two-wheel system is based on a point also made by Harned and Johnston (8), to the effect that, with four-wheel anti-lock systems, the driver may have to apply corrective steering action in order to maintain directional stability under some conditions.

With a two-wheel system, no special new driver skills or training should be necessary. Our position is that the advantage of early introduction of two-wheel control far outweighs any slight, eventual performance edge that might be obtained with the four-wheel system. This is not to say that four-wheel control is impractical or undesirable in the long run. We simply believe that at this time, two-wheel control represents the best solution from the viewpoints of both the public and the automotive industry.

## CONTROL PROGRAM

Although automatic skid control systems have been in widespread service in aircraft since about 1960, this technology has been of only limited value as a basis for the automotive problem because of basic dissimilarities in both the physics of the problem and the circumstances surrounding system use.

1. The tire friction mechanism differs significantly because the higher aircraft speed results in high pressure steam generation at the tire-road interface on wet surfaces and melting rubber on dry surfaces.

2. The required dynamic range of the aircraft system is much smaller because the pilot is required to judge the runway condition and make a manual selection of system response.

3. Runway surfaces are more consistent than public roads; hence, rapid changes in friction coefficient during a stop are unlikely, and large variations in friction coefficient between right and left wheels are unusual.

4. Since the maintenance of aircraft is very stringent, severe unbalance among braked wheels is unlikely. On the other hand, the dual servo brakes widely used in road vehicles will show large variations in brake coefficient from

wheel to wheel if they have been abused by overheating, glazing, or contamination. The need for assuring stability and good stopping distance with unbalanced brakes places another stringent requirement on the automotive skid control system.

5. Finally, aircraft brake control systems cost from $500-$1500 per controlled wheel while, to be feasible economically, the car system cost at the supplier's plant must be only a small fraction of this for the entire system.

In general, the problem of optimum automotive braking control represents an almost classic case of intractability. In mathematical terms, the car under extreme braking is represented by a tenth-order system of nonlinear differential equations with time-varying coefficients. The problem is made more difficult because quantitative data on the behavior of the tire-road system are almost totally lacking, except for the measurements reported by Harned & Johnston (8) and Meyer and Kummer. (6) Knowledge is sketchy at best of such factors as the difference, if any, between steady state and transient behavior, the influence of suspension dynamics, the effects of tire tread design and composition, and tire construction.

In such circumstances of insufficient data and understanding, the answer of the control engineer is to enclose the un-known elements of the system in a feedback loop. (Fig. 1). If the loop gain AB is sufficiently high over a broad band of frequencies and signal levels, the response of the system is governed only by the transfer characteristics of the feedback element.

Unfortunately, in some systems, and the car is one of these, the output response (in this case stopping distance and attitude) cannot be measured conveniently and fed back into the control system. Furthermore, in some cases, and again the car is one; the frequency and amplitude range of the available, easily sensed information occupy the same region as noise or unwanted signals. For example, there are suspension and tire resonant frequencies which cover the range from 1-22 cps, while the response of a wheel entering and recovering from a skid covers a time regime corresponding to a band width of about 10 cps.

One means of handling such a control problem lies in the use of the so-called model reference self-adaptive control technique (Fig. 2). The model is an idealized electrical analog of the system to be controlled. For example, it could represent a single wheel, unsuspended vehicle with an idealized brake system (no fade, grab, or hysteresis) and ideal tire-road friction characteristics. The brake command is fed into the model, which computes an ideal wheel speed

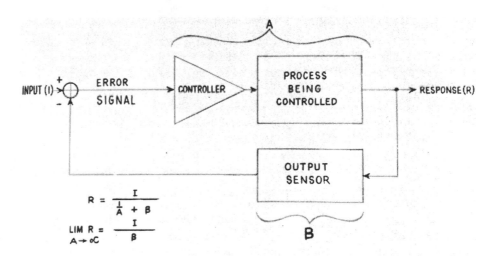

Fig. 1 - Typical feedback control system

$$R = \frac{I}{\frac{1}{A} + B}$$

$$\lim_{A \to \infty} R = \frac{I}{B}$$

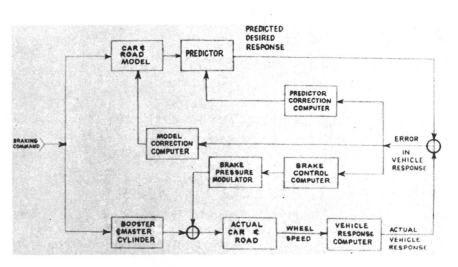

Fig. 2 - Model-predictive self-adaptive braking control system versus time characteristic against which the actual wheel speed is compared. The error signal is used to control the brakes in such a way as to cause the real and ideal wheel speeds to coincide. Meanwhile, the magnitude of the error signal is assessed, and its character is used to modify some of the model properties to improve the accuracy of its representation and to adjust the controller response.

Such methods are well-known in the control art and broadly applied in the aerospace field. (9, 10)

The application of these techniques in the automotive brake control system required a massive effort in analysis, computer simulation, and vehicle testing. In general, basic problems of system logic and structure were handled on the digital computer, while system tuning and tolerance establishment was done by both analog simulation and car testing.

## SYSTEM CONSIDERATIONS

The reasoning behind the selection of a two-wheel over a four-wheel anti-lock system has been discussed. Some discussion of the background for the choice of vacuum power rather than electrical or hydraulic power for the brake pressure modulator is appropriate, since the power steering system or the automatic transmission could have supplied oil under pressure, and the modern alternator electrical system could have furnished the required peak power. Vacuum power devices are intrinsically slow, have response characteristics which depend on piston position and displacement, and are bulky. Nevertheless, it was concluded that these disadvantages were outweighed by the much slower response and higher cost of an electrical modulator, and the coordination and installation engineering problems associated with a tie-in to the hydraulic steering or transmission systems. We also wished to retain flexibility to adapt the anti-lock system to cars without power steering or automatic transmissions. Having made this decision, we designed the control circuits for compatibility with modulator response characteristics which could be obtained within reasonable limits of complexity, cost, and size. The design of the present Sure-Track system assures that no performance has been sacrificed because of the choice of vacuum actuation.

More difficult choices are associated with alternatives among directional stability, stopping distance, and the accommodation of off-tolerance or deteriorated foundation brakes.

Since maximum lateral force capability is obtained with a freely rolling wheel, some compromise in directional stability is inherent in braking to any significant slip level.

The accommodation of lateral variations in road friction requires that the wheel on the high friction surface roll sufficiently freely to assure that unstable attitude changes cannot develop. Incorporation of this capability into the control system again requires some compromise if satisfactory "split coefficient" stability is to be maintained. Finally, if satisfactory stopping distances are to be achieved on high friction surfaces, both wheels must be driven to a relatively high slip level. With unbalanced brakes, it is necessary that the brake pressure be allowed to remain at a high enough level that the underbraked wheel achieves adequate slip. The other wheel will of necessity be overbraked. The system must be capable of adjusting for this condition without the risk of undue loss of stability with normal brakes under split coefficient conditions.

It is thus clear that the selection of system parameters involves a series of compromises among mutually contradictory requirements, so that the overall performance delivered inevitably implies a number of value judgments.

Where tradeoffs between stopping distance and stability were involved, we biased the system slightly toward stability.

Where the subjective "feel," that is, driver consciousness of system operation, conflicted with stopping distance or stability, we emphasized objective performance.

## COMPONENT EVOLUTION

This part of the paper deals with the evolution of major components of the Sure-Track brake system. Fig. 3 is a schematic of the system mounted on a vehicle chassis. Mechanically driven electro-mounted sensors located in each rear wheel generate a-c voltage pulses in proportion to wheel speed. These pulses are transferred through wiring to a solid-state electronic control module located under the glove box. The control module determines when a rear wheel skid is

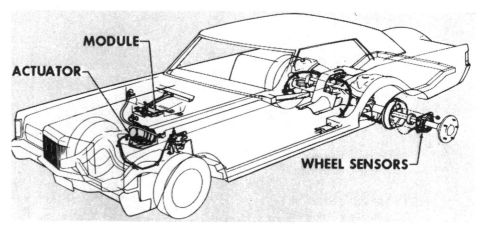

Fig. 3 - Sure-Track system components

imminent and signals the vacuum powered brake actuator. This unit, located on the right frame rail near the rear of the engine, releases the rear brakes, then reapplies them when the wheels spin up.

Fig. 4 shows early design sensors. One with teeth cut in the brake drum flange was discarded because of its vulnerability to damage. The next step was to put a rotating toothed ring and a stationary magnetic pick-up inside the brake. During this period of system development, when the control system was relatively unsophisticated, the performance of this sensor seemed to be satisfactory. However, when a production design was made, axle and brake tolerances and deflections became a problem. In addition, the sensor ring was vulnerable to damage in service handling. These reasons, together with the desire to obtain the simplicity and cost saving that could result from using only one sensor, led to the next phase being centered around transmission output sensing.

One of the several designs considered is shown in Fig. 5. This sensor was designed to attach to the transmission at the speedometer cable attaching boss, the speedometer cable being connected to the sensor. An integrated stator, rotor, and bearing permitted minimum clearance between the rotor and stator so that a good signal was generated. However, backlash and vibration between the sensor and the rear tires created electrical noise which the module sometimes read as skid signals. Lash between the speedometer drive gear and the rotor gear was particularly troublesome. Attempts to preload the gears with cam adjustments and spring-loaded flexible shafts were unsuccessful and caused excessive tooth wear.

At this point small size wheel sensors appeared to offer the best solution. They were designed to be attached as shown in Fig. 6. A special diameter was turned on the axle shaft outside of the wheel bearing on which the rotating member (rotor) was pressed. The wheel bearing outer retainer was replaced with a new, similar plate to which the stationary member (stator) was attached. The first design is shown in Fig. 7. The rotating member consisted of a steel ring, a rotating circular magnet of barium ferrite suspended in a rubber compound, and a soft iron outer ring with 160 teeth on its outer diameter. The stationary member con-

Fig. 4 - Early design sensors

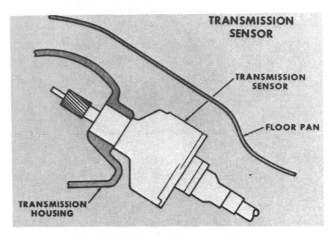

Fig. 5 - Transmission sensor design

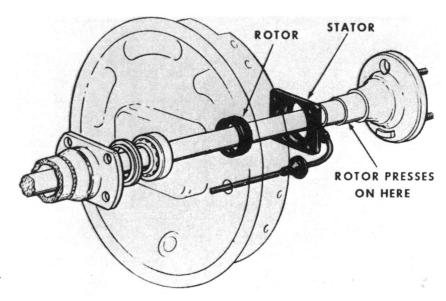

Fig. 6 - Axle shaft sensor installation

tained corresponding teeth and a coil. The design calculations indicated that the sensor would have sufficient output with the clearance required for production tolerances and axle deflections. However, the experimental components had much lower output than was expected, and unacceptably small tooth gaps were required for adequate signal strength. The rubber magnet also caused a problem due to distortion. Sensors set up with running clearance one day would have tooth interference the next day.

About this time the idea of a floating sensor was conceived. The first experimental design is shown in Fig. 8. The rotating member has 130 teeth and is pressed onto the axle shaft. The stator contains the coil and magnet and is held against the wheel bearing retainer by two leaf springs. The components are dimensioned so that the smooth rub rings contact each other before the teeth make contact. Rub ring contact causes the stator to shift position on the retaining springs so that it becomes self-aligning, thus providing for manufacturing variations and axle deflections.

It was found that this design was subject to damage during installation. If one nut was run up ahead of the others before

the axle shaft and bearing were completely seated, the bearing retainer might be bent so that the stator was significantly misaligned with the rotor. A design modification, as shown in Fig. 9, was made, incorporating two stand-off pads or feet

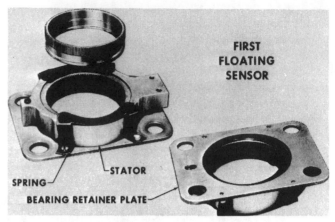

Fig. 8 - First floating sensor

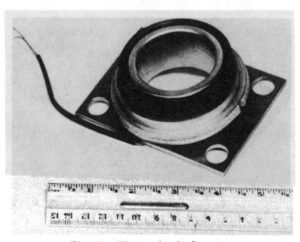

Fig. 7 - First axle shaft sensor

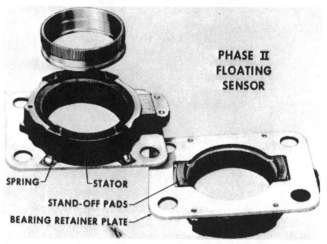

Fig. 9 - Phase II floating sensor

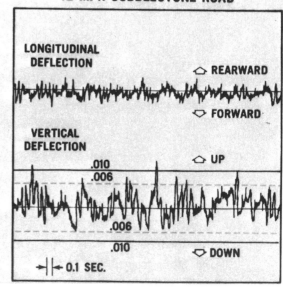

Fig. 10 - Typical tape showing axle deflection

on the rub ring so that they contacted the outer race of the wheel bearing and held the stator off the bearing retainer by 0.020 in. This change gave good stator-rotor alignment, even when the retainer was distorted. It had a second advantage of providing essentially constant friction contact with the bearing in spite of retainer corrosion.

Car tests were conducted over various road surfaces to determine the relative motion between the rotor and stator. A typical tape is shown in Fig. 10. It shows the vertical and longitudinal deflections of the axle shaft at the rotor with respect to the stator. In the static deflection chart shown at the left, note the shift of 0.005 in. in the center of rotation when the vehicle weight is put on the floor. With a fixed design sensor, clearance must be provided for this initial deflection, while the floating design merely shifts the stator to a new center. The dynamic portion of this tape, taken at 12 mph over a severe cobblestone road, shows a maximum deflection of ±0.0125 in., with most of the deflections less than ±0.006 in. Similar car tests were conducted on a variety of road surfaces and speeds. Other tests were made to determine the portion of time the rotating and stationary rub rings were in contact. These data were used to design laboratory tests which duplicated our rough road durability test procedure.

One of the test fixtures is shown in Fig. 11. With this fixture, three sensors can be tested simultaneously for rub ring wear. The rotors are pressed onto the test fixture spindles, shown near the bottom, which are turned the equivalent of 45 mph. The stators are bolted to face plates, which are cycled up and down by hydraulically powered cylinders in a programmed frequency and amplitude, based on car data from the rough road course. Fig. 12 shows the results of life tests on several materials. We would have preferred that the rotor be the more rapid wear member, but this was precluded by a combination of manufacturing and output signal considerations. The released stationary rub ring is molydisulphide coated, tuftrided sintered iron, and the rotor rub ring is mild steel, chrome plated, and molydisulphide coated.

As development was continued, the system became more and more sophisticated so that the sensor signal quality became more important. It was found that sometimes, because of insufficient difference between the velocity signal and the noise signal generated by axle deflections, the control module could not identify the skid signal. Several design modifications were built and checked for useful and noise signals. Some of the results are shown in Fig. 13. The control system is frequency-sensitive rather than amplitude-sensitive, so that the peak-to-peak amplitude of the signal is unimportant as long as the trace continues to cross the zero line. When it does not, as shown on the upper trace, the module reads it as wheel skid and fires the solenoid -- a false signal. Finally, two major changes were made. The teeth were reduced from 130 to 65 and the magnetic material was changed from Alnico 5 to Alnico 6. The latter also resolved a manufacturing problem, in that Alnico 5 frequently cracked.

With the floating sensors, the coils were wound on nylon bobbins and the wire ends were soldered to metal terminals, as shown at the top of Fig. 14. Later the lead wires were soldered to the terminals, then the stator components were put into the stator housing. With the initial design, the assembly was filled with a soft rubber-like potting. Because manufacturing equipment was not available, experimental sensors were filled with the potting material only at the terminal cavity. Several instances of wire failure were traced to motion within the potting material. Therefore, it was decided to encapsulate the coil and terminal in a more rigid material. Thermoset EL-798, a thermosetting epoxy, was selected. An encapsulated coil with its lead wires attached is shown on the bottom of Fig. 14. The sensor, developed as described, was released and is in production.

Another laboratory development and proof test fixture is shown in Fig. 15. It consists of two axle and driveshaft assemblies mounted in a test fixture. Weights, equivalent to the tire and wheel, are attached to each axle shaft flange. These and the brakes are then enclosed in the four environmental chambers shown. The propeller shafts are driven at the equivalent of 30 to 60 mph, while air-powered cylinders raise and drop the axles to produce g loads in a programmed pattern based on measurements taken on the Ford rough road

Fig. 11 - Sensor wear test

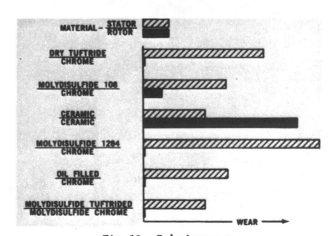

Fig. 12 - Rub ring wear

167

durability route. These shock-load and wear tests are conducted at ambient temperatures and at extremes ranging from -67F to +350F. During the ambient life test, a mixture of brake lining dust and Arizona test dust is introduced inside the brake drum. Sensor output is measured at 4 mph and 100 mph before and after the test. Rub ring wear is measured at the completion of the test.

The control module was originally packaged integrally with the actuator so that the module and actuator could be attached to a common mounting bracket and pretested as an assembly prior to delivery to the assembly plant. This arrangement proved to be undesirable because of the environmental extremes imposed on the module. In addition, nothing was gained by a proof test of this assembly. Therefore, it was decided to locate the module inside the passenger compartment, either inside the left kick panel or under the glove box, as shown in Fig. 16. The latter was released for production.

The development program revealed that it was relatively easy to design a module which performed satisfactorily down to 0.1 coefficient of friction. However, it required many hours of design considerations, computer analysis, and laboratory and car prove-out to develop a module which would self-adapt for road changes and give satisfactory performance on a complete spectrum of speed with friction coefficient from 0.05 up to dry concrete.

Performance requirements for the control module include input-output characteristics before, during, and after thermocycling at temperatures ranging from -67F to +200F over a 15-day period. The automatic equipment for checking performance characteristics is shown in Fig. 17. Other requirements pertain to vibration and impact resistance, established by measuring the loads imposed during car operation. It is interesting to note that the maximum impact load imposed on the control module occurred when the glove box door was slammed, which produced loads as high as 16.7 g. The specification we established requires satisfactory performance

after 3-1/2 hr of 30-40 g impacts of 10-15 millisec duration at the rate of 1/sec, and after smaller, varied, higher-frequency vibration loads for a period of 1 hr.

The brake actuator is vacuum-suspended, has a displacement-rod type hydraulic cylinder, and has a solenoid-operated vacuum-air valve. Fig. 18 is a schematic showing

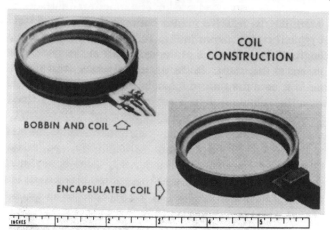

Fig. 14 - Coil construction

Fig. 15 - Sensor environmental test

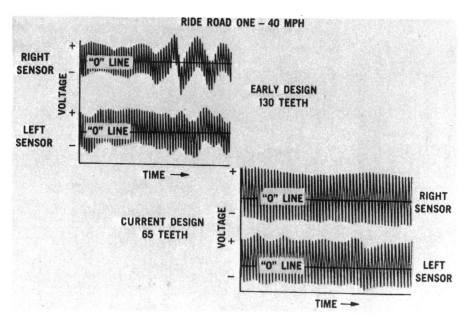

Fig. 13 - Sensor signal quality

168

how it operates. When a skid is imminent, a command from the control module causes the solenoid to close the vacuum passage to the right side of the diaphragm and to open it to atmospheric pressure. This causes the diaphragm and power plate assembly to move to the left compressing the return spring. Hydraulic pressure from the master cylinder acts on the end of the displacement rod and forces it to the left. The first short movement closes the check valve and isolates the master cylinder. Additional motion increases the volume for fluid, thereby reducing the hydraulic pressure to the rear brakes so that they release. When the wheels spin up, the control module causes the solenoid to revert to its normal position. Vacuum is applied to both sides of the diaphragm and the return spring forces the power plate, diaphragm, and displacement rod to the right. This restores line pressure and reapplies the rear brakes.

Fig 19 shows an early design actuator. During the development program several design improvements were incorporated. Among them were better sealing against corrosion for the solenoid, improved diaphragm clamping, and more free breathing. Even though no actuator was ever known to fail in the partially applied position, a position

sensitive switch was added to warn of loss of rear brakes in the event such a failure occurred.

The released actuator design is depicted in Fig. 20. Note that the power plate and diaphragm can move away from the displacement rod so that rapid cycling of the actuator cannot create a vacuum in the hydraulic system.

Originally the piston had straight sides and a spherical radius where it seats in the power plate. A taper was added to correct a bind problem that occurred after corrosion tests.

The early design actuator had pressure dump and build-up characteristics as shown at the left on Fig. 21. Computer studies indicated that ideal wheel slip could be more nearly attained with a lower rate of pressure build-up as the cut-off pressure was approached, particularly on high friction surfaces. Since car tests confirmed this conclusion, the actuator response was changed to that shown at the right on Fig. 21.

Fig. 16 - Module location

Fig. 17 - Module environmental test equipment

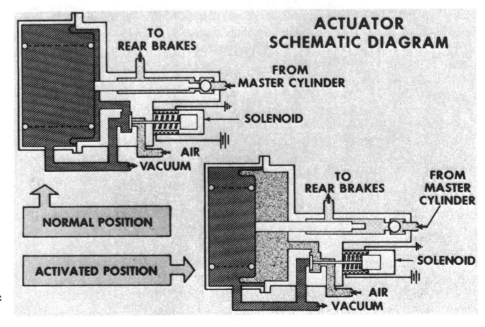

Fig. 18 - Actuator schematic diagram

169

The change in rate of pressure rise was achieved by the addition of a position-sensitive valve which restricted the vacuum passages.

Fig. 22 shows part of the test set-up for actuator validation. The test actuator is mounted in the environmental chamber shown in the foreground. A pneumatically driven master cylinder applies brake fluid pressure cycling from 0-1000 psi. On the tenth application, the pressure is held while the actuator is cycled ten times. This sequence is repeated for a total of 500,000 static and dynamic cycles at ambient temperatures. Similar cycle and soak tests are conducted at temperature extremes ranging from a low of -67F to a high of +280F. After each of these, the hydraulic and vacuum leakage and input-output response characteristics must be within prescribed limits.

Several locations were considered for the actuator. However, assembly plant processing problems dictated that the hydraulic system be attached to the chassis to permit brake system bleeding prior to front-end sheet metal decking. Therefore, the location on the right of the frame near the rear of the engine, as shown in Fig. 23, was released. A large

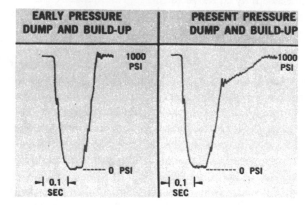

Fig. 21 - Actuator response

Fig. 22 - Actuator environmental test

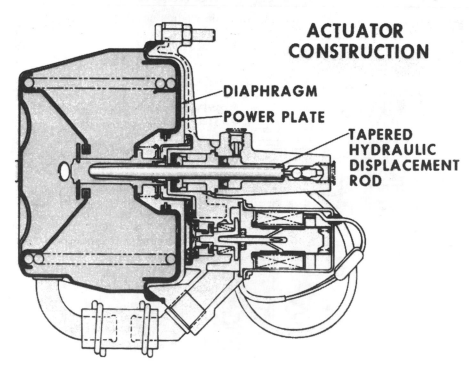

Fig. 19 - Early design actuator

Fig. 20 - Actuator construction

170

mounting bracket was developed cooperatively with assembly plant processing engineers to protect the actuator from damage during engine transmission decking and body decking.

## SYSTEM PERFORMANCE

The performance objective set for the system was to improve vehicle yaw stability, with overall stop distance no greater than obtained with four-wheel lock. The performance was determined by making a series of spike stops on each test surface with the system "off-on-on-off" and repeating the sequence until an average or statistical comparison was possible. On high friction surface tests, it was also necessary to test in a prescribed speed sequence; otherwise, results were unreliable because the base brake system had some kind of "memory" when repeatedly subjected to the same high speed, high friction stops. In fact, when high friction stops were being made on several successive days, the brakes were mildly burnished each night to overcome this phenomenon.

The performance data taken between January and November, 1968, was recently reviewed to see how many "spike" stops had been made. The result was an amazing 15,000 plus.

A car instrumented for performance test is shown in Fig. 24. Recording instrumentation is required to determine consistently the stop distance with reasonable accuracy. Fig. 25 is a representation of the data that must be taken. Vehicle velocity is measured with a 5th wheel. Hydraulic pressure is measured for two reasons; it shows if the system being tested affects the rate of brake application and it gives a starting point for measuring distance. When the data are reduced, stop distance is measured to the closest 0.5 ft using the point where the line pressure reaches 600 psi as the "start-of-the-stop." This distance is then corrected based on the square of the desired initial velocity divided by the actual initial velocity measured to within 0.2 mph. The need for this accuracy becomes obvious when we consider that on a stop from 60 mph an error of 1 mph in speed, 0.2 sec in point of application, and 1 ft in distance add up to more than a 10% error in stop distance.

During most development tests additional data were taken so that the performance of the system could be evaluated. A typical tape is shown in Fig. 26. The system is "off" (conventional brakes) at the top and "on" (Sure-Track system operational) at the bottom. In addition to the information previously mentioned these tapes show individual wheel velocity, vehicle deceleration, and the solenoid signal. Yaw is not shown here, but was usually measured by means of a gyroscope and recorded on the same tape. When appropriate, other data such as manifold or actuator vacuum, front brake line pressure, sensor signal, and engine rpm were taken.

Friction surfaces ranging from dry concrete down to about 0.05 were available on the Ford proving grounds. The low friction was obtained by using Hydrolube and water on a plastic sheet. The validity of development tests on this surface was confirmed by tests on ice at Dearborn, International Falls, Minn. and on the Olympic ice rink at Squaw Valley, Calif.

Almost any reasonable anti-lock system will show an improvement in vehicle yaw control, but it is extremely difficult also to reduce stop distances over wide ranges of speed and friction encountered in customer service. Fig. 27 shows a statistical plot of yaw during 60 mph stops on wet asphalt with conventional and Sure-Track brakes. Similar data for 20 mph on Hydrolube is shown in Fig. 28. While it is true that on a given stop any car may have relatively little yaw,

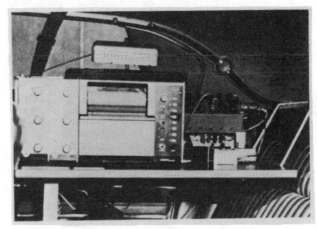

Fig. 24 - Car instrumented for performance test

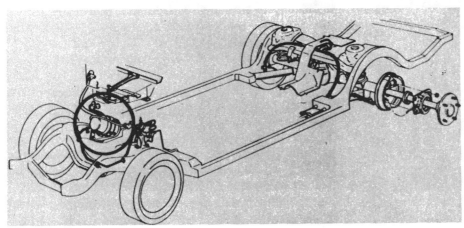

Fig. 23 - Actuator mounting

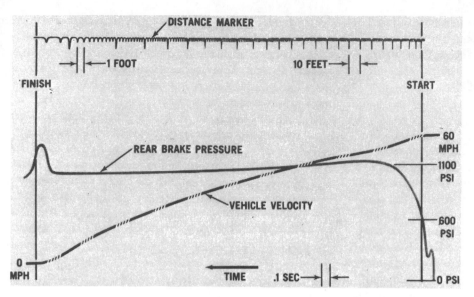

Fig. 25 - Stop distance measurement

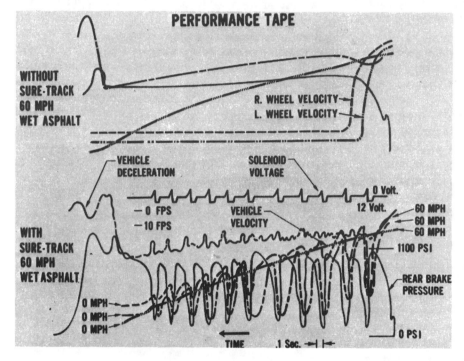

Fig. 26 - Performance tape

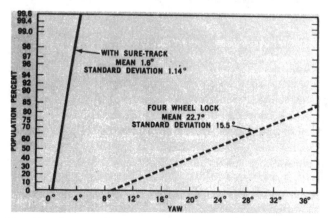

Fig. 27 - Improved yaw control-stopping from 60 mph on wet asphalt

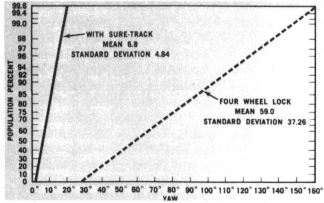

Fig. 28 - Improved yaw control-stopping from 20 mph on Hydrolube

these charts show that the Sure-Track system provides significant yaw reduction on an overall basis. Even more remarkable is the reduced yaw when stopping with high friction on one side of the car and low friction on the other. This significant improvement in yaw control is achieved with a system which also reduces the overall stop distance, compared to four-wheel lock, as shown in Table 1.

The upper portion of Table 1 shows average distances with 4-wheel lock and Sure-Track under a wide range of friction and vehicle speed conditions. Regular production tires were used on these tests, including determination of the coefficient of friction. The procedures previously described were used in determining the stop distances. As previously indicated, the yaw improvement due to Sure-Track is remarkable. On extremely low friction surfaces, such as ice, and high friction surfaces such as dry pavement, 4-wheel lock and Sure-Track stop distances are essentially equivalent. However on moderately low friction surfaces and at 60 mph on medium friction surfaces, the reductions in stop distance due to Sure-Track are impressive.

The lower portion of the table shows the benefit of improved vehicle stability when making panic stops with significantly different friction on the two sides of the car. This is equivalent to making panic stops with the two wheels on one side of the car on snow and the two on the other side on clear pavement. With 4-wheel lock the car goes out of control and yaws more than 90 deg in all such stops, while with Sure-Track yaw is significantly reduced even under these extreme conditions.

SUMMARY

The use of anti-lock braking system on aircraft led to efforts to adapt such a system to road vehicles, or to develop a unique system. Experimentation with several systems led to a set of requirements necessary for feasible vehicle anti-lock system development. Refinement of one particular system led to its introduction as an option early in the 1969 model year.

More than 15,000 on-the-road braking tests plus laboratory and computer analysis of preproduction units have assured attainment of system reliability, manufacturing feasibility, and substantially improved braking performance. Road tests have shown that yaw stability during panic braking is controlled to close limits under all conditions and that stopping distance is the same or less in most braking situa-

Table 1 - Performance Data

| Initial Speed mph | Average Friction | Mark III | | | | Thunderbird | | | | |
| | | Stop Distance, ft | | Yaw, deg | | Average Friction | Stop Distance, ft | | Yaw, deg | |
| | | 4-Wheel Lock | Sure-Track | 4-Wheel Lock | Sure-Track | | 4-Wheel Lock | Sure-Track | 4-Wheel Lock | Sure-Track |
|---|---|---|---|---|---|---|---|---|---|---|
| 10 | 0.07 | 47.6 | 48.3 | 4 | 3 | 0.05 | 66.0 | 68.6 | 5 | 3 |
| 18 | 0.07 | 152.8 | 148.2 | 6 | 4 | 0.05 | 197.0 | 200.3 | 10 | 6 |
| 20 | 0.21 | 62.6 | 55.4 | 4 | 2 | 0.21 | 63.6 | 58.3 | 7 | 2 |
| 30 | 0.16 | 187.1 | 163.8 | 6 | 2 | 0.19 | 186.8 | 170.6 | 11 | 3 |
| 20 | 0.83 | 15.9 | 16.8 | 1 | 1 | 0.71 | 18.7 | 19.4 | 2 | 1 |
| 40 | 0.66 | 80.4 | 80.8 | 7 | 2 | 0.56 | 91.7 | 91.7 | 4 | 2 |
| 60 | 0.51 | 233.4 | 213.6 | 29 | 2 | 0.47 | 251.7 | 234.6 | 23 | 2 |
| 20 | 0.80 | 16.5 | 16.9 | 2 | 1 | 0.83 | 15.9 | 16.0 | 2 | 1 |
| 40 | 0.74 | 71.8 | 71.4 | 5 | 1 | 0.75 | 70.4 | 71.6 | 8 | 2 |
| 60 | 0.72 | 164.1 | 166.8 | 15 | 2 | 0.71 | 166.4 | 163.5 | 15 | 2 |

Split Friction Performance With Sure-Track System*

| Speed, mph | Surface Friction | Mark III | | | Thunderbird | | |
| | | Yaw Angle, deg | | | Yaw Angle, deg | | |
| | | Minimum | Maximum | Average | Minimum | Maximum | Average |
|---|---|---|---|---|---|---|---|
| 25 | Right 0.85 Left 0.25 | 5 | 30 | 17.8 | 6 | 45 | 20.7 |

*Right side wheels on 0.85 friction surface; left side wheels on 0.25 friction surface with 4-wheel lock, yaw on split coefficient surface exceeded 90 deg in all cases.

tions, regardless of the coefficient of friction of the road surface or speed at the time brakes are applied.

REFERENCES

1. U.K. Patent 382,241 10/20/32 filed 5/26/32. Werner Mom, "An Improved Safety Device for Preventing Jamming of the Running Wheels of Automobiles when Braking."

2. U.S. Patent 2,914,359 11/24/59 filed 12/1/55. Gordon W. Yarber, "Anti-Skid Brake Control System."

3. U.S. Patent 2,930,206 3/22/60 filed 3/4/57. J. R. Steigerwald, "Skid Warning System."

4. U.S. Patent 3,245,727 4/12/66 filed 8/15/62. J. S. Anderson et al, "Anti-Skid Brake Control Systems."

5. U.S. Patent 2,906,376 9/29/59 filed 3/6/57. J. M. Zeigler, "Non-Skid Braking System Using Pre-Set Pulsing Action."

6. H. W. Kummer & W. E. Meyer. "Measurement of Skid Resistance." Special Tech. Publication No. 326, Symposium on Skid Resistance ASTM, 1962.

7. Final Report, Skid Control Performance Specification, GSA Contract No. Gs-00S-60857.

8. J. L. Harned and L. E. Johnston, "Anti-Lock Brakes." G.M. Engineering Staff, presented at G.M. Automotive Engineering Seminar, July 1968.

9. J. E. Gibson and J. S. Meditch, "A Class of Predictive Adaptive Controls." Technical Documentary Report No. ASD-TDR-61-28 f.i., Sept. 1965, Flight Control Laboratory, Wright-Patterson Air Force Base.

10. V. W. Eveleigh, "Adaptive Control Systems." Electro Technology, April 1963, pp. 80-98.

# The Chrysler "Sure-Brake"—The First Production Four-Wheel Anti-Skid System*

**J. W. Douglas**
Chrysler Corp.
**T. C. Schafer**
Bendix Corp.

*Paper 710248 presented at the Automotive Engineering Congress and Exposition, Detroit, Michigan, January, 1971.

IN 1966 CHRYSLER became very interested in testing and developing production designs of skid control brake systems. Research work had been in process since 1957 but not on production designs. Now various suppliers were moving forward in developing such devices, and it was obvious that it was just a matter of time before a system was sufficiently advanced to warrant development on a specific system with the objective of installation on a volume production vehicle. The potential benefits obtained by such devices were well known, the most important being directional stability and control during braking. It was recognized that a rear wheel system would keep the vehicle travelling in a straight line but it would have no steering capabilities. A four-wheel system, although having considerable complexity, would have the tremendous advantage of retaining steering control. A further potential advantage lay in obtaining shorter stopping distances on many surfaces where rolling friction was higher than sliding friction.

## BACKGROUND

Knowing the potential was one thing; how to utilize it was another. Some of the alternatives to be considered were:
1. Four-wheel or rear wheel only?
2. Mechanical inertia or electronic sensors?
3. Should the rear wheel accelerations be sensed at the wheels or at the drive shaft? If sensed at the drive shaft, which end would be best—the speedometer cable pick-up or the differential housing?
4. Should the modulator power source be vacuum, hydraulic, electric, or compressed air?
5. If vacuum power was used, could the inevitable large pressure reducers or modulators be packaged?
6. Could a single, combination power-brake booster and modulator be developed to provide satisfactory performance at an obviously lower cost?

---

────ABSTRACT────

The paper outlines testing, development, and operation of the first production four-wheel slip control system for passenger cars in the United States. The Chrysler Corp. calls the system "Sure-Brake," but it is more generally known as "anti-skid."

The first portion of the paper deals with considerations that led Chrysler into the Sure-Brake system, the philosophy behind the system, and a detailed explanation of its operation. The second portion deals with the development and testing of the system, leading to its release as an option on the 1971 Imperial.

The testing program introduced a new dimension to brake engineering. Before the advent of wheel slip control systems, many thousands of brake tests were conducted but were always terminated at the point of skid. These tests were also conducted mainly on black top or concrete roads. For the first time, thousands of stops were made at maximum deceleration on every available surface. The paper lists the results obtained and attempts to pass on some of the lessons learned in handling skidding vehicles.

Many of these questions were resolved by testing various suppliers' systems at the Chrysler Proving Grounds. It was finally decided that the system would be vacuum powered with an electronic logic controller and have an electromagnetic sensor at each wheel. Initially, no decision was made whether to develop a rear-wheel or four-wheel system, and so parallel programs were carried out for several months.

As experience was gained, it was concluded that to provide a truly advanced braking system, a four-wheel system was needed. The rear-wheel-only device, while generally keeping the vehicle traveling straight, had several deficiencies.

1. There was absolutely no steering control with the front wheels locked; no matter which way the driver steered, the vehicle would continue to travel straight.

2. The vehicle on a crowned or cambered road or under heavy crosswind conditions would tend to drift sideways, especially on a low-friction surface, where it could collide with a parked vehicle.

3. Front tires could be easily damaged on higher coefficient surfaces since they would lock under "panic type" braking situations.

4. Stopping distance was generally longer than that obtained by use of a four-wheel system.

5. Under split coefficient conditions, the vehicle would drift to the higher coefficient side and corrective action was impossible.

Having examined the problem, it was then possible to formulate a few ground rules. These were:

1. The vehicle must have good steering control and stability under all braking conditions.

2. Stopping distances must be less than that obtained by a skilled driver stopping in the minimum distance without skidding his tires.

3. The braking system must be unaffected by the weight distribution of the vehicle.

4. The system must be capable of installation on the normal production line as a vehicle option.

5. The system must be in production for the 1971 Imperial. Given the preceding ground rules, it was evident that:

1. It had to be a four-wheel system.

2. The sensing system could be either at each individual wheel or at each front wheel and one rear axle. It was finally decided to install one sensor per wheel.

3. The simplest system that could be developed in time had to be vacuum powered. Hydraulic power was definitely a possibility, but not at this stage.

The Bendix system offered the greatest potential of attaining these objectives. A joint test and development program was then instituted with The Bendix Corp. to reach the production date with an effective, thoroughly tested durable system.

To understand better how these requirements are met, it seems appropriate to discuss the basic tire-to-road relationship. These characteristics are depicted by the $M_u$ slip curve. Fig. 1 shows a typical plot of the retarding force characteristic and the lateral force characteristic.

The retarding force of the tire generally exhibits a rather sharp rise as wheel slip increases, peaking at some point, and

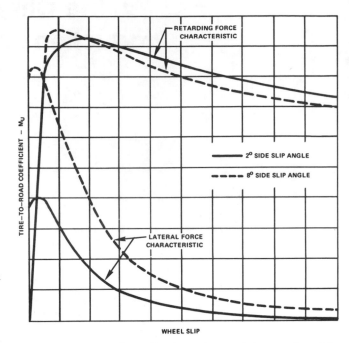

Fig. 1 - Typical plot of retarding force characteristic versus lateral force characteristic

then falling off as the wheel approaches a locked condition. This does not change appreciably with side slip angle, at least in the region most commonly encountered. Typically, then, the Sure-Brake system controls braking so as to stay at or near this peak, resulting in maximum retarding force and, thereby, reduced stopping distance. The lateral force characteristic varies considerably with both wheel slip and side slip angle. It also peaks as wheel slip increases, but at a lesser value than the retarding force, and then diminishes rapidly as wheel slip increases, as shown on Fig. 1. Since maximum retarding force and maximum lateral force cannot exist simultaneously the Sure-Brake system operates in such a manner as to provide maximum lateral force consistent with maximum retarding force. The lateral force characteristic provides vehicle stability in the case of the rear wheels and vehicle steerability in the case of the front wheels.

Fig. 2 is intended to clarify the action of the wheel as it is cycled back and forth about the peak. The upper line, $V_v$, represents vehicle speed and the two lower lines, $S_1$ and $S_2$, represent the limit of wheel speed as it cycles about the peak retarding force. $V_w$ shows the wheel oscillation between these limits.

Stability refers to the resistance to sideways movement or spinning of the vehicle, such as that which occurs when the rear wheels lock on a side-to-side unbalanced coefficient or in a curve. Controllability, to differentiate from stability, is primarily a front wheel function and refers to the ability to steer the vehicle under maximum braking conditions.

In order to achieve the previously mentioned objectives, the Sure-Brake system controls each front wheel individually and the rear wheels as a pair. This results in three control channels, each consisting of a pressure modulator, electronic con-

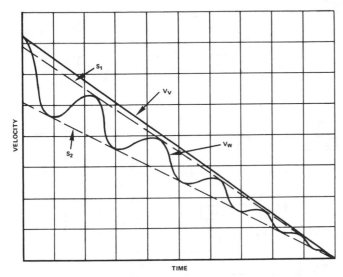

Fig. 2 - Wheel action, velocity versus time

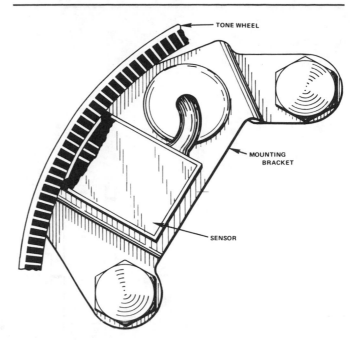

Fig. 4 - Front sensor

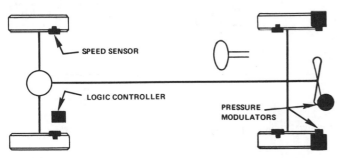

Fig. 3 - Hardware arrangement of Sure-Brake system

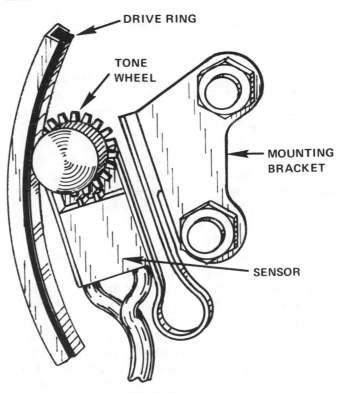

Fig. 5 - Rear sensor

trol circuit, and a speed sensor. Fig. 3 shows the hardware arrangement. In the case of the rear wheels, it is necessary to select which wheel will control that channel. A "select low" approach proved to be the most desirable scheme for good overall performance. This means that the rear wheel approaching the locked wheel condition first controls the operation of the rear channel.

Each of the three Sure-Brake channels senses the appropriate wheel speed and uses this intelligence as the input signal to the logic controller. The logic controller contains the program by which the system operates. When it determines wheel lock is imminent, signals are sent to the control valves on the appropriate modulator. The pressure modulator provides the muscle to adjust brake pressure so as to keep the wheel at or near peak retarding force.

The three channels of the Sure-Brake system operate independently of each other. The only interconnection is with respect to failsafe. The failsafe system is designed to negate function in all three channels if a Sure-Brake failure is detected in any one channel. When failure is detected, the brake failure light is illuminated and latched "on" until the ignition is de-energized. As discussed earlier, only those failures that produce reduced braking over that achievable with a standard brake system are detected and indicated.

## "SURE-BRAKE" COMPONENTS

SPEED SENSOR - The speed sensors are electromagnetic devices which continuously monitor each wheel speed. Due to design variances and deflection differences between the front and rear brakes, two sensor designs were implemented.

The front sensors shown in Fig. 4 are a "frictionless" type

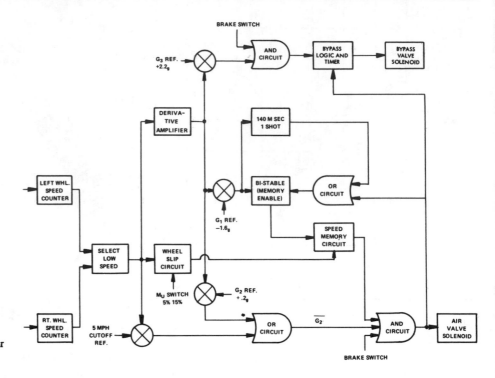

Fig. 6 - Functional block diagram for logic controller

design in that there are no wearing surfaces. The sensor, consisting of a magnet, coil, and toothed pole piece, is encased and mounted to the stationary knuckle. The sensor is opposed by the tone wheel, which is a large toothed-wheel fixed to the rotating disc and hub assembly of the front brake. All teeth in the tone wheel are encapsulated to prevent foreign objects from lodging in the voids and causing possible damage. The gap between the sensor and tone wheel is adjustable and set at 0.020 in. This arrangement yields a minimum output of 0.5 V peak-to-peak at 5 mph, and increases in voltage output as speed increases. The sensor output frequency is approximately 42 Hz/mph, which provides sufficient flow of information to obtain good system control.

In the case of the rear sensors, sizable deflections can occur between the rotating and nonrotating members of the rear brake and axle assembly. These large deflections made it impractical to use the "frictionless" approach. Therefore, a "friction drive" design was implemented, as shown in Fig. 5. The rear sensor provides the same voltage and frequency characteristics as the front sensor. The rear sensor and tone wheel are mounted on a spring loaded bracket, with a fixed air gap between the sensor and tone wheel. The tone wheel when assembled on the vehicle rides on the inside diameter of the rubber drive ring and can easily absorb run-out variances of 0.100 in.

The drive ring, being attached to the axle flange, rotates as the vehicle wheel turns, resulting in a corresponding rotation of the tone wheel. The tone wheel has a light knurl on the outside diameter to provide positive engagement between the tone wheel and drive ring.

## LOGIC CONTROLLER

The logic controller is the "computing module" or "brains" of the Sure-Brake system. It receives input signals from the speed sensors, processes these signals, and issues commands to the pressure modulator control valves when wheel lock-up is imminent. Fig. 6 is a functional block diagram for the logic controller. Fig. 7 shows the interrelationship between wheel speed, wheel acceleration, and brake pressure, and is included to clarify system operation further. Referring to these two figures, we will now explain the operation of the logic controller. The functional diagram shows the "select low" logic, referred to earlier, which applies to the rear axle operation only. The front wheel operation utilizes identical logic, except the wheel speed selection circuitry is not necessary.

Wheel speed enters the logic controller where it is processed and differentiated electronically. When the wheel deceleration signal reaches 1.6 g, the system memorizes wheel speed. This memory is maintained for approximately 140 ms. If during this time interval the wheel speed falls a predetermined amount below this memorized value when the brake is applied, the command is issued to decay or reduce brake pressure. If not, the memorized speed signal is cancelled. The percentage change in speed from the memorized value required to decay pressure is either 5 or 15%, depending on vehicle deceleration. Below approximately 16 ft/sec/sec vehicle deceleration, the 5% value exists; above 16 ft/sec/sec it is 15%. This adaptive feature was found necessary to obtain good performance under all tire-to-road conditions.

As brake pressure is reduced, the wheel ceases to decelerate

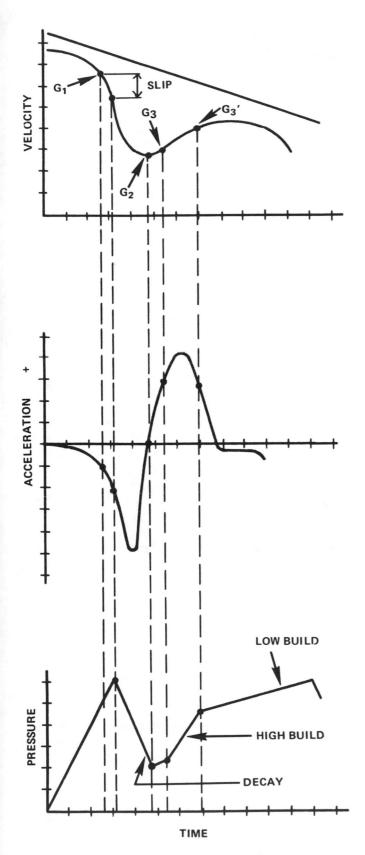

Fig. 7 - Interrelationship between wheel speed, wheel acceleration, and brake pressure

and starts to accelerate. When it reaches a slightly positive acceleration value, such as 0.2 g, pressure decay is discontinued and a low pressure build rate initiated. If the wheel acceleration exceeds a value of 2.2 g, the system builds pressure at a more rapid rate. This fast build-rate continues until the wheel acceleration falls below the 2.2 g level, at which time the system reverts to the low pressure build-rate until the wheel reaches incipient lock, when the cycle repeats.

The system incorporates a low speed cutout below 5 mph. This function is required since sensor output becomes unusable at low speeds.

## PRESSURE MODULATOR

The pressure modulator shown schematically in Fig. 8, provides the muscle to cycle brake pressure upon receipt of signals from the logic controller. In the normal de-energized position, the hydraulic check valve is held open by the displacement plunger, allowing brake fluid to pass freely from the master cylinder to the wheel cylinders. This plunger is spring loaded to withstand a minimum pressure of 1450 psi before retracting due to fluid pressure alone. During operation of the vehicle, the modulator is evacuated by use of engine vacuum. A vacuum check valve is incorporated in the vacuum inlet line to the modulator for the purpose of maintaining a high vacuum level in the modulator and to prevent gasoline fumes from entering the modulator on engine shutdown or during acceleration.

During Sure-Brake operation, the pressure modulator reduces and increases brake pressure through operation of the solenoid control valves. As brake pressure increases sufficiently to cause incipient wheel lock, the control system energizes both control valves. This causes the normally closed air valve to open and the normally open bypass valve to close. In this condition, atmospheric pressure is admitted to the front side of the diaphragm, creating a differential pressure. As this differential pressure increases, it overcomes the spring preload and retracts the pneumatic portion of the modulator. As the displacement plunger follows, the hydraulic check valve closes terminating communication between the master cylinder and the wheel cylinder. Continued movement of the displacement plunger creates additional volume in the wheel cylinder side of the brake system, causing brake pressure to decrease. As can be seen in the diagram, the displacement plunger is not fixed to the pneumatic piston, thus preventing the possibility of developing a negative pressure in the brake line, which could admit air into the brake system through the wheel cylinder seals.

After the brake pressure has been reduced sufficiently to allow the wheel to accelerate, the Sure-Brake system reverts to either "low build" or "high build," depending on the magnitude of wheel acceleration. High build is accomplished by de-energizing both valves, preventing the entrance of air through the air valve and opening a relatively large orifice between the front and back side of the pneumatic piston by use of the bypass valve. This causes rapid equalization of pressure across the diaphragm, resulting in rapid advancement of both the

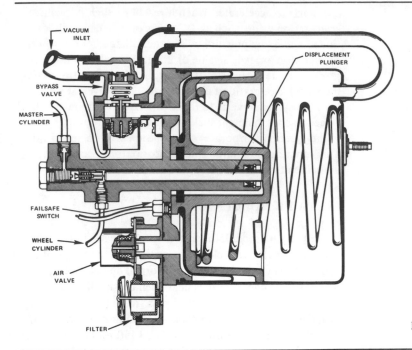

Fig. 8 - Pressure modulator schematic

Labels on figure:
VACUUM INLET
DISPLACEMENT PLUNGER
BYPASS VALVE
MASTER CYLINDER
FAILSAFE SWITCH
WHEEL CYLINDER
AIR VALVE
FILTER

pneumatic and hydraulic pistons due to the modulator spring, thereby increasing brake pressure at a rapid rate. If the wheel acceleration is low, the air valve is de-energized and the bypass valve is kept energized. A small bleed in the bypass valve short circuits the main orifice, allowing the differential pressure across the diaphragm to diminish at a slow rate. The resulting action builds brake pressure slowly.

The electrical switch on the modulator is part of the failsafe function. This switch is normally open when the modulator is in the de-energized position. When closed in the absence of a brake light signal, this indicates that the modulator has erroneously retracted. This switch signal results in deactivation of the system electronically and the subsequent illumination of the brake failure light.

To insure that the modulator receives frequent exercising to keep the interface between the displacement plunger and the seal lubricated, the system incorporates a "power-on transient." This function energizes the modulators for approximately 20 ms every time the engine is started with the brakes applied. This time interval is adequate to fulfill the intended exercise function with negligible vehicle motion on steep grades.

To prove out the Sure-Brake system, extensive testing was conducted. The following sections review this test activity and some of the associated problems.

## VEHICLE TESTING

Testing soon showed that the American passenger car is indeed a very forgiving vehicle and will not turn over under any provocation, providing it remains on a hard flat surface. Thousands of skid stops up to 80 mph have been made without tipping a car over although vehicles have often spun through as much as 390 deg.

One of the most useful development tools in the early stages was a split coefficient of friction surface consisting of a sheet of wetted stainless steel adjacent to a dry concrete road. This simulates conditions such as driving with two wheels on a gravel shoulder or on a patch of ice while the other wheels are on a high-friction surface. On hitting the brakes hard, the difference in tire drag produces a torque on the vehicle, pulling it towards the high-friction surface such that the vehicle turns crossways on the road. This action is extremely rapid and one for which the normal driver cannot correct with a conventional vehicle.

With Sure-Brake the vehicle experiences a pull toward the high-friction surface, but this pull can be easily corrected and the vehicle can be steered readily in any direction.

Steering control is one good measure of the usefulness of a wheel slip control system, but it is not always enough to steer out of an accident. The vehicle must also stop, and so stopping distance tests are conducted on the various surfaces at the Chrysler Proving Grounds at different speeds. We compared stopping distances under three conditions: locked wheel, Sure-Brake control, and a skilled driver and minimum stopping distance without locking wheels.

Table 1 gives the average values obtained from approximately 1000 stops. From the table, it can be seen that in all cases where we have comparative figures, the Sure-Brake control system provides shorter stopping distances than were obtainable by the skilled driver.

The skilled driver results reported in this paper were obtained by "seat of the pants" evaluation with no decelerometer or pressure gage aids.

Initially, stopping distance was measured by means of a gun firing a chalk pellet onto the pavement upon applying the brake pedal. The stopping distance was then measured by a tape measure and the distance recorded. This method was

Table 1 - Average Stopping Distance

| Speed, mph | Surface Condition | Locked Wheels, ft | Sure-Brake, ft | Skilled Driver, ft |
|---|---|---|---|---|
| 30 | Dry concrete | 42.2 | 43.3 | 56.2 |
| 60 | | 159.2 | 165.3 | 192.5 |
| 80 | | 306.9 | 306.1 | 345.1 |
| 30 | Wet concrete | 42.7 | 43.1 | 56.2 |
| 60 | | 199.8 | 176.5 | 215.2 |
| 80 | | 442.7 | 347.6 | 427.5 |
| 30 | Gravel | 52.4 | 64.5 | 76.3 |
| 60 | | 200.6 | 265.0 | 289.7 |
| 45 | Wet Jennite | 314.7 | 185.5 | 263.7 |
| 45 | Washboard road | 93.0 | 77.4 | 110.3 |
| 45 | Snow and ice | 255.2 | 226.9 | 260.5 |
| 60 | | 424.5 | 417.3 | 521.0 |

time consuming, requiring two men, and with the advent of cold or wet weather a very unpleasant job. The Chrysler Corp. Instrument Laboratory built fifth wheels with a null speed indicator to provide accurate, preset, initial test speed readings. A four-digit electronic counter with a "NIXIE" tube display is used for reading the stopping distance from the instant of the initial movement of the brake pedal to rest. This enabled the test to be conducted with one driver who could record the results without leaving the car. A further modification made by brake engineering was to build an air cylinder device which lifted the wheel clear off the ground should the vehicle rotate more than 60 deg while skidding. This innovation has saved many very expensive fifth wheels.

The biggest problem was the wide variation obtained in stopping distance results. To reduce this effect to a minimum, it is necessary to run 10 stops with and 10 stops without Sure-Brake control, using the average to determine the reduction in stopping distance. This variation exists equally with locked wheels and Sure-Brake control stops and appears to be due to driver variation in reading the vehicle speed indicator, the driver brake application rates, and the response in the power brake booster.

To facilitate gathering of data, the distance of each stop is recorded on a punched card along with other information such as vehicle weight, tire size, material, tire manufacturer, ambient temperature, surface, and speed. This information is used to compute average stopping distance and for determining subtle changes caused by revisions to various components.

During the course of this work, a certain amount of knowledge has been acquired in making hundreds of locked wheel stops from speeds of 60 and 80 mph without accidents or problems. Accordingly the following simple rules have evolved.

1. Seat belts are to be used at all times and are to be properly tightened.

2. Doors are to be locked.

3. Once the brake has been applied, it must be fully on until the vehicle comes to rest.

Items 1 and 2 are normal precautions all drivers should take.

Item 3 is not so obvious, and yet is the most important. Newton's first law of motion states that every body continues in its state of uniform motion in a straight line except in so far as it may be compelled to change that state by the action of some outside force.

If a vehicle has all four wheels locked, it may start to spin due to differences in tire-road friction values across the road width. It will still continue to travel along the highway virtually in a straight line and finally halt, usually on the highway and rarely more than a few feet over the shoulder. The vehicle tends to drift down the road camber, but should it not halt before actually running onto a gravel shoulder, the higher friction value of the highway will tend to pull it back onto the road. As long as the vehicle stops on the flat road, it cannot roll over.

If one examines many accidents in which vehicles have halted many yards from the highway or have crossed into oncoming lanes, it is apparent that in many cases the driver locked up the brakes. Then, upon starting to spin, the driver released the brakes. The effect of releasing the brakes causes the vehicle to charge off in the direction it is pointed, provided the vehicle has not turned beyond steering limits. Under such conditions high inertia forces are placed upon the driver, which can move him from the vehicle controls unless he is suitably restrained.

As mentioned before, with front wheels locked and rear wheels revolving, the vehicle will skid in a straight ahead position. Should the rear wheels lock with the front wheels still revolving, the vehicle is virtually uncontrollable and requires a high degree of skill to keep it within a traffic lane. Frequently a steering maneuver with this condition, especially on wet surfaces, is almost certain to spin the vehicle around.

## STOPPING DISTANCE TEST SURFACES

In order to develop the Sure-Brake system, hundreds of stops were made by many test cars on a wide variety of surfaces.

DRY CONCRETE - A vehicle can stop in a comparatively short distance with a maximum of noise and smoke with the wheels locked on dry concrete. Six locked wheel stops from 80 mph are usually sufficient to wear holes through new tires. Sure-Brake stops can be run continuously without flat spotting the tires, and the vehicle is always under steering control. The vehicle without wheel slip control may spin and halt facing the direction from which it came. The condition of the tire tread is unimportant. Worn smooth tires provide as short a stopping distance as new treads.

WET CONCRETE - The condition of the tread is most important. The stopping distances listed relate to new or partially worn tires. Bald tires would have greatly increased the stopping distance because of lack of drainage allowing a wedge of water to form. However, the comparison between locked wheel, Sure-Brake, and skilled driver would have been unchanged. The Sure-Brake stop would still be shortest.

WASHBOARD BUMPS - Initially, very low rates of deceleration were obtained on washboard bumps. These bumps are normally encountered on gravel roads, but at the Chrysler Proving Grounds a chatter bump stretch made of asphalt on the vehicle endurance road was used and provided a constant test surface.

The wheel or wheels bouncing upward tended to lose traction with the road, and due to brake torque would tend to lock. The sensors sent this intelligence to the logic controller which, in turn, reduced pressure to that wheel or axle, thus lowering the deceleration of the vehicle to a very low level. This problem was resolved by modifying the circuitry in the logic controller. To prove that the changes in the logic controller had corrected the problem, it was necessary to make hundreds of stops at maximum deceleration over the washboard surface road. The vehicles were stopped from speeds of 20-60 mph at 5 mph increments; 20 stops at each speed were made. It says much for the toughness of the vehicles that they were able to withstand such poundings.

JENNITE SEALED ROAD - Jennite is an asphalt sealer and has the characteristic of producing a very low sliding-coefficient of friction when wet and a high rolling-coefficient of friction when wet. This test surface represents conditions encountered in the summer on some asphalt roads when the surface becomes polished and oily during periods of dry weather, then the first downpour of rain produces a low-friction surface until the rain has washed off the slick film. The stopping distance of the car with Sure-Brake control is reduced by almost half compared with the locked wheel condition. This surface is one of the few in which a skilled driver pumping the brakes can indeed beat the stopping distance of the locked wheel car.

GRAVEL ROADS - It was found that the shortest stopping distance was obtained by hitting the brake hard, locking up all four wheels and plowing a deep groove in the road. During such a stop, there is, of course, no steering control but stopping distance is short.

There appears to be no way that a rolling wheel on top of gravel can produce as much resistance as one in a snowplowing situation. The vast majority of customers, however, could be better served with a system assuring steering control at all times on gravel and still providing decelerations greater than obtained without skidding wheels.

SNOW AND ICE COVERED ROADS - Table 1 lists average stopping distances and gives relative data for locked wheel stops and Sure-Brake stopping distance results, but the numbers themselves depend too much on variables such as depth of snow and temperature. In a group of 113 locked wheel stops, it was found that the stopping distance from 30 mph to rest varied 82-213 ft. It is certain that more tests would provide even wider results.

SPLIT COEFFICIENT OF FRICTION SURFACE - This facility readily duplicates two common situations. In the winter, it is not unusual to find patches of ice or snow on an otherwise bare dry road. Also many highways have gravel shoulders. Should a vehicle straddle such surfaces with two wheels on one side on a low-friction surface and the two others on the high-friction surface, a locked wheel stop will immediately rotate the car, sometimes through 180 deg. For

such a test, it is important that the brake not be released until the vehicle has halted; otherwise, the vehicle will charge off in the general direction it happens to be pointed. To reproduce this split coefficient of friction surface, a 80 ft × 3 ft sheet of stainless steel was laid down adjacent to a concrete road. With the stainless steel strip wetted, the friction values correspond closely to a sheet of ice.

With this facility, the effectiveness of the wheel slip control system was determined. Tests showed that a vehicle without a wheel slip control system will spin out. Further tests revealed that a rear-wheel only system will stop virtually in a straight line with 2 or 3 ft deviation. A four-wheel slip control system is capable of steering straight ahead or indeed in any direction, including lane changing.

Since a vehicle without an effective wheel slip control system will spin out as soon as sufficient hydraulic pressure has been delivered to the brakes, this split coefficient test proved to be an excellent screening tool for various systems offered.

## SYSTEM PROBLEMS

The majority of system problems stemmed from salt and water causing a leak path from the 12 V system to the millivolt circuits of the wheel control system. The solution was to isolate high-voltage connections from low-voltage ones and, where possible, install all connections in dry areas.

To resolve the problems, we used an eight channel light beam oscillograph to record the events occurring throughout the stop. The various readings taken during test stops were:

1. Master cylinder pressure.
2. Brake pressure.
3. Sensor output from each wheel, being proportionate to speed; this showed the action of each wheel.
4. Modulator air valve and bypasss valve actuation.
5. Internal signals from the logic controller, giving the acceleration of each wheel.

By examining these various signals, or absence of such signals, it was possible to identify the problem, and by making various changes, eliminate the trouble.

In addition, to keep it dry, the logic controller was moved from behind the right headlamp to the trunk. As always, this solution, while removing most of the problems, introduced several new ones. The sensor leads now had to run to the rear and were adjacent to the 12 V power supply lead; they began to pick up induced currents which, in turn, were relayed to the logic controller as erroneous signals. This caused the system to fire unnecessarily. This problem was finally solved by shielding the sensor wires. By this time the system was quite reliable, but on occasions a wheel would skid for no apparent reason. The drivers on the night shift were also finding this problem, but much more frequently. Due to its random malfunctioning, the problem was difficult to isolate and resolve. Finally, it was discovered that the problem was accentuated by the radio. The night shift drivers had the radio on, while the engineers trying to discover the failure turned it off to give their full attention to the problem. This particular problem was finally traced to the electronic voltage regulator which was feeding in "hash" to the electrical system which, in turn, caused the air valve to shut off prematurely and allowed the wheel to go to skid. A change was made to the electronic circuit which corrected this problem by filtering out these random spikes.

The other parts of the system were comparatively trouble-free, although some speed sensor failures were caused earlier by leaking bearing seals. This problem was corrected by improved sealing design.

The modulators which do the actual work of reducing or modulating the pressure have proved to be trouble-free whether on test cars cycled frequently or on endurance cars cycled rarely. Some corrosion was encountered on air valve springs and piston, but a change to stainless steel for these parts corrected that problem.

## STARTING ON STEEP GRADES

To ensure that the modulators do not deteriorate through years of inactivity, the modulators are cycled each time the vehicle is started with the brake pedal depressed. During a normal start cycle there is no problem; the duration of the cycle is so short with normal brake pressures that nothing happens. However, if one turned off the engine and applied the brake pedal several times to exhaust the power brake booster, considerable brake pedal pressure would be required to generate 350-400 psi, the pressure required to just hold the vehicle on the 32% grade. Under these circumstances, on restarting the vehicle the modulators cycled in the normal manner and the reduction in hydraulic pressure permitted the vehicle to roll downhill under the force of gravity until the engine started restoring the power brake booster to full effectiveness. The problem was corrected by logic controller circuit changes, thus cutting down the rollback to a maximum of 8 in. In comparison, a manual transmission car on stalling rolled 19 in. before the driver could stop it. It must be emphasized that with the parking brake applied or with normal vacuum levels in the power brake booster, no rollback of the vehicle takes place.

In addition to the normal brake engineering development program at the Chrysler Proving Grounds, endurance testing was carried out by other departments. The normal cycle consisted of driving for 50,000 miles over a wide variety of roads from concrete and asphalt to gravel, including washboard bumps, widely spaced bumps, smooth concrete, brine saturated gravel, loose stones, and a periodic pass through the splash trough. Speeds varied from 40 mph while traversing bumps to 100 mph around the high-speed oval. At the end of the 50,000 mile cycle, the vehicles were stripped down and all components carefully examined for wear or damage.

## TEMPERATURE TESTING

Most endurance work was done at the Chrysler Michigan Proving Grounds, but for the past three years vehicles have been running at the Chrysler Phoenix Test Station, where ambient summer temperatures frequently reach 110 F.

In order to obtain test data at various temperatures and altitudes, road trips were conducted both by Bendix and Chrysler Corp. In the middle of winter, trips have been made up north to Stephens Point, Wis., and Bemidji, Minn., where tests were conducted on the airport runway which had been wetted down to form ice. During the summer months, road trips were made to Death Valley, Nev., where stopping distance tests were conducted at 120 F ambient temperature to determine if the high temperatures coupled with vibration would show up any problems during Sure-Brake stops.

The road trips, while providing the greatest interest, only occurred after an extensive test program in the engineering test laboratories where all components were subjected to a wide range of temperatures and humidity conditions while cycling at peak loads. The modulators, for example, are cycled 225,000 times from -20 F to 200 F and special calibration tests are run at -40 F and 250 F.

## EFFECT OF VACUUM LEVELS ON PERFORMANCE

The modulators are vacuum powered and as such are dependent on the level of vacuum generated in the engine intake manifold. In general, the tendency of vacuum levels will be to drop during the next few years, so this has to be considered in the size of the units. A further consideration is the effect of altitude; the higher one climbs, the lower the vacuum level produced by the engine. In Michigan, at approximately 900 ft above sea level, vacuum levels of 24 in. Hg. are readily attained, but on top of Pikes Peak in Colorado, at 14,110 ft above sea level, 12 in. Hg. is a more usual figure. Tests were made descending the Pikes Peak at 20, 30, and 40 mph in which Sure-Brake stops were made at each 1000 ft vertical drop level when descending and, apart from a tendency for the system to cut out early and lock up the wheels prematurely towards the end of the stops, the systems functioned correctly each time. Since the road is a gravel road and like most mountain roads twists and turns and has steep grades, the correct functioning of the system at 40 mph became very important.

## ELECTROMAGNETIC INTERFERENCE (EMI)

Much work had to be done in the area to ensure that external sources of energy would not cause failure or unnecessary operation of the system. Sure-Brake stops were made as close as possible to radio and television towers in the Detroit area, "The Voice of America" station near Cincinnati, and at the National Bureau of Standards at Fort Collins, Col., where frequencies of 19 kHz to 25 mHz are used. No degradation in performance was noted at Detroit or Fort Collins but, while passing under the main supply cable at the "Voice of America" station, a momentary turning off of the system occurred.

Bendix conducted copius tests in the laboratories to cover even wider variations, using military specifications as the basis for the test.

## CONCLUSION

The four-wheel slip control system has made its debut on a low volume, luxurious vehicle, but everyone involved in its development is confident that eventually a four-wheel system will be standard equipment on all road vehicles. The performance advantages of such a system are now well-known. Stopping distances are frequently reduced, stability is preserved, and steering control is retained, making it possible for the driver to upgrade his performance during those times he is driving at the limit of road-to-tire adhesion. The Sure-Brake system will not necessarily help the driver who is not mentally alert and unable to react to an adverse situation. The driver must still steer away from accident situations and apply the brake early and hard enough to stop.

It is not possible to say what wheel slip control systems will be in general use in ten years time; the field is new and many approaches will no doubt be tried and rejected before arriving at one or two general designs capable of providing satisfactory performance without excessive size, complexity, or weight penalties. Combinations of functions may be achieved; the modulators could be designed to increase brake line pressure as well as reduce it, and thus eliminate the power booster. The wheel sensors could provide information to speed control and wheel spin control systems. Other uses for the sensors could be to provide odometer and speedometer input signals.

# Design and Development of an Hydraulic Powered Wheel Slide Protection System*

**Robin A. Cochrane**
Girling Ltd.

*Paper 720031 presented at Automotive Engineering Congress and Exposition, Detroit, Michigan, January, 1972.

ALTHOUGH BRAKE SYSTEM PERFORMANCE and tire adhesion characteristics have improved substantially in recent years, skidding, caused by wheel locking, continues to be a factor in a large percentage of road accidents.

Advances in the understanding of the stability and controllability of vehicles have led to better handling characteristics and improved steering geometries. However, the friction characteristics of the road surface—especially under wet or icy conditions—probably remain the greatest potential hazard to most drivers.

Currently used braking equipment can achieve high rates of deceleration under ideal conditions. However, because the coefficient of friction between the tire and road can vary considerably from wheel to wheel, there are many circumstances in which a vehicle will become unstable. An apportioning valve inserted into the rear brake line will go some way to obtaining more stable conditions, but it controls only the front to rear braking ratio of the vehicle in relation to axle load or pressure; it is still possible to lock either front or rear wheels, resulting in instability and loss of steering control.

Various countries have proposed legislation by which a vehicle will be required to meet certain braking standards without wheel locking, so that control may be improved. However, this must be satisfied without detriment to the performance that could be achieved in a vehicle not fitted with such a system.

A Wheel Slide Protection (WSP) system has been developed by Girling to enable a driver to make better use of available tire-to-road friction levels in emergency situations where invariably his reaction will be to apply excessive force to the brake pedal. The Girling system uses the technique of rapidly dropping the brake pressure on sensing excessive deceleration of any wheel. This is followed by a controlled reapplication of the pressure to hold it near to the value which gives maximum deceleration for as long as possible during each cycle.

Many companies around the world have experimented with systems which work on this principle, and some production cars are fitted with such devices. These generally use vacuum energy to modulate the brake pressure, with logic decisions being made electronically. The reasons for adopting this combination of technologies are:

1. Rapid response and flexibility of electronic devices.
2. Readily available vacuum energy source.
3. Low valve loadings required with vacuum.
4. Know-how existing on vacuum servo devices.

If a system is to be designed to function on more than one axle, and, in particular, if it must operate independently on

## ABSTRACT

A Wheel Slide Protection system is considered in which a hydraulic power source is used to modulate braking effort at each wheel of a vehicle. The reasons for selecting this type of power source are discussed, together with other parameters employed in the design of the system.

System operation is described with particular emphasis on the functioning of the hydraulic actuators and their response to electrical control signals.

Evaluation has been carried out on different vehicles over a wide range of surfaces. Certain observations are made as a result of these tests.

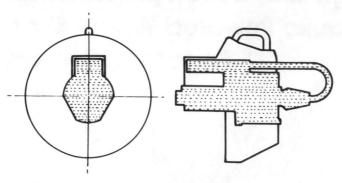

Fig. 1 - Comparison of vacuum and hydraulic actuator sizes

different wheels, then the use of vacuum energy to modulate brake pressure has certain restrictions:

1. Large size of vacuum units (Fig. 1).
2. Slow response (because a compressible medium is used).
3. Vacuum limitation can affect performance.

These problems become particularly critical if a system is operating on the front axle of a car. Therefore, it is desirable to use a more responsive system, with an energy source of a higher pressure level. Girling has therefore designed a system which contains its own energy source in the form of a central pump supplying brake fluid at 300 lb/in$^2$ (2070 kN/m$^2$) to individual actuators.

## DESIGN CONSIDERATIONS

For 10 years Girling has experimented with WSP systems for cars. These have generally been of the vacuum deboost type, or full-power dump systems, together with electronic decision making. Although successful vacuum systems have been developed for rear axle use, the limitations of such a system are acknowledged. A dump system depends first on fitting a car with a full-power braking system and this limits its application. For instance the WSP system could not normally be fitted as an option. Further, it would normally require acceptance of two new and unknown brake devices together on a new vehicle.

Work on the system described here was initiated in May 1969. In addition to the many requirements that have previously been published for such a system (1-3)* the following were considered to be important:

1. The system must be capable of operation on any individual wheel or axle.
2. It must operate with any known dual hydraulic brake system, including split calipers.
3. It must operate with any normal type of hydraulic brake actuation.
4. There must be no sacrifice of braking performance during partial failure of a dual system.

*Numbers in parentheses designate References at end of paper.

5. It must not be dependent on the engine running.
6. The system should be readily adaptable to different vehicle installations.

All these features have been achieved in the present Girling system.

The pressure level for the energy source was fixed at 300 lb/in$^2$ (2070 kN/m$^2$) for various reasons. The most important of these was that the design of gear-type pumps becomes more complex as pressures rise above this level. However, the reduction in actuator size for any increase in pressure becomes progressively smaller, while higher pressures mean less fluid volume transfer, which makes brake pressure control more difficult.

Consideration was given to other energy sources already on a car. Two particular sources were considered but dismissed— the power steering and engine oil pumps. Neither have sufficient excess capacity at idling speeds to power a 4-wheel system, and neither achieve feature 5. (The use of power steering on European cars is low and this would severely restrict the use of such a system for even rear wheel WSP.) The problems of system interaction and priority control further discourage the use of pumps already fitted to a car.

A system could be designed where an open-center loop, such as power steering, is used to charge accumulators, and this would overcome these limitations, but the cost of sufficient accumulators and priority control devices would be at least as expensive as using a separate pump, while taking up more underhood space. Every car has a battery which is readily capable of providing sufficient energy to power a WSP system, thus the most satisfactory solution was considered to be the use of a closed-loop system with a central electrically driven gear pump.

Such a system can provide constant output under any condition. It avoids problems of fluid contamination between mineral engine oil and vegetable brake fluid. It also simplifies detection of power source failure, or excessive valve leakage in the WSP system.

## DESCRIPTION OF SYSTEM

Fig. 2 shows the WSP system configuration used with a dual brake system. In this example, the use of split front calipers with unequal bores has also been shown. By connecting the smaller bores of the front calipers to the rear brakes as one half of the system, while the larger bores of the front calipers form the second system, it is possible to achieve a comparatively cheap brake system with high levels of performance during a partial failure.

Each wheel contains an inductive sensor. This produces a sinusoidal voltage signal which is fed back to a control module. The signal is converted to a d-c voltage level, which is proportional to wheel speed, and this, in turn, is differentiated to obtain wheel deceleration. When a threshold value is exceeded, a control circuit is switched which triggers an actuator in the brake line, causing the brake pressure to drop. The signal continues until the wheel speed increases. The actuator is de-energized at a wheel speed which corresponds to

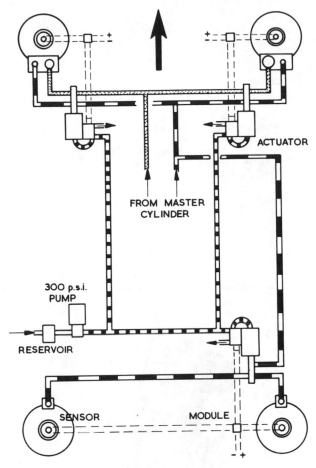

FROM MASTER CYLINDER

ACTUATOR

300 p.s.i. PUMP

RESERVIR

SENSOR

MODULE

Fig. 2 - System layout

a falling ramp voltage which was initiated at the moment of threshold deceleration. The pressure to either front brake is modulated independently by its own actuator, while experience has shown that a single actuator can satisfactorily control the pressure to both rear brakes.

The energy required to control the actuators is supplied by a small gear pump driven by a electric motor. This unit is rated intermittently to pump 0.6 U.S. gal/min (2.27 liters/min of brake fluid at 300 lb/in$^2$ (2070 kN/m$^2$) for short periods, relying on the large reserves of energy contained in the car battery as a power source even when the car engine is not running. A closed-center control loop is used so that the system may be held pressurized at all times without the pump running continuously. The pump is supplied by its own fluid reservoir, and the motor is switched as required by any actuator.

As the system is normally in a static condition for long periods, a checking system is incorporated to ensure that it is cycled regularly and that it will still function correctly. Continuous checks are also carried out, and any particular failure will cause the whole WSP system to revert to normal braking, while also giving the driver an indication of what has happened.

To reduce delays in operation, the actuators are always mounted as near to the brakes as is possible. The distance of the actuator from the supply pump is unimportant, as each actuator contains a reserve of fluid to maintain system pressure

at a constant level irrespective of switching and fluid flow delays.

## ACTUATOR

Fluid in the brake lines normally flows freely past the simple ball valve in the high pressure head of each actuator (Figs. 3 and 4). The valve is held open by a deboost plunger which in turn is held forward against brake pressure by a larger piston. During normal running, this piston is supported by the pressure output of the pump. If this piston is withdrawn, the plunger will follow under the pressure in the brake. First the valve closes off the line to the brake from the master cylinder, and then the volume of the brake system is expanded, causing the pressure to drop in the brake. Therefore, if the piston is moved, then the deboost plunger will follow it and cause the brake pressure to be dropped and reapplied irrespective of the driver's demands.

The 300 lb/in$^2$ (2070 kN/m$^2$) pressure also acts on a smaller piston to compress a large coil spring. This spring has two purposes. The first is to provide a mechanical loading onto the larger piston, and, therefore, the deboost plunger, if the pump should fail. In this situation a brake pressure of 1500 lb/in$^2$ (10,300 kN/m$^2$) may still be applied before the actuator self-limits. Girling generally designs for a pressure of 1500 lb/in$^2$ (10,300 kN/m$^2$) for 0.87 g (8.5 m/s$^2$) deceleration. The actuator will not, therefore, reduce this level, even if there is a failure in the WSP system. Under normal conditions, it will allow pressure to be applied to 2000 lb/in$^2$ (13,800 kN/m$^2$), which gives a high margin of safety even during brake fade. This margin is important when a WSP system is fitted on the front axle.

The second benefit of the spring and piston is that it forms a fluid accumulator of stored energy in each actuator. Immediately after the pressure drops in any actuator, a limit switch closes to start the pump running. The pump takes a finite time to reach working pressure, particularly since it may be situated at the far end of the car. During this time the accumulator maintains the pressure close to 300 lb/in$^2$ (2070 kN/m$^2$).

A small solenoid-operated control valve is fitted to each actuator and in its de-energized position maintains fluid to the backside of the large piston only, leaving the annular chamber ported to the reservoir at atmospheric pressure. On sensing an excessive wheel deceleration, the solenoid is energized by the control module. This causes the control valve to connect both sides of the large piston to the pump (Fig. 5).

The annular area of the piston is 35% larger than the area holding the brakes on. Pressure rapidly balances across the piston creating a large driving force to accelerate it away from the deboost plunger. However, the plunger must follow under brake line pressure and the result is that pressure at the brake is reduced.

The pressure continues to be relieved until the wheel speed increases and then the solenoid is de-energized. The control

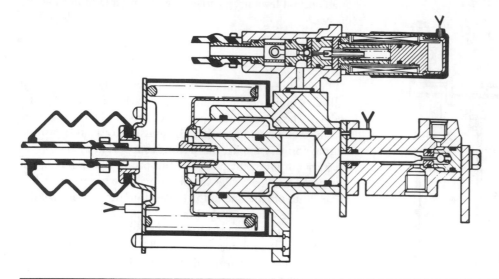

Fig. 3 - Actuator detail

Fig. 4 - Current actuator (left) and 300 lb/in$^2$ pump (right)

valve returns to its rest position and fluid that had passed into the annular chamber flows back to reservoir at a controlled rate, which, in turn, controls the reapplication rate of the brakes.

When the brake pressure again becomes excessive, the cycle is repeated. At the end of the stop, or when the maximum pressure at the brake is less than that required to lock the wheel, then the plunger will return to its rest position with the ball valve open.

If a dual braking system is used where the calipers are split, as in the example shown, then two parallel deboost plungers are incorporated in the expander head of the actuator. This provides simultaneous modulation of the pressure in both parts of the brake system, while keeping each part discrete. The self-limiting pressure levels are still maintained with this method of construction while variations in actuator design are kept to a minimum.

## PERFORMANCE TESTING

Initial trimming of the WSP system was carried out on an analog computer. Work that could have taken many months of track testing was carried out in a matter of days. A specially constructed computer is now used for development testing (Fig. 6).

A single actuator was first tested on the front axle of a Ford Cortina already fitted with a vacuum actuator on the rear axle. When this was proved to be satisfactory, a second actuator was added to permit independent front brake modulation.

The possible advantages of such a system were immediately shown under winter conditions. For instance, tests were carried out on an asphalt road with a light covering of snow. A corner which could just be negotiated at 35 mph (56 km/h) could be approached at up to 50 mph (80 km/h) and

188

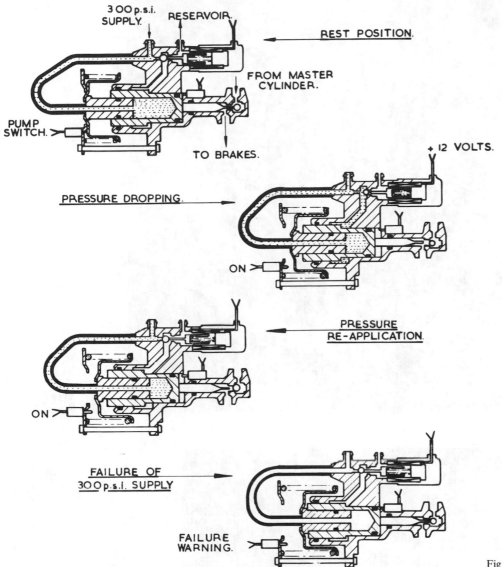

300 p.s.i. SUPPLY

RESERVOIR.

REST POSITION.

FROM MASTER CYLINDER.

PUMP SWITCH.

TO BRAKES.

+ 12 VOLTS.

PRESSURE DROPPING.

ON

PRESSURE RE-APPLICATION.

ON

FAILURE OF 300 p.s.i. SUPPLY

FAILURE WARNING.

Fig. 5 - Actuator operation

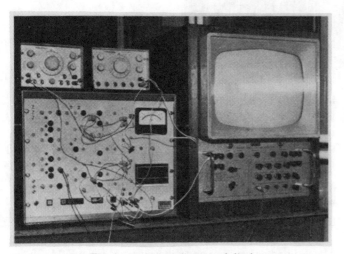

Fig. 6 - Analog computer and display

the car braked to a standstill in the corner without loss of control.

A full road test program was then begun, using cars fitted with three actuators (Fig. 7). Four surfaces of different friction levels have been used for assessment, ranging from a 300 ft (91.5 m) × 30 ft (9.2 m) epoxy resin strip giving a sliding friction of 0.14 when wet, to dry asphalt at 1.0 (Fig. 8). It is also possible to straddle test from high to low friction, or low to medium friction levels across test surfaces. Further test work has been carried out on washboard and paved surfaces, while winter testing has been carried out at environmental temperatures of -30C.

To reduce system variations, and, therefore, to simplify both development and production tasks, the minimum number of changes have been made in each installation. In fact, only two have been made:

Fig. 7 - Typical test vehicle installation

Fig. 8 - WSP track test surfaces

1. Varying actuator displacement to match brake system displacement.

2. Varying number of sensor teeth to maintain information rate at approximately 20 Hz/mph.

The first variation has been achieved without altering the low-pressure control section of the actuators, while the second allows for common modules to be used on all vehicles.

The work that has been carried out has shown the following:

STOPPING PERFORMANCE - Part of the reduction in stopping distance recorded on low friction surfaces is a result of

improved apportioning of the front-to-rear brake ratio. For instance, a vehicle with a front-to-rear ratio of 65:35 only reaches this ideal balance at high levels of deceleration. On a surface with a peak friction level of 0.25, the ideal ratio should probably be in the order of 55:45. A good driver in a car with a 65:35 ratio could at best achieve about 0.21 g (2.1 m/s$^2$) with the front wheels on the point of locking. The same car fitted with 4-wheel WSP should approach 0.25 g (2.5 m/s$^2$).

A poor driver, with or without apportioning, would lock all wheels in an emergency on such a surface, and would achieve a

Table 1 - Effect of Brake Ratio on Stopping Distance

Surface: Epoxy resin paint
Nominal friction level with locked wheels: 0.14
Nominal peak level: 0.25
Test speed: 35 mph (56 km/h)

| Car | Ideal Ratio, $0.25g(2.5 \text{ m/s}^2)$ | Actual Ratio, $0.25g(2.5 \text{ m/s}^2)$ | Theoretical Improvement, % | | Actual Improvement, % | |
|---|---|---|---|---|---|---|
| | | | GD/LW* | WSP/LW | GD/LW | WSP/LW |
| A | 57:43 | 60:40 | 41 | 44 | $-$VE | 32 |
| B | 61:39 | 67:33 | 39 | 44 | 28 | 28 |
| C | 54:46 | 73:27 | 24 | 44 | $-$VE | 25 |

*GD - Good driver.
LW - Locked wheels.

deceleration of only about 0.14g. (1.4 m/s²). WSP could therefore reduce his stopping distance by up to 44%.

Table 1 shows some practical examples. Cars A and B are reasonably well balanced and a good driver should approach 0.25 g (2.5 m/s²) deceleration while theoretically WSP should be capable of actually achieving a 0.25 g (2.5 m/s²) deceleration. Car C is badly balanced and even a good driver should be able to achieve only about 0.18 g (1.8 m/s²).

It is interesting, therefore, to see that it is impossible for a good driver to control the braking sufficiently well on this surface to prevent the wheels from locking on cars A and C. This is attributed to booster response time and hysteresis. The result is that WSP shows a very considerable improvement against any driver with these two cars. (The fact that the actual improvement in stopping distance appears to be about 15% below that theoretically possible is not an indication of the inefficiency of this system, but rather of the difficulty of measuring friction-slip characteristics.)

These results help to prove that it is irrelevant to compare test results from different antiskid systems fitted to different vehicles. Results can only be compared when expressed as percentage improvement over locked wheels and the tests are carried out with a constant friction-slip characteristic. This basic point is often forgotten or ignored, yet comparison can be very misleading.

COMBINED BRAKING AND CORNERING - The use of a single actuator on the rear axle results in some sacrifice of the ultimate braking performance during cornering. Braking when a car is near the limit of cornering adhesion has shown that the variation in modulated pressure level from outer wheel to inner wheel can be as much as 5:1.

A similar ratio could be achieved on the rear axle, but a single actuator sensing the slowest wheel must sacrifice about 80% of the peak braking possible on the outer wheel. In vehicle braking terms, this means that only 70-80% of the theoretically available performance can be achieved in this extreme condition. A good driver may be able to exceed this value slightly on a normal car, but only by cornering with

Table 2 - Effectiveness on Different Vehicles

Test speed: 40 mph (64 km/h)

| Surface | Nominal Locked Wheel Friction | WSP/LW,*% | |
|---|---|---|---|
| | | Car A | Car C |
| Epoxy resin paint | 0.14 | 32 | 25 |
| Bridport Macadam carpet | 0.30 | 22 | 18 |
| Dry asphalt | 0.85 | 6 | $-$3 |

*LW - Locked wheels.

either one or both of the inner wheels locked. A poor driver would probably lock all four wheels and the car would travel tangentially to the curve.

Use of individual modulation of the rear brakes would mean that all four tires could be near the limit of adhesion. By design, any WSP system must overbrake in order to discover the adhesion limit. This can produce a potentially dangerous situation by causing front or rear breakaway. On all the cars tested, this overbraking has led to rear breakaway. It is, therefore, preferable to reduce rear braking during cornering to improve stability. The adoption of a single actuator achieves this, while also reducing total system cost.

SYSTEM SOPHISTICATION - Various systems have been designed which depend on adjusting the pressure reapplication rate, or the electronic control module characteristics, in relation to the surface coefficient. The Girling WSP system depends on the rapid response characteristics of both electronics and hydraulics to reduce brake pressure rapidly after a wheel has exceeded the threshold deceleration. By acting in this way, pressure deviation from the ideal can be kept to a minimum while the need for complication in system logic is reduced.

Minimum pressure deviation means that a car will spend more time at maximum deceleration. The test results shown in Table 2 confirm the effectiveness of this system. Damage to vehicle and suspension, which can be caused by cyclic systems, is reduced by an efficient system which will hold the car at a nearly constant deceleration throughout a stop. Passenger discomfort is also reduced.

INSTALLATION VARIATIONS - The cars which have been used for development work have varied considerably in weight, suspension and steering design, etc. This has proved that there is no basic problem in cycling the front brakes independently if the pressure deviation is kept low. However, the steering feedback and vehicle stability vary considerably between vehicles, but at present insufficient practical information has been obtained to give any definite rules on suspension and steering design when associated with WSP.

## SUMMARY

A WSP system has been designed and developed, and tested on various cars. The results of this exercise have shown that such a system, with a high response rate, requires little adaption to achieve good braking performance when installed in different vehicles. However, the effects on handling and vehicle stability vary and there is still considerable investigation required in this area before the effect of WSP on different vehicles can be accurately predicted.

## ACKNOWLEDGMENT

The author wishes to thank the Directors of Joseph Lucas Ltd. and Girling Ltd. for permission to publish this paper.

## REFERENCES

1. Thomas C. Schafer, Donald W. Howard, and Ralph W. Carp, "Design and Performance Considerations for a Passenger Car Anti-Skid System." SAE Transactions, Vol. 77 (1968), paper 680458.

2. R. H. Madison and Hugh E. Riordan, "Evolution of Sure-Track Brake System." SAE Transactions, Vol. 78 (1969) paper 690213.

3. J. W. Douglas and T. C. Schafer, "The Chrysler 'Sure Brake' - The First Production Four Wheel Anti-Skid System." Paper 710248 presented at SAE Automotive Engineering Congress, Detroit, January 1971.

# CURRENT ABS
# TECHNOLOGY

*Theory*

# Excess Operation of Antilock Brake System on a Rough Road*

**M. Satoh and S. Shiraishi**
Honda R & D Company Ltd., Japan

*Paper C18/83 is published as part of **Braking of Road Vehicles 1983** and is reproduced by permission of the Council of the Institution of Mechanical Engineers© Institution of Mechanical Engineers 1983.

SYNOPSIS  Several types of antilock brake systems have so far been made available for practical purposes.  Of these systems, this report discusses the four-wheel-control antilock brake with particular emphasis on its braking performance on a rough road.  The report first explains why the new antilock brake system proposed here has adopted the "select-high" technique for the front wheels and the "select-low" technique for the rear wheels.  Then the performance of the braking system on a rough road is discussed in detail in Sections 2 and 3.  Finally Sections 4 and 5 compare the new antilock brake system with its predecessors based on the results of real vehicle tests.

## INTRODUCTION

The four-wheel-control antilock brake (4W-ALB) system, already installed in some production cars, can effectively curb the reduction of vehicle stability and loss of vehicle steerability resulting from locked-up wheels during panic braking.

Along with the above excellent capability, however, the existing 4W-ALB systems have a disadvantage such as the vehicle's stopping distance increases on gravel and other bumpy roads.  Preferably, further improvements should be made in these systems to eliminate this shortcoming.

Vehicles equipped with existing 4W-ALB's require a longer distance for stopping on a rough road probably because:
(1) Unlike a normal smooth road surface, a rough road provides the highest apparent adhesion coefficient between its surface and the vehicle's wheel treads when the wheels get locked; this means that locked-up wheels are more effective than controlled ones in achieving a shorter stopping distance on a rough road; and
(2) The control function of the 4W-ALB is unfavorably affected by wheel speed changes during braking on a rough road, subsequently resulting in system over-control which unnecessarily reduces braking torque.
    This report presents a comparative study on existing 4W-ALB's and a new system that can achieve better braking performance on a rough road by using advantageously or ravelling out the foregoing causes of ALB system tendency to worsen stopping capability on a bumpy road surface.

## 1   CONTROL TECHNIQUES FOR ALB

Three techniques are now available for controlling braking forces for the right and left wheels:
(1) Independent control technique, which controls braking forces for the right and left wheels independently;
(2) Select-low technique, which simultaneously controls the right and left wheels in response to a signal from either of them which is predicted to lock first; this implies that braking torque for the wheel on the other side may become lower than the desired level; and
(3) Select-high technique - this subsystem simultaneously controls braking forces for the two wheels in response to a signal from either of them which is predicted to be the least likely to get locked; therefore it may permit the other wheel to be locked up.

All existing 4W-ALB's use technique (1) or (2).  None of them have ever adopted technique (3) for the front or rear wheels presumably because vehicle stability and steerability may be affected by the possible reduction of a lateral force on the tire due to the tendency of this control technique to permit either the right or left wheel to get locked during full braking.

It should be noted, however, that wheel locking will always occur on the side which produces a smaller lateral force than the other.  In other words, the system may lock either the right or left wheel which less affects vehicle stability or steerability than the other.  On a split road surface with different adhesion coefficients ($\mu$) on the right and left sides, for instance, the wheel on the side with a lower $\mu$ may be locked during full braking.  When the vehicle is turning on an even road surface with a uniform $\mu$ on both sides, the system may lock the wheel on the inside of the curve which receives lighter load than the other.

Use of the select-high technique will therefore affect vehicle stability or steerability so slightly that no practical problem may arise.  On the other hand, this technique provides a very effective means of precluding an increase in stopping distance which may otherwise occur on a rough road due to system overcontrol.

The authors believe that the first consideration in a brake system is to provide the vehicle with sufficient stopping capability and stability. The following combination of techniques is recommended for the 4W-ALB system by the fact that general passenger cars, particularly front-drive ones, depend more heavily on the front wheels than on the rear for their stopping capability:

(1) Select-high technique, an effective means of curbing stopping distance extension on a rough road, for the front-wheel control subsystem; and

(2) Select-low technique, a most considerable means for the maintenance of vehicle stability, for the rear-wheel subsystem.

This combination of front and rear wheel control subsystems will be referred to as the "H-ALB" (high-selected antilock brake) in the subsequent paragraphs.

## 2 EXCESS OPERATION

The antilock brake system keeps monitoring wheel speed changes and automatically eases braking torque when any of the wheels is predicted to get locked. The reference standard for determining such system-actuating conditions is generally worked out by the following formulas:

$$\lambda > \lambda_0 \qquad (1)$$
$$\dot{V}_w < \dot{V}_{wo} \qquad (2)$$

Where $\lambda$ = slip ratio of the wheel;

$\dot{V}_w$ = circumferential acceleration of the wheel;

$\lambda_0$ = reference standard for $\lambda$; and

$\dot{V}_{wo}$ = reference standard for $\dot{V}_w$.

With vehicle and circumferential wheel speeds given as $V$ and $V_w$, respectively, it follows

$$\lambda = 1 - V_w/V \qquad (3)$$

Usually the reference standards are set at the the following levels:

$$\lambda_0 \simeq 10\% \qquad (4);$$
$$\dot{V}_{wo} \simeq -9.8 \sim -16 m/s^2 \qquad (5)$$

On a rough road, wheel speed varies at random, i.e., the so-called wheel speed pulsation occurs, due to the ever-changing contact between the tire tread and the road surface. Such wheel speed pulsation becomes more severe particularly when the brakes are being applied. This is because longitudinal vibration of the suspension may occur from varying braking forces due to changes in the way the tire tread contacts the road surface.

Fig. 2 gives some examples of the wheel speed pulsations recorded while a conventional test vehicle equipped with no ALB system was running on extremely rough road surfaces as shown in Fig. 1. Of these data, the values of $\lambda$ and $V_w$ were calculated when a force of 100N was put on the brake pedal on an extremely rough road. Such small force were selected to avoid causing any of the wheels to be locked up. The results are shown in Fig. 3 along with straight lines representing the reference standards, i.e., $\lambda_0 = 10\%$ and $\dot{V}_{wo} = 16 m/sec^2$.

The diagram indicates that the system-actuating conditions represented by formulas (1) and (2) above may sometimes occur on a rough road even if the driver steps on the brake so lightly that none of the wheels will get locked. This is why vehicles equipped with existing ALB systems unnecessarily ease braking torque on a rough road, making their stopping distance longer. In this report, such overcontrol of braking torque is defined as "excess operation" of the ALB system.

## 3 PRECLUSION OF EXCESS OPERATION

### 3.1 Adjustment of longitudinal suspension compliance

As noted earlier, wheel speed pulsation may be amplified by longitudinal vibration of the suspension assemblies on a rough road. This suggests that wheel speed pulsation could be curbed through adjustment of the longitudinal compliance.

Computer simulation was carried out to examine the effects of such adjustment (See Appendix). Fig. 4 shows some of the wheel speed pulsations simulated while the vehicle was supposed to be running on a rough road with its suspension compliance alternately set at three different levels, i.e., nominal level, halved, and doubled.

It was found that a significant change in the longitudinal compliance of the suspension assemblies hardly helps preclude excess operation of the ALB system as wheel speed pulsation remains virtually unchanged.

### 3.2 Use of select-high technique

Fig. 5 shows the probability density and probability distribution functions of slip ratio $\lambda(t)$, assuming that variable as shown in Fig. 3 changes at random process.

With the occurrence probability of the system-actuating conditions satisfying formula (1) given as $P$, the excess operation probability of the select-high system, given as $P_H$, is equal to the probability with which the right and left wheels simultaneously meet the conditions represented by formula (1). Providing that the mutual interference of the right and left wheels be left out of account, this can be formulated into:

$$P_H = P^2 \qquad (6)$$

If the values of $P$ and $P_H$ are obtained from Fig. 4, it follows:

o Rough road : $P = 0.068$, $P_H = 0.005$

o Extremely rough road : $P = 0.382$, $P_H = 0.146$

This means that the select-high technique can substantially curb excess operation of the ALB system on a rough road.

### 3.3 Means for setting the reference wheel speed

Some effective means must be provided somehow or other to find vehicle speed since slip ratio $\lambda$, a function of vehicle speed, represents an essential control factor in the ALB system as noted earlier.

From formulas (1) and (3), the reference standard for finding whether a wheel is predicted to get locked can be expressed by the following formula:

$$Vw < (1 - \lambda o)V \qquad (7)$$

Hence, defining Vs as follows, Vs represents the reference wheel speed.

$$Vs = (1 - \lambda o)V \qquad (8)$$

Therefore the following formula is considered equivalent to formula (1):

$$Vw < Vs \qquad (9)$$

This suggests, if vehicle speed is known, the braking torque can be properly controlled by slip ratio because the reference wheel speed can be determined. However, no practical means has yet been made available for directly measuring vehicle speed V during braking.

In general, therefore, wheel speed Vw is used to estimate vehicle speed V as shown in Fig. 6(a). In other words, this estimation is performed by using the memory C which stores the peak values of wheel speed and reduces it at a specified rate whenever the actual wheel speed goes down sharply. Assuming that output V* of this memory is approximate to V, the reference wheel speed Vs is set at a level equal to the result of multiplying this output by $(1 - \lambda´o)$. Where $\lambda´o$ represents the apparent reference standard for a slip ratio which is experimentally determined in adequate consideration of an error between V* and V.

Fig. 6(c) gives the values of Vs and $\lambda o$ obtained through computer simulation for the circuit in Fig. 6(a) using as an input the value of Vw; shown in Fig. 2, actually measured on an extremely rough road with the brake pedal force set at 100N. As is apparent from the diagram, this method allows $\lambda o$ to vary violently on a rough road, letting it decrease to a very low level locally, because of wheel speed pulsation that occurs on such a road surface. In other words, excess operation of the ALB system occurs more frequently, and the vehicle requires a longer distance for stopping.

Setting of $\lambda o$ involves an additional problem. Formula (7), worked out to determine the reference standard for control system actuation, can now be put into:

$$V - Vw > \lambda o V \qquad (10)$$

The left member of the formula represents the difference between wheel speed and vehicle speed, while the right member represents the reference standard for control system actuation depending on vehicle speed. If $\lambda o$ is set constant, the reference standard becomes very small at a low vehicle speed. This implies that even minor wheel speed pulsation makes the ALB system get into excess operation. Since the phenomenon appears only at a low vehicle speed, this disadvantage of the system in respect of stopping distance may be practically negligible.

However, the driver feels that brake system operation has worsened considerably. This unfavorable impression can be removed by slightly increasing $\lambda o$ at a low speed.

An attempt was made to work out a system improvement that can remove both drawbacks discussed above;
(1) great variation of the reference standard $\lambda o$ due to the estimation of vehicle speed from wheel speed by a simple method;
(2) worsened feeling of brake system operation caused by a decrease in the permissible amplitude of wheel speed pulsation at a low vehicle speed.
For this purpose, a charging resistor R was added to the memory C as shown in Fig. 6(b). The resistor serves as a one-way filter that works only when the memory is being recharged. Through a voltage drop in itself, the resistor also helps increase $\lambda o$ as vehicle speed decreases. Fig. 6(d) gives the results of simulating the operation of the improved system shown in Fig. 6(b) under the same conditions as for Fig. 6(c). The findings indicate that the resistor works to stabilize $\lambda o$ and increase it at a low vehicle speed, while substantially reducing the probability of the ALB's excess operation on a rough road.

### 4 COMPARISON OF STOPPING CAPABILITIES ON ROUGH ROADS

The stopping distances of three different test vehicles were compared to examine the braking performance. The test vehicles were:
(A) Equipped with an ALB consisting of an independent control subsystem for the front wheels and select-low subsystem for the rear - I-ALB;
(B) Fitted with another type of ALB comprised of a select-high subsystem for the front wheels and select-low subsystem for the rear - H-ALB; and
(C) Fitted with conventional brakes, no antilock system - N-ALB.

Of these vehicles, the ALB systems for I-ALB and H-ALB were both prepared with exactly the same control logic. The circuit shown in Fig. 6(b) was used in these ALB systems to set the reference standards for wheel speed.

The tests were conducted on three different road surfaces:
1. Rough road (See Fig. 1);
2. Extremely rough road (See Fig. 1); and
3. Gravel road.

Table 1 Stopping distance on rough road

| Road \ Vehicle | N-ALB | I-ALB | H-ALB | Initial speed |
|---|---|---|---|---|
| rough road | 25.0 | 26.0 | 25.5 | 67.5 km/h |
| extremely rough road | 16.5 | 27.6 | 19.0 | 50 km/h |
| gravel road | 37.3 | 50.3 | 43.6 | 75 km/h |

The results are as shown in Table 1.
Both H-ALB and I-ALB required a longer distance for stopping than N-ALB on any of the road

surfaces noted above. Compared with I-ALB, however, H-ALB achieved an appreciably shorter stopping distance, demonstrating that its system improvement produces favorable effects in this respect.

## 5 COMPARISON OF STEERABILITY ON SMOOTH, FLAT ROADS

As discussed earlier, use of the select-high technique for the front-wheel control subsystem may probably permit either one of the front wheels to be locked up during full braking. However, wheel locking will always occur on the side which affects vehicle steerability in a lower degree than the other, and H-ALB can maintain satisfactory vehicle steerability even during full braking. A series of tests were conducted to investigate this capability of H-ALB.

### 5.1 Braking performance test during turning on smooth, flat roads

Test conditions
   Road surface: Dry asphalt
   Initial vehicle speed: 80 km/hr
   Turning radii: 168m, 112m and 84m
   Lateral acceleration: 3.0, 4.4, 5.9 $m/s^2$
   Steering wheel: Fixed
   Force applied to the brake pedal: 882N
Measurement
   Position and heading of the test vehicle when coming to a stop
Test vehicles
   H-ALB and I-ALB

As is apparent from the test findings shown in Fig. 7. H-ALB deviated little from the specified course and kept heading along the course; it could maintain sufficient turning capability even during full braking. It was also found that both H-ALB and I-ALB can easily stop with their heading kept on the specified course if the driver is permitted to perform correction steering.

### 5.2 Evasive maneuverability test on low μ road surface

Test conditions
   Road surface: Wet, smooth, flat surface with
                 adhesion coefficient μ = 0.35
   Test course: As shown in Fig. 8(a)
   Force applied to the brake pedal: 882N
Measurement
   Distance ℓ between the test vehicle at the initial application of the brakes and an obstacle ahead was gradually reduced until it became too short for the vehicle to avoid a collision. Then the minimum ℓ in which the vehicle could avoid hitting the obstacle was measured.
Criteria
   Five rounds of test were conducted for each distance and the vehicle was considered to have sufficient evasive maneuverability for that specific distance if it could stop without touching the obstacle at any one of the five tries.

Test vehicles
   H-ALB and I-ALB

The test results are shown in Fig. 8(b). H-ALB proved to have evasive capability nearly equal to that of I-ALB; the difference between the two systems is only about 10 percent.

## 6 CONCLUSION

The following conclusion may be drawn from the foregoing discussion:
(1) The ALB system gets into excess operation due to wheel speed pulsation during braking on a rough road. This results in a longer stopping distance.
(2) Such excess operation of the ALB can be appreciably reduced through the use of an improved method for setting the reference wheel speed in combination with the application of the select-high technique to the front wheels which provide the greater portion of stopping capability.
(3) Use of this technique for the front-wheel control subsystem may permit during panic braking either one of the front wheels to be locked. However, the effects of such wheel locking on vehicle steerability is virtually negligible.

As noted above, the front wheels provide a major portion of stopping capability, while the rear ones play an important role in maintaining vehicle stability. The new 4W-ALB system uses the select-high technique for the front wheels and select-low technique for the rear. It may be assumed that this type of 4W-ALB can provide a new practical braking system particularly suitable for front-drive cars because of foregoing advantages.

## APPENDIX

The computer simulation used in the section 3-1 has been conducted by the co-operation with Dr. D. H. Weir, Mr. J. W. Zellner, Mr. G. L. Teper, et al.

This simulation is based on similar non-linear models and simulations developed before, for example Ref. 1 for trucks and buses and Ref. 2 for motorcycles.

Several parts of this simulation concerning this paper is summarized below.

### A Suspension Model

Each wheel has the following degrees of freedom relative to the sprung mass:

   Vertical motion, of the unsprung mass and suspension
   Longitudinal motion, due to suspension compliance
   Spin about the wheel spindle

Vertically, any or all of the wheels can bounce off the roadway.

The vertical suspension characteristics are modelled by a spring and a damper. The spring is represented by four different rate segments to provide for progressive rates and bump stops. The dampers are modelled by a two segment curve so that the compression and rebound damping can be different. The longitudinal compliance of the suspension is modelled by a four segment spring and a two segment damper, also.

## B  Tire Model

The program uses a modified ellipse tire model to compute the forces on the tire at the ground contact points. This model was originally presented in Ref. 3, and it has since been widely used in various forms. It takes into account the load normal to the ground, circumferential forces due to braking and acceleration, etc; and the interaction of them. The model is empirical, and it uses test data for a tire towed at various braking torques resulting in fractional rotational slip, normal loads, and for selected pavement conditions.

The required data include the radial force vs. tire deflection, the normalized locked wheel skid coefficient vs. normal load, and the ratio of circumferential friction coefficient to normalized skid coefficient vs. fractional rotational slip. Provision for tire damping is also included in the model.

For more detailed background information on the model the reader is directed to Ref. 3, and other pertinent literature.

## C  Roadway Model

The simulation provide for a generalized roadway bump profile. The front and rear tires encounter the bumps, with a time delay equal to the wheelbase divided by the forward velocity.

The tire can contact the roadway either at 2 points when encountering a bump, or with 1 point when rolling over the top of the bump or when operating a flat surface.

REFERENCES

1. Weir, D. H., Teper, G. L., et al.
   Analysis of Truck and Bus Handling.  Vol. II;
   Technical Report.
   DOT HS-801 153, Mar. 1974.

2. Weir, D. H., and Zellner, J. W.
   Motorcycle Handling.  Vol. II: Technical
   Report.
   DOT HS-804 191, May 1978.

3. McHenry, R. R. and Deleys, N. J.
   Vehicle Dynamics in Single Vehicle Accident -
   Validations and Extensions of a Computer
   Simulation.
   Cornell Aero. Lab. Inc., VJ-2251-V-3,
   Dec. 1968.

4. Leiber, H. and Czinczel, A.
   Antiskid System for Passenger Cars with
   a Digital Electronic Control Unit.
   SAE Paper, No. 790458.

5. Harned, J. L., Johnston, L. E. and Scharph, G.
   Measurement of Tire Brake Force Character-
   istics as Related to Wheel Slip (Antilock)
   Control System Design.
   SAE paper, No. 690214.

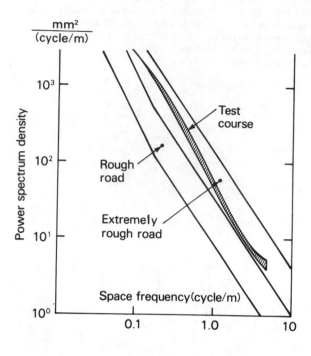

Fig.1 Characteristics of rough roads

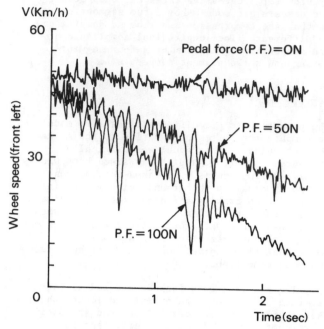

Fig 2 Example of the front wheel speed pulsations

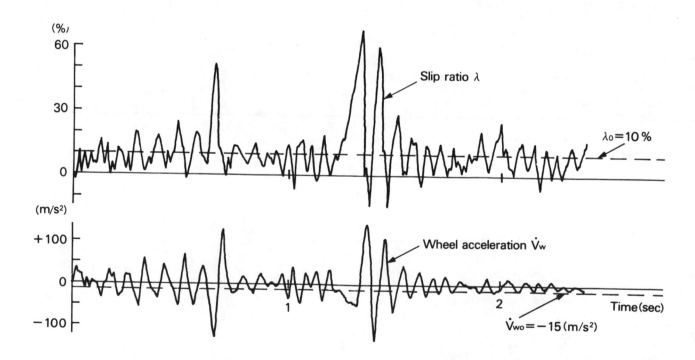

Fig 3 Pulsations of $\lambda$ & $\dot{V}_w$ over reference values

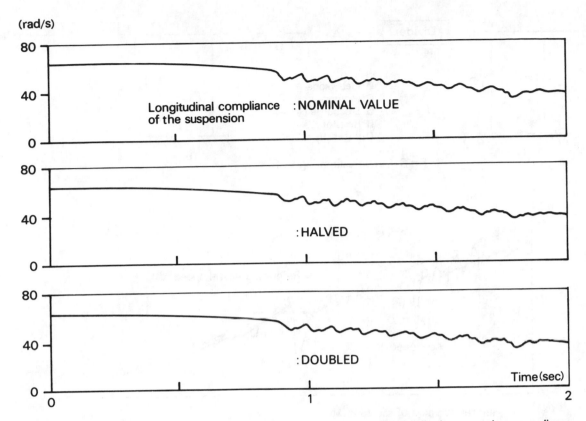

Fig 4 Simulation results of the front wheel speed with different longitudinal suspension compliance

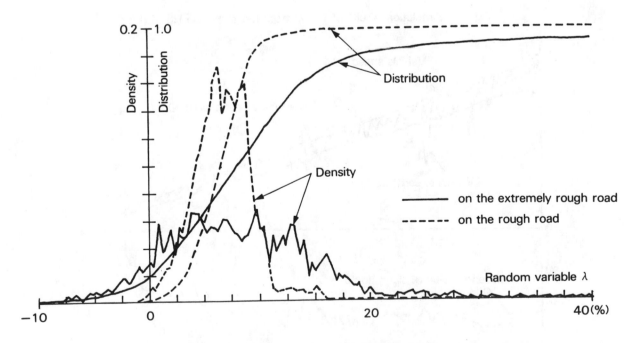

Fig 5 Probability of random process λ (t)

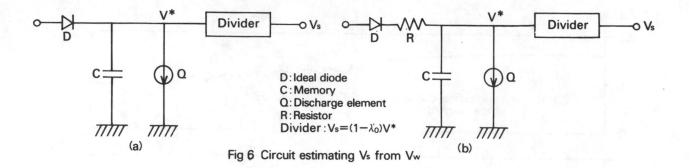

D : Ideal diode
C : Memory
Q : Discharge element
R : Resistor
Divider : $V_s = (1 - \lambda_0)V^*$

**(a)**　　　　　　　　　　　　　　　　　　　**(b)**

Fig 6　Circuit estimating $V_s$ from $V_w$

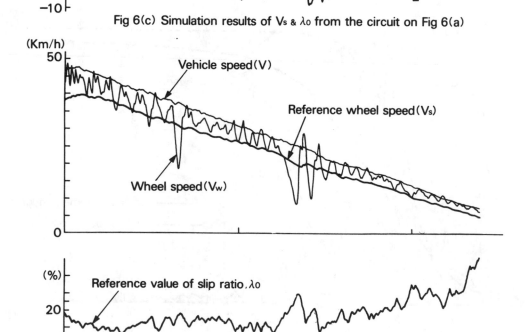

Fig 6(c)　Simulation results of $V_s$ & $\lambda_0$ from the circuit on Fig 6(a)

Fig 6(d)　Simulation results of $V_s$ & $\lambda_0$ from the circuit on Fig.6(b)

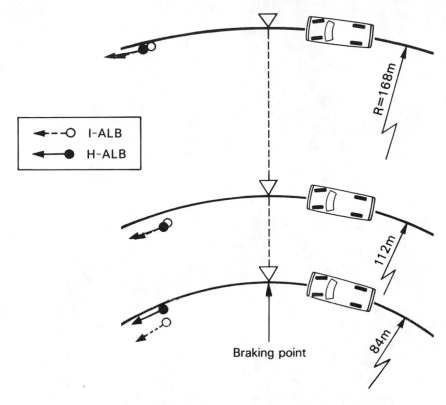

Fig 7 Test results on braking performance during turn

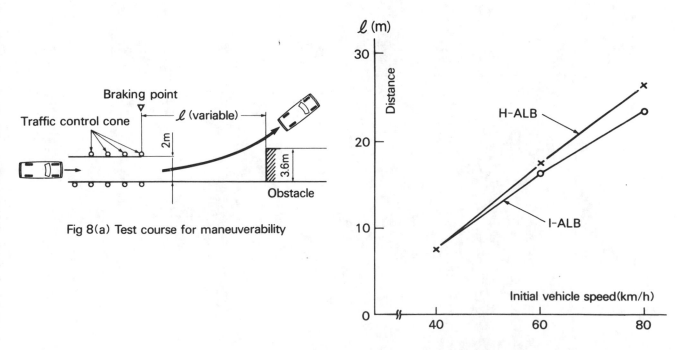

Fig 8(a) Test course for maneuverability

Fig 8(b) Test results on evasive maneuverability

# The Potential and the Problems Involved in Integrated Anti-Lock Braking Systems*

**H. Leiber and A. Czinczel**
Robert Bosch GmbH, Stuttgart, West Germany

*Paper C192/85 is published as part of **Anti-lock Braking Systems for Road Vehicles** and is reproduced by permission of the Council of The Institution of Mechanical Engineers© Institution of Mechanical Engineers 1985.

SYNOPSIS The paper discusses the potential and the problems involved in integrated anti-lock braking systems regarding an improvement of the braking performance as compared to separate anti-lock braking systems. The comparison shows that integrated systems have a great potential for improving the braking system as well as the anti-skid function and also for expansion with subsystems. However, integrated systems are not in all cases superior to good separate designs.

## 1 INTRODUCTION

Separate high-performance anti-skid systems for passenger cars have been in production since 1978 and by the end of 1984 Bosch had supplied some 500 000 systems. The Bosch ABS (Anti-Lock Braking System) is therefore already in widespread use, and customer attitudes toward the system have been exceptionally positive. The potential of the ABS for increasing traffic safety is recognized, for example, by the German Association of Automobile Insurers which recommends its members to grant a 10 % discount on fully comprehensive insurance for vehicles equipped with ABS.

In practice the Bosch ABS has shown itself to be exceptionally safe and reliable. The success of the system is based upon system reliability and upon the high-level performance of the controller.

Already in the seventies there were various anti-skid systems available in the USA for passenger cars and commercial vehicles. However, they did not gain any great popularity as a result of unsatisfactory performance and reliability. 3 different anti-skid systems are currently available in Japan.

At an early date Bosch presented the possible types of anti-skid system in the relevant technical journals. The basic possibilities are, firstly, separate systems which are functionally and structurally separate from the master cylinder and brake power-assist unit, and secondly, integrated systems which are functionally und structurally combined with the master cylinder and brake power-assist unit.

Integrated anti-skid systems have been under development for many years. The first system was recently introduced in the USA. Further systems are currently being prepared for production in Europa and the USA. It therefore appears appropriate to compare and evaluate the features of separate and integrated anti-skid systems. Fig 1 shows photographs of the separate and the integrated Bosch ABS.

## 2 METHODS OF PRESSURE MODULATION IN ANTI-SKID SYSTEMS

Efficient anti-skid systems are characterized by the fact that the pressures in the wheel brakes are automatically modulated as soon as the wheels show signs of locking as a result of excessive application of the brakes. Generally, pressure modulation involves successive pressure buildup, pressure reduction an pressure holding phases such that the wheels are controlled within a narrow slip range with optimum braking action.

This pressure modulation can be achieved by different principles which are shown in Fig 2 and 3. With the return-pumping principle (separate system) the ABS control valves are positioned directly between master cylinder and wheel brake cylinder. If automatic brake control calls for a pressure reduction in a wheel brake, the ABS control valve switches to the pressure reduction position which creates a connection between the wheel brake cylinder and the pump. The brake fluid taken from the wheel brake is pumped back to the master cylinders by the electrically driven return pump. With this principle, the brake circuits remain closed during pressure modulation.

The failure of a brake circuit does not affect brake boosting. The operation of the ABS in the intact brake circuit is maintained. The legally required emergency braking action is assured in the event of the failure of brake boosting. ABS operation is maintained insofar as the driver is able to produce the required brake pressure without brake boosting.

The three designs IIa, IIb and IIc show the principle of pressure modulation by changing the volume in the brake circuits (separate systems). Common to these designs is the fact that pressure modulation is performed by a plunger which changes the volume of the relevant brake circuit. The plunger is driven by a separate energy supply via the ABS control valves in the auxiliary circuit.

These principles correspond to the systems which have been built in the USA and which have also recently been used in Japanese systems.

This subject area was treated in detail at the 1984 Fisita Congress (1). For basic understanding, therefore, only principle IIa is explained. This design shows the plunger in the normal braking position; the ball-seat valve on the left is open. If the brake pressure is to be kept constant in the wheel brake, the plunger which is preloaded with a strong spring is pressurized with fluid from the accumulator via the ABS control valve until the ball-seat valve is closed. If a pressure reduction is required in the wheel brake, the accumulator pressure continues to be applied to the plunger so that the cylinder volume connected to the wheel brake (between ball-seat valve and stepped plunger) is enlarged. The pressure in the wheel brake is then relieved as a result of the enlarged volume.

As can be seen from the basic diagram, the main attractions of the return-pumping principle are its simple construction and its fail-safe design.

Fig. 3 shows pressure modulation by direct pressure supply into the brake (integrated system) circuits. This design shows one of the possibilities of functional integration. In this case, during operation of the ABS shown, the pressure of the brake power-assist unit is used for pressure modulation. According to the pedal force, the brake valve controls the servo pressure which is applied to the master cylinder pistons. Under partial braking the pressure in the closed brake circuits is transmitted to the wheel brakes via the open ABS control valves. If, during ABS operation, there is to be a pressure reduction in the wheel brakes, a corresponding quantity of fluid must be taken from the wheel brakes. If, during the next pressure buildup, this quantity of fluid were to be replenished from the master cylinder, the master cylinder pistons would advance by a corresponding amount, and after a few control cycles they would reach their end position. A further brake pressure increase would then be impossible.

During ABS operation, therefore, the pressure must be supplied directly to the wheel brakes from the servo circuit. This is accomplished by the pressure-supply valve shown on the right which is energized at the start of automatic brake control and establishes the connection between servo circuit and wheel brake. The separation of the circuits is guaranteed by two non-return valves.

The pressure supply must likewise be fail-safe thanks to a suitable sensor arrangement since, if a brake circuit fails, a pressure drop in the accumulator must be expected, which then necessitates the shutdown of the pressure supply.

If, for example, this valve fails during ABS operation, brake fluid is taken from the brake circuit for pressure modulation and, consequently, the brake master cylinder pistons are moved by the brake pedal as far as the limit stop. A further pressure buildup is then impossible, which, on an icy road surface, is virtually equivalent to the failure of the entire braking action.

Electrical monitoring of the solenoid valve alone is insufficient. Therefore, a pressure switch must be provided at the outlet of the pressure-supply valve.

The monitoring of the pressure supply is likewise of relevance with regard to safety. This necessitates a fail-safe sensor arrangement with monitoring circuit, which can be achieved through the use of a redundant pressure sensor with mutual monitoring.

Many systems additionally require the monitoring of the fluid level in the reservoir.

3 PEDAL CHARACTERISTICS WITH SEPARATE AND INTEGRATED ANTI-SKID SYSTEMS

Fig 4 shows pedal characteristics for separate and integrated anti-skid systems. This was also dealt with in detail at the previously mentioned Fisita Congress (1).

Graph A applies to plunger and return-pumping systems. At the respective lock-up pressure the brake pedal is arrested and superimposed with a slight oscillation. In the case of low lock-up pressure (e.g. black ice) the pedal is arrested at a short pedal travel whereas, in the case of high lock-up pressure (e.g. dry road surface), the pedal is arrested at a correspondingly long pedal travel.

Graphs B, C and D belong to integrated anti-skid systems. The characteristic feature of these systems is the safety zone which must be maintained so that, in the case of a defect, sufficient brake pressure can still be produced by way of the corresponding master cylinder piston travel.

In system B, a locking sleeve ensures that, during ABS operation, the master cylinder pistons are blocked directly before the safety zone. At low lock-up pressure this means that the pedal moves forward as far as the safety zone whereas, at high lock-up pressure, the pedal is forced back as far as the limit of the safety zone.

Until it nears the safety zone, system C has the same characteristic as system A. As soon as the pedal travel has reached the safety zone, the brake pressure is raised without further travel, which is linked to a corresponding increase of the pedal force. This is how, in this case, the safety zone is maintained.

In system D, the pedal undergoes a slight reaction (oscillation) at the start of ABS operation. In contrast to systems A, B and C, with this system the driver is able to bring the pedal into the position he desires, unimpeded by the ABS. The great advantage of this system is that the pedal characteristic is the same both under partial braking (i.e. without ABS operation) and under emergency braking (i.e. with ABS operation). Consequently, in critical situations in which the ABS is in operation, the driver is not disconcerted by an unaccustomed pedal characteristic.

4 SAFETY CRITERIA

Fig 5 summarizes the safety criteria for separate and integrated anti-skid systems.

Since, with the separate system, the brake circuits are always closed, not only under partial braking but also under emergency braking with ABS operation, the functions of the brake power-assist unit always remain strictly separate from those of the ABS.

This means that, as a basic principle, there can be no transfer of faults from the system comprising the closed brake circuits and the ABS to the brake power-assist unit or vice versa:

- Faults in the ABS components are not transferred to the brake power-assist unit. If, for example, a defective ABS valve were to remain permanently in the pressure reduction position, the accumulator chamber upstream of the return pump would accept a very small quantity of brake fluid with the result that there would be a slight lengthening of the pedal travel. The maximum braking effect would be virtually maintained.

- A failure of the brake power-assist unit does not affect the operation of the ABS. Insofar as the driver is able to produce the necessary lock-up pressure (e.g. on black ice or snow) without the aid of the brake power-assist unit, the ABS operates in the normal manner.

- The operation of the brake power-assist unit is not affected in the case of a circuit failure. Brake boosting in the intact circuit is maintained, and, since the operation of the ABS is also maintained, the driver is able to make full use of the braking effect remaining with the intact brake circuit. This also applies even if the failure of the brake circuit takes place during ABS-controlled emergency braking.

- The problems associated with a failure of the pressure supply under erratic changes of the adhesion coefficient will be indicated later when dealing with integrated anti-skid-systems. With the separate systems, the closed brake circuits guarantee, even in this case, that, should the pressure supply fail, the driver is able without hindrance to build up the brake pressure in accordance with the pedal force.

The summary shows that the separate ABS with its closed brake circuits offers a high degree of safety without the need for special monitoring devices in the hydraulic part of the system.

With the integrated anti-skid system featuring direct pressure supply, the brake circuits are open during pressure modulation: to reduce pressure, brake fluid is drained out of the wheel brakes into the reservoir and, to build up pressure, the drained brake fluid is resupplied via the pressure-supply valve.

During ABS pressure modulation, therefore, the brake circuits with the ABS control valves are connected via the pressure-supply valve to the pressure chamber of the brake power-assist unit and the pressure accumulator. This means that faults can be transferred from the brake circuits and the ABS valves to the brake power-assist unit and vice versa:

- If, during ABS operation, the pressure-supply valve fails to switch to the pressure-supply position, the master cylinder pistons move into their end position with the result that there is no longer any braking effect. This makes careful electrical and mechanical monitoring of the pressure-supply valve an urgent necessity. If the pressure-supply valve fails to switch, ABS operation must be switched off.

- In the event of a failure of the brake power-assist unit (leaking, pump failure etc.) the ABS must be switched off quickly enough so that there is still sufficient brake fluid in the pressure accumulator to permit some amount of residual boosted braking and, above all, to guarantee maintenance of the closed brake circuits for emergency piston operation. This requires careful monitoring of the accumulator pressure. If the accumulator pressure drops below a specified warning threshold, the pressure supply and the ABS must be switched off.

- Under partial braking, i.e. when the brake circuits remain closed, the failure of a brake circuit has no detrimental effect on the brake power-assist unit. During ABS operation, however, the failure of a brake circuit leads to the loss of brake fluid from the accumulator. If the accumulator pressure drops below the warning threshold, the ABS and the pressure supply must be switched off.

- Under certain conditions the failure of the pressure supply can lead to the loss of braking effect. If braking takes place with ABS on a high adhesion coefficient and there is then a change to a low adhesion coefficient, whereupon the pressure supply fails, it is then no longer possible to build up the brake pressure.

The study of the safety criteria for integrated anti-skid systems with direct pressure supply shows that the installation of sensors with appropriate monitoring circuits is necessary in order to make these systems fail-safe. This involves an outlay on equipment which separate anti-skid systems do not require.

5 EVALUATION OF SEPARATE AND INTEGRATED ANTI-SKID SYSTEMS

Separate and integrated anti-skid systems are to be evaluated with reference to essential features, the list of which, however, does not claim to be exhaustive (Fig 6). The separate system used in the comparison is the series-produced Bosch ABS.

Integrated systems can have different performance classes. A system of performance class III is rated better than the separate system only in installation size and installation outlay. As regards the safety criteria which have been treated in detail, the separate system fares well in the comparison. The evaluation of performance classes II and I shows the potential of integration whereby it should be noted that in class I a clear appraisal of all the listed features is currently not yet possible.

The positive evaluation of some features of the separate system calls for some explanatory remarks.

The safety criteria have been discussed at length. The better flexibility concerns the ease with which the hydraulic modulator can be housed in the engine compartment. Despite compact construction, integrated systems cannot in all cases be installed in a standard version. This therefore necessitates different packaging. Furthermore, one single hydraulic modulator can be used in a wide range of vehicles containing different brake master cylinder layouts for reasons of weight.

Consequently, it is necessary once again to provide different integrated units, with all the associated consequences as regards production and inventory management.

A comparison of separate and integrated anti-skid systems would be incomplete without an accompanying evaluation of the features of the corresponding brake power-assist units. Fig 7 shows the most important features of the vacuum power-assist units as well as of two hydraulic power-assist units, one of which is the long-term optimum solution and the other the minimum-demand solution which is almost ready for production.

The vacuum brake power-assist unit is of simple construction and highly reliable; it is, however, limited in performance. The current hydraulic brake power-assist unit is likewise limited in performance, whereas the long-term optimum concept has great potential for improving the braking system.

With the first two brake power-assist units, the failure of a brake circuit leads to sagging of the pedal because the rod-type master cylinder piston is rigidly connected to the pedal. In the optimum brake power-assist unit, piston and pedal are separate so that the pedal travel remains unaffected by a brake circuit failure.

With the first two systems, a ratio jump is not possible in the event of the failure of brake boosting due to the rigid connection between piston and pedal. With the optimum brake power-assist unit, the separation of piston and pedal makes it possible to implement a ratio jump which, as compared to the first two systems, allows greater retardation in the event of the failure of brake boosting.

Whereas, in the first two systems, the pedal characteristic is completely fixed by the volume capacity of the brakes and the master cylinder diameters, in the optimum system the pedal characteristic can, within limits, be freely selected with the aid of the characteristic of a simulator spring.

As compared to the vacuum brake power-assist unit, hydraulic brake power-assist units have shorter pressure build-up times. In conjunction with ABS the pressure build-up time of the first two systems is considerably lengthened by the throttling effect of the solenoid valves. With the optimum system, a solution is conceivable in which the throttling caused by the solenoid valves is allowed only during automatic brake control, with the result that there is a very short pressure build-up time for the initial braking phase.

Integrated anti-skid systems lend themselves to the idea of introducing a three-circuit braking system. Fig 8 shows such a system in which the three circuits comprise the two front wheels and the rear axle.

One of the advantages of this arrangement is the small reduction of the braking effect in the event of a brake circuit failure. The construction costs are lower than for the currently widespread diagonally-split braking system.

Fig 9 indicates the problems of the 3-circuit braking system as regards braking performance in the event of defects in the system. The failure of a front-wheel brake circuit leads to high yawing moments. Grahps A, B, C show the critical increase of the yawing moments in the sequence "circuit failure before braking", "during partial braking", "during emergency braking with ABS". Front-wheel brake-circuit failure necessitates a reduction of the yawing moment, in addition to a number of other measures, in order to maintain directional control of the vehicle. In order to achieve a reduction of the yawing moment the ABS valve of the intact front wheel brake circuit can be used. Brake pressure in the intact brake circuit can be immediately reduced when the brake circuit failure occurs and then gradually increased so that the driver has enough time to correct the gradually increasing yawing moment with the steering wheel.

To guarantee directional control in the event of a front-wheel brake circuit failure, the operation of the ABS must be upheld in the intact brake circuits.

A failure of the 3-circuit system's dynamic rear-axle brake circuit causes the failure of boosting.

## 6 PROSPECTS

Integrated anti-skid systems have great potential for improving the operation of the ABS and the braking system. Safety criteria are of decisive importance in any integration. The paper shows that integrated systems are not in all cases superior to good separate designs.

The facts which are indicated form the basis for the further development of the ABS, the end of which cannot be foreseen at present.

## 7 REFERENCES

(1) Heinz Leiber, Armin Czinczel: ABS-spezifische Bremskraftverstärker Fisita-Congress, Vienna, 1984, 845106

(2) H. Leiber and W. D. Jonner: Pressure Modulation in Separate and Integrated Antiskid Systems with Regard to Safety, SAE 840467

(3) H.-W. Bleckmann, J. Burgdorf, H.-E, von Grünberg, K. Timtner and L. Weise: The First Compact 4-Wheel Anti-Skid System with Integral Hydraulic Booster, SAE 830483

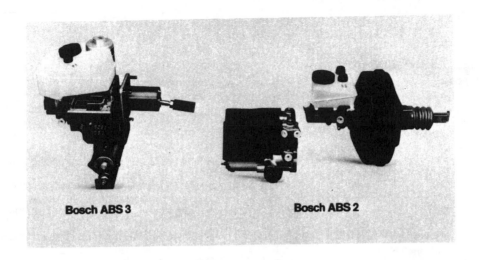

Fig 1    Comparison of separate and intergrated ABS

## I Return-pumping

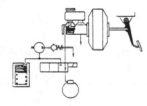

## II Change of volume of brake circuits (plunger system)

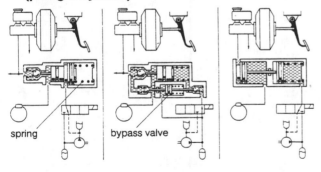

spring          bypass valve

Fig 2    Pressure modulation with ABS (part1)

fluid level switch

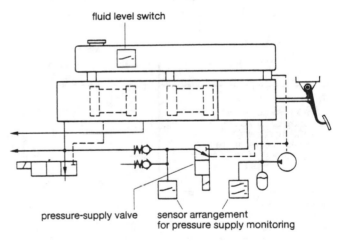

pressure-supply valve          sensor arrangement
for pressure supply monitoring

Fig 3    Direct pressure supply to brake circuit (integrated ABS)

211

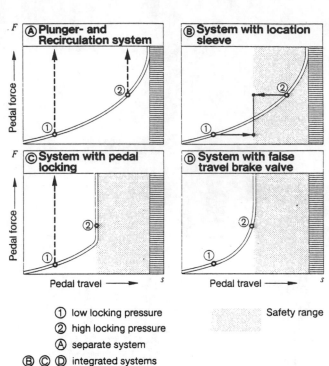

① low locking pressure
② high locking pressure
Ⓐ separate system
Ⓑ Ⓒ Ⓓ integrated systems
Ⓐ Ⓑ Ⓒ Systems with joint movement of master cylinder and pedal

Safety range

Fig 4   Pedal, characteristics of different hydraulic boosters with antiskid

**separate**   **integrated**

brake booster    brake booster

pressure-supply valve    sensor arrangement

● fault transfer from ABS components to brake booster

● fault transfer from brake booster to ABS

● brake circuit failure and effect on brake booster

● brake circuit failure during ABS operation

● pressure supply failure during certain braking conditions, e.g. sudden $\mu$-change

Fig 5   Safety criteria

| Characteristics | Evaluation | | | |
|---|---|---|---|---|
| | separate | integrated | | |
| | | III | II | I |
| ● ABS | | | | |
| – ABS functioning with brake circuit failure | + | – | – | + |
| – ABS functioning with brake booster failure | + | – | – | – |
| – Optimising of pedal reaction | – | – | + | + |
| – comfort of control | ○ | ○ | + | + |
| ● failure safety | + | – | – | |
| ● installation expenditure | ○ | + | + | + |
| ● installation volume | ○ | + | + | + |
| ● flexibility | + | – | – | |
| ● weight | ○ | ○)¹ | ○)¹ | |
| ● service – replacement costs | + | – | – | – |
| ● extension possibilities for subsystems, e.g. electronic traction control | –)² | + | + | + |

)¹ advantageous compared to great vacuum boosters

)² possible with addit. expenditure

Fig 6   Comparison of separate and integrated ABS (state of 1984)

| Characteristics | Vacuum-booster | Hydraulic booster | |
|---|---|---|---|
| | | min. | max. |
| ● brake system – pedal falling through with brake circuit failure | – | – | + |
| – transition jump with brake booster failure | – | – | + |
| – optimising of pedal characteristic – pressure build-up time | – | – | + |
| – pressure build-up time | – | –)¹ | + |

)¹ pressure increase throttled by ABS valve

Fig 7   Comparison of brake booster capacity

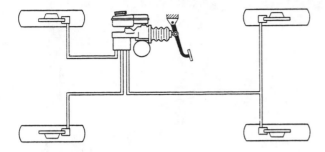

● small reduction of braking effect
  with brake circuit failure

● Reduced expenditure
  (compared to diagonal-split brake circuits)
  – lines

  – pressure reducing valves

  – ABS solenoid valves

  – bridging the pressure reducing valves
    during ABS operation

Fig 8   Advantages of three-circuit system

## Brake circuit failure

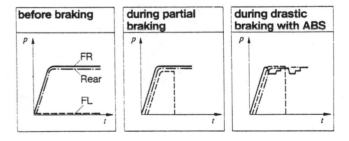

● substantial yawing moments during brake
  circuit failure of a front wheel require
  additional measures

● ABS operation is required to ensure sufficient
  stability after brake circuit failure

● rear brake circuit failure causes brake
  booster failure

Fig 9   Problems of three-circuit system

# Pressure Modulation in Separate and Integrated Antiskid Systems with Regard to Safety*

**W. D. Jonner and Heinz Leiber**
Robert Bosch GmbH

*Paper 840467 presented at the International
Congress and Exposition, Detroit, Michigan,
February, 1984.

## ABSTRACT

The antiskid systems which have been on the
market for some time are characterized by the
fact that they are separate from the brake
power-assist unit and are positioned between
the master cylinder and the wheel brakes
(separate configuration). At present, integrated antiskid systems are also being
prepared for launching on the market. In these
systems the hydraulic brake power-assist unit
performs the functions of brake boosting and
partly also of ABS pressure modulation.

The principles of ABS pressure modulation in
separate and integrated antiskid systems are
compared and questions concerning safety are
discussed.

With the separate ABS (plunger system, return
system) the brake circuits are closed, i.e.
when braking and also during ABS operation the
volume of brake fluid between the master
cylinders and the wheel brake cylinders is
closed and separated from the energy supply
of the hydraulic brake power-assist unit. If
a brake circuit fails during ABS operation,
the respective master-cylinder piston advances
into its end position. The energy supply is
unaffected. Special monitoring of the energy
supply is therefore not necessary.

With the integrated ABS there are two possible
principles, one with controlled master-cylinder
pistons and one with direct pressure supply into
the brake circuits. The controlled master-
cylinder pistons are arranged in parallel and
are moved to and fro on the primary side during
ABS operation by means of a control valve with
the result that the pressure in the wheel
brakes is modulated in accordance with the
control signals from the electronic control unit.

In the closed brake circuits between the master
cylinders and the wheel brakes there are merely
pressure-holding valves which do not remove any
brake fluid from the closed brake circuits. If
a brake circuit fails, the corresponding master
cylinder moves to its limit stop and the energy
supply is not affected, with the result that it
need not be specially monitored. However, the
elaborate design and the unconventional parallel
arrangement of the master cylinders speak
against taking this system into production.

The integrated ABS with direct pressure supply
into the brake circuits reverts to the proven
technique of the master cylinder with the pistons
arranged in a bore. Since the control valves
are between the master cylinders and the wheel
brakes, brake fluid must be supplied to the
wheel brakes directly from the brake power-
assist unit by way of a pressure-supply valve
during ABS operation. This prevents the master-
cylinder pistons from extending into their end
position. Thus, the brake circuits which are
closed for partial braking become open brake
circuits during ABS operation. The correct
functioning of the pressure-supply valve must
be monitored .If a brake circuit fails ABS
operation must be switched off promptly so that
the energy supply is not exhausted. The pressure
supply must also be monitored in order to
guarantee adequate braking capability under all
circumstances.

Finally, it should be noted once again that,
in contrast to separate systems, added attention
must be paid in integrated systems to the
possible faults described.

## INTRODUCTION

For some time now there have been antiskid systems on the market in which the antiskid system is separate from the brake power-assist unit and is positioned between the master cylinder and the wheel brakes (separate configuration).

At present, integrated systems are being prepared for launching on the market. In these systems the hydraulic power-assist unit performs the functions of brake boosting and also of ABS pressure modulation. The various concepts of integrated ABS brake power-assist units are very different in their pedal characteristics when the ABS comes into action and with regard to the possibilities of improving the properties of conventional brake systems.

The aim of this paper is to compare the principles of ABS pressure modulation in separate and integrated antiskid systems and to discuss questions of safety when braking.

## PRESSURE MODULATION IN SEPARATE ANTISKID SYSTEMS

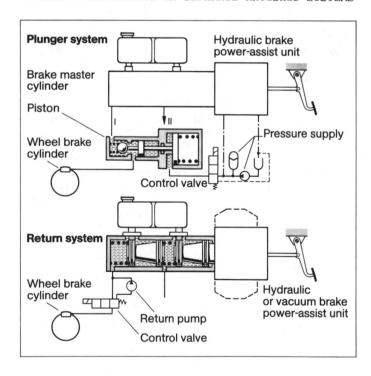

Fig.1: Pressure modulation on separate ABS

Fig. 1 shows 2 brake systems with separate ABS. They have closed brake circuits. In the case of partial braking and also in the case of emergency braking with ABS operation the volume of brake fluid between the master cylinders and the wheel brake cylinders is closed. If a brake circuit fails the corresponding master cylinder advances into its end position. The volume of brake fluid in the defective brake circuit is lost, but the energy supply remains intact. Special monitoring of the energy supply is therefore not necessary.

The top part of the figure shows the plunger system. The closed brake circuit I (master cylinder, brake lines, plunger, wheel cylinder) is shown in full.

Pressure modulation takes place in the auxiliary circuit by means of a control valve (in this example a 3/3 valve with the functions of pressure buildup, pressure holding, pressure reduction) which can apply metered pressure to the auxiliary piston from an hydraulic accumulator. If a wheel is about to lock, pressure is first of all supplied to the auxiliary piston so that the ball-seat valve closes. Consequently, the pressure in the wheel brake cylinder is kept constant. If, to avert the danger of a wheel locking, it is necessary to reduce the pressure in the wheel brake, further pressure is applied to the auxiliary piston. This causes the smaller plunger to move to the right so that the volume of the plunger space is increased. As a result, the wheel brake pressure decreases. The energy supply and the auxiliary piston space are separate from the plunger space. If the energy supply fails, ABS operation is no longer possible. This must be indicated by a warning lamp.

The bottom part of the figure shows the return system in conjunction with a vacuum power-assist unit or an hydraulic power-assist unit. Once again, the brake circuits remain closed for ABS. A return pump pumps the brake fluid taken from the wheel brakes during pressure reduction back into the master cylinders. If the energy supply of the brake power-assist unit fails, the ABS is unaffected and continues to operate. If the return pump fails, the pressure in the wheel brakes can no longer be reduced, with the result that ABS operation is no longer possible. However, the normal braking action is maintained.

# PRESSURE MODULATION IN INTEGRATED ANTISKID SYSTEMS

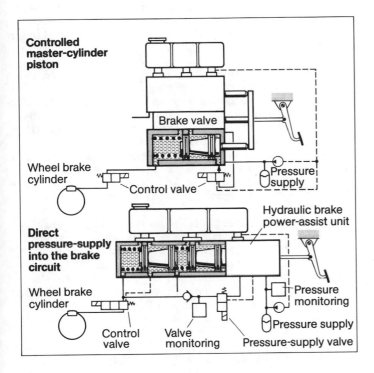

**Fig. 2: Pressure modulation on integrated ABS**

Fig. 2 shows principles of pressure modulation in integrated antiskid systems. The top part of the figure shows the principle with controlled master-cylinder pistons. The power-assist unit is characterized by the hydraulic actuation of the master cylinder pistons which are arranged in parallel and are actuated on the primary side by the pressure which is controlled by the brake valve.

During ABS operation this primary pressure is modulated by a control valve so that the master-cylinder pistons move to and fro, thus modulating the pressure in the wheel brakes in accordance with the control signals from the electronic control unit. In the closed brake circuits between the master cylinders and the wheel brakes there are two-position valves for implementing the pressure-holding function. These valves do not take any brake fluid from the closed brake circuits. If a brake circuit fails the corresponding master-cylinder piston moves to its end stop and the energy supply is not affected, with the result that the energy supply need not be specially monitored. If the energy supply fails, ABS operation is no longer possible. Emergency braking remains assured through the possibility of mechanical actuation of the master-cylinder pistons.

However, the elaborate construction and the unconventional parallel arrangement of the master cylinders (twin) are facts which speak against the production of this system.

The bottom part of the figure shows an integrated antiskid system with direct pressure supply to the brake circuits. This system reverts to the proven technique of the master cylinder with the pistons in a bore. Under partial braking the servo-pressure produced by the hydraulic brake power-assist unit is applied to the master cylinders which transmit the pressure to the wheel brakes. The brake circuits are closed in this case. The control valves are in the closed brake circuit.

Under emergency braking with ABS operation the closed brake circuits must be opened so that the pressure-supply valve can replace the brake fluid which has been returned from the wheel brake cylinders to the reservoir as a result of pressure reduction. The brake fluid needed for replacement is taken from the servo-circuit of the brake power-assist unit. If this volume of fluid were not replaced through the pressure-supply valve, the master-cylinder pistons would gradually move into their end positions, and a further pressure buildup would no longer be possible.

The safe and reliable operation of the brake system presupposes the correct functioning of the pressure-supply valve. Pressure supply must be guaranteed for ABS operation. On the other hand, if there is a failure in the energy supply or in a brake circuit, the pressure-supply valve must separate the servo-circuit from the brake circuits so that the emergency braking system is maintained. This therefore requires careful monitoring of the switching of this valve. How this can be done will be shown later.

Due to the opening of the brake circuits during ABS operation it is likewise necessary to carefully monitor the pressure supply. Fault criteria and technical approaches for the monitoring circuit will be indicated later.

## POSSIBLE FAULTS WITH INTEGRATED ABS

| Defect | Countermeasures |
|--------|-----------------|
| Failure of pressure supply, ABS not applied | ABS block, triggering of safety lamp and warning lamp |
| Failure of pressure supply, during ABS application | ABS block as soon as the accumulator warning pressure has been dropped below; provision of adequate piston residual travel |
| Circuit failure, ABS not applied | Warning display; observe the regulations for reservoir design |
| Circuit failure, during ABS application | Fail-safe monitoring of the accumulator pressure; blockage of both ABS and pressure supply, immediately warning threshold for the accumulator pressure is dropped below; provision of adequate piston residual travel |
| Pressure supply valve fails to switch in | Monitoring of the switch-in function of the pressure supply valve. In case of error detection, ABS block becomes effective |
| Pressure supply valve remains permanently switched-in | Monitoring of the ABS-block function |
| Safety lamp defective | The warning lamp informs driver of a failure of the brake system which includes possible failure of ABS. |
| Warning lamp defective | Redundant warning via safety lamp in case of critical pressure loss. |

Fig. 3: List of faults for integrated ABS with pressure supply

Fig. 3 shows the basic possible faults for the integrated ABS with pressure supply, and the countermeasures which are to be taken in the event of a fault. The impact on braking performance and the brake system are to be discussed in detail.

If the energy supply fails during partial braking (i.e. without ABS operation), the ABS must be disabled, and the safety lamp and the warning lamp must be energized. The failure of the pressure supply means that there is no brake boosting function with the result that braking takes place through the mechanical actuation of the pistons via the brake pedal (emergency piston operation). The pressure supply is ineffective.

If the energy supply fails during ABS operation, the ABS must be blocked as soon as a certain accumulator warning pressure is dropped below and there is no longer the guarantee that the spent volume of brake fluid can be replaced from the servo-circuit, in which case the master-cylinder pistons would move to their end stop.

To provide at least the legally required emergency braking action in the event of energy failure, it is necessary in this case to disable the pressure supply and thus also the ABS. The loss of the brake boosting and pressure supply functions may take place at long pedal travels, but at low pressure in the wheel brakes. For this reason, there must always be adequate piston residual travel during ABS control. This feature applies to all known integrated systems.

If a brake circuit fails during partial braking, the warning lamp must be energized. If, thanks to additional measures, it is possible to correctly identify which circuit has failed, the defective circuit can be cut off with the aid of the ABS control valves, and ABS operation can be maintained in the intact circuit. The otherwise possible maximum force can no longer be obtained, and there is a loss of brake fluid. However, the legally prescribed emergency braking action is assured through compliance with the regulations on reservoir design.

If a brake circuit fails during emergency braking (i.e. during ABS operation), this leads to a pressure drop in the accumulator because it is connected to the brake circuit via the servo-circuit and the pressure-supply valve. If the accumulator pressure consequently drops considerably, the maximum braking force is no longer obtainable. Unless special monitoring and corrective measures are taken, the failure of a brake circuit can lead to the failure of brake boosting.

Therefore, fail-safe monitoring must detect when the accumulator pressure drops below a warning threshold, and the ABS and the pressure supply must then be disabled. As already explained, there must be an adequate piston residual travel so that under these conditions it is possible to further increase the brake pressure.

To maintain ABS operation for as long as possible during critical braking, the fact that the fluid in the reservoir drops below a certain level should not in itself lead to the disabling of the ABS.

However, if the fluid drops below the warning
level during unbraked or partially braked
driving, the ABS can be switched off.

Let us assume the extreme case of the failure
of the energy supply during ABS-controlled
braking on black ice. In order, under these
circumstances, to produce sufficient pressure
for a 30% retardation without brake boosting,
it must be ensured - even if ABS control begins
on a high friction coefficient with a high
lock-up pressure level - that the position of
the master-cylinder pistons during ABS control
does not exceed a certain travel so as to
guarantee a residual actuation travel. This
can be achieved by various means. Reference
is made in this respect to (1).
(Publication will take place May 1984)

During ABS operation fluid must be fed directly
into the brake circuits so that the brake fluid
which is taken from the wheel brakes during
pressure reduction is replenished without the
master-cylinder pistons extending further. To
guarantee this, the position of the pressure-
supply valve must be monitored.

If the pressure-supply valve does not cut in,
the master-cylinder pistons will extend to
their limit stop during ABS pressure modulation.
A further pressure buildup is then no longer
possible. This is particularly critical if the
fault occurs on a low adhesion coefficient
which then changes into a higher adhesion
coefficient in the course of braking. To
preclude this possibility, the cut-in of the
pressure-supply valve must be monitored. If
a fault is detected, the ABS must be disabled.

If the pressure-supply valve stays on
permanently, the failure of a brake circuit will
be additionally accompanied by the failure of
the brake boosting function because the
accumulator is connected via the servo-circuit
and the pressure-supply valve to the failed
brake circuit. A total failure of the brakes
is then possible. The disabling function of the
pressure-supply valve must, therefore, likewise
be monitored. The simultaneous failure of a
brake circuit and the permanent operation of
the pressure-supply valve would be a double
fault. If both faults are safely monitored,
their simultaneous occurrence can virtually
be ruled out.

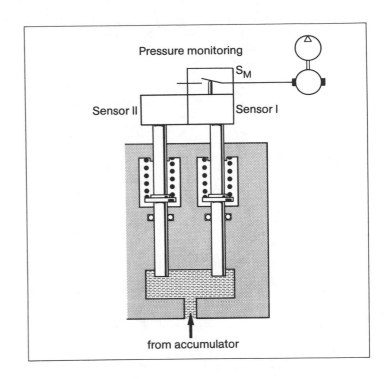

Fig. 4: Solution for fail-safe pressure
        monitoring

Fig. 4 shows a possible solution for fail-safe
pressure monitoring. If the energy supply or a
brake circuit fails, ABS operation must be
blocked as soon as the accumulator pressure
drops below a certain minimum pressure. This
is the only way of guaranteeing adequate braking
after the failure of a brake circuit or of the
energy supply. The minimum pressure should be
monitored redundantly with 2 independent systems.
Thus, for example, each of the two monitoring
systems can have an upper and a lower warning
threshold. If one of the upper warning thresholds of the two redundant systems responds,
the cutting-in of the pressure-supply valve and
the operation of the ABS are temporarily blocked.
After a hysteresis has been passed through,
the pressure-supply valve and the ABS are
enabled again.

Fail-safe monitoring can be achieved as follows: if the lower warning threshold of one system responds without the upper warning threshold of the other system having responded, this is detected as a fault. The ABS and the pressure-supply valve are then desabled. A corresponding logic circuit is presented in Fig. 5 for the expert.

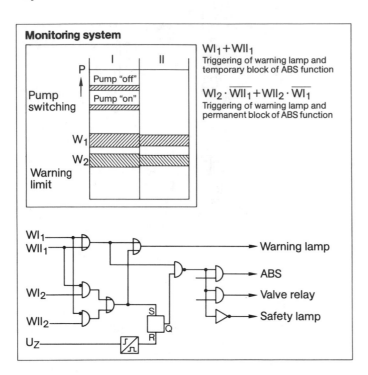

**Monitoring system**

$WI_1 + WII_1$
Triggering of warning lamp and temporary block of ABS function

$WI_2 \cdot \overline{WII_1} + WII_2 \cdot \overline{WI_1}$
Triggering of warning lamp and permanent block of ABS function

Fig. 5: Logic circuit for fail-safe pressure monitoring

For reasons of safety, there should be independent switch actuation, e.g. by means of 2 different switching plungers which are both under accumulator pressure, or, in the case of single actuation, it is necessary to choose a design principle which leads automatically to a warning in the event of failure. In this connection, it is appropriate to note that the seals of a switching plunger exhibit very different friction forces over the course of a long service life and also depending on the position of the switching plunger. For this reason, redundant actuation is important. If pressure monitoring failed and this were followed also by a failure of the energy supply (e.g. due to a defective pump), then partial braking would be unboosted. With ABS operation, however, there could be a total failure of the brake system.

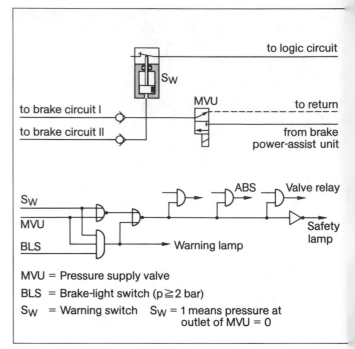

MVU = Pressure supply valve
BLS = Brake-light switch (p ≧ 2 bar)
$S_W$ = Warning switch  $S_W$ = 1 means pressure at outlet of MVU = 0

Fig. 6: Solution for monitored pressure supply

Fig. 6 shows a possible solution for monitoring the pressure-supply valve. Both positions of the pressure-supply valve, the on and the off position, must be monitored. Electrical monitoring of the pressure-supply valve (e.g. measuring of voltage across the valve coil) is not sufficient, but rather the mechanical-hydraulic operation of the valve must be monitored.

This can be done by means of a pressure switch which responds to the pressure downstream of the pressure-supply valve. As long as the pressure-supply valve is switched to off (i.e. no feeding in of brake fluid into the brake circuit) the pressure downstream of the pressure-supply valve must be zero; as soon as it is switched to on (feeding in) there must be pressure at its outlet. In a logic circuit the position of the pressure switch can be compared with the energization signals of the pressure-supply valve in order to monitor the correct operation of the pressure-supply valve.
If Sw = 0 and MVU = 0 (for explanations refer to fig. 6) the safety lamp is energized and MVU, ABS and the valve relay are disabled. If the AND condition from Sw, MVU and BLS (stop-lamp switch) is met, the safety lamp and the warning lamp are energized. MVU, ABS and the valve relay (VR) are disabled.

The functions of the pressure-supply valve and its monitoring circuit are checked each time the vehicle moves off. For this purpose, the pressure-supply valve is briefly excited and the agreement between electrical energization and pressure condition downstream of the valve is verified.

PROSPECTS

If the possible faults which have been described are monitored in the indicated manner, the integrated ABS will offer the following advantages over the separate ABS, given the selection of an appropriate brake power-assist unit:

. simple mounting

. free choice in design of pedal characteristics

. lower noise with optimum design

. improvement of braking performance (response behavior, no sagging of pedal in event of brake circuit failure)

The integrated systems are in the first phase of development. The system solutions shown are examples taken from the first development phase. The expected demand for antiskid systems, as well as the different requirements of the automobile manufacturers, will necessitate intensive development work to extend this existing basis.

REFERENCES

(1) H. Leiber und A. Czinczel: ABS-spezifische Bremskraftverstärker, Fisita-Kongreß 1984

(2) Leiber, H. und Czinczel, A.: Der elektronische Bremsregler und seine Problematik, ATZ 74 (1972) 7, pp. 269 - 277

(3) P. Müller and A. Czinczel: Electronic antiskid system - Performance and application, Fisita 1972, 3/92

(4) Heinz Leiber and Armin Czinczel: Antiskid system for Passenger Cars with a Digital Electronic Control Unit, SAE-Paper 790 458

(5) H. Leiber: Bremskreisaufteilungen und mögliche Antiblockiersysteme, Fisita 1982, 82 125

(6) Heinz Leiber und Armin Czinczel: Four Years of Experience with 4-Wheel Anti-Skid Brake Systems (ABS), SAE-Paper 83 0481

(7) H.-W. Bleckmann, J. Burgdorf, H.-E. von Grünberg, K. Timtner and L. Weise: The First Compact 4-Wheel Anti-Skid System with Integral Hydraulic Booster, SAE-Paper 830 483

# Brake Boosters Designed Specifically for Anti-Lock Braking Systems (ABS)*

**Heinz Leiber and Armin Czinczel**
Robert Bosch GmbH
Stuttgart, Germany

*Paper 845106 presented at the XX FISITA
Congress, Vienna, Austria, May, 1984.

## ABSTRACT

For some time now, integrated anti-lock braking
systems have been developed, which incorporate
the vehicle's brake booster to help modulate
pressure during ABS operation. These newer systems have a number of different features up for
evaluation here. It is the objective of this
paper to describe the criteria for selecting the
optimal ABS configuration, and also to reveal
potential for future development work on the
brake systems that would reflect vehicle-specific
requirements.

## INTRODUCTION

Anti-lock braking has become a well known concept
in braking technology, and is now recognised as a
significant contribution towards active safety.
Various anti-lock braking systems (ABS) have come
onto the market. These systems have a hydraulic
pump for ABS pressure modulation that works independently of the vehicle's brake booster; it is
connected with the vehicle's master brake cylinder
by brake lines. The systems differ in their performance; pedal feel and noise development during
ABS operation are also different.

For some time now, integrated anti-lock braking
systems have also been developed, which incorporate the vehicle's brake booster to help modulate
pressure during ABS operation. These newer systems have a number of different features up for
evaluation here. It is the objective of this paper to describe the criteria for selecting the
optimal ABS configuration, and also to reveal potential for future development work on the brake
systems that would reflect vehicle-specific requirements.

## POTENTIAL PRESSURE MODULATION WITH ABS

Effective anti-lock braking systems are characterized by automatic modulation of the pressure
levels in the wheel brakes, which sets in as soon
as the wheels threaten to lock up from excessively hard brake application. Pressure modulation
generally consists of a sequence of pressure-
buildup, pressure-reduction and pressure-holding
phases controlling wheel slip in a narrow range
for optimal brake action.

This pressure modulation can be accomplished
using any of several principles, as illustrated
in Figures 1 and 2.

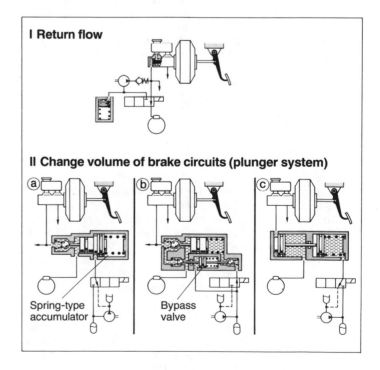

Fig. 1 - Different methods of pressure modulation
for ABS (Part 1)

Systems built on the principle of return flow have their ABS control valves located directly between the master cylinder and the wheel brake cylinders. Whenever the brake control unit requires pressure reduction in a wheel brake, the ABS control valve goes into its pressure-reduction position, which establishes a connection between the wheel brake cylinder and the pump. The brake fluid taken from the wheel brake is carried back into the master cylinder by the electrically driven return-flow pump. With this principle, the brake circuits remain closed during pressure modulation.

If a brake circuit fails, brake boosting is not impaired. The ABS function in the intact brake circuit is maintained. If the brake booster itself should fail, the legally required auxiliary braking action is still available. The ABS function is maintained as long as the driver is able to produce the brake pressure required for it without a brake booster.

Other systems work on the principle of pressure modulation by means of a change in the volume of the brake circuits, as shown in Examples IIa, IIb and IIc. These examples all incorporate a floating plunger to vary the volume of the respective brake circuit for pressure modulation. The floating plunger is controlled by a separate energy supply acting through the ABS control valves located in the auxiliary circuit.

Example IIa shows the floating plunger in its normal braking position: the ball seat valve located at the left is open. Whenever the brake pressure in the wheel brake is to be kept constant, the floating plunger is subjected to fluid pressure from the accumulator coming through the ABS control valve until the ball seat valve is closed. Whenever pressure reduction is required in a wheel brake, the floating plunger is subjected to accumulator pressure such that the cylindrical space communicating with the wheel brake (between the ball seat valve and the graduated floating plunger) is enlarged. The pressure in the wheel brake is then reduced in the enlarged space.

The brake circuits in Examples IIa, IIb and IIc also remain closed during pressure modulation. Example IIa incorporates a stiff return spring to ensure that the brake pressure produced in the master cylinder can never move the floating plunger by itself, thus keeping the ball seat valve always open during normal brake application. If the energy supply should fail, normal braking action is always maintained.

In Example IIb, the upper floating plunger is subjected to accumulator pressure when in its normal position. Whenever the pressure in the wheel brake is to be kept constant, the accumulator has to be cut off with the aid of the ABS control valve, and enough fluid has to be drawn off until the upper ball seat valve closes.

If a pressure reduction is required, more fluid must be drawn off, moving the floating plunger farther towards the right and increasing the cylindrical space communication with the wheel brake.

If the energy supply should fail, the floating plunger moves towards the right, closing the ball seat valve. In order to maintain normal braking action in this case as well, an auxiliary plunger standing under accumulator pressure is placed parallel to the first plunger; if the energy supply should fail, the auxiliary plunger is moved towards the left by a return spring, opening a ball seat valve. Then, the master cylinder and the wheel brake cylinder are connected with one another through this bypass, thus maintaining normal braking action. When the right ball seat valve closes, it locks the floating plunger in place, avoiding any undesirable increase in volume.

Example IIc requires no ball seat valve. Instead, a pair of plungers is moved towards the right by the pressure produced in the master cylinder, thus increasing wheel brake pressure. Whenever wheel brake pressure is to be kept constant, the ABS control valve is moved into its hold position. Whenever wheel brake pressure is to be reduced, additional fluid from the accumulator is fed to the right cylindrical space through the ABS valve. That moves the pair of plungers towards the left, reducing wheel brake pressure. If the energy supply should fail, normal braking action is maintained.

It is customary to monitor the position of the floating plunger in all plunger systems for reasons of safety.

The top of Fig. 2 shows an alternative in which the pressure modulation is done by controlling the master cylinder pistons. This alternative is characterized by having its brake valve positioned parallel to its master cylinder pistons. Whenever only partial brake pressure is applied, the primary ends of the pistons are subjected to pressure by the brake valve through the change-over valve, advancing the pistons. Control Bore R remains closed during partial application of brake pressure. There is no direct mechanical contact between pedal and piston. Pedal moveability is made possible by a simulator spring, the force of which acts on the brake valve. Movement of the master cylinder pistons is determined by the pressure/volume characteristic of the wheel brakes and the piston diameters.

Whenever the ABS control unit sets in calling for pressure reduction in the wheel brakes, the change-over valve is moved into its second position. That drains brake fluid off the primary end of the piston into the reservoir. The piston moves back, but no farther than enough for the left piston seal to uncover Control Bore R.

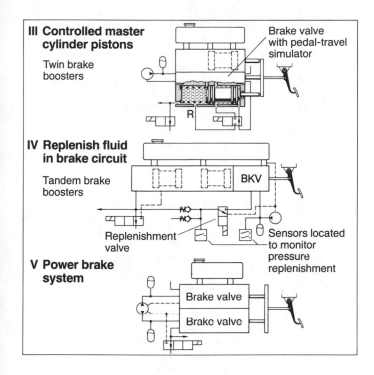

**III Controlled master cylinder pistons**

Twin brake boosters

Brake valve with pedal-travel simulator

R

**IV Replenish fluid in brake circuit**

Tandem brake boosters

BKV

Replenishment valve

Sensors located to monitor pressure replenishment

**V Power brake system**

Brake valve

Brake valve

Fig. 2 - Different methods of pressure modulation for ABS (Part 2)

The reason is that then brake fluid also flows out of the secondary circuit into the reservoir, causing the piston to come to a standstill. This is necessary so that the driver still can increase brake pressure further if necessary without having the brake rods make contact with the pistons.

When pressure has to be held, 2/2 solenoid valves are closed between the secondary ends of the pistons and the wheel brakes.

Under Example No. 4, the figure shows pressure modulation by means of replenishing fluid into the brake circuits. The brake valve controls the power-assist pressure impacting the master cylinder pistons as a function of pedal force. As long as the brake pressure to be applied is only partial, it is transmitted in the closed brake circuits to the wheel brakes through the open ABS control valves. Whenever the ABS control calls for a pressure reduction in the wheel brakes, a corresponding quantity of fluid has to be taken from the wheel brakes. If that quantity had to be replenished from the master cylinder for the next pressure buildup, the master cylinder pistons would advance by a corresponding increment, reaching their end positions after only a few control cycles. Then no further increase in brake pressure would be possible. That is the reason that during ABS operation brake fluid has to be fed directly to the wheel brakes from the power-assist circuit. That is the purpose accomplished by the replenishment valve on the right side, which is actuated whenever a brake control cycle begins,

establishing a connection between the power-assist circuit and the wheel brake. Two nonreturn valves keep the brake circuits separate.

The reliable functioning of the replenishment valve during ABS operation is essential for the safety of the brake system. That is the reason that a pressure switch is located at the output end of the replenishment valve. If the replenishment valve is not actuated, the pressure is zero; when the valve is actuated, the pressure must be at least as high as the pressure required to lock wheels on the particular friction combination of tires and road surface. It is better yet to monitor movement of the valve mechanically by means of a push rod. The position of the push rod can be compared with the control signals to the replenishment valve, such that the replenishment valve is monitored in both switching positions. The pressure supply also has to be made fail-safe by locating sensors accordingly, since it is reasonable to expect a pressure drop in the accumulator if one brake circuit fails, requiring shut-off of the pressure replenishment. More detailed treatment of the safety issues involved with this integrated anti-lock braking system is given in Reference (1)*.

At the bottom of the figure, the power brake system with anti-lock braking is shown. This system requires two separate hydraulic accumulators filled by a two-circuit pump. Two brake valves are also required. Of course, this brake system requires no additional equipment other than ABS control valves, because its brake circuits are open.

This system has the well known disadvantage that it produces no braking action if its pressure supply fails. This disadvantage constitutes a serious hurdle to be overcome before it could be applied on a large scale in passenger cars.

BASIC TYPES OF HYDRAULIC BRAKE BOOSTERS

There are two basic types of brake boosters relevant to this paper: those without a pedal-travel simulator and those with a pedal-travel simulator (refer to Fig. 3).

Brake boosters without a pedal-travel simulator have their brake valve connected directly with the brake pedal by means of the brake rod. The upper part of Fig. 3 shows the brake valve as movable inside a piston, which is impacted with brake pressure on its right side whenever the brake pedal is depressed. For more compact installation space, some manufacturers have the brake valve positioned outside the master cylinder axis, where it is moved by means of a lever. The piston surrounding the brake valve moves the brake rod piston by means of a rod. Pedal travel equals brake rod piston travel.

---

*) The numbers in parantheses refer to literature in the bibliography included at the end of this paper.

225

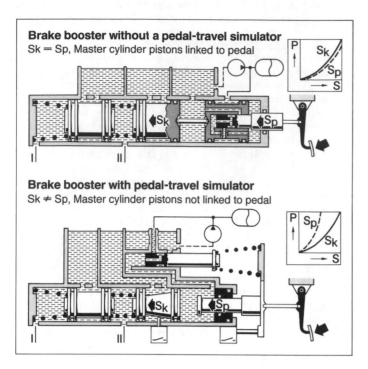

**Brake booster without a pedal-travel simulator**
Sk = Sp, Master cylinder pistons linked to pedal

**Brake booster with pedal-travel simulator**
Sk ≠ Sp, Master cylinder pistons not linked to pedal

Fig. 3 - Basic types of brake boosters (BKV)

With this system, it is not possible to specify freely a pedal-travel/force characteristic; instead, that characteristic is always a function of the brake's pressure/volume characteristic and the diameter of the master cylinder pistons. If one brake circuit fails, the pedal travel becomes excessive when the pedal is depressed.

With the second system, the brake valve is operated by the brake pedal by means of a simulator spring. The brake valve proportions brake fluid from the accumulator such that the resulting brake pressure impacting the brake valve from the left side is balanced off against the force of the simulator spring. Except for small movements to open the accumulator intake and return, the brake valve does not move; the fact that the simulator spring can be compressed makes it possible to move the brake pedal.

Brake pressure also acts upon the brake rod piston and brake rod, generating a resisting force which, together with the force of the simulator spring, equals the pedal force.

Fig. 3 shows the two systems subjected to partial braking pressure. The upper system exhibits rigid coupling between pedal travel and piston travel. With the lower system, one can see that the pistons have been displaced farther than the brake rod (and thus farther than the pedal as well). The diagrams at the right side of the figure display this difference graphically.

Movement of the brake pedal is decoupled from that of the brake pistons. The design engineer then is able to specify a pedal-travel/force characteristic freely, making him better able to satisfy desires that the automobile manufacturer may have. Then, piston diameters and piston strokes are determined by the pressure/volume characteristics of the brakes. Then, piston diameters can be dimensioned such that a driver can produce much higher brake pressures by moving the pistons with the brake rod directly in the case of an energy failure than he could with a conventional vacuum booster after an energy failure.

Travel of the stopper piston shown at the right side which guides the brake rod is limited by the brake pressure acting on the piston as long as the energy supply is intact, thus limiting pedal travel. If, however, there is an energy failure, this stopper piston becomes freely movable, so that the master cylinder pistons can be moved directly with the aid of the brake rod.

One advantage of this system is that the pedal does not travel excessively if one brake circuit fails, which has been known to shock drivers. Since a brake circuit failure is not noticed by the driver through the pedal, some device must be provided to warn the driver in such an event. In this system, such a device compares rod piston travel with brake rod travel. As long as brake circuits are intact, each increment of rod travel corresponds to a certain increment of rod piston travel. If one brake ciruit fails, the rod piston advances by a certain distance, thus disrupting the usual relationship between rod travel and rod piston travel. This can be detected with linear-displacement pickups attached to the brake rod and the rod piston respectively.

The travel measurements can also be used to alleviate impairment of the brake system caused by poor air-bleeding or vapor lock: Whenever the monitored piston slides ahead too far, additional fluid can be fed directly into the brake circuit for a short time (such as 100 ms), thus eliminating trouble.

Fig. 4 shows valve layouts designed to transmit the servo-pressure into systems with different brake boosters. The top brake booster has no pedal-travel simulator, whereas the lower one does. Let us take the worst possible scenario of an energy failure during an ABS-controlled braking maneuver on glare ice; in order to still be able to produce enough pressure for a 30 % deceleration without a brake booster, some provision would have to have been made - even if the ABS control set in at a relatively high wheel-locking pressure level while the vehicle was on a road surface with a high friction coefficient - for the master cylinder piston not to exceed a certain travel increment during the ABS control sequence, such that some residual actuation travel would still be available.

Thus, it is necessary to provide a safety margin for pedal travel in hydraulic brake boosters. Adding more fluid to the brake booster without a pedal-travel simulator like the one shown above would have to be done directly when the pistons reach a certain position, where they would have to be held with the aid of two shut-off valves. The correct piston position can be detected with a displacement pickup. If instead fluid were pumped in without holding these pistons in place, the pistons would be pushed ahead into the safety margin under the influence of pedal force. But that must be avoided.

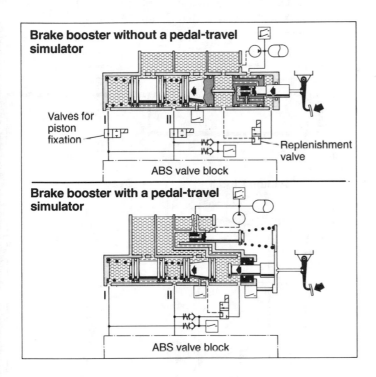

**Fig. 4 - ABS valve concepts used to match different types of brake boosters**

On the other hand, on the system with a pedal-travel simulator, shown on the lower half of the figure, no special shut-off valves are required. Since the brake pedal and piston are decoupled, the driver can increase brake pressure with the aid of the brake pedal without pushing the pistons into the safety margin. If the limit to brake booster action is reached, the ring-shaped stopper piston prevents further movement of the brake rod and brake pedal.

THE PEDAL CHARACTERISTICS OF DIFFERENT TYPES OF BRAKE BOOSTERS COMBINED WITH ABS

Fig. 5 shows the pedal characteristics of different kinds of brake boosters combined with ABS. At the upper left, the one for plunger and return-flow systems is shown. If the brakes are applied while the wheel-locking pressure is low (meaning a low adhesion coefficient), pedal force can be increased along the dotted line

above 1; however, pedal travel is limited to the value shown at 1. The pedal is locked in Position 1 because it is coupled with the master cylinder piston, the travel of which is in turn dependent upon wheel-locking pressure. If the wheel-locking pressure is high, the pedal is moved in accordance with the pedal-travel/force characteristic up to Point 2. But then again, it becomes no longer possible to increase pedal travel beyond that point.

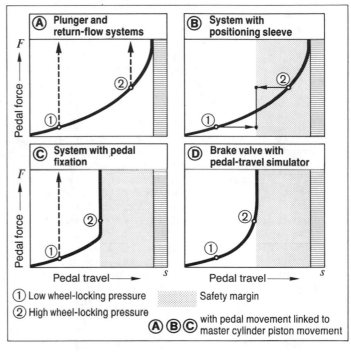

**Fig. 5 - Pedal characteristics of different types of brake boosters with ABS**

If pedal travel is increased further still, this happens along the dotted line above 2. The figure shows that the original pedal characteristic representing application of partial braking pressure is very different from the two dotted lines for ABS. The adaptation of this system to the driver is not optimal. Since the brake circuits are closed with these two systems it is not necessary to provide a safety margin in pedal travel.

At the upper right, the pedal characteristic for a hydraulic brake booster without a pedal-travel simulator is shown - a system incorporating a positioning sleeve to provide a safety margin in pedal travel.

With this system, pedal travel is also linked to piston travel. In order to provide the safety margin in pedal travel for the pistons, a positioning sleeve is activated to hold the pistons in a certain position when the vehicle is operated in the ABS mode.

When wheel-locking pressures are low, this means that the pedal offers little resistance in moving from Position 1 through to the safety margin for pedal travel as the brake control sets in, where it is then locked.

When wheel-locking pressures are high, this means that the pistons, and thus the pedal as well, are pushed back from Position 2 to the front limit of the safety margin for pedal travel. Here again it becomes apparent that the original pedal characteristic representing application of partial braking pressure is very different from the two finer lines for ABS braking maneuvers. During ABS braking maneuvers, which are often needed in critical driving situations, the driver will be lacking the pedal characteristic he is accustomed to, which can disturb and irritate him as described in Reference (1).

At the lower left, the system with pedal fixation is shown. The valve layout upon which this system is based was shown at the top of Fig. 4. Such a hydraulic brake booster without a pedal-travel simulator has its piston travel linked to its pedal travel. Whenever the pistons have advanced up to the safety margin for pedal travel, they are fixed in place by shut-off valves. When they are in this position, additional fluid is fed directly into the brake circuits through the replenishment valve, preventing any additional brake pedal travel. This characteristic is acceptable under conditions of partial brake application. In the lower pressure range, pedal movement contributes toward precise application of brake pressure, whereas in upper pressure range, the ability to apply pressure precisely is of less importance.

As soon as ABS operation begins, the shut-off valves are also activated and fluid is fed in directly.

When wheel-locking pressure is low (Position 1) the pedal characteristic departs radically from its original curve: Pedal travel cannot be increased by applying more force. When wheel-locking pressure is high, the pedal characteristic does not change from its original shape because the brake had no travel left originally.

At the lower right the pedal characteristic of the hydraulic brake booster with pedal-travel simulator is shown. With piston travel decoupled from pedal travel, it is possible to keep maximum pedal travel small enough by increasing the characteristic's progression that it is not necessary to provide pedal fixation for ABS operation. The pedal characteristic during application of partial braking pressure is identical with that during ABS braking maneuvers. The driver is given two parallel types of information on the braking force being applied: One is the resistance offered by the brake pedal and the other is pedal travel.

This way to inform the driver is optimal anthropometrically. The very situation in which ABS works are often those requiring increased concentration from the driver; then it is beneficial for him to have the same pedal characteristic he is accustomed to so that he is not distracted by the braking maneuver and can devote his entire attention to the traffic situation at hand.

The lower parts of Fig. 6 show a comparison between the pedal characteristics of the plunger system and those of a brake booster with a pedal-travel simulator. The upper part describes a hypothetical sudden transition in the road surface from a low to a high adhesion coefficient.

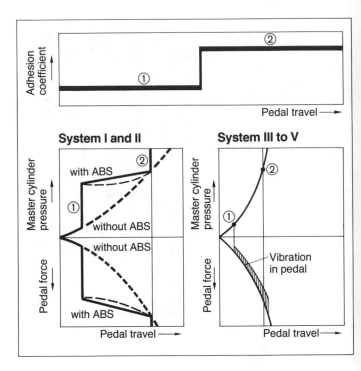

Fig. 6 - Pedal characteristics of different pressure - modulation systems

With the plunger system, brake control sets in at Point 1 in response to the low adhesion coefficient. Pedal travel can not be increased any further even if pedal force is increased. But as soon as the transition to the high adhesion coefficient takes place, pedal travel increases until steady-state brake control is reached on the road surface with the high adhesion coefficient at Point 2. During this time, depending on the driver's behaviour, pedal force may slightly increase or even decrease.

With the brake booster with a pedal-travel simulator, the brake control also sets in in response to the low adhesion coefficient. If pedal force is then increased further, pedal travel is also increased in accordance with the usual characteristic.

Pedal travel no longer depends at all on the brake control's cutting in, or on the adhesion coefficient prevailing between the road and the tires, but rather is now dependent upon pedal force alone in accordance with the usual pedal-travel/force characteristic.

During the brake control mode, there is slight vibration in the brake pedal, indicating to the driver that ABS is in action.

## CRITERIA FOR EVALUATING BRAKE BOOSTERS DESIGNED SPECIFICALLY FOR ABS

Now that we have had a look at brake boosters in their interaction with ABS, and have considered the problems involved with the various pedal characteristics, let us set up criteria for evaluating systems consisting of a brake booster and ABS components (see Fig. 7).

---

Pedal characteristics in cases of:
– ABS operation
– Circuit failure
– Fading
– Poor air-bleeding
– Vapor lock in a brake circuit

Braking action in cases of:
– Circuit failure without ABS
– Circuit failure with ABS
– Poor air-bleeding with ABS and
  alternating adhesion coefficient
– Vapor lock

Indication of a malfunction in cases of:
– Circuit failure
– Poor air-bleeding
– Vapor lock

Reliability of malfunctions being
indicated in cases of:
– Circuit failure
– Pressure supply

ABS operation in case of malfunctions:
– Circuit failure
– Pressure supply

---

Fig. 7 - Evaluation criteria for brake boosters specifically designed for ABS

When selecting a system consisting of a brake booster and ABS components, give special attention to the following:

ITS PEDAL CHARACTERISTIC in general, during ABS operation in cases of failure of a brake circuit, fading, poor air-bleeding and vapor lock in a brake circuit.

INDICATION IN CASES OF CIRCUIT FAILURE, poor air-bleeding and vapor lock.

BRAKING ACTION in cases of circuit failure with or without ABS, poor air-bleeding, vapor

lock, brake control on surfaces with alternating adhesion coefficients.

RELIABILITY OF A CIRCUIT FAILURE'S BEING INDICATED.

ABS OPERATION IN THE CASE OF A CIRCUIT FAILURE:

ADVANTAGES OF THE SYSTEM CONSISTING OF ABS COMPONENTS AND A BRAKE BOOSTER WITH A PEDAL-TRAVEL SIMULATOR.

Details have already been given on the major advantages of the integrated system consisting of ABS components and a brake booster with a pedal-travel simulator as regards its function in both normal braking and ABS modes. Let us just list here the advantages of this system in brief:

- Optimization of the pedel characteristic during ABS operation

- Indication of ABS operation in the form of slight vibration in the brake pedal

- Improvement in the brake system as a free side effect through such features as keeping the pedal travel from becoming excessive in the case of a brake circuit failure

- Higher brake pressure in the case of energy failure than with conventional systems

- Indication of malfunctions in the brake system (brake circuit failure, poor air-bleeding, vapor lock, for example)

- Potential for addition of subsystems such as electronic traction control incorporating brake management

- Potential for reducing costs, weight and installation space

- Pressure limitation in each brake circuit

## OUTLOOK

Anti-lock braking systems integrated with hydraulic brake boosters offer great potential for improving the function of the BAS components and that of the brake system as well. The systems described are examples taken from the first development phase. Given the projected demand for anti-lock braking systems and the different requirements of automobile manufacturers, one can expect this initial basis to be expanded by intensive development work.

REFERENCES:

(1) Leiber, H. and Jonner W. D.:
Pressure Modulation in Separate and
Integrated Antiskid Systems with Regard
to Safety, SAE 840467

(2) Leiber, H. and Czinczel, A.:
Der elektronische Bremsregler und seine
Problematik, ATZ 74 (1972) 7, pp. 269-277.

(3) Müller, P. and Czinczel, A.:
Electronic anti-skid system - Performance
and application, FISITA 1972, 3/92

(4) Leiber, H. and Czinczel, A.:
Antiskid System for Passenger Cars with a
Digital Electronic Control Unit,
SAE-Paper 790458

(5) Leiber, H.:
Bremskreisaufteilungen und mögliche Antiblockiersysteme, Fisita 1982, 82 125

(6) Leiber, H. and Czinczel, A.:
Four Years of Experience with 4-Wheel
Antiskid Brake Systems (ABS),
SAE-Paper 830481

(7) Bleckmann, H.-W., Burgdorf, J.,
von Grünberg, H.-E., Timtner, K. and
Weise, L.:
The First Compact 4-Wheel Anti-Skid System
with Integral Hydraulic Booster,
SAE-Paper 830483

*Systems*

# Antiskid System for Passenger Cars with a Digital Electronic Control Unit*

**Heinz Leiber and Armin Czinczel**
Robert Bosch GmbH
Stuttgart, Germany

*Paper 790458 presented at the Congress and Exposition, Detroit, Michigan, February, 1979.

AIMS FOR THE DEVELOPMENT OF ABS

- Based on a modular block system the ABS was to be developed for passenger cars and light weight trucks with a later version for heavy-duty trucks to follow.
- The electronic control unit was to be designed largely by using digital circuitry.
- The ABS was to feature maximum reliability.

- The wheel speed sensors were to be simple and rugged in design and most adaptable to a multitude of applications.
- The hydraulic unit was to be designed such as to allow for a straightforward installation without any changes in the existing brake system. Its operation was to be independent of other hydraulic systems in the car.
- The ABS was to offer high-performance, retain steerability and prevent swerv-

---

ABSTRACT

By introducing modern digital electronics, BOSCH succeeded in developing a high-performance antiskid system for passenger cars. A digital design approach was chosen since it allows for a greater degree of integration than an analog design.

This results in increased reliability. The hydraulic unit comprises three of four solenoid valves and a return pump driven by an electric motor. The system prevents vehicle swerving and maintains steerability while attaining remarkable gains in stopping distance at the same time.

As passenger cars developed, brake systems also improved steadily. Today almost optimal brake force generation has been achieved which assures stable braking on uniform roadways without any yaw movements as long as there is no excessive braking.

Excessive braking will result in wheel lock-up. In this case the vehicle is no longer steerable, may even swerve and in general the stopping distance will be longer than with optimal brake action.

Therefore, the objective of the antiskid system (ABS) is to reduce the brake pressure in the individual wheel should excessive braking occur so that the wheels generate maximum brake force without locking. Thus vehicle steerability is maintained and vehicle swerving prevented.

Although the first patents concerning antiskid systems date back to 1905, actual development did not begin until 1957. By 1966 the control problems experienced had been essentially solved. However, the introduction of high-performance antiskid systems, which also maintain steerability, failed in the following period due to their complexity and reliability problems with the electronic control unit which comprised numerous discrete electronic components. Simplified mechanical components and introduction of new technologies for the electronic control unit characterize the BOSCH high-performance ABS system for passenger cars.

233

ing while assuring at the same time optimal braking on uniform roads with constant friction, nonuniform roads with
sudden changes in friction coefficient
as well as different coefficients left
to right and on undulated roads.

## SOME FUNDAMENTALS OF THE DYNAMICS OF THE WHEEL BRAKED WITH ABS

### FRICTION FORCES BETWEEN WHEEL AND ROAD

The behavior of the brake force as a function
of wheel slip is shown in figure 1.

A criterion for the transmissable brake
force is the tire-to-road friction coefficient $\mu_B$. The coefficient rises steeply from
zero wheel slip and reaches its maximum at a
10 to 50% slip after which it drops with increasing wheel slip values.

Curve (1b) shows the side force coefficient $\mu_s$ which corresponds to the tire-to-
road friction coefficient $\mu_B$ of curve (1a).
With increasing wheel slip, the side force
coefficient $\mu_s$ drops and assumes its lowest
value when the wheel is locked. This low
value is only due to the angular position of
the slipping wheel which no longer has any
lateral guidance capability.

The control ranges of the ABS are marked
by hatched areas. The ABS must adapt itself
to a wide range of different road and driving
conditions.

The figure shows that, in general, shorter stopping distances can be achieved with
ABS as compared to excessive braking with
locked wheels. Only for curve (3) this does
not apply. The advantage of brake control
with ABS lies here - as with other road conditions - in maintaining stability and steerability of the car.

### SELECTION OF CONTROL VARIABLES

Basically, brake control uses circumferential wheel deceleration and acceleration,
and wheel slip as control variables. Today,
the following knowledge of control variable
selection is at hand:

### WHEEL SLIP AS THE ONLY CONTROL VARIABLE

Vehicle speed, for instance, can be
measured by the Doppler principle. As shown
in figure 1, optimum brake and side force
coefficients cannot always be utilized by
comparing the wheel slip with a preset slip
value.

### WHEEL CIRCUMFERENTIAL DECELERATION AND ACCELE-RATION AS THE ONLY CONTROL VARIABLE

This variable can be obtained by differentiating the angular velocity of the wheel.
On homogeneous roads with constant friction,
it is suitable as a control variable for
nondriven wheels. However, with the engine
engaged and driving in lower gear, the driven
wheels may attain high slip values without

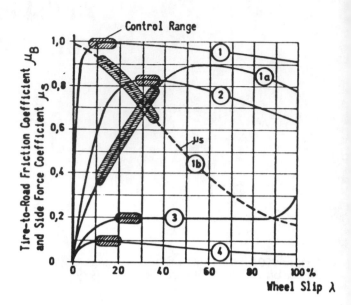

Fig. 1 - Brake and side force
coefficients versus slip

the angular wheel deceleration reaching a
preset reference value. With brake pressure
rising very slowly, these wheels may even
lock. Should a sudden decrease of the tire-
to-road friction coefficient occur, the non-
driven wheels may also attain high slip values
over a longer period. Therefore, circumferential wheel deceleration and acceleration
alone is not suitable as a control variable
either.

### COMBINATION OF CIRCUMFERENTIAL WHEEL DECELE-RATION AND ACCELERATION AND WHEEL SLIP AS CONTROL VARIABLES

By combining the two control variables
discussed above, namely wheel deceleration/
acceleration and a control variable corresponding to wheel slip, nearly optimal brake
control can be achieved. This corresponding
wheel slip variable is derived from a reference
speed which is obtained from the individual
wheel speeds. The process of generating same is
described later. This concept is used by BOSCH.

### SELECTION OF VALUE CONFIGURATIONS

The response of the ABS to varying road
and driving conditions is largely determined
by the way the brake pressure is varied. This,
in turn, is determined by the valve concept.
Before defining the valve concept different
solutions had been compared.

Figure 2 shows the variation of brake
pressure for different valve configurations.
The valve denoted by an arrow is activated
hydraulically, while all other valves are
solenoid valves.

Solution (A) comprises a hydraulic valve to limit the rate of pressure increase and a 3/2 - solenoid valve for pressure relieve. Slow and quick pressure rises and quick pressure rises and a quick pressure relieve are realized by this concept.

The small number of valves required here is advantageous. Definite disadvantages, however, are: Pressure increase during the wheel acceleration phase is critical on roads with low tire-to-road friction coefficients, because the wheels may be caused to decelerate prematurely. Wheel speed fluctuations and the consumption of brake fluid are high. Adaptation to sudden increases in friction coefficients, hysteresis of the brake and varying moments of inertia of the wheels with engine/transmission in gear is virtually impossible.

Solution (B) allows a pressure retention phase by applying an additional 2/2-solenoid valve. Stable brake control is possible with this configuration.

The large number of valves is disadvantageous. As with solution (A) there is no adaptation to positive step changes of the friction coefficient, to brake hysteresis and to varying moments of inertia at the driving wheels.

Solution C realizes slow and fast pressure rises also during the brake control cycle, pressure retention and quick pressure relieve. Hereby all requirements for a high quality antiskid system are met. In particular quick adaptations to positive step changes of the friction coefficient, to brake hysteresis and to varying moments of inertia of the driving wheels are possible. Of disadvantage is again the large number of valves required.

In Solution (D), only one 3/3-solenoid valve realizes all different pressure gradients of solution (C). Slow pressure increase is achieved by switching the valve between the first and the second position resulting in a stepped pressure increase. This valve meets all requirements. Only minimal electri-

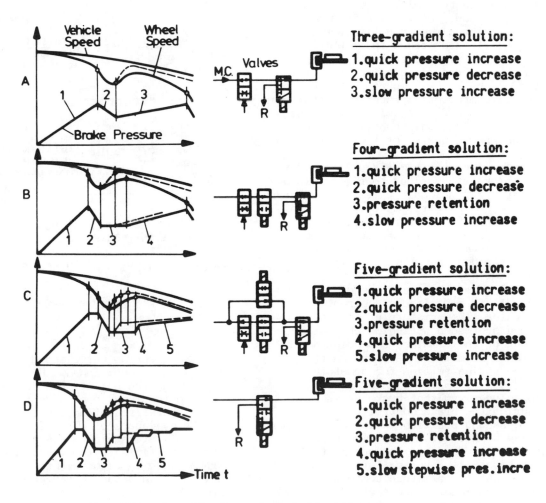

**Three-gradient solution:**
1. quick pressure increase
2. quick pressure decrease
3. slow pressure increase

**Four-gradient solution:**
1. quick pressure increase
2. quick pressure decrease
3. pressure retention
4. slow pressure increase

**Five-gradient solution:**
1. quick pressure increase
2. quick pressure decrease
3. pressure retention
4. quick pressure increase
5. slow pressure increase

**Five-gradient solution:**
1. quick pressure increase
2. quick pressure decrease
3. pressure retention
4. quick pressure increase
5. slow stepwise pres. incre

Fig. 2 - Fundamental behavior of brake pressure for different valve concepts

cal wiring is required. Stepped pressure increase, however, causes pressure vibration in the wheel brakes which necessitates special filtering techniques in the electronic control unit. In manufacturing this valve close tolerances must be observed.

Valve concept (D) was chosen for BOSCH ABS as a result of the above valve comparison. A functional description of this valve will be given later.

## SYSTEM DESCRIPTION

The ABS has been developed as a modular block system. Figure 3 shows one configuration for the conventional front-rear brake system with two wheel speed sensors for individual front wheel control. In order to minimize brake force differences at the rear axle on roads where different friction coefficients act on the right and left wheel, a common control for the rear wheels has been chosen.

The wheel speed sonsors measure the wheel speed and send sinusoidal voltage signals to the electronic control unit. By differentiating the wheel speed, the electronic control unit generates the circumferential wheel acceleration signal. The reference speed which corresponds approximately to the wheel speed at optimum brake force is obtained from the wheel speed signals. By comparing the control variables, circumferential wheel deceleration/acceleration and wheel slip with preset thresholds, the electronic control unit derives the command signals for the hydraulic unit.

The hydraulic unit comprises three (for front-rear brake systems) or four (for diagonal brake systems) solenoid valves and a return pump driven by an electric motor. Three positions in which brake pressure increases, is held constant or decreases can be assumed by the solenoid valve. It has three connections: the master cylinder, wheel brake cylinder and recirculation line.

Figure 4 shows the operation of the solenoid valve. It is energized by currents of different value. The armature of the valve incorporates a preloaded spring which limits the movement of the armature depending on the two different currents. In the non-energized position the spring retains the armature in the normal position $S_N$. The inlet is open and the outlet closed, so that brake pressure increases during the brake control cycle.

In the second position in which the valve is energized by half of the maximum current the magnetic force exceeds the spring force. In the position $S_I$ the spring force has a step increase so that the armature comes to a stop. In this position inlet and outlet are closed, holding the brake pressure at a constant level. In the third position the valve is energized by maximum current. The magnetic force exceeds

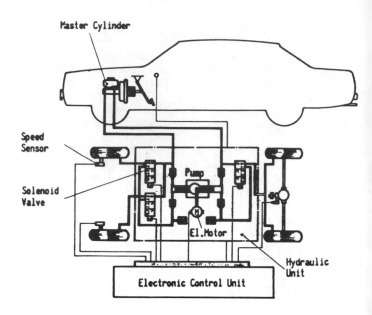

Fig. 3 - Schematic diagram of ABS

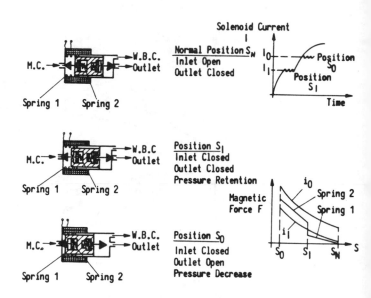

Fig. 4 - Functioning of 3/3 solenoid valve

the total spring force and drives the armature to its final position $S_0$. Here the inlet is closed and the outlet open, so that the pressure in the wheel brakes decreases.

The recirculation lines are connected to the cylinders of a small piston pump via two small storage chambers. The pump returns the brake fluid to the master cylinder if brake fluid is withdrawn from the wheel brake cylinders in order to decrease wheel brake pressure.

Figure 5 shows a picture of the ABS components on which, among other items, the two hydraulic units for the front-rear brake system and the diagonal brake system are shown. The latter incorporates four solenoid valves.

Figure 6 shows a block diagram of the electronic control unit. The signals provided by the wheel speed sensors activate a four-input amplifier designed as an integrated circuit (IC 1). Its output signals act on two integrated circuits each of them comprising several thousand transistors (ICs 2.1 and 2.2) in which the signal processing and the logic functions are realized in an all-digital manner. The output signals of these integrated circuits are the command signals for energizing the solenoid valves. They drive current regulators also designed as ICs. (ICs 4.1 and 4.2). According to the command signals, these circuits energize the solenoid valves with currents of different value.

Figure 7 illustrates the design features of the electronic control unit. A magnified view details a high density IC chip with a dimension of 6 x 6 mm.

To ensure the required system safety, the electronics and the electrical components of the hydraulic unit are checked for their proper functioning prior to driving the vehicle. While driving, the main components are continuously monitored. All checking is performed by a complex monitoring system also packaged in a highly integrated digital circuit (IC 3). The ABS system is deactivated in total as soon as a critical defect is detected. In that case, the basic brake system is still available while a signal lamp allerts the driver indicating that the ABS is not operational.

## TYPICAL CONTROL CYCLE

When activating the brake initially and when the tire-to-road friction coefficient decreases suddenly, high rates of pressure variation are required. A pressure decrease during the initial braking phase must be avoided. Once a control cycle has been initiated, the subsequent pressure increase must be considerably slower (by factor 5 to 10) in order to avoid hazardous resonances on the axle. Given a high tire-to-road friction coefficient and in consequence to the above we arive at a brake control cycle as shown in Fig. 8 and described as follows:

During initial braking the brake pressure in the wheel brake cylinder and the circumferential wheel deceleration increase. At the end of phase 1, the wheel deceleration passes a preset threshold $- a_1$. This, in turn, effects the solenoid valve to assume its second position, thus the brake pressure is maintained at a constant level. The brake pressure must not yet be decreased since the $- a_1$ threshold was already passed within the stable range of the tire-t-road slip curve. At the same time, the rate of increase of the reference speed is limited to a given value. The refe-

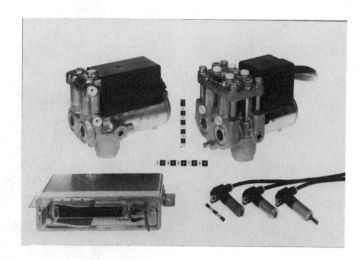

Fig. 5 - Photo of ABS components

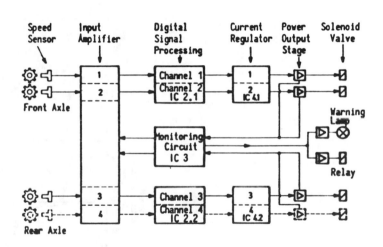

Fig. 6 - Schematic diagram of electronic unit with integrated circuits

rence speed is used to derive the $\lambda 1$ switching threshold. At the end of phase 2, the wheel velocity drops below the $\lambda 1$ threshold. As a result, the solenoid valve is switched to its third position causing the brake pressure to decrease for the time the circumferential deceleration of the wheel exceeds the $- a_1$ threshold. At the end of phase 3, the $- a_1$ threshold is again passed whereupon a pressure retention phase follows. Within a short time, the wheel circumferential acceleration has increased to such an extent that the $+ a_2$ threshold is passed. Hereby, pressure is still maintained at a constant level. At the end of phase 4, the wheel circumferential acceleration passes the relatively high $+ A$ threshold for example due to an increase of the tire-to-road friction coefficient. The brake pressure con-

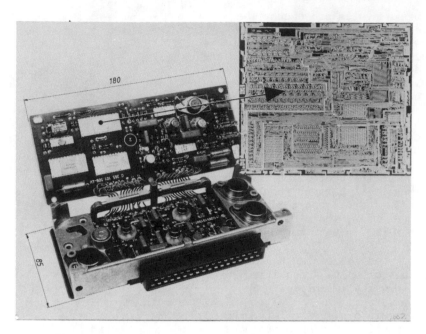

Fig. 7 - Photo of the electronic control unit

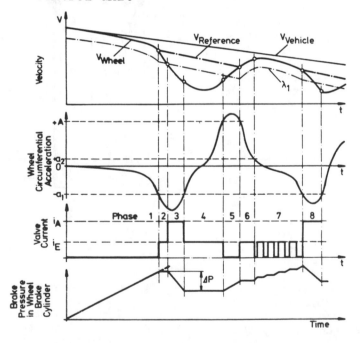

Fig. 8 - Control cycle of ABS

tinues to rise as long as the + A threshold is exceeded. During phase 6, the brake pressure is again maintained at a constant level. At the end of this phase, the wheel circumferential acceleration falls below the $+ a_2$ threshold. This is an indication that the wheel has entered the stable region of the tire-to-road slip curve and the brake force is below the optimal limit. By applying a pulse train to the solenoid valve between the first and second valve position, a stepped pressure build-up is achieved. This pressure increase is continued until the wheel circumferential deceleration drops below the $- a_1$ threshold (end of phase 7). This time, the brake pressure is decreased instantly, before a $\lambda 1$ signal has been generated. In contrast, during the first control cycle, the brake pressure was <u>not</u> to be decreased until a $\lambda 1$ signal developed in order to avoid a premature pressure decrease during initial braking in the stable region of the tire-to-road slip curve.

## SUMMARY

Utilizing advanced electronic technologies a high-performance antiskid system for passenger cars has been developed by BOSCH. Two well-known German car manufacturers are introducing the system presently, others will follow in the near future.

Among others, the system comprises two wheel speed sensors which measure the angular velocity of the front wheels. In order to minimize brake force differences of the rear axle on roads with different friction coefficients, acting on the right and left wheel simultaneous control for the rear wheels has been chosen.

Different valve concepts were evaluated to determine the best solution for a high-performance antiskid system. A valve was chosen which allows steady or stepped pressure rises, pressure retention and pressure decrease.

The hydraulic unit incorporates three (for front-rear brake systems) or four (for diagonal brake systems) of these solenoid valves and a return pump driven by an electric motor.

The electronic unit is mainly of digital design and contains a few integrated circuits. The antiskid system utilizes a large number of high level functions which result in a highly complex circuitry. The digital design was chosen because it allows for a greater degree of integration than an analog design.

To ensure optimal system safety the main components of the system are checked (self diagnosed) for their proper functioning prior to driving the vehicle. En route the main system components are continously monitored.

Upon detection of a critical defect the antiskid system is deactivated in total. While the basic brake system is still available a signal lamp allerts the driver indicating that the antiskid system is not operational.

# Four Years of Experience with 4-Wheel Antiskid Brake Systems (ABS)**

**Heinz Leiber and Armin Czinczel**
Robert Bosch GmbH, Stuttgart

**Paper 830481 presented at the International
Congress and Exposition, Detroit, Michigan,
February, 1983.

## ABSTRACT

Four years of experience with production line
installed 4-wheel antiskid systems (ABS) in
passenger cars were the motive for analyzing
and evaluating the results and the experience
gained. The performance, reliability and safety
in case of malfunction are the most important
reasons for the success of ABS. These points are
dealt with and future improvements are outlined.

## INTRODUCTION

Since October 1978 Robert Bosch GmbH has been
supplying the antiskid system for passenger cars
(ABS) to various automobile manufacturers. Over
200,000 passenger cars have now been equipped
with the ABS, and field experience as well as
customer reaction are very positive. After four
years of practical experience with the production
ABS it appears worthwhile to collect, analyze
and evaluate the experiences gained, to subject
the concept of the ABS to a critical review and
to draw conclusions as to possible improvements.

The success of the ABS in practice is dependent
on three main features:

- performance
- safety in case of malfunction
- reliability

## PERFORMANCE CHARACTERISTICS OF VARIOUS ANTISKID SYSTEMS

The performance of an ABS is decided by the
guarantee of

- steerability
- directional stability
- optimum retardation

As is well known, adaptive antiskid systems
are characterized by the fact that they control
the wheels within a narrow slip range in which
the maintenance of lateral control is guaranteed. The controlled front wheels thus have
guaranteed steerability while the controlled
rear wheels have guaranteed directional stability. Furthermore, the slip range selected is
usually the one in which the maximum braking
forces occur so that optimum retardation is
achieved.

These functions should be assured not only when
braking in a straight line, but also when braking while cornering and when braking on asymmetrical roadways having different right-left
conditions of grip. Disturbance factors, such
as the dragging moment of the engine or the
transition from asymmetrical to symmetrical
roadways, should also be mastered by efficient
antiskid systems.

The fulfilment of all these functions had
priority when the ABS was developed in cooperation with German automobile manufacturers.

Our extensive computer simulations and driving
tests with different antiskid systems had
already revealed at an early stage that only
ABS systems with 3 control circuits (individual control of the front wheels, joint control
of the rear wheels according to the "select-
low" method) can fulfil all these requirements
(1), (2)*). We feel that the success achieved
in the field and the positive reaction of
customers are due above all to the fact that
our ABS reliably performs the above-described
functions on all roadways and under all roadway
conditions.

---

*) Numbers in parentheses designate references
at end of paper

However, the technical layout involved with our present ABS gives cause time and time again to make critical comparisons with simplified antiskid systems.

The following pictures show the evaluation of possible antiskid systems which we have investigated to see how well they meet the requirements for steerability, directional stability and optimum stopping distance.

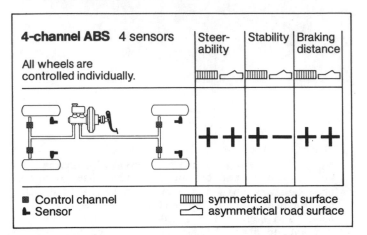

Fig. 1 - Evaluation of the 4-channel system with individual rear-wheel control

Fig. 1 shows the system with brake-circuit distribution front axle/rear axle, 4 wheel-speed sensors, 4 hydraulic channels and 4 control circuits. This means that all wheels are individually controlled. The brake pressure is individually metered to each wheel. While the steerability and stopping distance are optimal, the directional stability on asymmetrical split-grip roadways is not adequate since different braking forces are acting both at the front axle as well as the rear axle with the result that there is a large yawing moment. The typical driver is not able to compensate unequal breaking forces quickly enough with corrective steering.

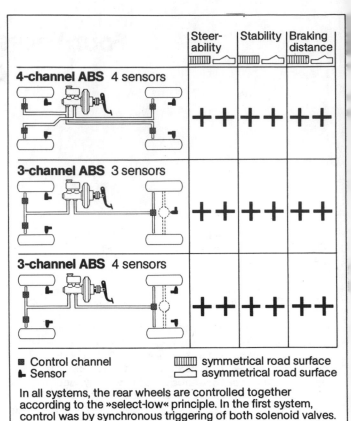

Fig. 2 - Evaluation of the Bosch-ABS

Fig. 2 shows the systems produced up to now by Bosch and fitted in vehicles of German manufacturers. Two hydraulic channels are used here as well on the rear axle, due to the fact that in this configuration the rear-wheel brakes belong to different brake circuits. Nevertheless, the rear-wheel ABS brake valves are controlled together in the "Select-low" operating mode with the result that the braking behaviour corresponds to that of the system described in the following.

This variant with 3 wheel-speed sensors and 3 hydraulic channels for a front-rear split braking system ideally fulfills all requirements as did the previously described system. Since the rear wheels are controlled jointly according to "select-low" and therefore with more or less the same braking forces, the undesired yawing moment on split-grip roadways is so small that it can be effortlessly compensated for with the steering. The same can be achieved with a 3-channel system with separate wheel-speed sensors on the rear wheels.

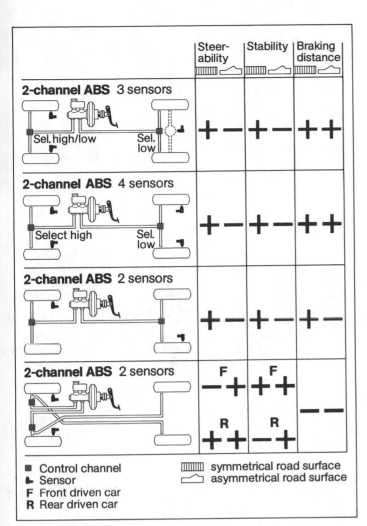

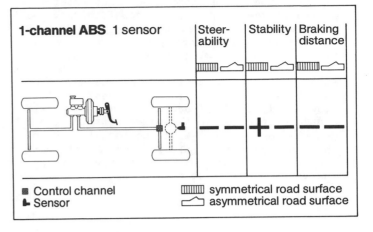

Fig. 4 - Evaluation of the 1-channel system

Fig. 4 shows the 1-channel system in which merely the rear wheels are controlled. This is done jointly according to "select-low". This, the simplest system is only able to guarantee directional stability when braking in a straight line. There is no steerability. The stopping distances are longer than in the case of optimum braking especially on wet roadways.

CRITICAL ROADWAY CONDITIONS AND CRITICAL DRIVING SITUATIONS

The deterioration in steerability and directional stability with the first three 2-channel systems was discussed in the last section.

Fig. 3 - Evaluation of the 2-channel systems

Fig. 3 shows different 2-channel systems. In the first of these systems with 3 wheel-speed sensors and 2 hydraulic channels there was provision for switching over from "select-low" to "select-high" on the front axle under severe split-grip conditions in order to also obtain sufficiently short stopping distances on such roadways. The problems of steerability and directional stability with this and the following systems are described in greater detail in Figs. 5 to 7. Under severe split-grip conditions this system suffered from an unreasonable deterioration in steerability and directional stability with the result that this system cannot be recommended.

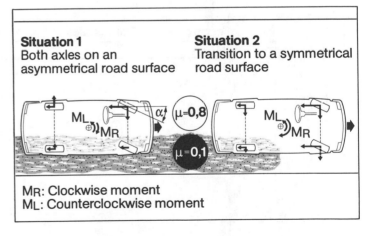

Fig. 5 - Critical steering behaviour with 2-channel ABS

Fig. 5 shows the problems of steerability and directional stability for these systems when moving from asymmetrical split-grip conditions to a symmetrical roadway with good grip.

In situation 1 the right-hand front wheel is locked at the lower friction coefficient. The front wheels are at the steering angle $\alpha$ with the result that the yawing moment as a result of the unequal braking forces is compensated for by the steering moment. As soon as the front axle comes onto the symmetrical roadway, the full braking force builds up suddenly at the locked, right-hand front wheel. This leads to a sudden increase in the moment turning the vehicle to the right with the result that in that instant the vehicle follows the steering angle. In this critical situation at high speeds the vehicle cannot be controlled.

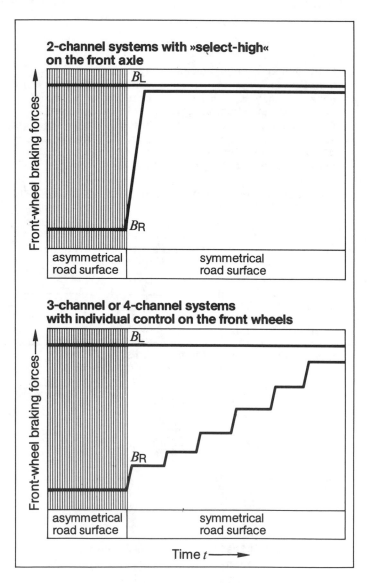

Fig. 6 - Characteristic curves of the front-wheel braking forces on 2-channel systems and on 3 or 4-channel systems

Fig. 6 shows the time curve of the front-wheel braking forces $B_L$ (left) and $B_R$ (right). On the asymmetrical split-grip roadway the braking force of the right-hand front wheel is much lower than that of the left-hand front wheel. Upon the transition to the symmetrical roadway with good grip, with the 2-channel system the braking force of the right-hand front wheel rises virtually instantaneously to the value of the left-hand front wheel braking force so that there is a sudden increase in the moment turning the vehicle to the right.

The braking pressure curve for the previously described 3-and 4-channel systems for the same braking maneuvre is shown in the lower Figure.

With these systems in which the brake pressure in the front wheels is individually controlled, this critical driving situation cannot occur because upon the transition to the symmetrical roadway the brake pressure is, starting out from the low level corresponding to the low friction coefficient, slowly built up in steps so that the driver has sufficient time to return the steering wheel to the straight-ahead position.

The second system is identical regarding its braking behavior. The only difference being that separate speed sensors are fitted to the rear sheels.

The third of these systems represents the so-called "diagonal system" in which two wheel-speed sensors are arranged diagonally and each control one axle with their own valve unit. As with the last systems, we feel that under severe split-grip conditions there is an unreasonable deterioration in steerability and directional stability, at least in the case of towing vehicles. Furthermore, compared with optimum braking the stopping distances are considerably longer if the more heavily loaded axle is controlled according to "select-low".

The last of these systems can in principle only be used for vehicles with a diagonally split braking system. The front wheels are individually controlled with two wheel-speed sensors and also determine the brake pressure of the diagonally opposite rear wheels. This system fails when the clutch is out on a slippery roadway; as a result of the dragging moment of the engine the rear wheels lock on rear-wheel-drive vehicles whereas, in the case of front-wheel-drive vehicles, the rear-wheel brakes are virtually pressureless, i. e. have no braking effect.

Wet roadways are characterized by a considerable difference between the maximum friction coefficient and the sliding coefficient (see Fig. 7 top right) when braking from high road speeds on such surfaces.

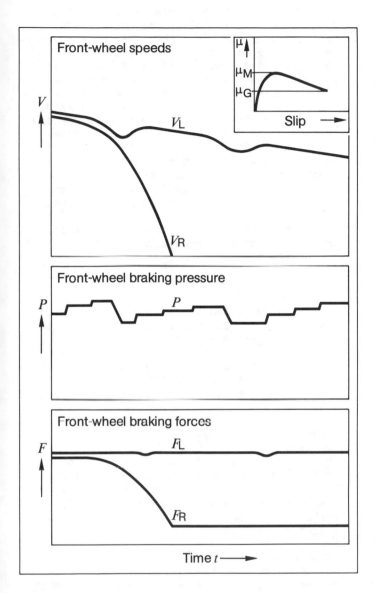

Fig. 7 - Control of the front wheels according to "Select-high" on wet road surface

Fig. 7 shows at the top the front wheel speeds and the brake pressure common to the two wheel brakes in the 2-channel systems which control the front wheels according to "select-high". Inevitable variations in the friction coefficient cause one of the front wheels to lock and, owing to the large difference between the friction coefficients for the controlled front wheel and the locked front wheel, there is an abrupt yawing moment which is very difficult to control particularly when braking from high speeds.

In our view these investigations show that such 2-channel systems cannot be recommended, but that separate control of the front wheels guarantees the best and safest braking performance.

## SAFETY IN CASE OF MALFUNCTION

The operation of the safety circuit is decisive with regard to safety in the case of a malfunction. The current, high safety level was reached by constantly expanding and improving the safety circuit from the concept stage through the test and trial phase down to a computer-controlled fault simulation.

Channel separation in the controller prevents centralized faults. The safety circuit employs reasonableness checks, it checks whether the occurrence of signals and their combination is logically correct, and whether the on-times of signals are physically possible. Apart from checking the controller signals, the safety circuit also monitors the sensor systems, the pump motor, valve coils and relays, as well as the wiring harness.

Before the vehicle moves off and after it has been standing still, a test cycle is initiated which checks the system. Control cycles are simulated and an identity comparison is made between the two controller modules. This is important because this involves the testing of circuitry components which are inactive during normal driving, i. e. when the ABS is inoperative, and whose failure would only become noticeable when the ABS comes into operation.

An important part of the test cycle is the self-test of the safety circuit. Faults are simulated and the reaction of the circuit is analysed. Valve coils, pump motor and relay are included in the test before the vehicle moves off. Should the monitoring or test circuit detect a fault, the system is switched off redundantly and a warning lamp lights up. The normal vehicle brakes, though, remain operational.

As mentioned later under "Prospects", we shall use the possibilities offered by large-scale integration to improve both system operation well as the safety circuit. In mid 1983 we shall be taking a revised ABS into series production. The new ABS will contain, among other things, 2 digital large-scale integrated circuits. In addition to the 2-channel signal processing and logic circuits, these large-scale integrated circuits each contain a complete safety circuit for the entire ABS. There is, therefore, a complete, redundant safety circuit. Since there is also redundant monitoring the possible failure of one of the safety circuits cannot interfere with the safety of the system in the event of a malfunction.

RELIABILITY

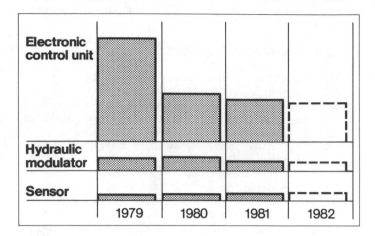

Fig. 8 - Field-fault failures of ABS components
         - comparison

Fig. 8 shows the comparative field failure-rate
of the ABS–system components within the warranty period. In the course of the four years
of production it has been possible to continuously reduce the already low failure rate
for all components. This is only possible thanks
to careful quality assurance which is described
below.

The quality of the future controllers is assured
as early as in the development phase. Each individual component is subjected to a type test
which guarantees the suitability of the component for the environment specific to motor
vehicles. The subsequent test bench endurance
test using a quick-motion method demonstrates
the stability of the components for the required
service life. The complete ABS electronics are
then subjected to a clearance-for-production
test. This includes mechanical and climatic
tests as well as laboratory and in-vehicle investigations into possible system disturbances.
Finally, the complete system is subjected to an
extensive large-scale trial in Bosch vehicles
and at the vehicle manufacturers. At the same
time, programs are devised for early failure
detection. These are subsequently incorporated
into the production process.

The ABS controllers also undergo several tests
during production. Assembly is followed by insu-
lation testing, fully automatic trimming and
a functional test of the units. After the
soldering of the trimming resistors and after
the visual examination some of the controllers
are subjected to a "burn-in" as a production
check. This "burn-in" means that the controller,
while switched on, is subjected to various
temperature cycles. This is a proven method of
early failure detection.

A route card is kept on the production process
and the test results of the final inspection
are documented. Random sample tests are constantly made during the individual production
phases.

The component failure concentration will be
dealt with briefly. Until the end of 1979 the
majority of failures on the electronic controller were due to the voltage regulator which
was replaced at the beginning of 1980 by an
improved type. As regards the hydraulic modulator the principal faults were damage to an
O-ring during assembly and to slight leaking
of the pump piston. These faults were remedied
through a modified assembly process and by
changing the design of the pump piston. The
main faults with the wheel-speed sensor are
open circuits between plug and cable and between cable and wheel-speed sensor which can
occur with unfavorable  routing of the cable.

To the best of our knowledge, not a single
critical situation arose as a result of one of
these defects.

PROSPECTS

The Bosch ABS was developed to the production
stage in almost 20 years. Its development can
be broken down into three phases:

- Solution of the technical problems of control.
- Achieving the high reliability and safety
  indispensable for parts of the brake system.
- Reduction of manufacturing costs so as to
  be able to offer a low-cost ABS; this phase
  is still in progress.

The experience gained from the 4 years of production have been used to revise and improve
the ABS components. This revised ABS will be
taken into series production in mid 1983.
Particularly on the hydraulic modulator, many
components have been made more production-
oriented and the weight has been reduced. The
electronic controller contains several functional improvements: the operating limit of the
ABS on ice has been extended until the vehicle
comes to a standstill, and the lateral stability, the stopping distances and the quality of
control have been further improved. There is
further integration of the electronic components. Thus, for example, the digital part has
been reduced to two identical large-scale
integrated circuits which both include a two-
channel controller and a safety circuit.

The total circuitry with 60 elements is accommodated on one printed-circuit board so that
there is a more than 50 % reduction in the
number of elements compared with the previous
production model.

In view of the great expense for the hydraulic part of the system, we have been looking for a solution for several years in order to make the hydraulic part smaller, lighter and more economical.

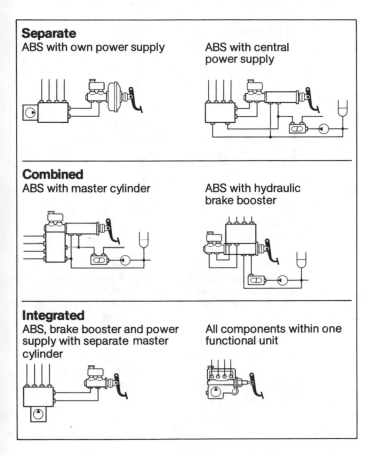

Fig. 9 - Possible antiskied systems

In considering the system comprising brakes and ABS, it is possible to make a distinction between separate, combined and integrated designs (see Fig. 9). The adding of a separate ABS to the brake system is a convenient possibility when first introducing the ABS so that no modifications are necessary to the existing brake systems. In the case of combined systems it is possible to structurally combine the ABS with the master cylinder or with the brake booster. This first step towards integration still proves complicated as regards pipework or internal routing. In integrated systems the booster and energy supply can be combined with the valve block - the normal master cylinder can then be used - or the brake booster is combined with both master cylinders and the reservoir and ABS valve block are integrated. This, the most compact solution requires approximately the same installation space as the present-day vacuum booster and is flexible as regards mounting. We are currently working on

this structurally and functionally integrated system so as to make a decisive step forward regarding cost and size.

The hydraulic brake booster is utilized here in two ways by means of a new principle of brake-power boost and ABS pressure-modulation. At the same time, the following brake-system functions can be improved in this system:

- If the pressure supply fails, or one of the brake circuits, the brake pedal does not push through to the floor
- In comparison to conventional brake boosters, up to 50 % higher brake pressures are available in case of failure of the pressure supply. Normally, the brake pedal has an unpleasantly hard feel when the brake booster fails. This is avoided in this system.
- If the pressure supply fails, this does not result in a failure of the circuit.

The integration of brake booster and ABS offers essential advantages as regards weight and volume.
Compared with the conventional system comprising vacuum booster and ABS, the weight of the integrated hydraulic system could be reduced by half.

This concept represents the Bosch objective. Due to the high innovative content in integrated systems, it is necessary to plan a long period of testing in order to achieve extremely safe and reliable systems.

It is intended that this system should also find its way into middle-class vehicles so that the increase in active safety through the ABS can be of benefit to the majority of passenger cars.

Fig. 10 - Photo of the integrated system

Fig. 10 shows a prototype of this integrated system.

## SUMMARY

During the past four years, more than 200,000 passenger cars have been equipped with the 4-wheel antiskid system (ABS) from Bosch. The experiences gained under practical everyday driving conditions are presented, analyzed and evaluated. Of major importance for the success of ABS are its performance, its reliability and the safety in case of malfunction. The performance of the 4-wheel ABS is compared with that of simplified systems. In doing so, it becomes apparent that the 2-channel systems which have appeared up to now have such functional disadvantages that their introduction and use appears to be inadvisable. The practical increase in safety to be gained with the simple rear-axle antiskid systems seem to be so small that, in our opinion, they cannot be recommended. Using the operational criteria steerability, directional stability and braking distance, it is demonstrated that, under given road-surface conditions, the 2-channel systems can become critical from a safety point of view.

Reliability is dealt with by comparing various components with respect to their failure rate within the warranty period. The failure-concentration points are outlined. In order to guarantee the reliability of the electronic control unit, extensive quality assurance tests have been introduced during production. The safety in the case of malfunction, is guaranteed by a special safety circuit which automatically checks and monitors itself and all the important components in the ABS before start-off and every time the vehicle is at a standstill. If a fault is detected, the ABS is switched off. During actual driving, all the important components of the ABS are continuously monitored and if a critical fault is detected by the safety circuit this switches off the ABS.

In the middle of 1983, a constructionally modified ABS will come onto the market with a much reduced weight. The possibilities now offered by the increased component density in large-scale integrated circuits means, among other things, that the safety circuit could be designed, in two large-scale circuits, to be fully redundant.

Finally, it is demonstrated how we are aiming, in the middle-term, at achieving widespread application of ABS even in passenger cars of the medium-price range. We intend to achieve this goal by integrating the ABS, both functionally and constructively, in a hydraulic brake power assist unit. A prototype of this system upon which development work is being carried out, is shown in the photograph.

## REFERENCES:

(1) Leiber, H. und Czinczel, A.:
Der elektronische Bremsregler und seine Problematik, ATZ 74 (1972) 7, S. 269 - 277

(2) Burckhardt, M., Glasner von Ostenwall, E. Ch. und Krohn, H.:
Möglichkeiten und Grenzen von Antiblockier-systemen, ATZ 77 (1975) 1, S. 13 - 18

(3) Leiber, H. und Limpert, W. D.:
Der elektronische Bremsregler, ATZ 11 (1969) 6, S. 181 - 189

(4) R. H. Madison, H. E. Riordan:
Evolution of Sure-Track Brake System, SAE 690213, January 1969

(5) Schafer, T. C., Howard, D. W. and Carp, R. W.:
Design and Performance Considerations for a Passenger Car Anti-skid System, SAE 690058, 1968

(6) Reinecke, E. und Weise, L.:
A new Anti-skid Concept for Commercial Vehicles, with an Optimal Solution Concerning Active Safety and Costs, Report 6 - 7, Fisita Tokio 1976, p. 6 - 55/63

(7) Heinz Leiber and Armin Czinczel:
Antiskid System for Passenger Cars with a Digital Electronic Control Unit, SAE 790458, 1979

# Rear Brake Lock-Up Control System of Mitsubishi Starion*

**Satohiko Yoneda**
Passenger Car Engineering Center
Mitsubishi Motors Corp.
**Yasuo Naitoh**
Himeji Works
Mitsubishi Electric Corp.
**Hideo Kigoshi**
Automotive Products Division
The Nippon Air Brake Corp.

*Paper 830482 presented at the International Congress and Exposition, Detroit, Michigan, February, 1983.

## ABSTRACT

This paper describes a new type Rear Brake Lock-up Control System which is mounted on MITSUBISHI STARION. This system meets all the requirements of the ECE R13, Annex 13, gives a high braking stability regardless of the road conditions, and minimizes the body vibration and the pulsative reaction on the brake pedal to achieve a comfortable feeling during the brake application. It would be expected that the adoption of this sytem could contribute greatly to the reduction of car accidents caused by the rear wheel lock-up.

MITSUBISHI MOTORS CORPORATION developed jointly with Mitsubishi Electric Corporation and The Nippon Air Brake Co., Ltd. a new type Rear Brake Lock-up Control System which was placed on the market by MMC in Europe in April, 1982, in Japan in May and in the U.S. in September, as one of the major product features on newly-introduced MITSUBISHI STARION.

The MITSUBISHI STARION has a high-powered engine producing a maximum car speed of more than 125 mph. The Rear Brake Lock-up Control System has been adopted in this car to support its high vehicle performance.

This system meets all the requirements of the ECE R13, Annex 13 and has been approved in Europe as an anti-lock device.

## DEVELOPMENT OBJECTIVE

When driving on a rainy day, any driver will be inclined to drive slower than on a fine day. This should be mainly because the driver has no confidence on the brake performance of the car he drives. However, if a car can give full-powered running only on a fine day, that car cannot claim to be a highpowered one. If a reliable brake operation is insured at all times, the car could be driven at a higher speed than a car of comparable engine power but with a less reliable brake system.

The MITSUBISHI STARION required a powerful and reliable brake system to fully achieve its high power. This car is equipped naturally with disc brakes on both front and rear wheels and for additional safety in braking, a new type Rear Brake Lock-up Control System was developed.

It is well established that a conventional anti-lock system is very effective to insure safety when braking a car, while the system has not yet been widespread in the market probably because of high cost penalty among others. In view of this, we started development of a control system for two rear wheels that would meet the following performance requirements.

(1) To attain far higher braking stability than that of a car without a conventional anti-lock system and to keep a stopping distance shorter than or at worst the same as that of a car not equipped with the system, regardless of the road conditions.

(2) Not to cause a pulsative reaction on the brake pedal, car shake or shock that gives the driver an uncomfortable feeling or an uncertainty when the system is operating.

(3) Not only to insure high reliability under any of car-driving environments but to cut off the system in case of any failure in the system, so that the regular brake system function comes into work.

(4) To function under the vehicle speed of from 5 mph to 140 mph regardless of the transmission gear position,

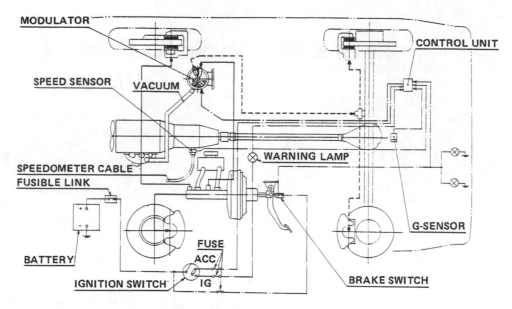

**Fig 1 System configuration**

the brake pedal force and the brake pedal depressing speed.

## SYSTEM DESCRIPTION

Fig. 1 and 2 show the configuration of the Rear Brake Lock-up Control System and the appearance of its major components, respectively.

Two sensors are provided, one is the speed sensor, mounted at the speedometer cable take-out of the transmission case, to generate the wheel velocity signal and the other is the G-sensor mounted in the luggage compartment, to generate the vehicle deceleration signal in the car longitudinal direction. The control unit calculates the wheel deceleration and the slip ratio between the road surface and the tires through the wheel velocity signal and the vehicle deceleration signal. Thus, the control unit determines the brake fluid pressure control mode and gives a pressure control command to the modulator.

The modulator, that cuts off the brake fluid line between the master cylinder and the wheel cylinder to control the fluid pressure of the wheel cylinder, is operated by the vacuum pressure, supplied from the engine intake manifold. This modulator is equipped with two solenoid valves, a release solenoid valve and a build-up solenoid valve, which control the wheel cylinder fluid pressure in three modes of RELEASE (REL), FAST BUILD (FB) and SLOW BUILD (SB) according to the pressure control command from the control unit.

A warning lamp is located in the instrument panel and lights up in case of any trouble occurred in the system.

## CONCEPT OF THE CONTROLS

The wheels under the control should ideally be decelerated at all times with the slip ratio that gives maximum coefficient of friction (hereafter denoted $\mu$max) between the road surface and the tires. As the slip ratio that gives $\mu$max changes continuously depending on the road surface and the tires conditions, common approach in an anti-lock system is to control the wheel deceleration with the slip ratio in the vicinity of the one corresponding to $\mu$max. Achieving this on all roads is, however, difficult unless the vehicle velocity can be sensed very accurately.

On the other hand, if the wheels are controlled at the slip ratios corresponding to two extreme cases such as wheel lock-up and almost being free, not only a large change in the vehicle deceleration is produced, but the engine will stop

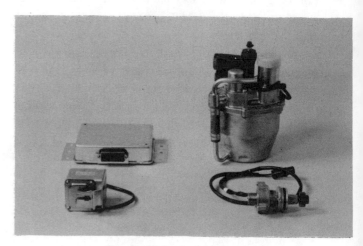

**Fig 2 Major components**

quite readily on a slippery road when the manual transmission is in-gear state.

In our Rear Brake Lock-up Control System, in order to decelerate the wheels with a slip ratio in the vicinity of the one corresponding to $\mu$max, the target slip ratio is not fixed but instead is varied appropriately at every control cycle. Namely, the timing of the pressure control command is varied according to the minimum slip ratio value of each control cycle. Fig. 3 shows an example of the control cycle. In the first control cycle, the control is made with the slip ratio that is assumed to be in the vicinity of the one corresponding to $\mu$max because no data is available for the calibration. In the second control cycle, the release timing is varied according to the minimum slip ratio registered during the first control cycle. This calibration is made at every control cycle (excluding the first control cycle).

In this manner, the peak of the $\mu$-slip curve is constantly followed by this system (adaptive control), making it possible to shorten the stopping distance and improve the braking comfort.

GENERAL DESCRIPTION OF COMPONENTS

SPEED SENSOR – There are two types of speed sensors that are employed in a conventional anti-lock system, one is the wheel sensor that directly senses the wheel rotational speed and the other is the propeller shaft sensor that senses the mean rotational speed of R.H. and L.H. wheels.

For our Rear Brake Lock-up Control System we adopted the propeller shaft sensor which is mounted at the speedometer cable take-out of the transmission case in order

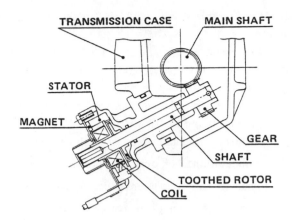

Fig 4  Speed sensor

to protect the sensor from the external environment (boucing stone, mud, etc.) and to give good accessibility for maintenance.

Fig. 4 shows the construction of the speed sensor, which is a variable reluctance type consisting of a magnet, a coil, a stator and a toothed rotor. The toothed rotor rotates via a gear that is driven by the speedometer drive gear in the transmission, and induces 60-pulse/rev. in the coil. The air gap between the stator and the toothed rotor is 0.6 mm and an output voltage of approximately 0.9 volts peak-to-peak is generated when the car is running at 5 mph.

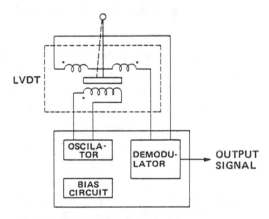

Fig 5  Configuration of G-sensor

G-SENSOR – The G-sensor is a linear accelerometer that generates a voltage in proportion to the longitudinal deceleration of the vehicle. The G-sensor linearly senses the longitudinal deceleration of the vehicle in a range of coefficient of road surface friction from 0.1 to 1.0.

Fig 5 shows the configuration of the G-sensor. The G-sensor is so constructed that the core of the LVDT (Linear Variable Differential Transformers) supported by leaf springs moves according to the magnitude of the vehicle deceleration. When the G-sensor is subjected to the deceleration, the core is displaced by the amount corresponding to the deceleration. The displacement of the core is sensed by the demodulating

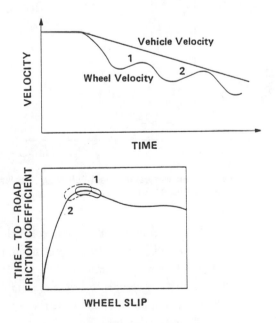

Fig 3  Control cycle

circuit. The G-sensor is filled with oil to dump the mechanical vibration.

The G-sensor has a resolution of 0.01 g or better and a response time of approximately 100 ms.

CONTROL UNIT – The control unit provides three kinds of pressure control commands of RELEASE (REL), FAST BUILD (FB) and SLOW BUILD (SB) and it selects the optimum control command among the three based on the result of computation. For optimum pressure control command timing, it uses two third-order active filters with different cut-off frequency to compute the two wheel velocity signals $\omega_1$ and $\omega_2$. The vehicle velocity, which cannot be measured, is assumed in the control circuit accord-

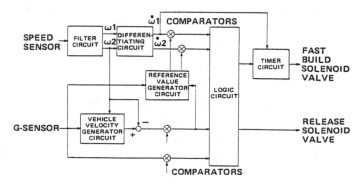

**Fig 6  Block diagram of control unit**

ing to the wheel velocity signals supplied from the speed sensor and the information from the G-sensor. The control unit has a circuit to compensate for the error caused by the inclination of the G-sensor to the road surface as well as a circuit for response time compensation.

Fig. 6 shows the block diagram of the main circuit in the control unit and Fig. 7 shows an example of the control cycle.

The REL mode is set at the point of time $t_1$ when both of the two wheel deceleration signals computed by the two wheel velocity signals have reached the threshold deceleration. Namely, the REL mode is set when it is assuredly sensed that the brake torque has clearly reached the friction torque caused by the friction between the road surface and the tires. Resetting of this mode takes place at the point of time $t_2$ when the wheel deceleration computed from $\omega_1$ has decreased to the switching threshold. This is to prevent the brake torque from becoming too small relative to the friction torque. Then, the brake fluid pressure stops falling while transition to the SB mode takes place, followed by gradual build-up of the wheel cylinder fluid pressure. In the SB mode, the slip ratio in the vicinity of the one corresponding to $\mu$max is effectively utilized and at the same time, the

fluctuation of the wheel cylinder fluid pressure during braking is minimized for better braking comfort. During the time period of the REL mode, the wheel rotation changes from deceleration to acceleration. When the wheel acceleration $\dot{\omega}_1$ reaches the FB threshold (point $t_3$), the FB mode is selected. The duration time of FB mode is determined by the magnitude of subsequent wheel acceleration. Namely, the FB mode is intended to provide good correspondence in case where the brake torque has become too small as compared to the friction torque or where the friction torque has increased significantly due to friction change between the road surface and the tires. (due to the car proceeding from a low adhesion coefficient road to a high one.)

In this manner, the control unit provides controls according to the wheel deceleration and acceleration signals. In addition, the deceleration and acceleration threshold are corrected according to the G-sensor output and at the same time, they are corrected at every control cycle based on the

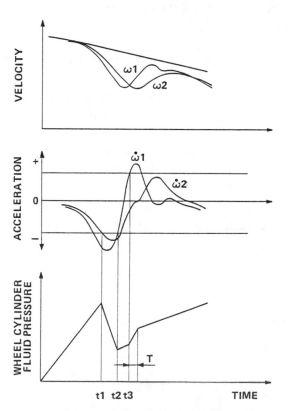

**Fig 7  Interration ship between wheel velocity, acceleration and  wheel cylinder fluid pressure**

slip ratio as described in concept of the controls. Through these modification the pressure control command timing is varied, making it possible to decelerate the wheels with the slip ratio in the vicinity of the one corresponding to $\mu$max, and to secure high braking stability even on the surface of which $\mu$max is not definite, such as on a gravel road.

The logic circuit not only selects the control command

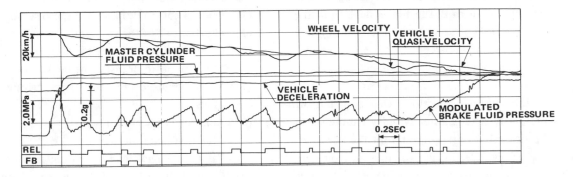

Fig 8  Osillographic recording of the system operation

based on the wheel deceleration and the acceleration signals and the slip ratio, but also discriminates the true signals from the noisy signals, supplied from the speed sensor and the G-sensor resulting in effective prevention from erroneous operation by using a timer and a sequential circuit.

Through these controls, high braking stability is insured at all times regardless of any changes in the factors that determine the wheel behavior such as the road adhesion, the transmission gear position and the vehicle velocity.

Fig. 8 shows the vehicle quasi-velocity, the wheel velocity, the master cylinder fluid pressure and the modulated brake fluid pressure.

MODULATOR – Fig. 9 shows the sectional views of the modulator.

The modulator consists of a choke valve to limit the build-up rate of the wheel cylinder fluid pressure, a check valve to shut off the brake fluid line between the master cylinder and the wheel cylinder, a plunger to change the displacement of the fluid chamber connected to the wheel cylinder for controlling the wheel cylinder fluid pressure, a vacuum piston to drive the plunger and two solenoid valves.

The choke valve limits the wheel cylinder fluid pressure build-up rate to 24.5 MPa/s, when the master cylinder pressure build-up rate is too high. This is effective in improving the braking capability when a panic braking is applied. Fig. 10 shows the construction and characteristics. So long as the master cylinder fluid pressure remains below 4.9 MPa, the master cylinder fluid pressure is transmitted to the fluid pressure control section as it is. When the pressure reaches 4.9

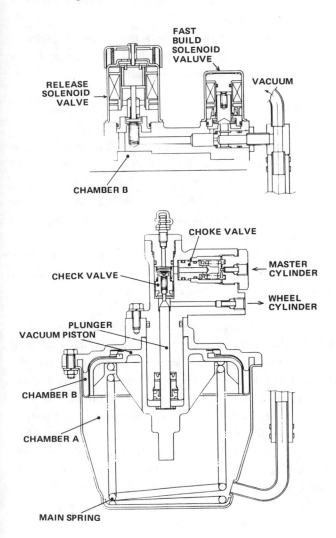

Fig 9  Modulator

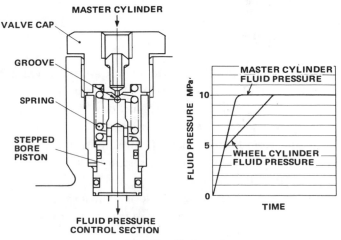

Fig 10  Choke valve

253

MPa, the stepped bore piston overcomes the spring, getting into contact with the valve cap. The brake fluid from the master cylinder to the fluid pressure control section can flow only through the groove of the stepped bore piston so that the fluid pressure build-up rate in the control section becomes lower.

The check valve has two functions. One is to shut off the fluid pressure between the master cylinder and the wheel cylinder and the other is to prevent a abrupt fluid pressure build-up in the wheel cylinder in case where a fluid pressure higher than the pressure existing at pressure shut off is required, such as where the car is proceeding from a low adhesion coefficient road to a high one. When the release and fast build solenoid valves are not in operation, the fluid pressure circuits on the master cylinder side and the wheel cylinder side are at the same pressure level as if there were no Rear Wheel Lock-up Control System.

When the release solenoid valve is actuated by pressure control command from the control unit, the atmosphere is introduced into the chamber B, and the vacuum piston and the plunger overcome the main spring force, moving to close the check valve and increasing the volume of the brake fluid line on the wheel cylinder side, which in turn reduces the brake fluid pressure. Maximum fluid volume change produced here is approximately 1.8 cc. When the release solenoid valve becomes inactive later, the chamber B is connected via the orifice to the vacuum source, causing gradual decrease of the pressure in the chamber B. As a result, the vacuum piston and the plunger are forced back by the main spring, and then the wheel cylinder fluid pressure builds up gradually. If, under this condition, the fast build solenoid valve is actuated, the passage bypassing the orifice is opened to fast-decrease the pressure in the chamber B down to the pressure in the chamber A. As a result, the return speed of the vacuum piston and the plunger increases, causing fast build-up of the wheel cylinder fluid pressure. It takes less than 20 ms to switch the modes.

## FAIL-SAFE FUNCTION

Each part and component of this system is well attended to EMI (Electro-Magnetic Interference) and the control unit has various diagnosis functions. The system is so designed that whenever any fault is detected by the diagnosis logic, the Rear Brake Lock-up Control System stops functioning itself and the regular brake system takes over, and the warning lamp alert the driver.

The self check diagnosis logic functions no matter whether the car is running or not, so long as the ignition key is in the ON position.

This function checks the following items.
(1) Open circuit to or inside the speed sensor
(2) Short circuit of the G-sensor
(3) Open circuit to or inside the G-sensor
(4) Open circuit to or inside the modulator
(5) Short circuit of the transistors driving the solenoid valves
(6) Abnormal deceleration signal
(7) Abnormal slip ratio signal
(8) Fail in the logic circuit operation
(9) Open circuit of stop lamp switch wiring and all stop lamps blown.

In addition, the following functions are provided.
(1) Startup check function to check functioning of the modulator by placing the ignition key in the ON position with the brake pedal stepped on and actuating the modulator momentarily to check for its operating sound.
(2) Lamp check function to cause the warning lamp to light for 3 seconds when the ignition key is placed in the ON position.
(3) Function to stop the Rear Brake Lock-up Control System functioning in the event of abnormally flow power voltage to the control unit.

## THE EVALUATION TESTS

The evaluation tests were conducted on every conceivable type of a road surface from a high coefficient of adhesion to a low one including a split road surface, with the emphasis placed on (1) braking stability, (2) braking comfort under system operation, (3) durability of system components and (4) electro-magnetic compatibility.

Actual running tests were conducted not only in Japan but also in West Germany and the Netherlands in winter season to successfully demonstrate its highly practical performance.

The car equipped with this system gives higher braking stability than a car without the system, naturally, at hard braking regardless of the road surface conditions, but the effect of the Rear Brake Lock-up Control System is most markedly seen on a split road surface. On this type of road surface, a car without this system often ran out of control but when equipped with this system, the same car could be brought to stop, sliding only slightly toward the dry road surface. (Fig. 11) Thus, it was demonstrated that this system is highly effective on a road surface partially frozen.

Fig. 12 compares the stopping distance of cars with and without the Rear Brake Lock-up Control System. Thanks to this system the stopping distance by 5% on a dry surface, 6% on a wet surface and 13% on a slippery surface has been reduced. This data indicates that this system controls braking in such a manner that the braking takes place always in the vicinity of the slip ratio at which the coefficient of friction is the largest, regardless of the transmission gear positions, on any road surfaces from a low coefficient of adhesion to a high one. Even when the car rushed from a low adhesion coefficient road to a high one or vice versa, the wheel locking was positively prevented by adequate computer control, securing the stability.

As for the car's longitudinal shake caused by the operation of this system, tests were repeated with the initial target of less than 0.05 g. Through the adoption of the linear G-sensor, three-control modes by two solenoid valves, improvement of the computer control logic and the adoption of the pressure control valve, the car shake was reduced to as low as 0.04 g. Thus, with this system, good braking comfort is achieved, with minimized car shake and little brake pedal

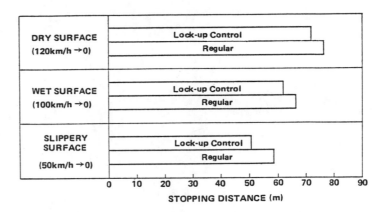

Fig 12 Comparison of stopping distance

pulsative reaction.

As for the durability of the system components, a wide range of road tests covering a high speed durability, a rough road durability and an irregular road durability were conducted and their high durability was demonstrated.

Concerning EMI, this system was tested in an anechoic chamber to check for erroneous operation in a strong electric field and the braking test while a radio set carried on the car was transmitting were conducted. Various improvements were incorporated, based on the results of these tests and finally sufficiently electro-magnetic compatibility was established.

CONCLUSION

This system shows following characteristics.

(1) This system attain far higher braking stability than a car without the system and keeps a stopping distance shorter than or comparable to that of a car not equipped with the system regardless of the road conditions.

(2) Drivers do not have a feeling of discomfort or uncertainty when the Rear Brake Lock-up Control System is operating.

(3) A driver accustomed to slippery road could be able to derive from this system the performance that is very close to that of the four-wheels control system.

Through widespread use of this system, it would be expected that traffic accidents resulting from the rear wheel lock-up could be significantly reduced.

**With the Rear Brake Lock-up Control System**

**Without the Rear Brake Lock-up Control System**

Fig 11 Evaluation test on the split road

# The First Compact 4-Wheel Anti-Skid System with Integral Hydraulic Booster*

**H.-W. Bleckmann, J. Burgdorf, H.-E. von Grünberg, K. Timtner, and L. Weise**
Alfred Teves GmbH

*Paper 830483 presented at the International Congress and Exposition, Detroit, Michigan, February, 1983.

## ABSTRACT

The major step in promoting vehicle safety in the past decade is undoubtedly the advances made in developing automatic anti-skid systems. The authors ' company, Alfred Teves GmbH, has made a significant contribution with an intensive 12 year anti-skid program. After successful presentation of the Mk I System in the late 70's, further progress led to the Mk II Anti-Skid System, which has attracted wide attention. It will see production in 1984 in advanced European vehicles.

The main feature of this new system is its integral and highly compact design, using a single hydraulic medium, namely brake fluid.

A hydraulic booster, master cylinder, energy supply with electro-pump and accumulator, and a solenoid valve block form the assembly.

Another key element is the first passenger car use of microprocessors which double processing of all information and mutually check every step. These microprocessors are also highly adaptable to program changes, without affecting the hardware.

This new Anti-Skid design has unique features which will set a standard for all future systems, especially as the demand for anti-skid spreads from today's upperclass vehicles to small and medium size vehicles, and the very significant safety advances are made available to the broadest range of the motoring public.

AS MOTOR VEHICLE DEVELOPMENT ADVANCED in the first half of this century, basic design challenges were overcome, allowing engineers to turn their attention to improving the brake system. The 20's saw the early development of hydraulics to replace rods and cables as a way of quickly and uniformly actuating four brakes, not two.

Those drum brakes left a lot to be desired, for they could cause unpredictable pulling left to right, quite apart from the problem of front to rear balance. By today's standards, these systems were most unsatisfactory, and improvements became mandatory as traffic volume and speed increased.

The next major breakthrough came in the 60's with the advent of fade-free disc brakes, whose chief advantage was their ability to predictably and uniformly convert the driver's pedal effort into brake torque. Traffic safety then made a great step forward, and at the same time the effect of front-to-rear brake balance took on a new meaning. Brake systems were then designed for shortest brake distance and maximum stability, usually in that for all road conditions, the rear brakes should not lock-up before the fronts. This basic fact is expressed in most of today's International Braking Regulations.

However, even these latest systems are not ideal, in that dependent on road surface and the brake force applied, the wheels may lock up. When this happens, the driver can lose control of the vehicle, for a locked wheel cannot absorb any cornering or lateral forces, and steerability will be lost.

It has been demonstrated in recent years that to further improve active road safety, an anti-skid system should be introduced which will allow the driver to achieve maximum deceleration while maintaining full steering control.

After early theoretical considerations and tests in the late 30's and early 40's, it took until the mid-70's before a satisfactory system could be put into limited production, initially at very low volumes.

The authors' company has set itself the target of making this extraordinary improvement in vehicle safety available in larger volumes to as broad a section of the motoring public as possible, in the form of a cost-effective, lightweight compact anti-skid system.

At the same time the concept should not be too futuristic, in order to retain the well-tried properties of today's brake systems. Above all, the reliable function under all possible emergencies must be guaranteed. With these goals in mind the Teves Mk II Anti-Skid System has been developed, the description of which is the subject of this paper.

After a brief review of the hydraulic booster, the historical development of anti-skid systems is covered; with the aid of a system analysis, the Mk II concept was arrived at. A description of the hydro-mechanical group with its features and advantages follows.

Because the electronic controller is a major unit of the Anti-Skid System a full section is devoted to it, particularly the first passenger car application of microprocessors with whose help a highly sophisticated safety and self-monitoring concept has been achieved.

THE HYDRAULIC BRAKE BOOSTER

With the introduction of disc brakes especially in higher weight vehicles, brake power-assist units became mandatory. These reduced the driver's pedal efforts, and are typically vacuum boosters, which have proved to be simple and cost-effective throughout the world.

As other safety-related features began to be introduced, such as hydraulic power steering and self-levelling suspensions, the application of hydraulic brake boosting became obvious, particularly as a hydraulic energy source, e.g. steering pump, was already available. Such units were put into production in the USA in the 70's. In Europe the Teves H31 System (Fig. 1) has been fitted to the 7-Series BMW's since 1977.

Fig. 1 - Dual-Medium H31 Hydraulic Booster

As a logical further development of this H31 System, the compact self-contained H21 System has been developed, with the clear aim of being easily upgraded into an automatic anti-skid system. Thus, the key advantages of weight and cost optimization have been achieved. Fig. 2 shows an H21 booster which is scheduled for production in the coming years, and which is the subject of a separate 1983 SAE-Paper.

Fig. 2 - Compact H21 Hydraulic
Booster

MECHANICAL AND HYDRAULIC DESIGN OF
THE ANTI-SKID Mk II SYSTEM

BRIEF DEVELOPMENT BACKGROUND OF AUTOMATIC
WHEEL-LOCK PREVENTION DEVICES ON PASSENGER
CARS - Already in the 30's, engineers were
trying to prevent the problem of skidding
wheels on rail and road vehicles, for it was
realized that the maximum peripheral and lateral forces can only be transmitted when the
wheel is rolling. With a locked wheel, the
vehicle cannot be steered, and furthermore,
the sliding wheel could develop a flat spot
quickly enough to destroy the function of the
wheel itself.

At the same time, however, efforts were
fruitless, because mechanical sensors don't
provide sufficient information about the
rotational behavior of the wheels and mechanical control units are not flexible enough.
Only with the progressive development of
electronics with high speed and sophisticated
control algorithm, the solution to the problems have come within reach.

After 1966, electronic controllers were
being developed by several firms, using the
analog technique. For experimental vehicles
the results were quite satisfactory, but to
achieve the necessary series production reliability standards under all operating conditions, further steps were needed.

These units are nowadays referred to as
the "1st Generation", as shown in Fig. 3,
and were successfully used during a fleet
test of Volvo vehicles.

Fig. 3 - Teves Anti-Skid
1st Generation

The period 1973-1979 saw the development
of digital controllers. Because of the complex
data processing involved, highly integrated
custom-made digital circuits were needed,
whereby the peripheral circuitry to interface
with the digital information processing was
using discrete electronic components.

On the hydraulic side two development
trends emerged:

- Normal vacuum-boosted brake systems
  received an additional hydraulic control
  unit which contains an electro-pump and
  solenoid valves; the pump is only
  activated if at least one wheel starts
  to lock up. These "add-on" systems are
  currently in production on some German
  cars.
- In brake systems which have a hydraulic
  booster the control unit acts directly
  on the brake circuits by means of a
  "plunger piston". Fig. 4 shows an anti-
  skid system using this principle.

For weight and complexity reasons, this
latter concept was not put into production,
although its function met all demands. Therefore, it was decided to replace the "plunger-
piston" with the "dynamic flow-in principle"
on the hydraulic, and to make full use of

259

microprocessors with more extensive integration of the peripherals on the electronic side.

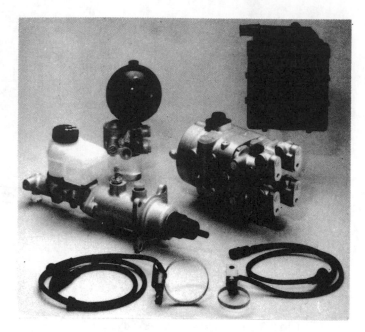

Fig. 4 - Teves Anti-Skid 2nd Generation with Hydraulic Booster H31

All hydraulic components are grouped together, and the high performance Teves Anti-Skid Mk II System shown in Fig. 5 is the result of this effort.

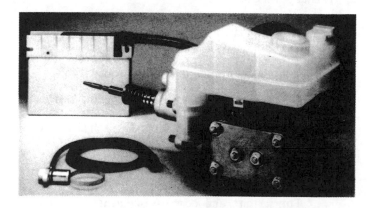

Fig. 5 - Teves Anti-Skid Compact Unit Mk II System

Compared with earlier designs the reduced size of the individual components can be seen, and the simplified electrical and hydraulic connections to install the unit in the vehicle are obvious. This compact cost-effective unit, with integral energy source for normal braking and anti-skid actuation takes the place of the conventional vacuum booster/master-cylinder assembly. It uses only brake fluid as a working medium and the previous need for mineral oil brake boosting, as in Figs. 1 and 4, is no longer required with the problem of medium separation eliminated.

REVIEW OF SYSTEM ANALYSIS AND EVALUATION WHICH LED TO THE DECISION TO DEVELOP THE Mk II ANTI-SKID - As the add-on system was introduced in selected upper-class vehicles on a limited scale, Teves continued to work on an advanced system which will cover a wide range of vehicles into the late 80's. System analysis showed that advantages must be provided in the following areas:

- Compact size to limit weight, cost, assembly operations on the unit itself and simple vehicle installation.
- Upgrade the regular brake system with a hydraulic booster.
- 3-Circuit hydraulic brake actuation.
- Dynamic flow to the brakes during anti-skid control.
- Modular design for anti-skid and hydraulic booster.
- Compatible with rear disc or drum brakes.

By fulfilling these requirements the system has proven acceptable to all customers by retaining many well proven features of today's brake technology.

These key concepts are expanded upon as follows:

- By designing the unit to fit in essentially the same envelope as today's vacuum booster and combining the functions of energy source, accumulator, booster, dual master-cylinder and anti-skid valve block in one compact unit, weight, cost and complexity are optimized. Fire-wall mounting is the same as today's vacuum booster with multi-pin electric connectors and no extra hydraulic connections.
- The dynamic flow-in principle is only feasible for use with a single hydraulic medium, namely brake fluid, and a high pressure level applicable to the vehicle's brake system demands. This results in the added advantages of low weight, reduced reaction times and excellent pedal feel. The hydraulic accumulator provides energy reserves for a larger number of brake actuations, should an emergency arise.
- The 3-Circuit approach was chosen to match the brake circuits with the control circuits (for passenger cars the basic concept is individual front wheel and select-low rear axle control). For this only 6 solenoid valves are required, whilst for diagonal circuits 8 would be needed. The rear-axle control only needs one hydraulic line, and where needed, a brake proportioning valve. A further advantage of the 3-Circuit split is shorted

stopping distances in the event of a circuit failure. The rear brakes receive dynamic flow from the hydraulic booster chamber. In this way the conventional master cylinder can be dimensioned to supply the front brakes alone with short pedal travel.

- The 3-Circuit system can, however, only be applied to vehicles whose steering geometry will compensate for the effect of possible loss of one front brake circuit. This condition already is taken care of in all vehicles designed with diagonally split brake systems from the outset. In fact, the ability of a vehicle to accept large variations in front brake effectiveness side to side, which occur when braking on different surfaces with intact anti-skid control, must be a key design aim when a vehicle is specified to have an anti-skid system.

The following table shows the achievable decelerations for comparable vehicles in the case of different circuit failures of the three system layouts:

| Circuit Split | Normal-Decelerations Intact System | Achievable Decelerations with Circuit-Failures | |
|---|---|---|---|
| Front-Rear | 100 % | Rear : | 80 % |
| | | Front: | 33 % |
| Diagonal | 100 % | 50 % | |
| 3-Circuit System | 100 % | Rear : | 80 % |
| | | Front: | 73 % |
| | | (1 Wheel) | |

- The energy source is fully self-supporting. An electromotor driven high pressure pump charges a gas filled accumulator with upper and lower limit switching. This concept anticipates future engine developments whereby vacuum availability is limited, or with the latest cut-off modes, no vacuum exists unless an extra reservoir or pump is provided. In the Teves design the pump runs relatively often for short periods to supply the service brake system, which is advantageous for anti-skid system reliability. In contrast the pump for an "add-on" system runs only when anti-skid operates, in other words very infrequently, so that pump malfunction will not be noticed on normal braking situations. The pump is designed to provide the necessary booster pressure up to max. 180 bar (2610 psi). The motor consumption at that pressure, dependent on

vehicle size, is 120 or 180 Watt. Vehicles with GVW up to 1600 kg (3500 lbs.) can use the 120 watt motor; up to 1900 kg (4200 lbs.) the 180 watt unit.

- Both sizes are designed for the pressure level of European brake systems. For heavier vehicles, e.g. some US models with larger foundation brakes and greater fluid displacements, appropriate system enlargement is needed.

The dynamic flow-in principle and the use of the same energy supply for brake boosting and anti-skid control, influence the anti-skid function in the event of failure of one of the service brake circuits. A partial cut-out of the front axle anti-skid can also occur. In every case the rear axle anti-skid circuit ensures the necessary lateral stabilizing force and if one front brake circuit fails that wheel can nevertheless still be effectively steered. The driver is warned with the help of an ingenious combination of the brake fluid level warning device and the pressure warning switch.

DESCRIPTION OF THE MECHANICAL AND HYDRAULIC SYSTEMS OF THE ANTI-SKID Mk II - As the preceding explanation showed, the Mk II System was the clear solution to follow especially because the complete anti-skid function could be provided by simply adding a valve block to upgrade the H21 hydraulic booster.

Fig. 6 shows the actuating unit plus a symbolic illustration of the wheel sensors, electronic module and electrical/hydraulic connections.

Fig. 5 is a photograph of the actual components, with the accumulator nested in a recess in the fluid reservoir, and the electromotor pump unit below the master cylinder and behind the valve block. The brake pedal is connected to the input pushrod at the left of the picture.

To further explain the principle of operation, a hydraulic circuit diagram is shown in Fig. 7. The diagram is divided into five sections, shown as individual modules:

- H21 hydraulic booster
- dual-master cylinder for normal anti-skid applications
- fluid reservoir with level-indicators warning switches and filter
- energy supply group
- anti-skid valve block

As can be seen, the rear brake pressure is taken direct from the pedal-effort proportional booster chamber: in this way a dynamic circuit is created for the rear axle, in contrast to the static front brake circuits, one per wheel, which are supplied from the conventional dual master cylinder.

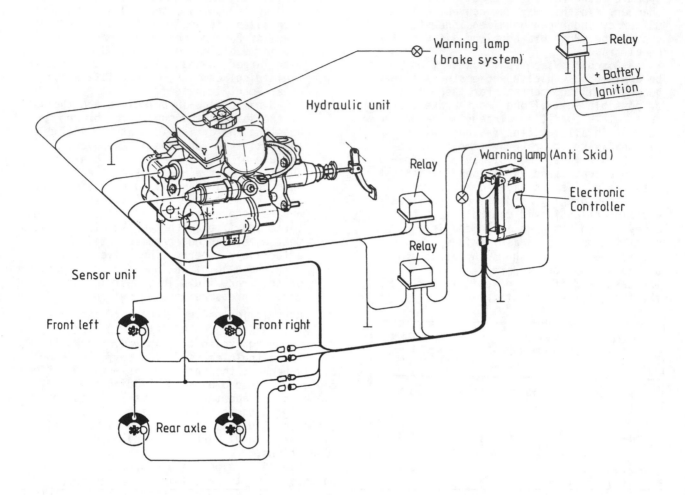

Warning lamp (brake system)

Relay

+ Battery
Ignition

Hydraulic unit

Relay

Warning lamp (Anti Skid)

Relay

Electronic Controller

Sensor unit

Front left

Front right

Rear axle

Fig. 6 — Symbolic Illustration of Teves Anti-Skid Mk II System

In this three-circuit brake system, each circuit has one pair of solenoid valves:

- inlet valve SO (open without current)
- outlet valve SG (closed without current)

Appropriate electrical signals control brake cylinder pressure to each front wheel, and for both rear brakes together, on the SELECT-LOW principle.

Each valve pair controls three phases:

- pressure increase phase
- pressure maintenance phase
- pressure decrease phase

During the anti-skid pressure modulation in the front brake circuits, the fluid volume required is taken from the dynamic circuit via a main solenoid valve, the pressure being pedal-effort proportional, i.e. a dynamic flow

into the static circuits takes place.

The following paragraph describes the function for normal brake actuation, without anti-skid.

The accumulator is held at a pressure level between 140 and 180 bar (2030 psi and 2610 psi). When pressure falls below 140 bar (2030 psi), the pressure switch (20) cuts in the electro-pump (15) until the 180 bar (2610 psi) pressure is again reached.

When the brake pedal is depressed, the control valve (2) in the hydraulic booster is activated by the scissor-lever mechanism (1), and a pressure proportional to the pedal effort is introduced into the annular chamber, and via the open SO valve (9) into the rear brakes. At the same instant the dynamic pressure moves the booster piston to the left, and after the central valves close in each master cylinder

262

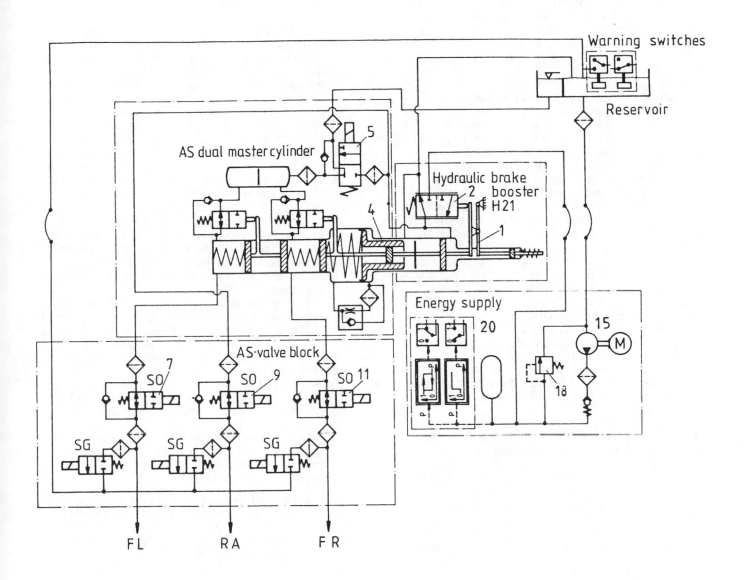

Fig. 7 - Hydraulic circuit diagram of the Mk II System

the front brakes are actuated via the two open SO valves (7 and 11). If the pedal is further depressed, the master cylinder actuating rod contacts the locating sleeve (4) and moves it to the left; the spring load on the sleeve is very low and has virtually no effect on the actuation characteristics.

For the driver the pedal feeling is particularly good, thanks to the precision of the hydraulic booster's spool-type control valving, and because only the front brake actuation has a direct link with the driver. The booster piston has low-friction seals which are not continuously under accumulator pressure, due to the special layout of the control valving.

When an anti-skid control is needed, the system works as follows - if there is imminent brake-lock tendency during an application

on one of the vehicle's wheels, which has been detected by the electronic sensors, the appropriate SO valve is closed, and the SG valve opened, so that the pressure in the wheel brake can be lowered by creating a connection direct from the brake circuit to the fluid reservoir. At the same time the main valve (5) is activated and the dynamic circuit joined hydraulically with the static circuits.

To reapply pressure to the wheel brakes, the SG valve closes, and the SO valve opens to allow the necessary volume of fluid to flow from the dynamic circuit, through the main valve (5), over the periphery of the master cylinder primary seals to the caliper or wheel cylinder in question.

Because the chamber ahead of the location sleeve (4) is also subject to the dynamic pressure during anti-skid modulation, the sleeve is forced to the right abutment and limits the operating stroke of the booster/master cylinder piston, thus giving feed-back to the driver's foot. Once the pedal position is initially located, only a slight vibration will be felt at the brake pedal.

When the anti-skid modulation begins with low pedal efforts, e.g. on icy or wet roads, the booster piston, and consequently the brake pedal, advance progressively until contact with the sleeve (4) is made.

If modulation begins with higher pedal efforts, such as on dry roads, the brake pedal is urged rearwards by the location sleeve. In this way there is always sufficient master-cylinder displacement available, should there be a system failure during an anti-skid stop.

On completion of an anti-skid phase, either during a stop or at the end of a stop, the main valve (5) switches back, interrupting the dynamic flow-in and reconnecting the annular chamber behind each master cylinder piston primary seal with the reservoir. Also, the location sleeve function has equal pressure on either side again, allowing normal three-circuit braking to continue.

A description of the safety and warning concept of the hydro-mechanical portion is given below.

During every anti-skid modulation the location sleeve acts to limit the pedal travel, thus guaranteeing that an adequate pedal stroke is provided to actuate the master cylinder circuits in the event of malfunction.

Should the accumulator pressure fall below the safety limit, there is a partial cutout of the anti-skid function, i.e. anti-skid actuation of the static master cylinder circuits is prevented, as is switching off the main valve (5). Rear axle anti-skid is retained, and the driver is warned via the anti-skid and the brake warning lamps. Even then, a series of brake applications with reduced boosting of front and rear brakes is possible.

If the reservoir fluid level drops, the fluid level sensor lights up the brake warning lamp. A further drop in level, indicating leakage, activates a second fluid level sensor, causing partial cutout of the anti-skid function as mentioned above.

In the event of loss of boosting, the full pedal travel is available for manual operation of the static master cylinder circuits. The vehicle can be stopped with an acceptable deceleration and pedal effort.

Eletric motor malfunction and insufficient charging of the accumulator is indicated by the warning lamps.

## ELECTRONIC CONTROLLER

FUNCTION AND DESIGN OF THE ELECTRONIC CONTROLLER - The controller processes the four wheel-sensor signals, and converts their frequency information into values which correspond to wheel speed and acceleration, from which approximations of vehicle speed and slip of each wheel are established.

Appropriate to the rotation behavior, determined by the above information, the electronic controller then generates solenoid valve command signals, which switch the main valve, and 3 pairs of inlet and outlet valves (one pair for each of the 3 hydraulic control circuits).

These valve signals reduce the brake pressure, hold it constant, or increase it.

Additionally, the electronic controller has the key task of monitoring its own function and the electro-mechanical components in the system.

In the event of a failure or temporary disturbance, the controller switches automatically to a condition which has no detrimental effect on vehicle control. In every case, at least the normal service brake function is guaranteed.

This fail-safe system is achieved through redundant information processing in two fully independent channels, in conjunction with plausibility criteria.

The controller circuitry is mounted on a single printed wiring board in a metal casing and is divided into the following main groups:

- 2-channel redundant data processing using two identically programmed microprocessors. If differing processing outputs are detected in the 2 channels, the controller switches off, and reverts to the conventional service brake operation.
- peripheral circuitry for the preprocessing and interfacing of input and output signals. This circuitry consists chiefly of custom-made circuits.

SAFETY CONCEPT - REDUNDANCY - Fig. 8 shows the main electronic circuits, and the safety concept in a simplified form, namely redundancy by doubling the data processing set-up.

The 4 pre-processed sensor input signals are fed into 2 identical function blocks (microprocessors 1 and 2). As both microprocessors receive the same set of input information, they must both produce the same internal and external signals, as long as there are no errors.

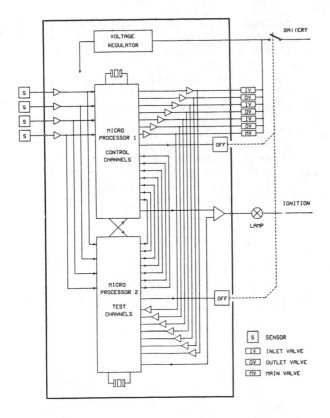

Fig. 8 - Block diagram - Electronic
Controller

These sets of signal combinations are independently compared by both microprocessors, whereby the external valve signal combination of microprocessor 1 is "diverted" via the valve drivers to the valve solenoids, and from there over a level-monitoring circuit into microprocessor 2. In this way, the function of the valve current loop is included in the "safety by redundancy" concept.

In order to monitor the valve loop in its two logic states during normal driving without anti-skid actuation, both microprocessors periodically produce short test pulses at the valve outputs, to which the valves do not mechanically react.

In the event of different signal combinations, the two microprocessors independently cut out the current supply to the electronic controller.

PLAUSIBILITY CRITERIA - Naturally, the redundancy concept will only recognize failures in the p r o c e s s i n g of the information. The physical c o n t e n t of the information, usually from one information source, e.g. a wheel sensor, can on the other hand only be verified against plausibility criteria.

Such criteria are:

● availability of sensor signals
● continuity of the signals
● comparison of the 4 different
  sensor signals

The formal monitoring procedure per these criteria is part of the information processing, and can therefore be carried out redundantly.

This has the following practical effect: an erroneous sensor information (sensor disturbance) will not lead to a cut-off of the controller, because the error signal is produced simultaneously in both microprocessors, and thus the redundancy is not disturbed.

An error signal of this type causes the controller to be inhibited only partially and temporarily.

If, however, an erroneous sensor information is incorrectly processed, the controller is cut off by the redundancy monitoring, and the system reverts to normal brake actuation.

SECOND SAFETY LEVEL - In the previous section, the different failure mechanisms between the incorrect processing of information (defective redundancy) and incorrect content of the information (e.g. sensor disturbance) were discussed. Also, it was mentioned that a second failure reaction will be effective if the first failure reaction cannot be effective.

In order to differentiate between these two failure mechanisms, the expressions "failures" and "disturbances" are defined as follows:

● A failure encompasses all events in which the redundant information processing in at least one of the two comparators leads to different results; in other words, essentially component defects.
● Disturbances are all events which lie outside the limits given by the plausibility criteria.

Between these two mechanisms there are interactions which represent a "second safety level".

If we differentiate between the terms "switch-off" (cut the current supply to the controller) and "partial inhibiting" (to inactivate appropriate valve signals) the following mechanisms make up the second safety level.

● If the controller cannot respond correctly to a failure due to a further defect, it will respond as if a disturbance had occured.
● If the controller cannot respond correctly to a disturbance, it will respond as if a failure had occured.

Usually, when a failure is signalled, the controller switches off, but with a disturbance it is partially inhibited. In some special cases of disturbances the partial inhibition as defined above is overruled and the controller directly switches off, when applying the plausibility criteria clearly points to a component defect, e.g. a missing sensor signal.

Electromagnetic interference at extremely high field strengths represents another special case. Such a temporary disturbance can affect the synchronization of the two microprocessors, and upset the redundant signal processing. The controller would respond as if a failure had occured, and switch off.

Fig. 9 lists the various reactions of the controller for failures and disturbances in the first safety level.

| Failure mode | Controller status | Effect |
|---|---|---|
| Controller component defects | Controller switched off | Conventional brake function and warning indication |
| Sensor disturbance | Controller temporarily and partially inhibited | At least conventional brake function and warning indication during disturbance |
| Sensor or valve failures (broken or shorted cables and coils) | Controller switched off | Conventional brake function and warning indication |
| Electric power failure | Controller switched off | Conventional brake function and warning indication (no warning at main power failure) |
| Hydraulic power failure | Controller temporarily and partially inhibited | At least conventional brake function and warning indication during failure (possibly without boosting) |
| RFI with extreme field strength | Controller switched off | Conventional brake function and warning indication |

Fig. 9 - Anti-Skid System Mk II
Failure Modes and Effects

CONTROL CONCEPT - Many methods of controlling brake pressure in anti-skid systems have been described in technical literature, so that details need not be covered in this paper.

For the Teves Mk II System, using an electric pump for hydraulic energy supply, the main additional objective is to minimize hydraulic energy consumption during the anti-skid phase.

The lowest consumption is realized at the smallest pressure amplitudes, and lowest anti-skid control frequency.

The demands of lower control frequency on the one hand, and fastest reaction to sudden road surface variations on the other, are not contradictory. Both demands can be accommodated if the brake pressure can be controlled fast enough and at the same time with sufficient precision.

The rapidity of system response is chiefly influenced by high speed solenoid valve switching, shortest data-processing time and high pressure gradients.

The precision of pressure control is largely governed by the initially unknown, and within a broad range variable parameters of the "control plant" where, however, greater pressure gradients make the precise control of the pressure more difficult.

These problems are depicted in the control-loop diagram shown in Fig. 10. For simplicity, a 4 -wheel vehicle is assumed, with only one controlled wheel.

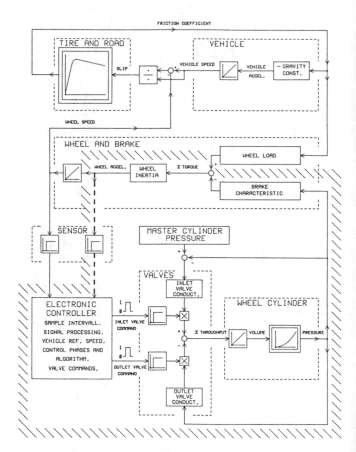

ASSUMPTION ... ONLY ONE CONTROLLED WHEEL

Fig. 10 - Simplified control loop

The part of the control loop which can be directly affected is surrounded by a shaded border, in which all elements outside the electronic controller are considered as the control plant.

Because the electronic controller processes the wheel speed signals from the wheel sensor and derives the wheel acceleration values indirectly, the main control variable "wheel acceleration" in Fig. 10 is shown as a dotted line.

In this way, the control plant approximates to a plant with "integral action" and "dead-time-response" (delay).

The proportional part of the plant (gain) varies within broad limits, and is initially undefined. This proportionality factor depends among others, on the following effects:

- road condition, master cylinder pressure, wheel cylinder pressure, lining coefficient of friction, temperature, axle loading, wheel inertia, etc.

The dead-time in the control plant is determined by the response time of the solenoid valves, the number of teeth on the sensor ring, and the computing time of the electronic controller, and is approximately constant.

For the high pressure gradients required, the dead-time must not be ignored, on account of which the electronic controller has a "predictor structure". This predictor structure is, however, only feasible if the initially unknown proportional part of the control plant can be determined by an adaptive procedure in the electronic controller.

MICROPROCESSORS PROVIDE FLEXIBILITY - Once the basic problems of microprocessor programming related to processing time and the safety concept have been solved, the microprocessor represents an almost ideal means of applying a controller structure to other basic vehicle configurations.

Prototype vehicles can be fitted with the original hardware of the electronic controller, whereby the fix-program microprocessor (with ROM) is replaced by a programmable unit (with EPROM).

In the same way, future knowledge and improvements will be possible without circuit changes and without repeating hardware testing.

The comprehensive "breadboard" units for prototype systems will no longer be needed.

SUMMARY

After important improvements on all parts of the brake system, it proved indispensable in the past decade to concentrate on developing an anti-skid safety during a braking maneuver.

The main challenge was to allow all drivers, including those less experienced to be able to brake a vehicle in an optimum way when confronted with an emergency situation. This meant intervention in the driver's control of a brake actuation, with the target of maintaining directional stability with the shortest stopping distance and allowing the driver to have steering control of his vehicle.

In order to make this major advance available to as broad a spectrum of the motoring public as possible, Alfred Teves has set itself the target of bringing into large scale production a light, compact, cost-effective anti-skid system, designated Anti-Skid Mk II. This system's design function and safety philosophy has been described in this paper. As explained, the compact hydraulic portion consists of a dual master cylinder, a hydraulic high pressure booster, an electro-hydraulic energy source with accumulator, a fluid reservoir and an integral anti-skid valve block.

The whole assembly replaces the usual vacuum booster/master cylinder assembly and is simply connected to the regular system and the vehicle's wiring harness.

The electronic controller embodies the first use of microprocessors for passenger cars, and includes sophisticated safety features. The electronic elements are mounted on a single printed circuit board, the whole assembly being protected from environmental influences.

As this paper illustrates, the Teves Anti-Skid Mk II is the most advanced, cost-effective and reliable unit available on the world market.

# Performance of Antilock Brakes with Simplified Control Technique*

**Makoto Satoh and Shuji Shiraishi**
Honda R & D Co., Ltd.

*Paper 830484 presented at the International
Congress and Exposition, Detroit, Michigan,
February, 1983.

ABSTRACT

The four-wheel controlling antilock brake
system is considered as an effective safety
device because of its capability to help a
driver to maintain vehicle stability and
steerability during panic braking even on a
slippery road surface.

This report deal with a simplified control
technique which simultaneously controls right
and left wheels on each front or rear axle.
Both front wheels are controlled in response to
a signal from the front wheel with the least
slip, while both rear wheels are controlled in
response to a signal from the rear wheel that
has the greatest slip.

A series of tests proved that this
technique ensures vehicle steering ability
even during panic braking. On a gravel and
other rough roads, this system provided shorter
stopping distance compared to other four-wheel
antilock systems. It has been generally
assumed that stopping distance extension on
such roads is only one disadvantage of the
four-wheel antilock brake system.

There are two types of antilock brake
systems now in use. One is a rear wheels
controlling system and another is a four
wheel controlling system. The former system
can greatly improve straight line stability
during panic braking, but it does not serve
to preclude loss of vehicle steerability
resulting from front wheel locking. The
latter, on the other hand, additionally allows
to keep vehicle steerability during panic
braking. Generally, the antilock brake is
most frequently needed and demonstrates its
capability on slippery roads such as wet,
snowy or icy roads. On such a road, since
a vehicle always requires a very long distance
for stopping, capability to maintain adequate
steerability even during panic braking is
an important factor in precluding an accident.
This is exactly why the four wheel antilock
brake system is more excellent and desirable
than the other.

However, the four wheel antilock brake
system also has disadvantages which are
stopping distance extension on a rough road
and system complexity, size, weight and cost.

This report discusses a new four wheel
antilock brake system that employs "select
high" technique for the front wheels and
"select low" technique for the rear wheels.
This new brake system is capable of retaining
sufficient vehicle stability and steerability
even during panic braking. This system also
improves stopping distance extension on a
gravel or other rough roads, and its structure
can be simplified.

## 1. CONTROL TECHNIQUES FOR ANTILOCK BRAKES

Three techniques are available now to
control braking torque for the right and left
wheels:
(1) Independent control technique, which
    independently controls braking torque for
    the right and left wheels;
(2) Select low technique, which simultaneously
    controls the right and left wheels in
    response to a signal from either of them
    which is predicted to get locked first;
    this implies that braking force for the
    wheel on the other side may become lower
    than optimum level; and
(3) Select high technique; which also
    simultaneously controls braking torque for
    both right and left wheels, contrast with
    select low technique, in response to a
    signal from either of them which is
    predicted to get locked later; therefore
    it may permit the other wheel to get locked.

Existing four wheel antilock brakes use
technique (1) or (2). None of them have ever
adopted technique (3) in passenger cars
presumably because vehicle stability or

steerability may be affected by the possible reduction of a lateral force on the tire due to the tendency of this technique to permit either the right or left wheel to get locked during full braking.

However, wheel locking may occur only on the side which produces a smaller lateral force than the other. In other words, the system permits the locking of either the right or left wheel which less affects vehicle stability or steerability than the other. On a split road surface with different adhesion coefficient μ on the right and left side, for instance, the wheel on the side with lower μ may get locked. When the vehicle is turning on an even road surface with a uniform μ on both sides, another instance, the system may permit locking the wheel on the inside of the curve which receives lighter load than the other.

Therefore applying the select high technique may affect vehicle stability or steerability so slightly that no practical problems may arise. On the other hand, this technique may provide a very effective means for precluding the stopping distance extension on a gravel or other rough roads as described later.

## 2. CAUSES OF STOPPING DISTANCE EXTENSION

Vehicles equipped with existing four wheel antilock brakes require a longer distance for stopping on a rough road probably because:
(1) Unlike a normal smooth road surface, a rough road provides the highest apparent adhesion coefficient between tire tread and road surface when the wheel is locking; this means that locked-up wheels are more effective than controlled one in achieving a shorter stopping distance on a rough road; and
(2) The control function of the antilock brakes is unfavorably affected by wheel speed changes during braking on a rough road, subsequently resulting in system's excess operations which unnecessarily reduce the braking torque.

The former is an inevitable disadvantage of antilock brakes but the latter can be considerably improved by applying the select high technique.

## 3. EXCESS OPERATIONS

The antilock brake system keeps monitoring wheel speed changes and automatically eases braking torque when any of the wheels are predicted to get locked. The standards for determining such system-actuating conditions are generally worked out by the following formulas:

$$\lambda > \lambda_0 \qquad [1]$$
$$\dot{V}w < \beta_0 \qquad [2]$$

Where
$\lambda$ = slip ratio of the wheel;
$\dot{V}w$ = circumferential acceleration of the wheel;
$\lambda_0$ = reference standard for $\lambda$;
$\beta_0$ = reference standard for $\dot{V}w$

And $\lambda$ is defined with vehicle and circumferential wheel speeds given as $V$ and $Vw$ respectively, as follows

$$\lambda = 1 - Vw/V \qquad [3]$$

Usually, the reference standards are set at the following levels:

$$\lambda_0 \simeq 10\% \qquad [4]$$
$$\beta_0 \simeq -9.8 \sim -15m/s^2 \qquad [5]$$

On a rough road, wheel speed varies at random, i.e., the so-called wheel speed pulsations occur due to the ever-changing contact between the tire tread and the road surface.

Such wheel speed pulsations become severer particularly when the brakes are being applied. This is because longitudinal vibration of the suspension may occur from varying braking forces due to changes of the tire tread contacts with the road surface.

Fig. 2 gives some examples of the wheel speed pulsations recorded while a conventional test vehicle equipped with no ALB systems was running on an extremely rough road surfaces as shown in Fig. 1. The values of $\lambda$ and $\dot{V}w$ were calculated using the data when a force of 100N was put on the brake pedal. Such small force was selected to avoid causing any of the wheels to get locked up. The results are shown in Fig. 3 along with straight lines representing the reference standards, i.e., $\lambda_0=10\%$ and $\beta_0 = -15m/sec^2$.

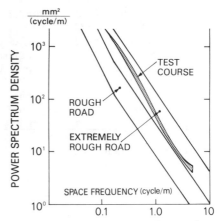

Fig. 1: Characteristics of rough roads

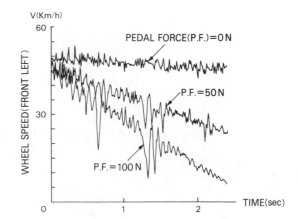

Fig. 2: Example of the front wheel speed pulsations

The diagram indicates that the system-actuating conditions represented by formulas [1] and [2] above may sometimes occur on a rough road even if the driver steps on the brake so lightly that none of the wheels will get locked. This is why vehicles equipped with existing ALB systems unnecessarily ease braking torque on a rough road, making their stopping distance longer. In this report, such overcontrol of braking torque is defined as "excess operations" of the ALB system.

Additionally, in the case of the independent technique, this excess operations may occur for only the wheel on the rough surface when brakes are applied on a split road with a rough surface on one side and smooth one on the other, so that the braking force for the right and left wheels becomes out of balance. Therefore the vehicle may sometimes tend to be pulled to one side on such a road if the independent technique is applied for front wheel.

## 4. APPLING OF SELECT HIGH TECHNIQUE

Fig. 4 shows the probability density and probability distribution functions of slip ratio $\lambda(t)$, assuming that variable as shown in Fig. 3 changes at random process.

With the occurrence probability of the system satisfying formula [1] given as P, the excess operation probability of the select-high system, given as $P_H$, is equal to the provability with which the right and left wheels simultaneously meet the conditions represented by formula [1]. Providing that the mutual interference of the right and left wheels can be ignored, this can be formulated into:

$$P_H = P^2 \tag{6}$$

If the values of P=and $P_H$ are obtained from Fig. 3 it follows

o Rough road  :  P  = 0.068,
$\qquad\qquad\qquad$ $P_H$ = 0.005
o Extremely  :  P  = 0.382,
$\quad$ rough road  $\quad$ $P_H$ = 0.146

This means that the select-high technique can substantially curb excess operations of the antilock brake system on a rough road.

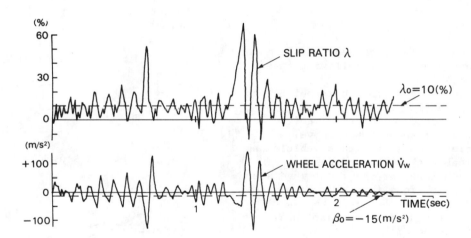

Fig. 3: Pulsations of $\lambda$ & $\dot{V}w$ over reference values

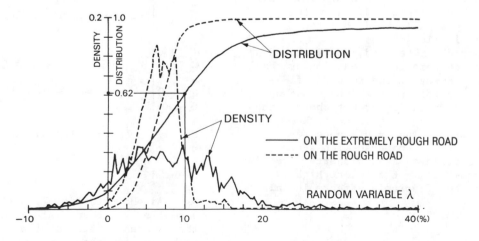

Fig. 4: Probability of random process ($\lambda$)

271

## 5. SIMPLIFIED CONTROL TECHNIQUE

With above examination result, the following aspects have been carefully considered.

(1) The most important function of the brake system is to provide the vehicle with sufficient stopping capability and stability.

(2) The rear brakes provide a relatively minor portion of stopping capability in passenger cars particularly front-drive-cars; and

(3) The rear wheel's rotating conditions play a very important role in maintaining vehicle stability.

As a result, the following combination of techniques will be recommended for the four wheel controlling antilock brake system.

- o Select high technique, an effective means of curbing stopping distance extension on a rough road, for the front wheel;
- o Select low technique, a most suitable means for the maintenance of vehicle stability, for the rear.

This new system requires only one control subsystem for each front or rear brake, accordingly the system can be simplified in size, weight and cost.

## 6. PERFORMANCE

To examine the above concept, a series of tests were conducted on the following three vehicles.

(a) The vehicle named N-ALB: this vehicle has no antilock devices.

(b) The vehicle named H-ALB: this vehicle has the antilock brake proposed above.

(c) The vehicle named I-ALB: this vehicle has the antilock brake with independent control technique for the front wheels and select low technique for the rear.

The H-ALB and I-ALB have exactly the same control logic for judgement whether wheels are likely to get locked or not.

### 6-1 Stopping Distance

As shown in Fig. 4, both H-ALB and I-ALB achieved a shorter stopping distance than N-ALB on the dry asphalt road, but the difference was little.

On the wet concrete road, H-ALB showed a stopping distance some 10 percent longer than that of I-ALB. However, the results of H-ALB are considered satisfactory as the stopping distance was more than 10 percent shorter than that of N-ALB. On the rough road both H-ALB and I-ALB required a longer distance for stopping than N-ALB. Compared with I-ALB, however, H-ALB achieved a much shorter stopping distance, demonstrating the select high technique produces favorable effects in this respect.

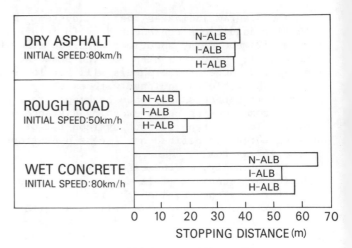

Fig. 5: Stopping distance

### 6-2 Stability Test on Rough/Smooth Split Surface

Test method: Yaw angles of the three test vehicles were compared when they came to a stop under full braking with the steering wheel fixed in the straight-ahead position on the split road surface shown in Fig. 6. The initial speed was set at 50 km/hr.

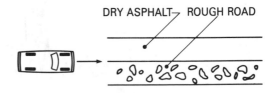

Fig. 6: Rough/smooth split road

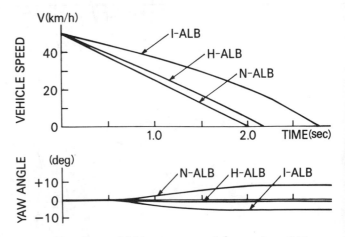

Fig. 7: Stability on rough/smooth split

### Results

As shown in Fig. 7, H-ALB was subject to a smaller yaw than I-ALB apparently because of the favorable effects of the select high technique for the front wheels. However, it may be assumed that I-ALB also has practically satisfactory capabilities since it can stop the vehicle under specified conditions with only an 8 or 9 degree yaw.

## 6-3 Stability Test on Dry/Wet Split Surface

Test method: Exactly the same test as described in 6-2 was conducted on the three vehicles with the initial vehicle speed 80 Km/hr on a split road surface shown in Fig. 8.

## Results

As is apparent from Fig. 9, both I-ALB and H-ALB proved to maintain sufficient stability with only small yawing during full braking. With N-ALB, all wheels locked and the vehicle was naturally subject to a very large amount of yaw, this did not decrease appreciably even with corrective steering.

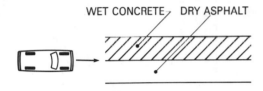

Fig. 8: Dry/Wet split road

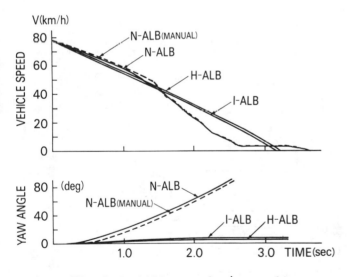

Fig. 9: Stability on dry/wet split

"N-ALB (Manual)" in Fig. 9 shows the behavior of this vehicle when corrective steering was performed to help stabilize it. The data indicates that the vehicle was out of control during full braking.

The apparent rapid slow down of vehicle (N-ALB) speed during braking was due to the measuring method used in this test which recorded only the speed along the longitudinal axis of the vehicle; it does not mean that the test vehicle actually slowed down rapidly.

## 6-4 Braking Performance Test During Turning

Test method: The position and heading of each test vehicle were recorded when it came to a stop under full braking applied while it was turning with a given radius at a given speed on a dry asphalt road with the steering wheel fixed in that turning position. The test was conducted with the initial speed set at 80 Km/hr and turning radius set at three different lengths, i.e., 168m, 112m, and 84m. These radii, in the decreasing order, are equivalent to an initial lateral acceleration of 0.3g, 0.45g, and 0.6g, respectively.

## Results

The test findings are as shown in Fig. 10. The white and black circles indicate where the vehicles stopped, while the arrows show their heading.

| Radii | Lateral accel. |
|-------|----------------|
| 168m ................... | 0.30g |
| 112m ................... | 0.45g |
| 84m ................... | 0.60g |

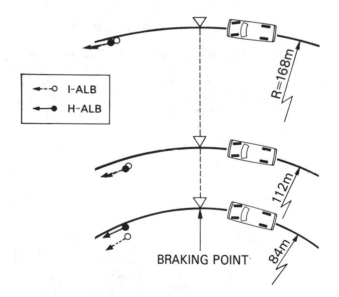

Fig. 10: Test results on braking performance during turn

As is apparent from the diagram, H-ALB deviated little from the specified course and kept heading tangentially; it could maintain sufficient turning capability even during full braking.

It was found that both H-ALB and I-ALB can easily stop with their heading kept on the specified course if the driver is permitted to perform corrective steering.

## 6-5 Evasive Maneuverability Test Under Full Braking on Wet Concrete Surface

The evasive maneuverability of the test vehicles were compared during full braking to examine their steerability on a slippery road. N-ALB was excluded from this test because it was not able to complete the maneuver during full braking.

Test method: Full braking was applied at a given point on a test course described in Fig. 11 while the vehicle was running at a specified speed. Distance ℓ between the braking point and the obstacle was gradually reduced until it became too short for the vehicle to avoid a collision. Then the minimum ℓ in which the test vehicle could avoid hitting the obstacle was determined. Five rounds of test were carried out for each distance, and the vehicle was considered to have sufficient evasive maneuverability if it could stop without touching the obstacle at any one of the five tries.

## Results

Fig. 12 shows the test findings. H-ALB, although liable to lock one of the front wheels, proved to have evasive capability nearly equal to that of I-ALB. It was found that the difference between the two systems was only about 10 percent.

## 7. DESCRIPTION OF EXPERIMENTAL SYSTEMS

Fig. 13 shows the general construction of H-ALB used in this test. The four-wheel-control antilock system with the select high technique for the front wheels and select low technique for the rear wheels has an electronic control

unit that receives a signal of wheel speed from a sensor attached to each wheel. Then the control unit regulates the brake pressure modulators through solenoid valves.

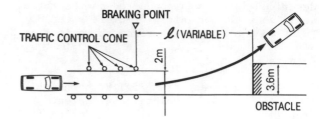

Fig. 11: Maneuverability test course

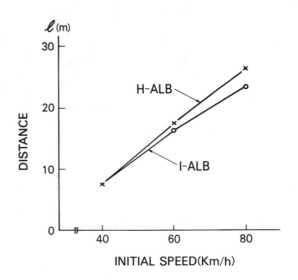

Fig. 12: Maneuverability

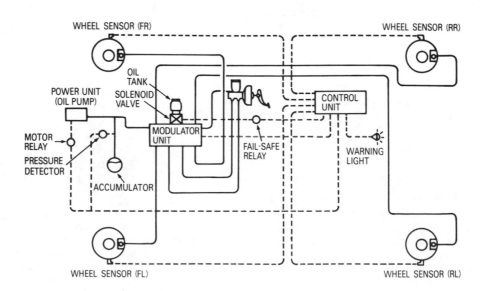

Fig. 13: General construction

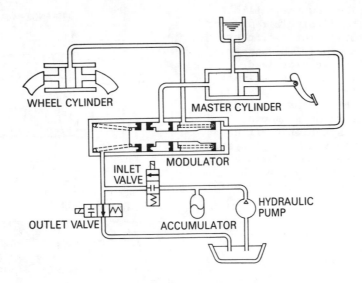

Fig.14: Modulator

To ensure high responsiveness, the system uses a hydraulic control modulator as shown in Fig. 14 which can respond to a command from the control unit with only about 5 ms delay.

Fig. 15 presents the inside structure of the control unit. Wheel speed signals from each wheel sensor are first sent to the select circuit which picks out signals on either the most likely to lock (select low rear system) or least likely to lock (select high front system).

Based on the results of calculating the selected signals, the decision circuit gives the solenoid valve a command to actuate it. Since highly accurate control of hydraulic pressure

is required in a four-wheel antilock brake, the control unit is designed to regulate the solenoid valve in five different modes:
(1) Rapid increase of pressure
(2) Gradual increase of pressure
(3) Maintenance of pressure at a constant level
(4) Gradual decrease of pressure, and
(5) Rapid decrease of pressure

Meanwhile, I-ALB has separate calculation and decision circuits for each of the front wheels to control them independently. However, the control logic of these circuits is exactly the same as the system used for H-ALB.

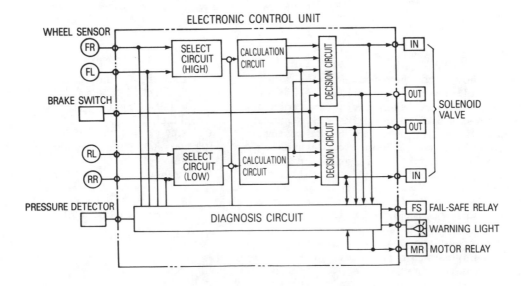

Fig.15: Inside construction of control unit

275

Fig. 16 and 18 describe how the antilock brake system operates. The first diagram gives an example of system operation for the front wheels when running on a course with sharply differing surface conditions as described Fig. 17. The wheels get locked alternatively depending on the road surface. Fig. 18 shows system operation when the brakes are applied during straight-ahead driving on a smooth, iced road with an extremely low adhesion coefficient.

As shown in Fig. 18 the select high system seldom let either one of the front wheels get locked when the vehicle is running straight ahead on an even, uniform road surface on both sides.

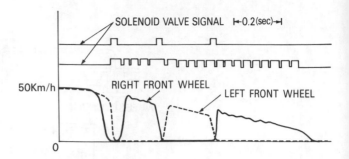

Fig. 16: Example of system operation (1)

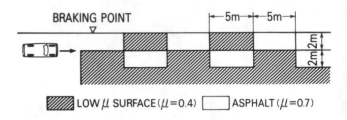

Fig. 17: A road with changing μ

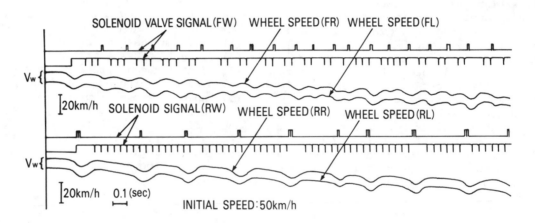

Fig. 18: Example of system operation

## 8. CONCLUSION

As noted above, the new four-wheel antilock brake system utilizes the select high technique for the front wheels and the select low technique for the rear wheels. The foregoing tests and analysis found that this antilock brake system is subject to only a small reduction in braking force on a rough road, while it can maintain sufficient vehicle stability and steerability during panic braking. System overcontrol for the front wheels on a rough road is undesirable as these wheels play a major role in providing stopping ability. The loss of braking force on a split road surface due to the installation of a select low subsystem for the rear wheels is not significant since the rear wheels provide a minor portion of stopping capability. This suggests that the new system is particularly suitable for front-drive cars which depend more heavily on the front wheels for their stopping capability.

## REFERENCES

[1] Leiber, H. and Czinczel, A.
"Antiskid System for Passenger Cars with a Digital Electronic Control Unit"
SAE Paper, No. 790458.
[2] Harned, J.L., Johnston, L.E. and Scharph, C.
"Measurement of Tire Brake Force Characteristics as Related to Wheel Slip (Antilock) Control System Design"
SAE paper, No. 690214.

# Evaluation Criteria for Low Cost Anti-Lock Brake Systems for FWD Passenger Cars*

**W. R. Newton**
Lucas Industries Inc.
**F. T. Riddy**
Lucas Girling

*Paper 840464 presented at the International
Congress and Exposition, Detroit, Michigan,
February, 1984.

ABSTRACT

   Nearly all of the small and medium sized passenger cars manufactured throughout the world
are now front wheel driven.  The dynamic weight distribution and consequent braking systems
of these cars has enabled the development of a two channel anti-lock brake system, known as
the Lucas Girling Stop Control System, that achieves the major characteristics of much more
complex three or four channel systems at a substantially lower cost.  This paper describes the
Stop Control System, its main component, typical performance achievable, and proposes practical assessment criteria.

IT IS NOW generally accepted that road vehicle anti-lock braking systems would, if widely
adopted, make a significant contribution towards increasing road safety.  All of the systems
currently fitted to European passenger cars employ four wheel electronic sensing and controlling
means and electrically powered hydraulics, as shown in Figure (1), making them inherently expensive and, therefore, only economically viable on higher priced cars, where the additional
cost to the purchaser is usually less than 10% of the vehicle base model selling price.

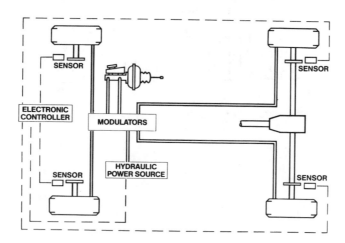

FIGURE (1) TYPICAL 3 OR 4 CHANNEL ANTI-LOCK BRAKING SYSTEM

The performance of these systems is such that stopping distances are better than even the most experienced drivers can normally achieve on most road surfaces and, more importantly, the car is stable and can be steered at all times without skidding through overbraking.

## THE CASE FOR LOW COST ANTI-LOCK

There is no doubt that these excellent systems will be succeeded by other less expensive electronic systems but, even if the most optimistic predictions are realised and full electronic anti-lock system costs reduce by 50%, they will still not be priced low enough to be adopted on those small cars that constitute more than half of the European passenger car market, as shown in Figure (2), as well as a significant proportion in other industrialised areas of the world. This vehicle sector is usually the most price sensitive so, if it is not to be denied the additional safety of anti-lock brakes, then much lower cost systems must be devised.

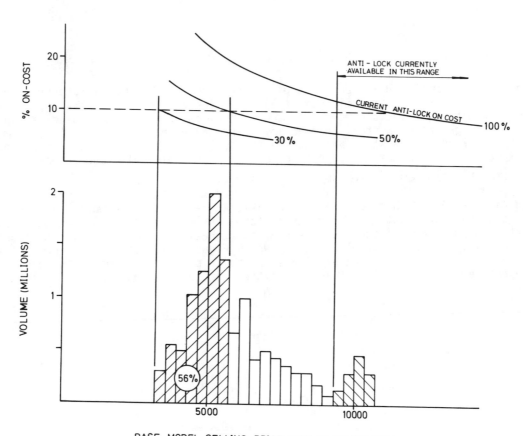

FIGURE (2) PRODUCTION VOLUME OF EUROPEAN CARS 1982

Most of these small cars are front wheel driven and feature dual, diagonally split braking systems, as shown in Figure (3) in order to satisfy the half-system failed braking requirements together with steering geometry that ensures vehicle stability in this condition.

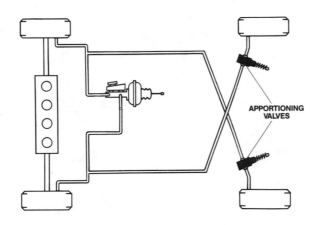

FIGURE (3) TYPICAL BRAKING SYSTEM FOR FRONT WHEEL DRIVE CARS

These features together with the European legislated requirements that front brakes must lock before rears, have enabled the development of much simpler brake anti-lock systems, as shown in Figure (4), that do not require all four wheels to be sensed and controlled independently, resulting in considerable cost reduction. Such a system does, however, place new demands on the apportioning valves and fast response, low hysteresis valves are essential.

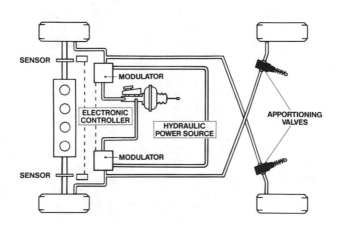

FIGURE (4) 2 CHANNEL ANTI-LOCK BRAKING SYSTEM FOR FWD CARS

Figure (5) shows the transmission powered Lucas Girling Stop Control System (referred to later as SCS) which further reduces cost to a level where anti-lock can be considered for even the lowest price front wheel driven car. A full description of the Lucas Girling SCS modulators appears later.

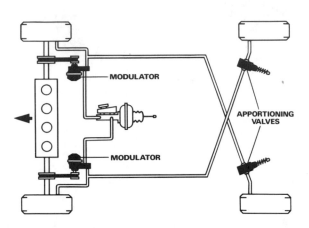

FIGURE (5) THE LUCAS GIRLING SCS STOP CONTROL SYSTEM

SCS OPERATING PRINCIPLES

The operation of SCS on a uniform road surface is obvious as each front brake must, by regulation, always approach locked wheel condition before rears, but is independently sensed and prevented from locking by its modulator, it follows that each diagonally opposite rear brake is similarly prevented from locking. The overall braking efficiency is therefore determined by the efficiency of the anti-lock modulators that directly control the front brakes which typically account for more than 80% of the total braking force, and the amount of under-braking of the rear wheels that is necessary to ensure that fronts always lock first.

Operation of SCS on non-uniform road adhesion surfaces or when braking whilst cornering is less obvious. Firstly, consider panic braking whilst both left wheels are on dry road and both right wheels are on ice. This is a situation that would almost certainly result in a normally braked vehicle spinning out of control. With SCS functioning the right front wheel on ice begins to anti-lock first, thereby ensuring that the left rear wheel on dry road does not lock. In fact, the resulting underbraking of this rear wheel optimises its lateral adhesion capability and improves vehicle stability in the same way that full electronic systems underbrake one wheel when they "select low" on the rear axle. However, with SCS the right rear wheel on ice is over-brakes and allowed to lock since it contributes very little to either braking or stability. The result is that the vehicle remains stable and can be steered even under panic braking conditions. Secondly, consider panic braking whilst cornering to the right. In this case lateral weight transfer occurs, tending to roll the car to the left, so that the loads on both left wheels increase and the loads on both right wheels decrease. As far as braking is concerned, this has the same effect as the non-uniform adhesion condition previously described. When the brakes are applied, the right front brake will begin to anti-lock first, thereby ensuring that the heavily laden left rear wheel does not lock.

Again, underbraking of this wheel optimises lateral adhesion and improves stability in the same way that the "select low" principle is applied to four wheel controlled electronic systems. Both front wheels are, of course, independently sensed and prevented from locking so that full steering control is maintained.

From the foregoing description it will be appreciated that SCS is also effective in overcoming less demanding conditions such as high to low or low to high adhesion changes.

As far as performance is concerned first priority is given to stability and steerability, which are considered to be the major benefits of anti-lock systems. The establishment of performance criteria for anti-lock systems which embrace stability and controllability in addition to adhesion utilisation and straight stopping distance is needed. If these criteria are to be realistic it is necessary to consider a normally braked car in a panic braking situation as the baseline for assessment.

When travelling along a straight road most drivers will overbrake in an emergency situation causing at least the front wheels, and probably all four wheels to lock. The stopping distance is then determined by the locked wheel adhesion value, but more importantly, the driver is unable to steer to avoid the obstacle, resulting in an accident. This situation is even more likely when the road surface is slippery or non-uniform.

The present position when panic braking whilst cornering or attempting to change lanes is even more serious when loss of steering due to locking the front wheels causes the vehicle to continue in a straight line with probably more disastrous consequences.

PERFORMANCE BENEFITS OF SCS

The beneficial effects of SCS in these conditions are clearly shown in Figures (6) and (7) where a typical small car with SCS is directly compared with an identical car having standard brakes.

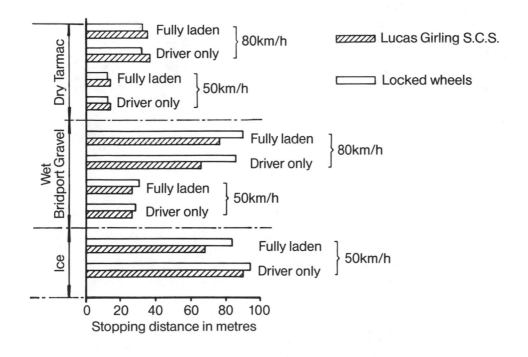

FIGURE (6) COMPARATIVE STRAIGHT LINE STOPPING DISTANCE
LUCAS GIRLING SCS AGAINST LOCKED WHEELS

This diagram compares stopping distances when the brake pedal is pressed with a force exceeding 70 kg - a normal reaction to an unexpected emergency. It can be seen that the stopping distance on ice for the SCS car is between 5 and 16 metres shorter. Similarly, on wet bridport gravel the stopping distance for the SCS car is between 2.5 and 20 metres shorter, depending on speed and condition of loading. The slight increase in stopping distance for the SCS car on dry tarmac was due to the very high locked wheel adhesion value under these conditions. Although the SCS car stopping distances are generally shorter, its main advantage is its ability to be steered by the driver at all times.

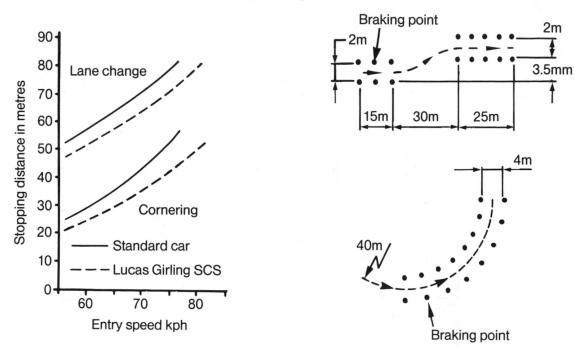

**Comparative cornering and lane change stopping distances**
**Lucas Girling SCS against standard car**

FIGURE (7) COMPARATIVE CORNERING AND LANE CHANGE STOPPING
DISTANCES LUCAS GIRLING SCS AGAINST STANDARD CAR

This diagram illustrates the main advantage of an anti-lock system - the ability for even an average driver to control the car whilst braking hard and changing direction.

The upper part of the diagram shows a car changing lanes according to the standard ISO Chicane Test at varying speeds. In this case the standard car stopping distances are the best attainable by a very experienced test driver without overbraking and locking the wheels which would make the lane change maneuvre impossible. Even so, the SCS stopping distances, which are achievable by any driver, were between 3 and 9 metres shorter depending on speed.

The lower part of the diagram shows a car negotiating a 40 metre radius and then braking from various speeds...Again the standard car stopping distances are the best attainable by a very experienced test driver. In this case the SCS stopping distances, which are achievable by any driver, were between 4 and 12 metres shorter depending on speed.

These results illustrate the additional vehicle safety that even a low cost anti-lock system can provide. Figures (8) - (12) show the construction and operation of the Lucas Girling SCS Modulators used during these tests.

THE LUCAS GIRLING SCS MODULATOR

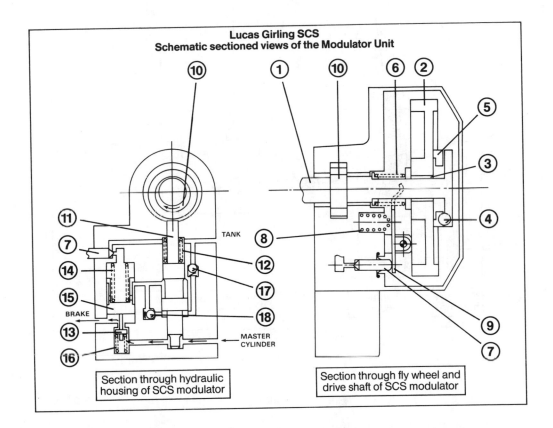

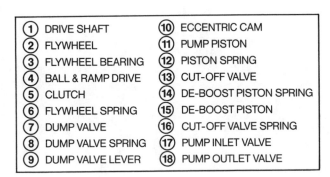

| | |
|---|---|
| ① DRIVE SHAFT | ⑩ ECCENTRIC CAM |
| ② FLYWHEEL | ⑪ PUMP PISTON |
| ③ FLYWHEEL BEARING | ⑫ PISTON SPRING |
| ④ BALL & RAMP DRIVE | ⑬ CUT-OFF VALVE |
| ⑤ CLUTCH | ⑭ DE-BOOST PISTON SPRING |
| ⑥ FLYWHEEL SPRING | ⑮ DE-BOOST PISTON |
| ⑦ DUMP VALVE | ⑯ CUT-OFF VALVE SPRING |
| ⑧ DUMP VALVE SPRING | ⑰ PUMP INLET VALVE |
| ⑨ DUMP VALVE LEVER | ⑱ PUMP OUTLET VALVE |

FIGURE (8) COMPONENTS OF THE LUCAS GIRLING SCS MODULATOR

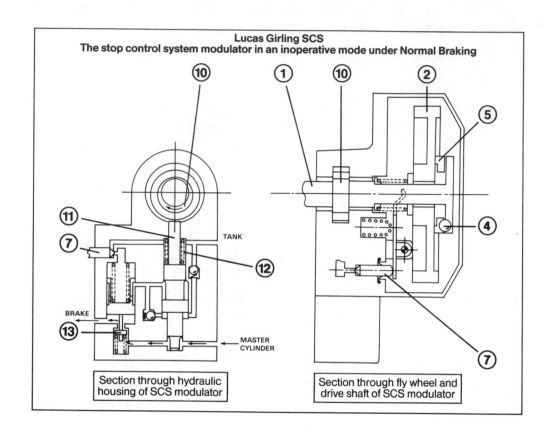

FIGURE (9) THE SCS MODULATOR DURING NORMAL BRAKING

    During normal braking the master cylinder is connected to the brake through the open Cut Off Valve (13) and the Dump Valve (7) is enclosed.  The Shaft (1) which rotates at a multiple of wheel speed, drives the Flywheel (2) through the Ball and Ramp (4) and Clutch (5).  The Pump Piston (11) is held out of engagement with the rotating eccentric (10) by a spring (12). In this condition the brake system operates in a conventional manner.

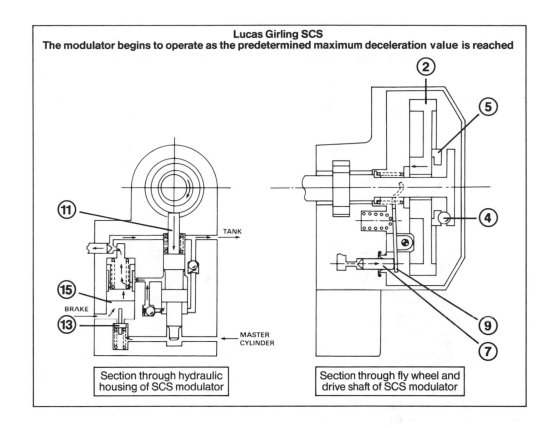

**Lucas Girling SCS**
**The modulator begins to operate as the predetermined maximum deceleration value is reached**

Section through hydraulic housing of SCS modulator

Section through fly wheel and drive shaft of SCS modulator

FIGURE (10) BRAKE PRESSURE REDUCTION

If deceleration of the front wheel, and therefore the drive shaft, exceeds a predetermined maximum value then the flywheel (2) through its inertia, overruns the ball and ramp (4). The axial movement which this causes then opens the dump valve (7) through the lever (9) and therefore connects the deboost-piston to tank.

This causes the de-boost piston (15) to move, which thus relieves the brake line pressure and also closes the cut-off valve (13) isolating the master cylinder from the brake.

At the same time the line pressure from the master cylinder causes the pump piston (11) to be pushed against the rotating eccentric but no pressure can be generated as the dump valve (7) is open to tank.

The flywheel is then decelerated at a controlled rate through the braking effect of the clutch (5).

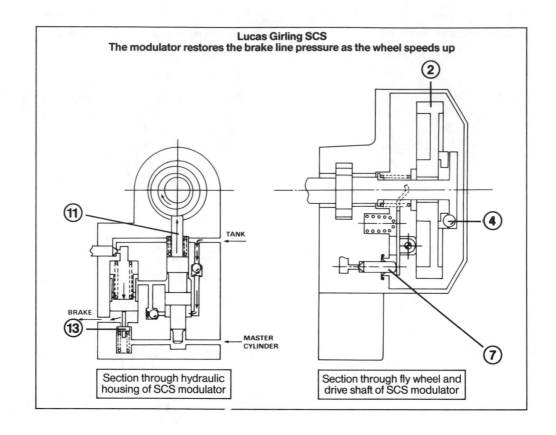

**Lucas Girling SCS**
**The modulator restores the brake line pressure as the wheel speeds up**

Section through hydraulic housing of SCS modulator

Section through fly wheel and drive shaft of SCS modulator

FIGURE (11) BRAKE PRESSURE INCREASE

The pressure reduction resulting from Figure 10 releases the brake and allows the wheel to accelerate to the speed of the still decelerating flywheel (2).

At that point the ball and ramp (4) runs back into its "start" position and the dump valve (7) closes, which causes the pump piston (11) to develop pressure and the clutch to accelerate the flywheel (2) at a controlled rate. Through this process the brake pressure is increased until either the maximum deceleration point is reached again and the disengaged sequence is repeated, or the master cylinder pressure is restored without "wheel lock". In the latter case the cut-off valve (13) opens again, the pump piston (11) is disengaged, and the system is back in a "normal braking" condition. (See Figure 9.)

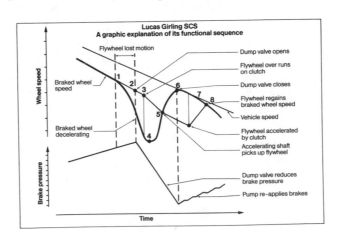

## FIGURE (12) LUCAS GIRLING SCS FUNCTION SEQUENCE

1. The braked wheel angular deceleration exceeds the maximum deceleration threshold set by the flywheel sensor. The flywheel thus overruns and expands the ball and ramp against the threshold spring.
2. When "lost motion" is taken up, the flywheel further expands the ball and ramp activating the dump valve and reducing the brake pressure.
3. The ball and ramp is fully expanded and the flywheel begins to overrun the clutch. At this point the dump valve is fully open.
4. The reduction in brake pressure permits the road wheel to recover and thus accelerate.
5. The accelerating road wheel reaches the flywheel speed. The threshold spring starts to move the flywheel back to close the ball and ramp and the dump valve.
6. The dump valve closes and the pump re-applies brake pressure.
7. The ball and ramp is fully closed so that the flywheel is accelerated by the faster rotating road wheel via the clutch.
8. The flywheel regains road wheel speed and the pump continues to re-apply brake pressure.
9. Rapid response capability enables the Lucas Girling SCS Modulator to operate at more than five cycles per second.

SUMMARY

This paper has described how front wheel driven cars fitted with the Lucas Girling Stop Control System are considerably more stable and controllable in panic braking situations than similar cars with standard brakes.

In the future there will, no doubt, be other low cost anti-lock systems aimed at the small and medium passenger car market. It is now time to establish realistic evaluation criteria using similar normally braked cars in panic braking situations as the baseline for assessment. In this way the added safety of anti-lock brake systems can be made economically available to all classes of passenger cars.

# A New Anti-Skid-Brake System for Disc and Drum Brakes*

**Heinrich Schürr and Adam Dittner**
FAG Kugelfischer KGaA

*Paper 840468 presented at the International
Congress and Exposition, Detroit, Michigan,
February, 1984.

## ABSTRACT

The principal requirements on
an anti-skid device (steerability,
stability on symmetrical and asymmetrical road surfaces, stopping
distances) are well known and not
part of this representation.

The development of the FAG
Anti-Skid-Brake system (ASBS) was
preceeded by other activities on
the mobile sector, from which the
most important construction units
could be taken, or from which at
least the requirements to be expected in vehicle application were known.

The knowledge from these developments
in the field of hydraulics and automatic control technology and the experience in brake hydraulics were the base
line for the development of an Anti-
Skid-Brake system, which started in
1980.

The today's stage of development
is a prototype system (mechanic-hydrau-
lic and electronic) running in some
prototype cars of european car manufacturers.

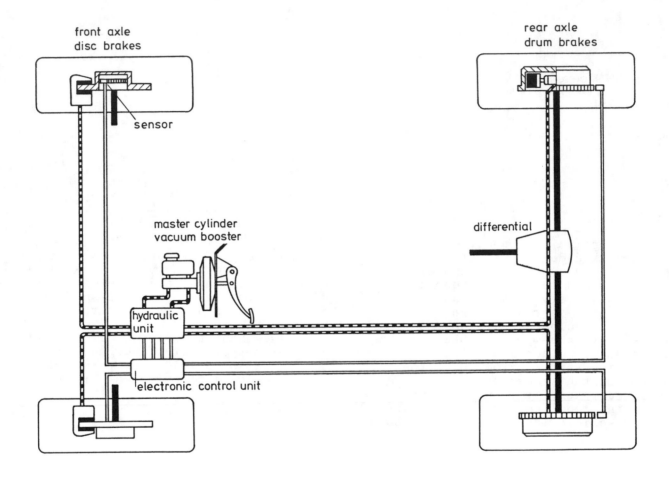

front axle
disc brakes

sensor

rear axle
drum brakes

master cylinder
vacuum booster

differential

hydraulic
unit

electronic control unit

Fig. 1 - 4-channel add-on system

THE SYSTEM CONSISTS OF the hydraulic unit, the 4 sensors on the wheels and the electronic control unit (Fig. 1). The hydraulic unit basically contains a pressure reservoir, a pilot valve, 4 control valves, 4 stepped pistons each with a displacement measuring device and an oil reservoir. Furthermore a piston pump with an electric motor mounted onto it (Fig. 2).

The hydraulic unit has 6 brake line connections: 2 inputs from the master cylinder and 4 outputs to the wheel cylinders and is not built together with the master cylinder to form one unit and can therefore be placed anywhere (in the motorspace). Of course it is also possible to combine the hydraulic unit with the master cylinder to one unit.

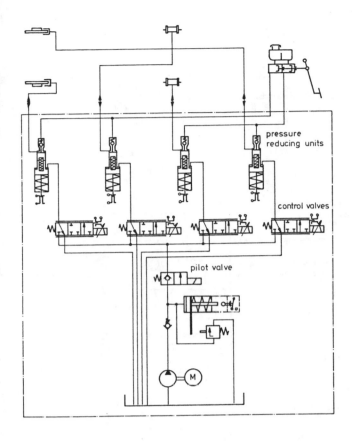

pressure reducing units

control valves

pilot valve

M

Fig. 2 - Hydraulic circuit diagramm

The base of the FAG Anti-Skid-Brake system (ASBS) is a spring-loaded pressure reducing unit with a stepped piston and an annexed displacement measuring device, which together with an analog control valve (servo-valve) forms a closed electro-hydraulic control loop (Fig. 3).

By means of control valves - in this case direct operated servo-valves (Fig. 4 and 5) the spring-loaded stepped pistons in the pressure reducing units are actuated. The driving principle - the pilot piston of the control valve is translatorily shifted by a spindle - allows such high shifting forces that mechanical jamming, caused by dirt particles in the oil, is practically excluded. For this reason this principle of valve is especially suited for the application in the mobile sector and patented by FAG in the most important industrial countries.

By a suitable input of a nominal value into the control loop it is possible to change the position of this unit by a defined value. The change of position of the stepped piston in the pressure reducing unit causes a proportional change of the pressure in the brake system only by change of volume of a controlled stroke volume, i.e. no ejection or refill of brake fluid.

This kind of control allows to work with constant pressure gradients for pressure increase and decrease. The advantage of this type of pressure modulation is, that vibrations of the drive and steering elements, especially on frontwheel driven vehicles, are greatly minimized compared to other known systems. At the same time drum brakes of commercial size (most types have a considerable hysteresis) can be controlled same as disc brakes (Fig. 7 and 8).

With the start of ASBS the pilot valve opens. The accumulator serves at this moment for a fast oil delivery during the run-up of the pump. The control valve is actuated and the stepped piston in the closed control loop is retracted. The ball valve above the stepped piston closes and the pressure in the wheel cylinder is reduced. The stepped piston works in its optimum position (at constant $\mu$). The part of the piston with the small diameter is divided, therefore no underpressure in the brake system independently from the position of the spring-loaded stepped piston. The 4 stepped pistons and the piston of the accumulator are acted upon by spiral springs. When a control period is over, the spring pushes the stepped piston back into its rest position, the ball valve above the stepped piston opens and the normal brake system is restored.

The advantage of this type of pressure reduction is that there is absolut no pulsation on the brake pedal during a control procedure.

In case of power failure during a control period the control valve is pushed back by the spring into its rest position and by this the annulus area of the stepped piston is opened to the reservoir. Even if the control valve jams the spiral spring will push the stepped piston back into its rest position because of leakage oil through the control valve back to the reservoir.

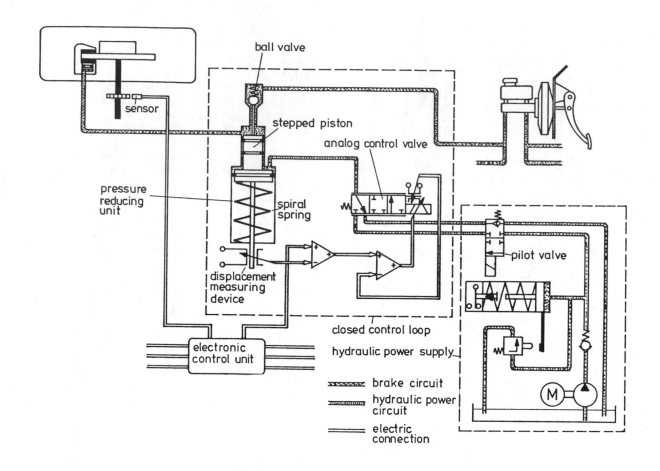

Fig. 3 - Simplified functional diagramm
only one controlled wheel

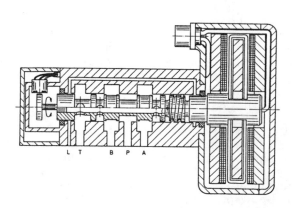

Fig. 4 - Cross section of the direct
operated servo-valve
(ServoRot)

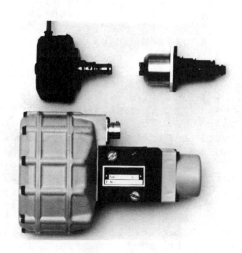

Fig. 5 - Below standard ServoRot.
Above on the left simpli-
fied analog control valve;
next to it the pilot valve

Fig. 6 - Photo of the present
prototype system

Fig. 7 - Sensor on the front axle

Fig. 8 - Sensor on the rear axle

Fig. 9 - Fitting position of the
FAG-ASBS in the motor-space
of a prototype car

The position of the control valves
can be checked at any time. The spiral
spring in the accumulator permits by very
simple means (limit switch) a statement
whether pressure fluid is available when
ASBS is started. Thus a sure determination of the filling is possible. Due to
weight reduction of the hydraulic unit
also a hydropneumatic accumulator is
applicable as well.

The spiral springs under the stepped
pistons ensure that an excessive pressure
increase will be impossible. At pressures
over e.g. 145 bar (2103 psi) the ball
valve above the stepped piston will
close.

The required electronic monitoring
of the control components is very simple,
because the state of the closed control
loops and the position of the valves can
be checked in a very simple way by the
displacement measuring devices, also outside a brake period. All necessary parameters can be controlled even if the vehicle is not moving.

The present prototype system (Fig. 6
and 9) is an add-on-system and has 4 individual control channels. Of course a
select-low control is possible without
any problems. The drive does not necessarily require brake fluid. Mineral oil can
be used just as well. If the vehicle is
already equipped with a powersteering,
electric motor, pump and reservoir can be
omitted.

Fig. 10 - Control panel in a
prototype car

The control panel in the prototype
cars (Fig. 10) allows the following
kinds of control:
- Braking without ASBS (OASBS)
- 4-channel system individual rear-
  wheel control (4 sensors) (4CIWC)
- 3-channel system select-low rear-
  wheel control (4 sensors) (3CSLC)
- 2-channel system rear axle only
  with individual wheel control
  (2 sensors) (2CRIC)
- 1-channel system rear axle only
  with select-low wheel control
  (2 sensors) (1CRSC)

CONTROL OF THE UNDRIVEN WHEEL

For the development of this type of
control it has been taken into consideration that any given wheeled vehicle
can be decelerated in the plane only by
means of the maximum coefficient of adherence between road and wheel. In optimum conditions the maximum value is
about 1. This means, that a vehicle can
be decelerated at best by -9,81 m/s$^2$.

If one wheel is overbraked by the
driver due to the characteristic of adherence the deceleration is higher than
-9,81 m/s$^2$. As soon as this information
is given the anti-skid-brake system is
activated. Through the output of threshold (3) the ramp generator (4) is started. This ramp generator (4) now gives
a constant releasing speed to the closed
control loop of the pressure reduction
unit (Fig. 11). The reduction continues

until the deceleration drops below the
value of threshold (3) of -9,81 m/s$^2$.
At this point the braking force on the
undriven wheel is too high by a value
which corresponds to the moment of inertia of the wheel, independently from the
actual adherence coefficient between
road and wheel and from the hysteresis
of the brake unit. With the supplementary release (5) a new position in release direction is given with a constant
releasing speed to the closed control
loop by the ramp generator (4). Because
of the almost linear connection between
displacement and pressure reduction it
is possible to change the position of
the closed control loop so that the resulting change of the braking force is
higher than the force which results from
the product of the moment of inertia and
the acceleration due to gravity. As soon
as the closed control loop has reached
its position, the ramp generator (4)
will be held at a constant value for a
certain time. Thus the closed control
loop and consequently also the stepped
piston is held in its actual position.
The actual braking force acting on the
undriven wheel is now lower than the
possible accelerative force resulting
from the road. If during the waiting
time threshold (3) will be activated
again due to a decrease of the adherence
coefficient, there is a absolut priority
for a new release. After the waiting
time the brake pressure and consequently
also the braking force are increased
again with a constant rate of pressure
rise via the ramp generator. If threshold (3) is exceeded again, the above
described control cycle will again be
initiated.

Typical settings used for an european middle class vehicle for pressure-
decrease-gradient, pressure-increase-
gradient and waiting time are:
- 1400 bar/s pressure decrease
- 450 bar/s pressure increase
- 200 ms    waiting time

In case of a sudden change of the
adherence coefficient with a resulting
short locking of the wheel, the electronic deceleration limiting device (2) ensures a continuous control at very low
angular velocities of the wheel.

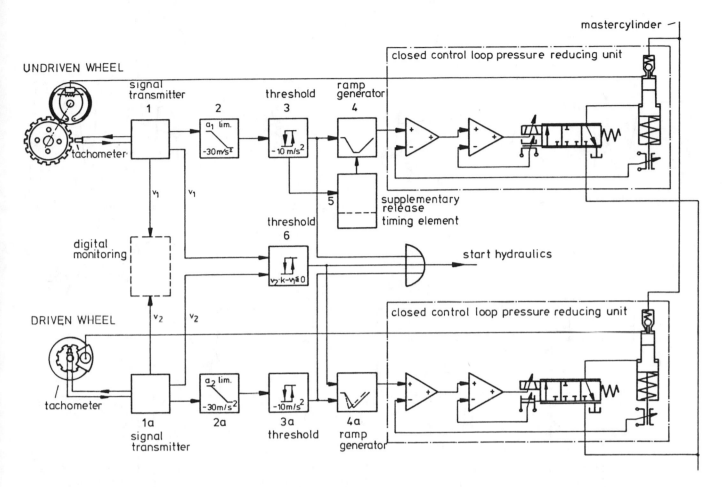

Fig. 11 - Simplified closed control loop for a driven and an undriven wheel

CONTROL OF THE DRIVEN WHEEL

Through the signal transmitter (1a) and threshold (3a) the ramp generator (4a) is started after an overbraking and the connected pressure reducing unit is displaced. This is done at first with a definitely lower pressure-decrease-gradient, than on the undriven wheel. If the value drops below threshold (3a) a pressure increase takes place without an interposition of a supplementary release and a timing element.
- 380 bar/s pressure decrease
- 380 bar/s pressure increase

On a nearly uniform surface this control is sufficient to decelerate the driven wheel within a stable slip range.

In case of a considerable decrease of the adherence coefficient, the control would not work, i.e. the wheel would slip more and more and would finally be locked. To avoid this possibility a superior closed control loop (6) is neces-

sary, in which the nominal value represents the actual speed of the undriven wheel.

At a certain difference between the angular velocity of the driven wheel and the angular velocity of the undriven wheel, which is used as a reference value, the ramp generator (4a) switches to a definitely higher pressure-decrease-gradient.
- 1500 bar/s pressure decrease

This will take place until the fast release is stopped by the closed control loop (6). The following control cycles are now determined again by threshold (3a) and by the ramp generator (4a) with a lower pressure-decrease-gradient.

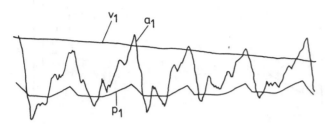

UNDRIVEN WHEEL

constant coefficient of friction asphalt dry

Fig. 12 - Recording of typical
measurement

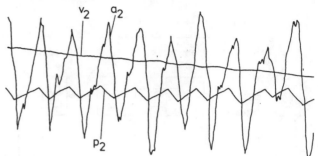

DRIVEN WHEEL

constant coefficient of friction asphalt dry

Fig. 13 - Recording of typical
measurement

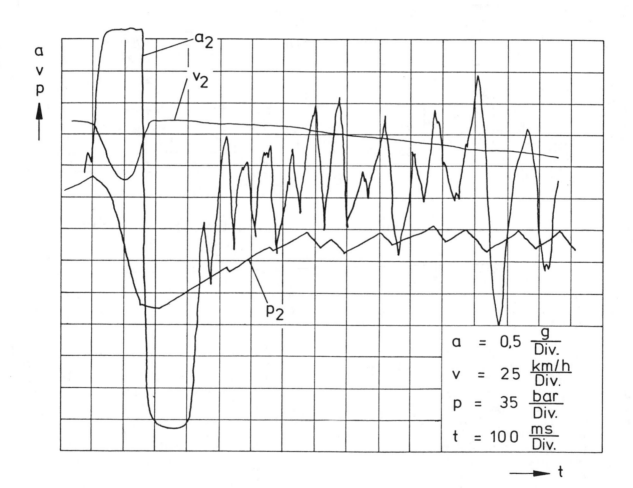

| | | |
|---|---|---|
| a | = 0,5 | $\frac{g}{Div.}$ |
| v | = 25 | $\frac{km/h}{Div.}$ |
| p | = 35 | $\frac{bar}{Div.}$ |
| t | = 100 | $\frac{ms}{Div.}$ |

Fig. 14 - Control procedure of a driven wheel
sudden decrease of the adherence coefficient

Fig. 12 and 13 show typical measurements of a controlled brake procedure on a uniform surface.

a = wheel acceleration
v = wheel velocity
p = brake pressure

Fig. 14 shows a controlled brake procedure of a driven wheel with a sudden change of the adherence coefficient. At a certain difference between the angular velocity of the driven wheel and the angular velocity of the undriven wheel the brake pressure is lowered with a definitely higher gradient. In the example the pressure is lowered down to zero. After the difference of the wheel velocities has fallen below the certain amount the following control cycles are determined again by lower pressure gradients. The brake pressure now corresponds to the actual surface.

| road condition | vehicle speed km/h | stopping dist. m | ∅ deceleration m/s² | Kind of control |
|---|---|---|---|---|
| asphalt wet | 150 | 137,3 | 6,32 | OASBS |
| | 147 | 120,8 | 6,90 | 4CJWC |
| | 153 | 133,8 | 6,75 | 3CSLC |
| | 150 | 163,3 | 5,31 | 2CRJC |
| | 150 | 163,8 | 5,3 | 1CRSC |
| | 79,5 | 31 | 7,86 | OASBS |
| | 80 | 33 | 7,48 | 4CJWC |
| | 80,6 | 34,67 | 7,23 | 3CSLC |
| | 80,8 | 37,92 | 6,65 | 2CRJC |
| | 80,1 | 37,3 | 6,63 | 1CRSC |
| low friction (simulated ice conditions) | 51 | 89,5 | 1,12 | OASBS |
| | 50 | 81 | 1,19 | 4CJWC |
| | 49,3 | 80,2 | 1,17 | 3CSLC |
| | 50,4 | 85,4 | 1,14 | 2CRJC |
| | 49,9 | 85 | 1,13 | 1CRSC |

Fig. 15 - Stopping distances reachable with different kinds of control

DESCRIPTION OF THE CONTROL OF
AN UNDRIVEN WHEEL

Symbols used:

| | |
|---|---|
| $m_A$ | proportionate weight of the vehicle = const. |
| $v_F$, $\dot{v}_F$ | speed, acceleration of the vehicle |
| $F_{Br}(t)$ | braking force |
| $r$ | radius of application of braking force |
| $\Theta_R$ | moment of inertia of the wheel = const. |
| $R$ | radius of the wheel |
| $F_R$ | frictional force = const. |
| $w$, $\dot{w}$ | angular velocity, angular acceleration of the wheel |
| $\mu$ | adherence coefficient = const. |
| $g$ | acceleration due to gravity |
| $A$ | translatory acceleration of the wheel (related to R) |
| $M_{Br}(t)$ | braking moment |
| $m_{red}$ | reduced mass transferred into radius of gyration R |

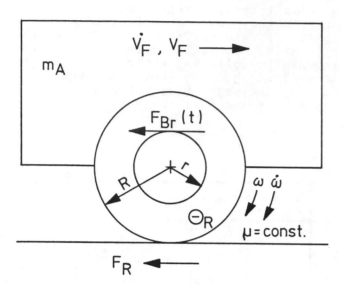

Fig. 16 - Equivalent mechanical modell of an undriven wheel

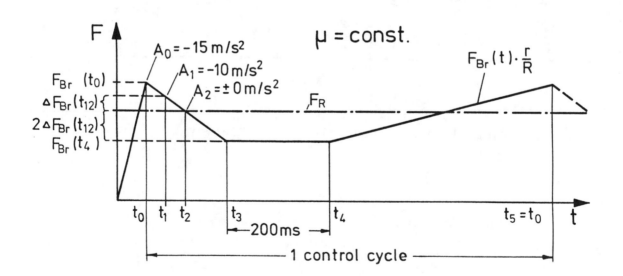

Fig. 17 - Forces acting on the undriven wheel

Example:    $m_A$ = 250 kg          $\mu$ = 0,8 = const.
            $R$ = 0,28 m           $m_{red}$ = 12,27 kg ($\Theta_R$ = 0,962 kgm$^2$)
            $g$ = 10 m/s$^2$        $r$ = 0,1 m

## BEHAVIOUR IN MOTION OF THE EQUIVALENT MECHANICAL MODEL

Braking moment:

$$M_{Br}(t) = F_{Br}(t) \cdot r$$

Friction moment:

$$M_R = F_R \cdot R = m_A \cdot g \cdot \mu \cdot R$$

Dyn. moment:

$$M_{Dyn} = \Theta_R \cdot w = m_{red} \cdot R^2 \cdot w$$

$$= m_{red} \cdot R \cdot A$$

From balance of moments:

$$(1) \quad F_{Br}(t) = \frac{R \ (m_A \cdot g \cdot \mu - m_{red} \cdot A)}{r}$$

$t = t_o$: Activation of the anti-skid-brake system by e.g. an acceleration - threshold
$A_o = - 15 \ m/s^2$.
Start of pressure release.
Constant pressure gradient.

Braking force:

$$F_{Br}(t_o) = 6115,34 \ N$$

$t = t_1$: Reaching of acceleration - threshold
$A_1 = - 10 \ m/s^2 = g$.
Start of additional release.

$$3 \cdot \Delta F_{Br}(t_{12})$$

Braking force:

$$F_{Br}(t_1) = 5943,56 \ N$$

$t = t_2$: Acceleration of the wheel
$A_2 = + 0 \ m/s^2$.
Continuation of constant pressure gradient. The deceleration of wheel and vehicle now results from the adherence coefficient.

Braking force:

$$F_{Br}(t_2) = 5600 \ N$$

$$\Delta F_{Br}(t_{12}) = 343,56 \ N$$

$$\Delta F_{Br}(t_{12}) \cdot \frac{r}{R} = 343,56 \ N \cdot \frac{0,1 \ m}{0,28 \ m}$$

$$= 122,7 \ N$$

The braking force at time $t_1$ is always too high by a value which results from the product

$$m_{red} \cdot g = \Delta F_{Br}(t_{12}) \cdot \frac{r}{R}.$$

This value is always the same, independently from the road surface, because the moment of inertia of the undriven wheel does not change as on the driven wheel (depending of the gear transmission). ($\Theta_R = m_{red} \cdot R = const.$)
Every further reduction of the braking force causes necessarily an inversion of the sign of the acceleration of the wheel. This further reduction of the braking force should be at minimum

$$2 \cdot \Delta F_{Br}(t_{12})$$

to reduce the slip between wheel and vehicle during the waiting time at low adherence coefficients.

$t = t_3$: Start of waiting period (constant braking pressure).

Waiting time $t_w$ = 200 ms

Braking force: $F_{Br}(t_3) = 5354,6 \ N$

Acceleration of the wheel at $t_3$:
from (1) $A_3 = + 7,14 \ m/s^2$

The wheel is accelerated from time $t_3$ by $+ 7,14 \ m/s^2$ from its instantaneous speed to the speed of the vehicle until a slip coefficient is reached which corresponds to the relation between frictional force $F_R$ and instantaneous braking force $F_{Br}(t_3)$.

$t = t_4$: Start of the increase of braking force.
The pressure in the braking system is now increased with a defintely lower gradient than the gradient of the preceeding pressure reduction. This takes place until acceleration threshold $A_o = - 15 \ m/s^2$ is activated again, a new reduction of the braking force is initiated and a new control cycle starts.

$t = t_5$   Acceleration threshold

$= t_0$:   $A_0 = -15$ m/s$^2$ is reached again.
New start of pressure reduction.

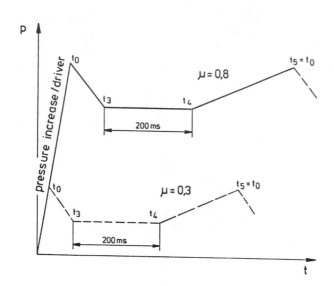

Fig. 18 - Control cycle at different coefficients of
friction

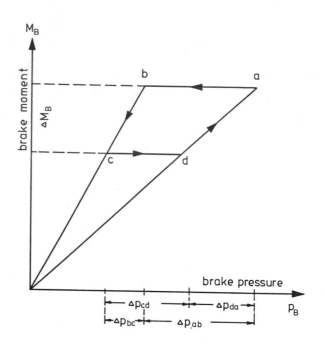

Fig. 19 - Relation of brake moment
and brake pressure

## COMPARISON OF ANALOG AND DIGITAL PRESSURE MODULATION ON A DRUM BRAKE

Fig. 19 shows the relation of brake moment and brake pressure of a drum brake with mechanical hysteresis used on the passenger car sector. At every necessary reduction of the brake moment at first the brake-pressure-difference pab has to be run through before a brake moment reduction is started. On the opposite side, during the following pressure increase, at first the brake-pressure-difference pcd has to be passed through before a brake moment increase is initiated. Increasing the brake moment causes a rise of the disturbing hysteresis. On nowadays drum brakes 50 % hysteresis are usual, which can only be lowered by design changes of the braking device to approximately 20 %.

Fig. 20 shows the advantage of the FAG-ASBS with analog pressure control opposite to a digital pressure modulation. The example shows a production size, unmodified drum brake on a passenger car and describes the necessary brake pressure reduction after a change of the coefficient of friction. Corresponding to the pressure reduction and the hysteresis of the brake unit the brake moment follows. It is obvious, that the optimum brake moment corresponding to the new coefficient of friction is reached earlier and with a higher precision using the analog pressure reduction than using the digital. The same behaviour demonstrates the subsequent pressure increase. The reason for this behaviour is the hysteresis of the drum brake.

Using the analog pressure modulation the brake moment, after running the brake-pressure-difference pab, follows proportional every further pressure reduction.

Due to this fact it is possible to adapt the actual brake moment corresponding to the control parameters with a high precision to the optimum brake moment.

Using the digital pressure modulation the reachable control quality is limited by the finite switching time of the solenoid valves, because the actual brake moment compared to the optimum brake moment is still too high or already too low.

Making a limit consideration it is evident, that with a given hysteresis of the drum brake, a comparable control quality can only be realized with infinite small switching times of the solenoid valves.

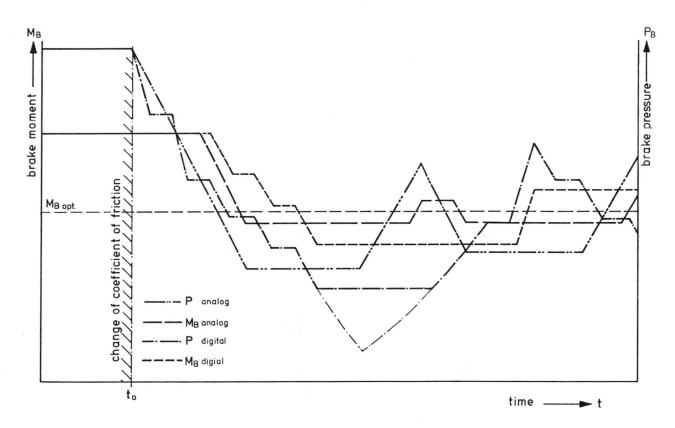

Fig. 20 - Comparison of digital and analog pressure modulation

## SUMMARY

The major aim of this development was to design a system which is able to control disc brakes as well as drum brakes, because on the passenger vehicle sector the number of cars having mixed brake systems is world wide a multiple compared to those cars working with 4 disc brakes. This problem was solved by using an analog controlled hydraulic circuit with constant pressure gradients for pressure modulation by change of volume of a controlled stroke volume. This kind of control has the advantage that the pedal feeling is excellent (absolut no pulsation), vibrations of the drive elements are minimized and therefore the system is especially suitable for middleclass vehicles. Due to the control principle (closed control loop) all components can be tested outside a brake procedure, even if the vehicle is not moving.

# Cost-Benefit Analysis of Simplified ABS**

**Peter Hattwig**
Volkswagenwerk AG
Wolfsburg, Germany

**Paper 850053 presented at the International
Congress and Exposition, Detroit, Michigan,
February, 1985.

## ABSTRACT

Anti-skid brake systems (ABS), as offered
in German cars, have attained a high technical
standard, but make up between 5 and 10 % of the
price of the vehicle for the customer. Investigations are therefore being made to establish
whether it is possible to have an ABS concept
with a simpler, more reasonably priced design,
which would nevertheless provide effective vehicle braking controlling.

To this end, on the basis of the Audi 100
system, a cost/benefit analysis is made for
several systems, which feature a smaller number
of control channels and other components. One of
the systems is a completely new concept not as
yet published in the relevant literature. This
particular system works similar to an amplifier
and diagonal control and is referred to as "multiple amplifier control". This control system
ensures that the vehicle keeps a high degree of
directional stability and steerability and good
deceleration under various driving and braking
conditions.

## COST-BENEFIT ANALYSIS BASED ON THE AUDI 100 SYSTEM

From the beginning of production anti-skid
brake systems have had a high technical standard
approaching the limit of what is currently feasible.

In Germany, however, these systems cost
about $ 1,000 as an option. In view of this high
price, the question arises whether it is possible to produce a less expensive ABS, even if the
system is less sophisticated. This system must
in the final analysis still be better than no
system at all.

The aim of this investigation is therefore
to discuss whether any cost optimized concepts
exist which still offer effective brake control
for the vehicle, without impairing the braking
characteristics.

In order to assess whether or not a simplified system is viable, a simple cost-benefit

analysis will be carried out, which is intended
to give an indication of how layout and benefits
are matched.

The considerations are based on the
Audi 100 and 200 (in the USA the Audi 5000 and
Audi 5000 turbo). The optional ABS is an add-on
to the existing braking system, without altering
this system. The additional expenses for the ABS
are assessed as 100 %, while the braking system
without the ABS has a cost index of zero.

The benefits are assessed in the same way.
The braking behavior with the ABS is such that
physical possibilities are largely exploited.
However, this does not mean that no further
improvements could be made to the ABS. The additional benefits of the ABS in the Audi 100 are
defined as being equal to 100 %, while without
the ABS the braking performance is by definition
equal to zero.

On the basis of these two values - the
100 % costs and the 100 % benefits - the
Audi 100 system has a cost-benefit ratio of 1,
while according to this method, no ratio can be
determined for the simple braking system without
the ABS.

In the opinion of the author a simplified
anti-skid brake system is only viable if the
cost-benefit factor is at least approximately 1
or better.

## THE ADDITIONAL COSTS CAUSED BY THE ABS

The Audi 100 ABS is a separately installed
system, from Robert Bosch, Germany (1)*. Fig. 1
shows the hydraulics for one wheel.

The hydraulics consist of two main components, which are added to the brake circuit
between brake master cylinder and the wheel
brake:
a)   The control elements, in this case a
     3-way/3-position valve, and
b)   The power supply, which in this case is a
     pump with a separate electric motor and a
     small accumulator.

* Numbers in parentheses designate references at
  end of paper

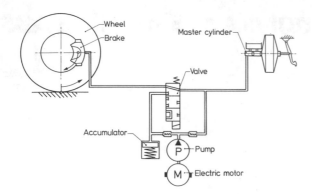

Fig. 1: Hydraulic system Bosch ABS
Valve in normal position

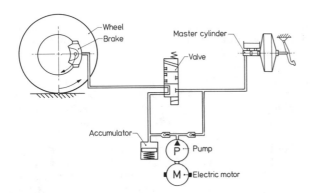

Fig. 2: Hydraulic system Bosch ABS
Valve in activated position

Under normal braking conditions without a control system the pressure can act directly on the brake via the normally open valve port. If the wheel begins to lock because the braking moment is greater than the frictional moment produced between the tire and the road surface, then the valve switches into its third position, as shown in Fig. 2. The brake fluid is first taken up by a small intermediate accumulator, but is then immediately pumped out and returned to the valve inlet side. This returning of the fluid is felt as a pulsing sensation through the brake pedal. When there is no longer any chance of the wheel locking and the balance has been restored, the valve opens rapidly and releases the metered pressure for the brake again. The valve also has an intermediate position, in which the pressure can be held constant and which is essential for control purposes.

Fig. 3 shows that vehicles with diagonally split brake circuits are fitted with 4 final control elements and 1 power supply unit (but with a dual circuit pump and two small accumulators).

In addition there is an electro-magnet sensor for each wheel, which measures the speed of the wheel, as well as an electronic control unit.

In the Audi 100 the front wheels are con-

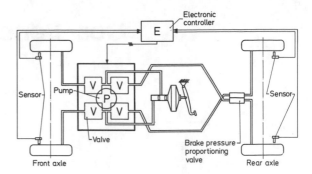

Fig. 3: 3-channel control with diagonally split brake circuit

trolled individually and the rear wheels jointly, according to the select low principle.

The vehicle has 3 independent control circuits. Thus the system is defined as a 3-channel control system.

The hydraulic operating principle of the Bosch ABS system is known as the return pumping principle. This is used as the basis for the cost analysis of most of the simplified systems which are mentioned in this paper.

The 100 % costs set for the 3-channel control system are made up of 55 % hydraulics, 27 % electronics and electrical components and 18 % sensors. The installation cost for the individual components is already included in these figures.

THE SAVINGS RESULTING FROM THE ABS

The additional cost must be calculated against the savings made due to the following points
a) When making an emergency stop on dry asphalt the wheels do not lock and there is no localized tire wear (flat spot).
b) Increased safety reduces the accident risk and there is less or no accident repair cost.

Neither of these points can be assessed quantitatively with any degree of certainty, and therefore neither of the savings is considered in the analysis.

THE BENEFITS OF THE ABS

To assess the benefits of the ABS, the braking action in a properly adjusted vehicle without the ABS will be studied:
If the actuating forces are sufficient then initially the front wheels will lock. The vehicle can no longer be steered. With the ABS both front wheels are controlled within a slip range where the lateral control is still sufficient to maintain vehicle steerability. The vehicle steerability is assessed at 35 %.

If the actuating force is further increased then in a vehicle without the ABS the rear wheels will also lock. With a uniform road surface the vehicle will be aimed in a straight forward direction and will skid in the direction of travel. With an asymmetrical (split friction) road surface a moment is produced about the vertical axis, which causes the vehicle to spin. With the ABS the vehicle can usually be kept stable even by the average driver. The ABS therefore provides better vehicle stability when braking on asymmetrical road surfaces and permits emergency braking while cornering. The directional stability is also assessed at 35 %.

Wheel slip is normally controlled so that the highest coefficient of friction between the tire and the road surface is extensively exploited. This makes the braking distance with the ABS shorter in many situations than without the ABS. The braking distance is assessed at 30 %.

## THE CALCULATION OF THE BENEFIT

The systems described subsequently are examined and presented in shortened form according to the following outline:

The steerability can be calculated from the lateral force of the front wheel. It is assumed either that the vehicle travels in a curve and that the loads on the outer wheel and the inner wheel have a ratio of 70 to 30 or that the vehicle travels on an asymmetrical road surface and that the friction coefficients are in the relation 70 to 30. Under this condition, different control principles will result in different lateral forces during braking in a curve or on split friction. The optimum can be attained on the front wheels by means of an individual control (100 % = 35 points). If, however, a wheel locks, as is the case with the select high control, the lateral forces corresponding to the assumptions are slighter (70 % = 21 points).

It must be taken into consideration, however, when calculating the steerability that the steering behavior depends not only on the lateral force of the front wheels, but is also influenced by the lateral force of the rear wheels. In order to be able to calculate the steerability of a vehicle, it is necessary to introduce a correction factor which is multiplied with the lateral force of the front axle. In the ideal case, the correction factor is 1, when the steering behavior of the vehicle is not changed by the control, and the correction factor is smaller than 1, when the vehicle understeers by the chosen control principle.

In order to be able to evaluate the stability of the vehicle, the lateral forces of the rear wheels are calculated. It is again assumed that the lateral forces with individual control are in the relation 70 to 30 (100 %). The lateral force with select low control is even more favorable (125 %). It is therefore evaluated with the highest number of points (35 points).

In order to calculate the braking distances, four different road conditions were assumed: ice (share 45 %), loose snow on a firm surface (20 %), wet asphalt (20 %) and dry asphalt (15 %). The typical friction-slip-graphs served as a basis for the calculation. The brake force distribution corresponds to that of a

front wheel drive vehicle with a load sensitive brake pressure proportioning valve. Regarding the calculation it was furthermore assumed that 70 % of the vehicles are driven empty and 30 % loaded.

According to this schedule the benefits of the 3-channel control system achieve a total of 89 points. As was defined earlier, this value is now taken as the 100 % value.

The individual points for the criteria are given in Table 1 and 2 at the end of the paper.

## 3-CHANNEL CONTROL WITH 3-CIRCUIT BRAKING SYSTEM

Fig. 3 shows that two final control elements are required for the rear wheels despite the fact that both wheels are jointly controlled according to the select low principle. Axially split brake circuits, where there is one brake circuit for the front wheels and one brake circuit for the rear wheels, would allow one final control element to be omitted for the rear wheels. Such a brake circuit configuration is not usually possible for front-wheel drive vehicles for reasons of axle load distribution.

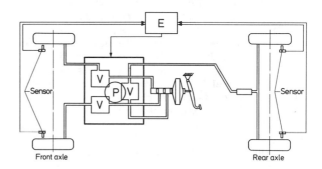

Fig. 4:   3-channel control with 3-circuit braking system

An alternative is offered by a 3-circuit braking system in which each front wheel and both rear wheels have a separate brake circuit as shown in Fig. 4. Not only does this save one final control element, but also a brake pressure proportioning valve. On the other hand there is the additional cost for a 3-circuit brake master cylinder to be considered. In practice there is the problem of uneven brake pull when one of the front brake circuits fails. If no additional problems result, then this arrangement would give a cost saving of 5 %. As the function remains unaltered, the cost-benefit ratio increases to 1.05.

Experiments with 3-circuit braking systems are being carried out in connection with the next generation of the ABS, the "Integrated System" (2).

## 2-CHANNEL AXLE CONTROL

A simplified system which has been available from a Japanese manufacturer for the last two years is the 2-channel axle control system (3). This system controls the front axle according to the select high principle and the rear axle according to the select low principle.

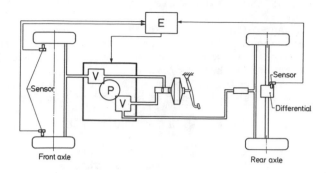

Fig. 5:   2-channel axle control with axially split brake circuit

This system offers little advantage for diagonally split brake circuits as the number of final control elements required is still the same. In contrast, the number of valves required is halved for axially split brake circuits, as shown in Fig. 5. This results in costs of 77 % since with rear-wheel drive a common sensor can also be fitted to the differential.

To work out the benefits, the braking performance must be examined. Select high control always results in one front wheel locking after a short time, even on a homogeneous road surface, as the vehicle is not usually symmetrically laden. Instead of the possible 35 points the steerability was rated at 16 points, as the vehicle heavily understeers when braking. As the braking distances also result in a loss of points, according to the calculation the system benefits are 71 %. This gives a cost-benefit factor for 2-channel axle control with axially split brake circuits of 0.92, which firstly is less than 1 and which also has further disadvantages. Tire cost must also be added to the total cost of the system, as emergency braking on dry asphalt could damage or ruin them.

On the other hand this system still has potential for further development, for example locking of the second front wheel when braking on a dry road surface can be prevented by the addition of simple control means.

2-CHANNEL FRONT WHEEL CONTROL

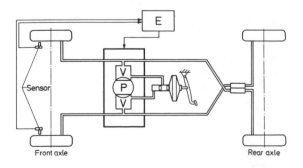

Fig. 6:   2-channel front wheel control with diagonally split brake circuit

Another method of reducing cost is to have only one final control element for each of the diagonally split brake circuits. There is the possibility of only fitting sensors to the front wheels, i. e. only the front wheels can then be controlled. The rear wheels are diagonally connected to the front wheels, see Fig. 6. This fact requires the vehicle brakes to be adjusted in such a manner that they are geared to the rear axle's tendency to underbrake, which naturally impairs the braking distance and the steerability as the vehicle understeers. Overbraking and locking of the rear wheels should be avoided, as this is more dangerous than braking without the ABS. If the driver was to make an emergency stop, without the ABS, the rear wheels would initially lock and usually after a short interval the front wheels as well. The vehicle might only be unstable for a few tenths of a second. If the front wheels were controlled and the rear wheels locked, then the vehicle would be permanently out of control. This system has not been assessed, because of the as yet unsolved problems.

2-CHANNEL DIAGONAL CONTROL

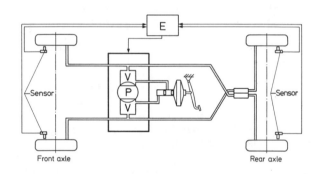

Fig. 7:   2-channel diagonal control with diagonally split brake circuit

For this type of control all 4 wheels are fitted with sensors, see Fig. 7. Each diagonal circuit is controlled individually according to the select low method. This means that when the brakes are correctly set only the front wheels are controlled, but if the brakes are not correctly set then the rear wheels are controlled first. The problem of overbraking of the rear wheels would be eliminated with this type of control.

The last two control systems described both have further shortcomings. For example: In the case of front-wheel drive vehicles, braking with the engine engaged results in a braking moment by engine torque acting on the front wheels. The control unit cannot differentiate between these two moments and reduces the pressure until the correct total moment is achieved. As there is no engine braking moment acting on the rear wheels, a lower braking pressure than that which the wheels can transmit is set by the control unit. This results in the rear wheels being considerably underbraked on a slippery road surface and at worst there would be no pressure on the rear brakes at all.

With 2-channel diagonal control this effect can be lessened by periodically switching one of the diagonal circuits to select high, but this can result in stability problems when cornering.

The assessment results in a cost value of 88 % and a benefit value of 88 %, which results in a cost-benefit factor of 1.0.

1-CHANNEL DIAGONAL CONTROL

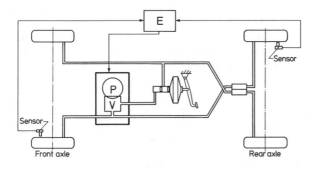

Fig. 8:    1-channel diagonal control with diag-onally split brake circuit

Halving the 2-channel diagonal control system results in the 1-channel diagonal controller, see Fig. 8. In this case, a normally-braked diagonally split brake circuit is the braking system available, which in some respects could even be seen as an advantage. The illus-trated version would, however, present insur-mountable problems when cornering or travelling on an asymmetrical road surface. Imagine the vehicle travelling into a left-hand turn under emergency braking and the front inner wheels are under no load: It would then be almost impos-sible to steer the vehicle, because the front outer wheel is locked. The situation is worse when travelling into a right-hand turn: If the rear inner wheel is under no load then the ve-hicle would oversteer because there would be almost maximum lateral control at the front and almost no lateral control at the rear. The 1-channel diagonal controller is not assessed because of this disadvantage.

1-CHANNEL REAR AXLE CONTROL

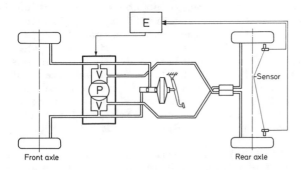

Fig. 9:    1-channel rear axle control with diag-onally split brake circuit

As early as at the end of the sixties rear wheel control systems were available from the large American automotive companies, and for the last three years they have also been available from a Japanese automotive company (4).

Fig. 9 shows that two final control ele-ments are still required for the rear wheels with diagonally split brake circuits, as well as the complete power supply unit, even though it is somewhat smaller. The costs have been esti-mated at approximately 64 % of those for the 3-channel system. Only stability is considered when determining the benefits: When braking in a straight line on a symmetrical road surface the vehicle handles just like a properly adjusted vehicle without the ABS: It skids straight for-ward. The additional directional stability does not improve the braking characteristic when cornering, in fact it impairs them because the tendency just before the front wheels lock is toward heavy understeer. The only time a rear axle controller is of benefit is when driving on an asymmetrical road surface, but even then if there is a major difference between the coeffi-cient of friction for the left and right wheel the rear axle controller will not prevent the vehicle from spinning.

Instead of 35 possible points only 20 points were given for the stability, because the vehicle cannot be controlled in a curve.

A total benefit value of 26 % was assessed, which includes a few points from theoretically slightly improved braking distances. This yields a cost-benefit ratio of only 0.41.

SIMPLE AMPLIFIER CONTROL

In comparison with the 3-channel system the simple control concepts explained up to now have had fewer final control elements and – with one exception – fewer control channels. If the pos-sible measures for cost reduction are taken still further, then one has to consider control-ling the amplification of the braking force. In this case only one final control element is left, which acts on all four wheels.

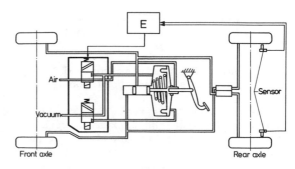

Fig. 10:    Simple amplifier control
           Valves in normal position

In order for such a system to function at all, the amplifier must be able to work in two directions. 2 air valves would be required with a vacuum amplifier, as shown in Fig. 10. The valves are switched so that air passes into the control neck via one of the valves and a vacuum is produced on the front side by the other valve. Under normal braking air is passed to the rear side of the diaphragm which causes the plunger rod to move forward. If the braking control unit is to cut in, both valves are switched so that the air and the vacuum on either side of the diaphragm are exchanged, which causes the amplifier to be cut out.

The next question to be answered is which wheel or wheels can or should be controlled. Under no circumstances may the wheel or wheels which lock first be controlled, as this would cause a catastrophic lengthening of the braking distance when travelling on an asymmetrical road surface or when cornering. Only the wheel which locks last may be controlled, and on the assumption that the front wheels are locked before the rear wheels, then this would be a rear wheel. This corresponds to select high control, which is less favourable as far as stability is concerned than select low control. Furthermore the regulating speed is too slow by a factor of between 5 and 10 for good control, but may still be adequate for a simplified control system.

In the cost-benefit analysis the costs were worked out to be 43 % and the benefits to be 16 %. This gives a cost-benefit factor of 0.37, which makes the system equally as poor as the 1-channel rear axle controller mentioned earlier.

MULTIPLE AMPLIFIER CONTROL

The last system to be mentioned is - in simplified terms - a combination of amplifier control and diagonal control and has not yet been discussed in the relevant literature.

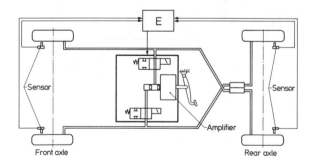

Fig. 11:  Multiple amplifier control
          Valves in normal position

The technical requirement is an amplifier which works forwards and backwards with a sufficiently fast regulating speed, e. g. an electrical amplifier, which is simply represented as a black box in this case.

The complete system is shown in Fig. 11: There is a normally open 2-position valve at the outlets of the brake master cylinder for both brake circuits, which can block off the pressure to each of the diagonal circuits. The other components are four sensors and an electronic control unit.

Several fundamental ideas are contained in the system:

1st fundamental idea - One of the two diagonal circuits is blocked by a valve as soon as adequate pressure has built up at the front wheel when braking, while the other diagonal circuit is controlled according to the select low method, i. e. with a correctly adjusted braking system only the front wheel is controlled by the amplifier modulating the pressure according to the control requirements. The blocked-off diagonal circuit is not affected by this control sequence and may initially lock. This principle works when travelling in a straight line and on a homogeneous, uniform road surface. In this case the lateral control of one of the front wheels and one of the rear wheels is adequate to maintain directional stability and reasonable steerability.

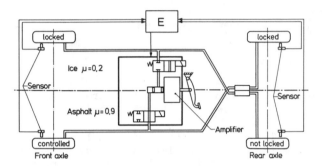

Fig. 12:  Multiple amplifier control
          Vehicle on asymmetrical road surface

2nd fundamental idea - It is not arbitrary which of the diagonal circuits is controlled, but rather that circuit whose front wheel locks second. This principle has advantages when travelling on an asymmetrical road surface, as shown in Fig. 12. On the right-hand side ice is illustrated and on the left-hand side asphalt. In this case the right-hand side front wheel would be the first to lock. As a result of this the valve would then cut off the pressure to this diagonal circuit and also prevent a further increase in pressure in the associated rear wheel, which is running on the asphalt side. The shutting of the valve causes this wheel to be underbraked and supply a considerable lateral force, in a similar manner to select low control. The next wheel to lock is the right-hand side rear wheel, where no controlling action is taken as this wheel contributes very little to lateral control. The last wheel which is in danger of locking is the left side front wheel, which is then controlled and which is responsible for the steerability of the vehicle. The control unit recognizes that the vehicle is travelling on an asymmetrical road surface when the right side rear wheel locks, and can adjust the control of both diagonal circuits accordingly.

This principle offers a satisfactory steering characteristic and adequate directional stability when braking on asymmetrical road surfaces or during cornering.

3rd fundamental idea - The fact that one wheel can lock when braking on dry asphalt, as originally mentioned, was seen in a negative light when studying the 2-channel axle controller. For this reason a control process was developed whereby both control channels can be simultaneously controlled by means of one control unit.

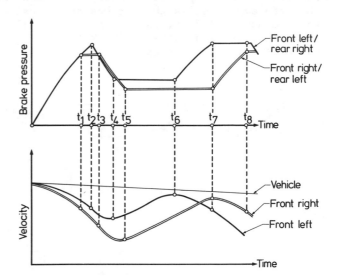

Fig. 14: Multiple amplifier control Control function

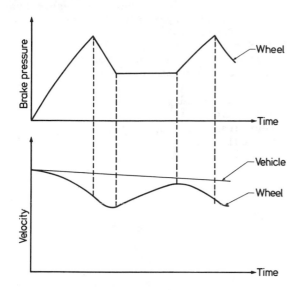

Fig. 13: General control function (schematic)

Fig. 13 shows schematically a normal control sequence. In the diagram the pressure and speed of a wheel are plotted against time; this representation was simplified for clarity. Such a control sequence consists of phases where the pressure is decreased and then increased again, which are called active phases, and also phases where the pressure is held constant which are called passive phases. The special feature of the control process is that when one of the control channels is in a passive phase the controller goes over to the other control channel by switching both of the valves, and if required then affects control with the help of the control unit.

Such a control system is shown in Fig. 14. It is assumed that the two sides of the road surface are slightly different: The left side has a coefficient of friction of 0.9 and the right-hand side a coefficient of friction of 0.8.

At time $t_1$ a deviation occurs at the right side front wheel, the pressure is only shut off and the wheel can continue slowing down. At time $t_2$ the left side front wheel locks; the control unit reduces the pressure in this diagonal circuit and at time $t_3$ it reduces the pressure in the second diagonal circuit, until at time $t_4$ the left side front wheel is accelerating again, the valve for this diagonal circuit shuts and only the pressure in the other diagonal circuit decreases, until at time $t_5$ the right side front wheel is also accelerating again.

At time $t_6$ the pressure in the front left/rear right diagonal circuit is increased again

until time $t_7$ when a new deviation occurs, which should trigger a drop in pressure. At exactly the same time, however, an opposite control command is given by the right side front wheel, i. e. a command to increase the pressure, as the wheel has now reached a stable braking condition again. Two basic rules apply to this case:

Rule 1: If the two brake circuits have opposite control requirements, i. e. to increase the pressure and to decrease the pressure, then when travelling on a symmetrical road surface in a straight line the command to increase the pressure has priority over the command to decrease the pressure. This is to prevent an increase in braking distance.

This control philosophy ensures that one of the front wheels is always being optimally controlled, while the other wheel is running under higher slip conditions and is not quite so well controlled, but nevertheless adequately enough for travelling in a straight line and on a uniform road surface.

Rule 2: If the two brake circuits have opposite control requirements then when travelling on an asymmetrical road surface or when cornering the command from the front wheel which has the higher coefficient of friction or which has the higher load acting on it has priority. The controller recognizes this condition because the rear wheel in this diagonal circuit locks.

This control philosophy ensures that the vehicle has a high degree of directional stability under various driving and braking conditions as the rear wheels are controlled in a manner similar to the select low method. It also ensures a high degree of steerability, better than with select high control, and normally good deceleration, which lies somewhere between the wheels being locked and being fully controlled.

The benefits were rated at 79 %. The costs are somewhat more difficult to estimate as a completely new component is being introduced in the form of the amplifier. Under the assumption that the amplifier is three times as expensive as a vacuum amplifier, the costs would total 74 %. This gives a cost-benefit ratio of 1.07, which is slightly greater than 1.

REFERENCES

(1) Heinz Leiber and Armin Czinczel:
    Four Years of Experience with 4-Wheel Anti-skid Brake Systems (ABS)
    SAE 830481, 1983
(2) H.-W. Bleckmann, J. Burgdorf, H.-E. von Grünberg, K. Timtner, and L. Weise:
    The First Compact 4-Wheel Anti-skid System with Integral Hydraulic Booster
    SAE 830483, 1983
(3) M. Satoh, S. Shiraishi:
    Performance of a Simplified Control Technique for Antilock Brake
    Kyoto, 1982
(4) Satohiko Yoneda, Yasno Naitoh, Hideo Kigoshi:
    Rear Brake Lock-up Control System of Mitsubishi Starion
    SAE 830482, 1983

| | Without ABS | 3-Channel Control Dia. Circuit 4 Sensors | 3-Channel Control 3 Circuit 4 Sensors | 2-Channel Ax. Control Ax. Circuit 3 Sensors | 2-Channel Dia. Control Dia. Circuit 4 Sensors | 1-Channel Rear Ax.Con. Dia. Circuit 2 Sensors | Simple Ampl. Contr. Dia. Circuit 2 Sensors | Multiple Ampl.Control Dia. Circuit 4 Sensors |
|---|---|---|---|---|---|---|---|---|
| Sideforce front | 0 | 35 | 35 | 24.5 | 35 | — | — | 30 |
| Correction factor | — | 0.85 | 0.85 | 0.65 | 0.90 | — | — | 0.85 |
| Steerability | 0 | 30 | 30 | 16 | 32 | -2 | — | 26 |
| Stability | 0 | 35 | 35 | 35 | 31 | 20 | 11 | 31 |
| Braking distance | 0 | 24 | 24 | 12 | 15 | 5 | 3 | 13 |
| Sum Benefit | 0 | 89 | 89 | 63 | 78 | 23 | 14 | 70 |
| Benefit (converted to a 100 %) | 0 | 100 | 100 | 71 | 88 | 26 | 16 | 79 |

Table 1: Benefits of all systems

| | Without ABS Dia. Circuit | 3-Channel Control Dia. Circuit 4 Sensors | 3-Channel Control 3 Circuit 4 Sensors | 2-Channel Ax. Control Ax. Circuit 3 Sensors | 2-Channel Dia. Control Dia. Circuit 4 Sensors | 1-Channel Rear Ax.Con. Dia. Circuit 2 Sensors | Simple Ampl. Contr. Dia. Circuit 2 Sensors | Multiple Ampl.Control Dia. Circuit 4 Sensors |
|---|---|---|---|---|---|---|---|---|
| Benefits | 0 | 100 | 100 | 71 | 88 | 26 | 16 | 79 |
| Costs | 0 | 100 | 95 | 77 | 88 | 64 | 43 | 74 |
| Benefit / Cost | un-defined | 1 | 1.05 | 0.92 | 1 | 0.41 | 0.37 | 1.07 |

Table 2: Cost-benefit ratio of all systems

SUMMARY

The benefits and the cost/benefit ratio of all simplified systems, which have no serious disadvantages, are summarized in Tables 1 and 2. Assuming that a 3-channel control, similar to the Audi 100 system, has a ratio of 1, it is noticeable that two systems have a low ratio of cost to benefit. These are the systems which only control the rear axle or the rear wheels.

The further systems have a ratio which is near to 1, but only two of them have a cost saving of about a quarter of the total cost. These are:

o   The 2-channel axle control which is only advantageous in vehicles with axially split brake circuits and
o   The multiple amplifier control which is also feasible in connection with diagonally split brake circuits. This system offers a great potential for simplifying of ABS without essential disadvantages.

# Anti-Lock Brake Systems for Passenger Cars, State of the Art 1985**

**Hans-Christof Klein**
Alfred Teves GmbH

**Paper 865139 presented at the XXI FISITA
Congress, Belgrade, Yugoslavia, June, 1986.

ABSTRACT

Over the course of the past decade, several four-wheel, automatic anti-lock device (ALD)
systems have been developed and gone into production. And this has evidenced the marketabi-
lity of this product, something that automotive
engineers had never deemed necessary. In this
report, a specially developed systematic approach is employed to compare ten ALD-supported
brake systems with respect to their particular
characteristics. Suitable measurement results
are not available for providing an objective
performance comparison. The report concludes
with an outlook on future developments.

THE INTRODUCTION OF AUTOMATIC ANTI-LOCK DEVICES
(ALD) into regular production represented a major
step in the direction of improving active driving
and road safety. Today, this is an indisputed
fact. The results of investigations by an insurance company in West Germany (HUK), as well as
other studies, both in Germany and abroad, have
shown that it has been possible to avoid roughly
7 % of all traffic accidents involving passenger
cars and to mitigate the consequences of another
10 to 15 % by means of anti-lock braking systems
/22/*. The positive response on the part of the
general public and the market itself, also
confirms the correctness of the decision, made
some 20 years ago, to commence development of
ALD-supported braking systems on a broad-scale,

*Numbers in / / designate references at end of
 paper

incorporating all of the opportunities offered
by the rapidly developing field of electronics.
The increasing number of patent applications and
publications on this subject suggest the effort
and intensity with which the work is being pur-
sued. Innumerable alternatives were designed and
reviewed, only to be rejected again, until the
solutions that were later to gain market acceptance were then culled and implemented. And this
trend is by no means over.

With the knowledge that we enjoy today, it
is possible for a more efficient, systematic
analysis and evaluation of ALD-supported braking
systems to be performed in a simpler manner. However in order to do this, it is necessary for
two major prerequisites to be satisfied, i.e.

o A uniform morphological classification
  principle

o Uniform evaluation criteria that are
  accepted by all parties involved.

A suitable system of characterization is
discussed in the following section of this State
of the Art Report. Unfortunately, it is not possible to provide a quantitative evaluation of
the various anti-lock braking systems here, as
this would far exceed the scope of this report.
With respect to the evaluation criteria that
must be taken into consideration, reference is
made to a prior publication by the author /1/,
as well as to the older principles that are discussed in /2/ through /6/. It should, however,
be noted here that the subjective perception of
the driver must always be taken into account, in
addition to objective evaluation criteria.

In the third section, the anti-lock braking
systems that are available on the market today,
as well as a selection of characteristic development alternatives, are compared with one another
in accordance with the previously discussed characterizing features. Neither very old developments (such as the Maxaret device, for example)

nor ideas that have thus far been presented only in the form of patent specifications are included. With respect to the state of the art in each case, reference is made to the publications by Ostwald (1964) /7/, Burckhardt (1975) /8/ and Cardon et.al. (1976) /9/.

## THE CHARACTERIZING FEATURES OF AN A L D BRAKING SYSTEM

There are two basic ways in which different system solutions can be compared with one another:

A <u>Comparison of the effects,</u>
of the performance, of the overall system which must be measured and analyzed under identical conditions and in accordance with uniform aspects.

B <u>Comparison of the system concepts,</u>
of the equipment design, the circuitry and the functional interaction, with respect to specific characteristics.

Since it is not possible to provide anywhere near an objective comparison of the effects merely on the basis of previously published data, the second approach (B) has been employed in this report. Moreover, as a result of the experience that has thus far been gathered, a knowledge of B alone, permits conclusions to be drawn with respect to the basic performance of a system. However the oversimplified PLUS-MINUS evaluation concept that can sometimes be encountered has not been employed in this report (cf. /1/).

The automatic anti-lock device is, at least on a functional plane, an integral constituent of the brake system. Consequently, a system for characterizing ALD-supported braking systems must also include the basic brake system. This is taken into consideration by the system of classification and characterization that is employed below. Three characteristic areas, each with three further subclassifications, produce a 3 x 3 characteristic matrix (TABLE 1).

| 1. | BRAKE SYSTEM LAYOUT | 2. | INFORMATION COLLECTING & PROCESSING | 3. | BRAKE FORCE MODULATION |
|---|---|---|---|---|---|
| 1.1 | BRAKE CIRCUIT AND CONTROL CHANNEL LAYOUT | 2.1 | TECHNOLOGY | 3.1 | MODULATION PRINCIPLE |
| 1.2 | BRAKE FORCE DISTRIBUTION AND BASIC LAYOUT | 2.2 | SENSORS | 3.2 | INTERFACE LOGIC CONTROL/MODULATOR |
| 1.3 | BRAKE SYSTEM ACTUATION | 2.3 | LOGIC | 3.3 | ENERGY SUPPLY |

Table 1  Characteristic areas employed for characterization of anti-lock brake systems for passenger cars

The three characteristic areas denote the following:

1. Brake system layout
   *The basic brake system*

2. Information collection and processing
   *The nerves and brain of the ALD*

3. Brake force modulation
   *The muscles of the ALD*

The briefest system description can be provided by means of the information contained in the diagonal elements (1.1-2.2-3.3) of the characteristic matrix. Supplementary characteristics are described by the two remaining elements of the second diagonal (1.3-3.1). And finally, taken together, all nine groups of characteristics permit unambiguous characterization of an anti-lock braking system.

Although it is not possible to treat all of the branches of the characteristic matrix here, the basic aspects will be discussed on the basis of typical examples.

Let us first discuss Characteristic Area 1. Within the scope of this discussion, it will first be necessary to treat the three possible system modes: NB, CB and EB (cf. Appendix for explanation), as well as the results of the related safety and failure effect analysis.

| NUMBER OF ANTI LOCK CONTROL CHANNELS | | BRAKE CIRCUIT LAYOUT | | | | |
|---|---|---|---|---|---|---|
| | WHEELS[1] | 12–34 | 13–24 | 1–2–34 | 124–123 | 12–1234 |
| | 1 | 2 | 3 | 4 | 5 | 6 | 7 |
| 1 | 4 | 1–2–3–4 | S 5, S 10[2] | S 5, S 10 | | | |
| 2 | 3 | 1–2–34 | S 2, S 9 | S 2, S 9 | S 9 | | |
| 3 | | 12–34 | S 7 | | | | |
| 4 | 2 | 13–24 | | S 1, S 6 | | | |
| 5 | | 14–23 | | | | | |
| 6 | | 34 | S 8 | | | | |
| 7 | 1 | 12 | | | | | |
| 8 | | 13 oder 24 | | | | | |
| 9 | | 1234 | | | | | |

▢ SERIES    ⌐⌐⌐ SOON TO GO INTO SERIES

▨ COMBINATION NOT PERMITTED BY FUNCTION AND/OR COSTS

1) INDEXING AS PER TABLE A1    2) S 1...S 10 SYSTEM DESIGNATION AS PER TABLE 5

Table 2  Brake circuit and control channel layout

Table 2, which belongs to Characteristic Group 1.1, illustrates the interaction that exists between the brake circuit layout and the number and layout of ALD control channels. The heavily outlined fields indicate solutions that are in series production today, while the open white fields denote room for future alternatives. In this connection, it should be noted that, due to its unacceptable performance, the single-channel control concept is not considered to have any chance of success in Europe. Moreover, it can also be seen from Table 2 that, with the introduction of ALD, the brake circuit layouts shown in the last two columns in the table which, although more sophisticated, offered increased advantages in the EB mode, were discarded.

The other two groups in Characteristic Area 1 break down as follows:

1.2  BRAKE FORCE DISTRIBUTION AND BASIC DESIGN

- Basic design $\Phi_G$ of the BFD, including non-adjustable pressure valves
  (Lock-up sequence, adhesion coefficient utilization $\varepsilon_\Phi$ in the NB mode, hysteresis while passing through an actuation cycle)
- BFD controlled on the basis of a theoretical, stipulated, fixed interaction of at least one variable parameter (e.g. ALB)
- BFD adaptively controlled, with optimized adhesion coefficient utilization being the target function

1.3  BRAKE SYSTEM ACTUATION

- Brake system actuation in the NB mode
  (Principle, nature and magnitude of boost effect)
- Brake pedal characteristic in the NB and EB modes
  (Force, travel, application, hysteresis, change between NB and EB modes)
- Brake pedal behaviour in the CB mode
  (Pulsation, non-recurring positioning, locked, uninfluenced)
- External energy in the NB and EB modes
  (Type of energy, intermediate energy storage)

Since it is not necessary to employ the same external energy (cf. Appendix for definitions of terms) in both the NB and CB modes, although this does determine the design of the brake force modulator, energy supply in the CB mode is treated separately under 3.3. Characteristic Group 3.3 breaks down as follows:

3.3  ENERGY SUPPLY

- Type of energy
- Energy source
  (Nature, power-source, -performance and -dissipation, failure possibilities)
- Energy storage
  (Accumulated energy, short-time output)
- Energy transmission medium

The significance of the question of energy storage is frequently all too gladly overlooked. If, for example, the brake system of a middle class vehicle is controlled between 60 and 80 bar at 10 Hz, this represents an average net power requirement of 125 W. If the pressure build-up phase amounts to 35 ms, the net performance demand during this period then increases to approx. 360 W. Should, while the brakes are being applied, this vehicle pass a friction co-

efficient threshold which necessitates that the braking pressure be increased from 20 to 80 bar within a period of 50 ms, this means that the momentary net power demand amounts to 750 W. TABLE 4 (cf. Page 4) provides a comprehensive overview of the entire complex of external energy.

The final example deals with Characteristic Group 2.2, Sensors, which breaks down as follows:

2.2  SENSORS

- Measured property
- Type of signal (analog, digital)
- Physical measurement principle
  (Nature, measurement range, resolution, signal to-noise ratio, ambient conditions)
- Number and location of sensors in the anti-lock braking system

In addition to the information pertaining to rotary wheel motion that is picked up by the wheel sensors, a number of further physical parameters also play a significant role in the optimum implementation of brake pressure control processes. Not in the least as a result of cost and reliability considerations, the aim is to employ the smallest possible number of sensors. TABLE 3 shows the factors that determine the minimum number of wheel sensors required. In addition to the parameters that have already been discussed in conjunction with Table 2, the location of the drive axle is also an influencing factor (Table 3, Column 3). Certain two-channel systems necessitate the employment of rear-axle load sensitive brake force reducing valves (Table 3, Line 5).

| BRAKE CIRCUIT LAYOUT | NUMBER OF CONTROL CHANNELS | DRIVE AXLE | SENSOR ON WHEEL | | | | SENSORS FOR | | | TYPICAL SYSTEM DESIGNS[1] |
|---|---|---|---|---|---|---|---|---|---|---|
| | | | 1 | 2 | 3 | 4 | $F_H$ | $\ddot{x}$ | | |
| 1 | 2 | 3 | 4 | 5 | 6 | 7 | 8 | 9 | 10 | 11 |
| 1   I | 3 | RA | + | + | + | + | | | | S 2, S 5, S 9 |
| | | | + | + | + | | | | | S 2 |
| 2   X | 3 | FA | + | + | + | + | | | | S 2, S 5, S 9 |
| 3   Y | | RA | + | + | + | + | | | | S 9 |
| | | | + | + | | + | | | | |
| 4   I | 2 (12–34) | RA | + | + | + | + | | | | S 7 |
| | | | + | + | + | | | | | |
| 5   X | 2 (13–24) | FA | + | + | + | + | + | | | S 8 |
| | | | + | + | | | + | | | S 1, S 6 |
| 6   I | 1 (34) | RA | | | + | | | + | | |
| 7   X | | FA | | | + | + | | | | |

[1] System designation as per Table 5    $F_H$ – Rear axle load    $\ddot{x}$ – Longitudinal vehicle deceleration

Table 3  Number of sensors and layout

The inductive wheel sensors that are being employed virtually exclusively today operate with on the order of 90 - 100 pulses per wheel revolution and, depending upon the speed of the vehicle, supply signals with pulse repetition frequencies of from (0)...30...3000 Hz or with signal clocking of from 33...0.33 ms. The signal clocking path depends only upon the wheel slip and amounts to approx. 20 mm for the unlocked wheel, representing a virtually constant path frequency of 50/m.

| EXTERNAL ENERGY At least partially effecting actuation of the friction brakes | | ACTUATION IN THE NB-MODE | | | ACTUATION IN CB-MODE | | ACTUATION IN EB-MODE Stored external energy | | |
|---|---|---|---|---|---|---|---|---|---|
| | | Muscle power, not assisted by external energy | Muscle power, assisted by external energy | Muscle power controls external energy | External energy | Energy storage | Alone | Tog. with Muscle power | Alone |
| | 1 | 2 | 3 | 4 | 5 | 6 | 7 | 8 | 9 |
| 1 NONE | Driver's muscle power | | | | | S2 | | | |
| 2 MECHAN. | Kinetic energy of moving vehicle | SB | SB | | S6 | | | | |
| 3 MECHAN. | Stored mechanical energy | | | | | S1, S5 | | | |
| 4 HYDRAUL. | OC hydraulic system | | | | | | | | |
| 5 HYDRAUL. | CC hydraulic system | | S2, S9, S10 | | | S5, S7, S9, S10 | | S9, S10 | S9, S10 |
| 6 PNEUMAT. | Eff. pressure <1 bar | | S1, S2, S5 S6, S7 | | S1 | S1 | | S1, S2, S5 S6, S7 | S1, S2, S5 S6, S7 |
| 7 PNEUMAT. | Eff. pressure >1 bar | | | | | | | | |
| 8 ELECTR. | | | | | S2, S5, S7 S9, S10 | | | | |

☐ SERIES ▪▪▪ SOON TO GO INTO SERIES ▒ FOOLISH OR INVALID COMBINATION    SB – SELF-ENERGIZING FRICTION BRAKE    S1...S10 – SYSTEM DESIGNATION AS PER TABLE 5

Table 4  External energy for brake system actuation

## TYPICAL EXAMPLES ON ANTI-LOCK BRAKING SYSTEMS

Within the scope of this report, ten anti-lock braking systems were analyzed under the aspects that were developed in Section 2. To the extent that dependable information was able to be obtained, the major results are summarized in TABLE 5. As can be seen from this table, four- and single-channel solutions were also tracked, in addition to the three- and two-channel ALD control concepts that predominated. In addition to vacuum and hydraulic energy sources, electricity is gaining significantly as a source of external energy in the CB mode.

Except in only one case, microelectronics are employed for information collection and processing. Almost without exception, the analog system processing that was initially utilized has - aside from individual tasks - been replaced by digital processor technologies. Four- and three-, as well as two-channel control concepts, should have four, however not less than three, wheel sensors as information sources. If only the two front wheels are equipped with wheel sensors, special provisions must be made in order to ensure compliance with the legally required lock-up sequence and adhesion coefficient utilization factor. Wheel rotation angle transducers with single- or double-pole inductive signal pick-ups have now become the exclusive design for wheel sensors.

The development of the basic control strategies can be deemed to have been completed. Even though, understandably enough, hardly anything is published by manufacturers regarding their specific details, it can be assumed that the same physical criteria are being employed as guidance and target parameters. Three-channel systems control the brake pressures of the front wheel brakes individually and those of the rear axle in accordance with the select-low principle.

In contrast, the two-channel System S7 (table 5) controls the front wheel brakes jointly in accordance with the select-high principle. More significant differences than in the case of the control philosophy can be clearly seen under closer observation of the tracked safety concepts. Bleckmann and Weise /17/ define five safety levels which, together with systematic failure effect analysis /19/, could serve as the basis for uniform evaluation of overall system safety.

Various principles can be utilized to modulate the brake force. Only brake pressure modulation is employed in modern automatic anti-lock device systems. It is effected either directly by means of high-speed solenoid valves (response time 3...8 ms) or with the aid of plungers that are actuated by means of external power. Control of the fluidic external power, in turn, is effected by means of high-speed solenoid valves.

| LINE NUMBER | ANTI-LOCK SYSTEM MANUFACTURER/ DESIGNATION | BRAKE SYSTEM LAYOUT | | | QUANTITY | | | PRESSURE MODULATION PRINCIPLE | EXTERNAL ENERGY FOR CB MODE | INTERMEDIATE ENERGY STORAGE IN CB MODE | NUMBER OF INDEPENDENT ASSEMBLIES | | REMARKS REFERENCES |
|---|---|---|---|---|---|---|---|---|---|---|---|---|---|
| | | BRAKE CIRCUIT LAYOUT | EXTERNAL ENERGY FOR NB MODE | NUMBER OF BFD VALVES | CONTROL CHANNELS (WHEELS) | PRESSURE MODULATORS | WHEEL AND OTHER SENSORS | | | | MECH. | SENSOR AND ELECTR. | |
| COLUMN NUMBER | | 4 | 5 | 6 | 7 | 8 | 9 | 10 | 11 | 12 | 13 | 14 | 15 |
| E1–E6 | LEGEND OF ABBREVIATIONS | I, X, Y | Vacuum; Hydraul.; Electr.; Pneum. | ..F Fixed setting; ..V variable change-over point; ..A adaptive BFD; ..AR Depending upon vehicle | 4...1; (...) wheel indices | 4...1 | ...1; +; ...1 | Plunger; Valve, switching; Vp Valve, proport. | Mechan.; Vacuum; Hydraul.; Electr.; Pneumat. | Mechan.; Vacuum; Hydraul.; Electr.; Pneumat.; Driver's muscles | Not including connectors and fasteners | | C (...) Corporate publication (Year of publication); /.../ cf. References; NI No topical information; II Insufficient information |
| S1 | AUTOMOTIVE PRODUCTS UK LOCKHEED – ANTILOCK | X | V | 2F/2V | 2 (13-24) | 2 | 2/(2+1) | P/V | V | M/V | 6 | 3/4 | C (1985) II |
| S2 | BOSCH ABS 2 | I, X | V, H | AR | 3 | 3 | 3, 4 | V | E | D | 2 | 4...5 | C (1985) /5/, /10/ |
| S3 | BENDIX | | | | | | | | | | | | NI |
| S4 | DELCO MORAINE EBCS | | H | | | | 4 | | E | H | 1 | 5 | C (1985) II |
| S5 | FAG KUGELFISCHER ASBS | I, X | V | – | 4 | 4 | 4 | P/Vp | E | M/H | 2 | 5 | C (1983) /11/ |
| S6 | GIRLING LUCAS SCS | X | V | 2V | 2 (13-24) | 2 | 2+1 | V | M | – | 4...5 | (1) | C (1985) /12/, /13/ |
| S7 | HONDA 4w ALB | I | V | | 2 (12-34) | 2 | 4 | P/V | E | H | 4 | 5 | C (1983) /14/ |
| S8 | MITSUBISHI ASBS | I | V | – | 1 (34) | 1 | 1+1 | P/V | V | M/V | 2 | 3 | C (1982) /15/ |
| S9 | TEVES ABS–MK II | Y, I, X | H | AR | 3 | 3 | 4 | V | E | H | 1 | 5 | C (1985) /16/, /17/ |
| S10 | WABCO | I, X | H | – | 4 | 4 | 4 | V | E | H | 4 | 5 | C (1985) II (/18/) |

Table 5  Anti-lock braking systems and their characteristic features

The specific characteristics of various automatic anti-lock device systems are discussed below.

**BOSCH ABS 2 (System S2, Table 5), Fig. 1**

The following features are especially characteristic of this well-known, four-wheel ALD system, which has been in production since 1978:

o Can be retrofitted to an existing brake system (add-on system)
   - Operates with vacuum and hydraulic boosters
   - ABS hydraulic unit installed in piping system, between booster and wheel brake

o Brake force reduction through direct drainage of brake fluid from the wheel cylinder circuit in question into a reservoir S (Fig. 1c) having a defined, limited volume

o Employment of the recharging concept: In each brake circuit, a pump driven by a common electric motor returns the brake fluid from the reservoir to the master cylinder against the effective actuation force (driver's muscles and booster)
   - This causes the brake pedal to pulsate in the CB mode
   - The maximum brake pedal actuation force determines the maximum pump delivery pressure
   - The resulting change in the pressure-travel characteristic of the master cylinder necessitates the employment of central valves instead of vent ports

o The employment of 3/3-way valves (Fig. 1d) means that only one valve is required per hydraulic control channel

o Employment of at least three single-pole, inductive wheel speed sensors

o Digitalized electronic controller with customer largescale integration (CLSI) circuits for logic and peripheral, separate safety-check circuit, all circuitry now contained on one board.

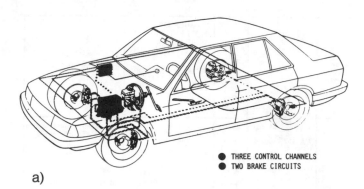

a)

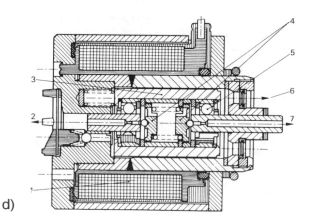

b)

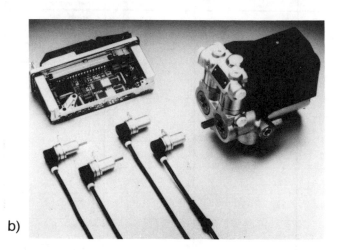

c)

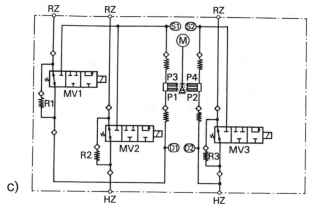

d)

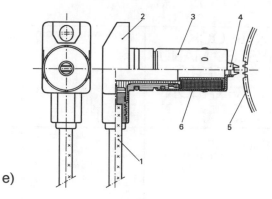

e)

Fig. 1  Bosch ABS 2 (System S2, Table 5)
a) Detached vehicle installation concept
b) Components, 1985 development status
c) Functional schematic of the I-3 hydraulic unit
d) 3/3-way hydraulic solenoid valve
e) Single-pole inductive wheel speed sensor

FAG KUGELFISCHER ASBS (System S5, Table 5), Fig. 2

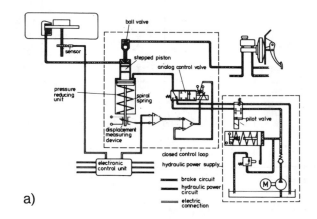

a)

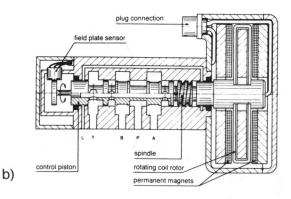

b)

Fig. 2  FAG Kugelfischer ASBS (System S5, Table 5)
a) Simplified functional schematic of a control channel
b) Principle of the directly-controlled servo valve with position return feature

This solution, too, employs the add-on principle. Four control channels are provided; pressure modulation is effected in accordance with the plunger principle, by means of valves in the hydraulic power circuit.

The following special characteristics deserve emphasis:

o The electric position feed back of the plunger (Fig. 2a)

o The employment of 3/3-way servo valves with rotary drive and electric gate return (Fig. 2b)

The same company also developed the motorcycle ALD system that was presented at the 1985 IAA in Frankfurt (Fig. 3). As the only non-passenger-car automatic anti-lock device, it has been included in this report in order to call attention to the direct electro-hydraulic modulation principle with the aid of a plunger actuated by a linear electric motor, which is employed in this system.

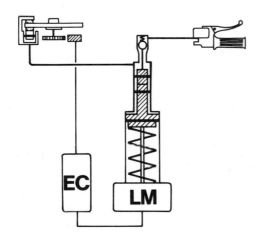

Fig. 3   Schematic representation of the FAG Kugelfischer motorcycle ALD system /20/

LUCAS GIRLING SCS (System S6, Table 5), Fig. 4

This system, which was developed for vehicles with front-wheel drive and X-pattern brake circuit layouts, is characterized by its lack of any electrical and electronic systems. However to date (December, 1985) there has not been any indication of system monitoring.

The mechanical interlinkage of the relative motions between flywheel 2 (Fig. 4b) and drive shaft, which actuates the fast-action brake pressure drain valve directly with the aid of a ball/ramp/slip clutch mechanism, can be classified as being analog signal processing. External mechanical energy is taken from the moving vehicle to build up the pressure again, the fact that the pressure can not be built up again while the wheel is stationary is not viewed as being a handicap.

Thus, a two-channel control concept that handles the front wheels individually has been implemented with two modulators associated to the two front drive wheels. The need for especially careful balancing of the brake force distribution between the directly controlled front wheel and the follow-along rear wheel has already been pointed out in this report.

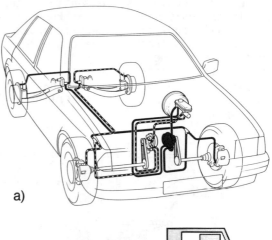

a)

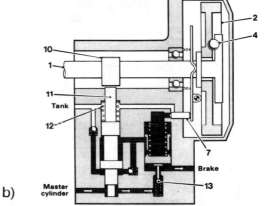

b)

Fig. 4   Lucas Girling SCS (System S6, Table 5), for vehicles with front wheel drive and X-pattern brake circuit layouts
a) Detached installation in the vehicle; pressure modulator driven from front wheel drive shaft by means of toothed belt
b) Simplified cross sectional representation of the mechanically driven brake pressure modulator

It should be added that an electronically controlled, four-channel ALD system with high-pressure hydraulic accumulator is discussed briefly in an L.G. publication /21/.

HONDA 4wALB (System S7, Table 5), Fig. 5

In this system, too, a two-channel, four-wheel automatic anti-lock device control concept has been implemented. The wheel cylinder pressure is modulated by means of four displacing plungers, which are controlled on the external hydraulic energy side by means of 2/2-way hydraulic valves (Fig. 5b). In both the NB and CB modes, the volumes of the wheel and master cylinders are coupled, which not only causes that changes in volumes produced by the ALD cycle to be noticeable on the pedal, but also means that pressure modulation must be effected against the pressure of the driver's foot (Fig. 5a).

The fact should also be mentioned that this system is the only one thus far known in which the front axle control strategy is based upon the select-high principle, while the rear axle is controlled in the customary manner, in accordance with the select-low principle.
(Note: Select-high does not mean that it would be impossible for one of the interlinked wheels to lock up).

The average velocity of the rear wheels is sensed by a signal pick-up at the transmission inlet (Fig. 6b). Attention should be called to the employment of a longitudinal deceleration sensor (Fig. 6c) for determination of a reference parameter.

At the IAA (Frankfurt, Sept. 1985) and the Tokyo Motor Show (Tokyo, Oct. 1985), there were indications of further independent Japanese ALD developments. Unfortunately, however, it was not possible to learn any engineering details.

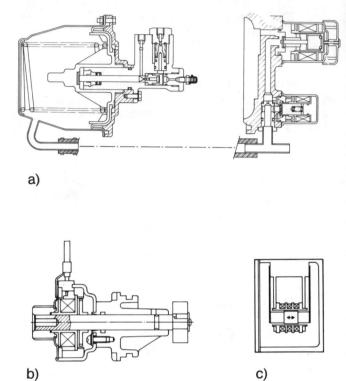

a)

b)                                    c)

Fig. 6  Mitsubishi ASBS (System S8, Table 5)
    a) Simplified cross sectional representation of the vaccum-actuated brake pressure modulator with vacuum solenoid valves
    b) Installation of the rear axle velocity sensor
    c) Longitudinal deceleration sensor

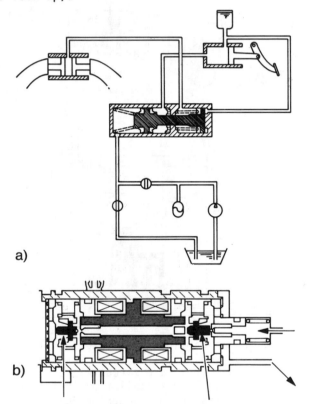

a)

b)

Fig. 5  Honda 4wALB (System S7, Table 5)
    a) Simplified functional schematic of a control channel
    b) Basic layout of two 2/2-way hydraulic solenoid valves in a cartridge-type housing

MITSUBISHI ASBS (System S8, Table 5), Fig. 6

This single-channel, rear-axle automatic anti-lock device continues the tradition of a development that is characterized by the names AP-Lockheed and Kelsey-Hayes, among others. It has a vacuum-actuated displacing plunger with cut-off valve, which is controlled on the vacuum side by means of a 3/2-way and a 2/2-way solenoid valve (Fig. 6a). Thus, the master cylinder and the wheel cylinders are separated in the CB mode.

TEVES ABS MK II (System S9, Table 5), Fig. 7

This four-wheel automatic anti-lock device system is the only ALD system that is being installed on a full-series basis, i.e. as standard equipment in entire vehicle model series, since its introduction in the United States and in Europe. It is characterized by the following engineering features:

o  Brake system actuation and ALD pressure control are combined in a compact assembly, the integrated hydraulic unit
    - Simplest possible installation in the vehicle, e.g. in place of a vacuum booster (Fig. 7a)

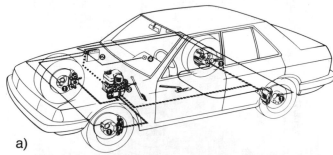

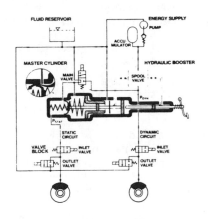

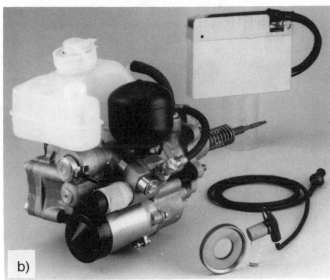

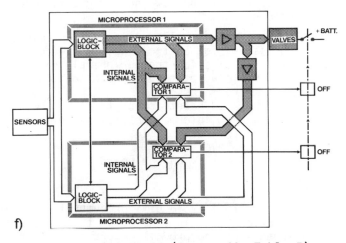

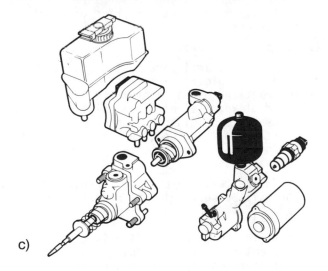

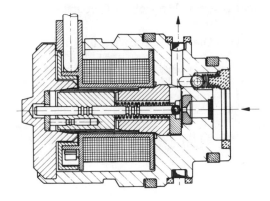

Fig. 7   Teves ABS MK II (System S9, Table 5)
a) Installation of the compact, integrated brake actuation and ALD in
the vehicle
b) Components, 1985 development status
c) Modular design of the hydraulic unit
d) 2/2-way hydraulic solenoid valve
e) Simplified functional schematic of
the hydraulic unit
f) Simplified block diagram of the
electronic controller

- External hydraulic lines only to the wheel
brakes
- Modular design (Fig. 7c) affords flexibility in adapting the system to the requirements of a wide variety of vehicle models
and concepts

o Hydraulic CC booster
- Shortest possible response time; sensitive,
hysteresis-free pressure control
- Energy storage in hydraulic accumulator

o Common energy supply for booster and ALD control
- Monofluidic high-pressure accumulator system for brake fluid
- Single-circuit hydraulic pump, driven by
an electric motor

319

Independentl of the speed of the internal combustion engine
$P_{max}$ determined by upper accumulator charging pressure
Minimum total energy demand over service life of vehicle

o Direct brake pressure modulation in the CB mode through simple, reliable 2/2-way hydraulic solenoid valves (Fig. 7d) in the brake circuit
  - Pressure build-up by means of dynamic flow-in of hydraulic energy from the booster into the brake circuits (Fig. 7e)
  - A failsafe positioning mechanism ensures the effect of the emergency brake system in the event of an external energy failure and retains the brake pedal in a defined position, i.e. no brake pedal pulsation

o Employment of four two-pole inductive wheel speed sensors

o Electronic controller, incorporating microprocessors, for control and safety logic, as well as CLSI's for the peripheral
  - Flexible adaptation through the employment of microprocessors
  - Integrated safety concept:
    Redundant information processing through redundant design (Fig. 7f)
    Safety level E /17/
  - All circuitry on one board

## OUTLOOK

In the life cycle of a product, the market introduction stage is considered to have been concluded when the product enters the growth phase. Various ALD-supported brake systems have now been evidenced to have reached this status. More and more comprehensive penetration of the market can be expected. By year-end 1985, over one million passenger cars had been equipped with modern, four-wheel automatic anti-lock device systems. The present annual production output - and the trend is rising - amounts to far in excess of 500,000 units. Nevertheless, the engineering development path continues on unabatedly.

The state of the art that has been attained with four-wheel, automatic anti-lock device systems is very high, especially if the cost factor is taken into consideration, in addition to function, safety and reliability. A cost/benefit ratio of better than UNITY can be deemed to have been achieved! The "dream" trend of further development is the one that has been denoted OT (overall target) in the cost/benefit diagram (Fig. 8):

o A further increase in the benefit, or technical content, accompanied by significant cost reduction far in excess of any possible rationalization effects!

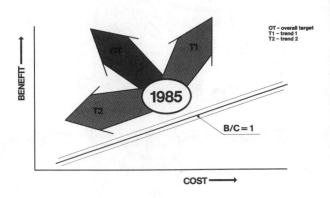

Fig. 8 Anti-lock development trends
Legend: OT - overall target
        T1 - trend 1
        T2 - trend 2

This ambitious objective will only be able to be achieved if further advances - or decisive breakthroughs - are also made in other areas of technology that have a bearing on automatic anti-lock device systems.

At present, two realistic trends of development can be observed:

TREND 1 (T1)

o Increased engineering performance (technical content), however at the price of marginally higher cost.

Examples of this are:

  - Automatic anti-lock devices for four-wheel-drive vehicles
  - Traction control systems (TCS)
  - Adaptive, i.e. electronically controlled brake force distribution
  - Keep stopping function
  - Integral self-diagnostics and maintenance instructor

All of these tasks have either already been solved or will be solved in the near future.

TREND 2 (T2)

o Here, the primary objective is to reduce costs substantially, while maintaining performance at the present level or accepting minor degradation (although, of course, still above Line B/C = 1).

Over the medium term, progress will be made primarily in the structural integration of components and assemblies, following the example set by the TEVES MK II system. Furthermore, large-scale integration (LSI) circuits will grow together on very large-scale integration (VLSI) chips. The higher application temperatures that are expected to result from the employment of

GaAs semiconductor technology would permit the control electronics to be grouped together with the hydraulic unit for installation in the engine compartment.

It remains to be seen whether the overall cost/benefit ratio will be able to be further improved during this time frame through multiple use of existing assemblies in combination with adjacent vehicle systems.

Consistent use of the auxiliary energy source required for the ALD function for normal brake application, as well, and its employment for the TCS function will further increase the utilization factor of the high-pressure energy system, e.g. of the TEVES ABS MK II, however it will still fall far short of the design limits to its capacity. The expandability of the basic system in terms of electronic hardware and software, as well as modular hydraulics, would suggest the integration of further functions, e.g. controlled engine exhaust braking, speed control, control of the torque distribution on four-wheel drive vehicles, etc.

And, finally, over the medium term an answer can also be expected to the question of whether simplified automatic anti-lock device systems with limited capabilities might be able to meet all present-day legal requirements in vehicles that are tailored to the needs of a specific clientele and if such systems will be accepted by the market.

It is always very risky to predict long-term trends. In this case, a forecast of this type also depends to a considerable extent upon the pace of progress in other relevant areas of technology (TABLE 6).

nents. It would therefore not be unreasonable to assume that future developments will also produce an electronically controlled service brake system (ECB) - sometimes termed "brake by wire". A system of this type would also permit numerous supplementary functions to be integrated with the greatest of ease (Fig. 9).

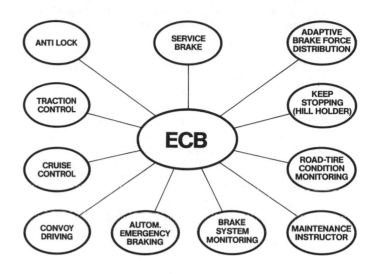

Fig. 9  Electronic controlled brake system (ECB) Integrated subsystems

Of course, this ECB would not serve as an isolated system in the vehicle, but would communicate with other on-board electronic systems, such as the central chassis system, the engine & powertrain management, the central body system and the central information system (Fig. 10).

---

**SENSORICS (ET + NT)**

- **VEHICLE DYNAMICS**
  - CHASSIS, SUSPENSION
  - BRAKE-SYSTEM
  - (LEGISLATION)
- **BRAKE-SYSTEM ACTUATION (NT)**
- **FLUIDICS (NT)**
- **SAFETY POWER-BUS (NT)**

- **SEMICONDUCTOR COMPONENTS**
  - SWITCHING TIME (of binary element)
  - STORAGE DENSITY
- **SIGNALPROCESSING**
  - ALGORITHM
  - CIRCUIT ARCHITECTURE
- **OPTOELECTRONICS (NT)**
- **SAFETY SIGNAL BUS (NT)**

**FLUITRONICS (NT)**

| **MECHANICS-HYDRAULICS** | **ELECTRONICS** |
|---|---|

Table 6  Relevant technology trends
Legend: ET - existing technology for anti-lock brake systems
NT - new technology for anti-lock brake systems

One thing, however, is certain: The symbiosis of microelectronics and hydraulics has brought forth a key technology for passenger car anti-lock braking systems. Each and every vehicle on the road that is equipped with one of these systems demonstrates the reliability of its electro-hydraulic and electronic compo-

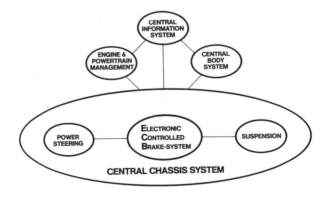

Fig. 10  Advanced automobile system control tech

At this point, the author's particular appreciation is expressed to the following companies for their support of this work through the provision of information and illustrative materials: Bosch, FAG Kugelfischer, Girling, Honda Germany, Mitsubishi Germany, Teves and Wabco.

# REFERENCES

/ 1/ Klein, H.-Ch.: Bewertung von Brems- und Regelkreisaufteilungen sowie der Bremskraftverteilung im Hinblick auf den Einsatz von ABV-Systemen.
VDI Conference on "Automotive Engineering, Sub-Areas and Methods", Bad Homburg, 1985, VDI Report No. 553.

/ 2/ Schlegel, H.: Wirtschaftlichkeitsanalyse von Sicherheitskomponenten im Personenwagenbau.
ATZ 74 (1972), pp. 465/469.

/ 3/ Fiala, E.: Regelungskonzepte für Fahrzeuge und Verkehrstechnik.
XXth FISITA Congress, Vienna, 1984, Paper No. 845027.

/ 4/ Seiffert, U.: Sicherheitstechnische Anforderungen an das Automobil der Zukunft.
Gfs Seminar, Ingolstadt, May, 1984.

/ 5/ Leiber, H., Czinczel, A.: Four Years of Experience with 4-wheel Antiskid Brake Systems (ABS).
SAE Paper 830481

/ 6/ Hattwig, P.: Cost-Benefit Analysis of Simplified ABS.
SAE Paper 850053

/ 7/ Ostwald, F.: Entwicklung der Blockierverhüter für Kraftfahrzeuge.
Automobile-Revue, No. 40, 1964

/ 8/ Burckhardt, M. et.al.: Möglichkeiten und Grenzen von Antiblockiersystemen.
ATZ 77 (1975), pp. 13/18.

/ 9/ Cardon, M.E. et.al.: Development and Evaluation of Anti-Lock Brake Systems.
SAE Paper 760348.

/10/ Leiber, H. et.al.: Antiblockiersystem (ABS) für Personenkraftwagen.
Bosch Technical Reports, No. 7 (1980), pp. 65/94.

/11/ Schürr, H., Dittner, A.: A New Anti-Skid Brake System for Disc and Drum brakes.
SAE Paper 840468.

/12/ Newton, W.R., Riddy, F.T.: Evaluation Criteria for Low Cost Anti-Lock Brake Systems for FWD Passenger Cars.
SAE Paper 840464.

/13/ v. Fersen, O.: ABV für Mittelklassefahrzeuge. ATZ 87 (1985), pp. 468/472.

/14/ Satoh, M., Shiraishi, S.: Performance of Antilock Brakes with Simplified Control Technique. SAE Paper 830484.

/15/ Yoneda, S. et.al.: Rear Brake Lock-up Control System of Mitsubishi Starion.
SAE Paper 830482.

/16/ Bleckmann, H.-W. et.al.: The First Compact 4-Wheel Anti-Skid System with Integrated Hydraulic Booster.
SAE Paper 830483.

/17/ Bleckmann, H.-W., Weise, L.: The New Four-Wheel Anti-Lock Generation - A Compact Anti-Lock and Booster Aggregat and an Advanced Electronic Safety Concept.
Conference on "Antilock Braking Systems for Road Vehicles", London, Sept., 1985.

/18/ Leffler, H. et.al.: Ein neues Anti-Blockier-System (ABS) für leichte Nutzfahrzeuge und PKW mit integriertem hydraulischen Bremskraftverstärker
XXIst FISITA Congress, 1986, Belgrade. Paper No. 865140

/19/ Wobben, D.: Sicherheitsaspekte in der Kfz.-Hydraulik. Conference on "Hydraulics in Passenger and Commercial Vehicles", Technology House, Essen, Nov., 1985.

/20/ -: Neues Antiblockiersystem, Elektromechanik. Motorrad, 1985, No. 20, p. 20.

/21/ Rath, H.: Entwicklungstendenzen im Internationalen Automobilbau - Auswirkungen auf die Bremsanlage der 80er Jahre.
Conference on "Brake and Clutch Engineering for Motor Vehicles", Technology House, Essen, Nov., 1983.

/22/ Danner, M.: Mehr Eigen- und Mitverantwortung sind notwendig.
Versicherungswirtschaft, 3/1986.

/23/ Leiber, H., Czinczel, A.: Erste Prinzipversuche mit einer multigeplexten elektronischen Bremsreglerschaltung.
Entwicklungslinien in der Kraftfahrzeugtechnik, 1976, pp. 593 - 602, Verlag TÜV Rheinland.

Definitions and abbreviations

ABS                 - Anti-lock braking system
ALB                 - Automatic load-dependent brake
                      force regulator
ALD                 - Automatic anti-lock device
Basic BFD $\Phi_G$  - Brake force share of the rear
                      axle when all vehicle brakes are
                      applied with equal hydraulic
                      pressure (determined as a result
                      of structural brake design and
                      the μ-values of the friction
                      materials employed)
BF                  - Energy transmission medium,
                      brake fluid
BFD                 - Brake force distribution,
                      --distributor
Brake circuit       - All conjointly operating compo-
                      nents of a BS between BS actua-
                      tion and wheel brake whose func-
                      tion is jointly dependent upon
                      the lack of a primary fault.
BS                  - Brake system
Control             - That portion of the BS that is
channel               controlled from an output of the
                      control logic in the CB mode
EBD                 - Electronically controlled adap-
                      tive brake force distributor
Energy supply       - Subsystem of a BS, supported/
circuit               actuated by external power, from
                      the energy source drive and the
                      fluid reservoir, if applicable,
                      all the way to the BS actuation
                      interface (booster, brake valve,
                      brake pressure modulator)
External            - Energy that is taken from Vehi-
energy/power          cle Subsystem I (moving vehicle,
                      running engine) and is made
                      available at System Interface II
                      in the form of auxiliary energy
                      for servo systems
Numerical           - Systematic numerical indexing
indices               system for vehicle components
                      and controls that occur fre-
                      quently in calculations and
                      diagrams as per TABLE A1.
Mode                - The operating mode of a brake
                      system
OB Mode             - Unactuated BS
AB Mode             - Automatic actuation of the en-
                      tire BS or a portion thereof
                      which is not initiated by the
                      driver of the vehicle (e.g. in
                      conjunction with TCS)
CB Mode             - Controlled braking operation,
                      i.e. braking with the service
                      brake system and active ALD
EB Mode             - The driver of the vehicle brakes
                      with the emergency brake system
                      (in the event of a primary fault
                      in the BS)
NB Mode             - Normal braking operation, i.e.
                      braking with the service brake
                      system; ALD inactive

MUX                 - Multiplex operating
L-MUX               - Multiplex operating of logic
                      channels
M-MUX               - Multiplex operating of modulated
                      channels
S-MUX               - Multiplex operating of signal
                      channels
Primary fault       - Any individual fault that can oc-
                      cur in the BS which causes mal-
                      function of a brake circuit, ex-
                      ternal power support, an external
                      energy circuit, the entire ALD or
                      any portion thereof. A minimum
                      braking effect must be ensured in
                      the event of a primary fault. A
                      primary fault must be able to be
                      identified and indicated to the
                      driver (monitoring and warning
                      facilities)
Select-             - Interlinking a plurality of input
average               signals in the control logic in
                      such a manner that the average of
                      these signals is dominant in de-
                      termining the output signal
Select-high         - Interlinking a plurality of input
                      signals in the control logic in
                      such a manner that the highest
                      value of these signals is domi-
                      nant in determining the output
                      signal
Select-low          - Interlinking a plurality of input
                      signals in the control logic in
                      such a manner that the lowest
                      value of these signals is domi-
                      nant in determining the output
                      signal
TCS                 - Traction control system
                      (Start-up assist not only)

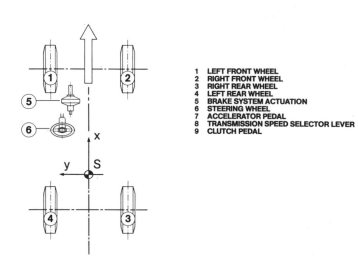

1  LEFT FRONT WHEEL
2  RIGHT FRONT WHEEL
3  RIGHT REAR WHEEL
4  LEFT REAR WHEEL
5  BRAKE SYSTEM ACTUATION
6  STEERING WHEEL
7  ACCELERATOR PEDAL
8  TRANSMISSION SPEED SELECTOR LEVER
9  CLUTCH PEDAL

Table A1  Numerical indices for chassis compo-
          nents and driver controls that are of
          relevance for the brake system

# A New Anti-Lock Braking System for Passenger Cars and Light Commercial Vehicles, with Integrated Hydraulic Brake Booster**

**Heinz Leffler and Erwin Petersen**
WABCO Westinghouse Fahrzeugbremsen GmbH
**Brian Shilton**
Clayton Dewandre Holdings Ltd.

**Paper 865140 presented at the XXI FISITA
Congress, Belgrade, Yugoslavia, June, 1986.

## ABSTRACT

THE PAPER DESCRIBES A NEW CONCEPT OF HYDRAULIC BRAKE BOOSTER which incorporates an internal "ratio-change" feature. This allows highly effective vehicle decelerations to be achieved under a variety of brake circuit failure conditions, whilst optimizing brake pedal force and travel characteristics under both normal service braking and secondary braking conditions.

Different dual-circuit system configurations can be easily accomodated, and the number of anti-lock control valves and control channels remains unchanged regardless of the complexity of the brake system split.

The ABS control philosophy with MIR (modified individual regulation) of the front wheels, and individual regulation of the rear wheels is recommended for all drive-train configurations, it not being necessary to switch off the ABS on all-wheel-drive vehicles, even when the differentials are locked.

The concept of the booster/ABS system enables a drive slip control system to be incorporated.

## GENERAL

The primary task of an anti-lock braking system is to maximise the lateral guiding forces of the braked vehicle, and thus to ensure even under full braking, stability and steerability of the vehicle within the available physical limits.

In addition, a major function of ABS is to optimize the utilization of adhesion and thus vehicle deceleration and to minimize the braking distance.

This is only possible when ABS acts on every wheel of the vehicle.

These requirements apply on both straight roads and bends, with tyre to road adhesion conditions which may be high or low, and which may vary both laterally and in direction of motion.

Scientific tests have proved that the use of ABS can lead to a considerable reduction in the number of accidents and to their severity (1)*.

In addition to a possible reduction in the number of accidental deaths and the avoidance of serious injuries, the use of ABS can also result in considerable reductions in material damage.

The first introduction of a modern electronic ABS for passenger cars was by DAIMLER-BENZ (2) and BOSCH (9) in 1978. This 3-channel ABS is added to a virtually unchanged braking system, and has an electrically driven pump as its own source of power.

The 4-channel ABS first introduced in 1981 for air-braked commercial vehicles by DAIMLER-BENZ (3) and WABCO is also an addition to the basic braking system but uses the existing air pressure system as its power source.

In 1985, FORD and ATE introduced the first integrated 3-channel ABS for passenger cars (6). The components are combined with those of the hydraulic booster brake system, the mono-fluid brake booster and ABS using the same power source.

---

*Number in parentheses designate references at end of paper.

The following text deals with a 4-channel ABS developed by WABCO Westinghouse, which is especially suitable for heavy passenger cars and light commercial vehicles with servo assisted brake systems, because of its special characteristics. A hydraulic monofluid brake booster operating on the closed-centre (CC) principle provides the basis of this integrated system.

## THE WABCO HYDRAULIC BOOSTER

GENERAL - The hydraulic booster used in hydraulic brake systems of passenger cars, lightweight buses and commercial vehicles functions on the CC-principle, supplied by a high pressure source of fluid from a pump and an accumulator. The main components of the booster are the master cylinder with a power valve and at least one servo-cylinder. The integration of rapid response solenoid valves into the booster provides the WABCO-hydraulic-ABS.

ENERGY SUPPLY - The hydraulic pressure is provided by a pump which is driven by an electric motor powered by the vehicle battery. A pressure switch controls the electric motor to maintain the operating pressure within the required range. A pressure limiting valve protects the system. The dimensions of the pump are such that, when the vehicle starts, any prior loss in accumulator pressure is quickly compensated in addition to the fluid consumption which occurs during ABS-braking.

The energy supply could alternatively be provided by a mechanically driven hydraulic pump.

DESIGN OF THE BOOSTER - The actual booster consists of a master cylinder, power valve, one, two or several servo-cylinders and a number of solenoid valves for ABS control. These components can be integrated in different ways in one or more housings.

Having all components completely integrated in the housing of the compact booster is an advantage, because of the small number of hydraulic connections and external pipes. Separate units (Fig. 1) are more suitable for vehicles with restricted installation space in the area of the brake pedal unit. Both units require less space than a comparable vacuum servo. With the exception of the booster chambers in the servo-cylinders which have to be bled independently, the entire system is designed to be self-bleeding.

BOOSTER OPERATION - Application of the brake pedal and the resulting movement of the push rod operates the master cylinder, closing its recuperation valve. Simultaneously a captive spring moves the booster spool valve through its free travel. Hydraulic pressure subsequently developed by the master cylinder acts upon the booster power valve, resulting in an accompanying built up of pressure in the booster circuit.

By choosing the appropriate valve and reaction piston diameters, the required pressure ratio characteristics of the unit can be achieved.

SERVO CYLINDER OPERATION - The master cylinder pressure and the booster pressure is built up in hydraulically separated chambers of the servo-cylinder. The check valve which connects the master cylinder and the brake wheel cylinder is open in the off

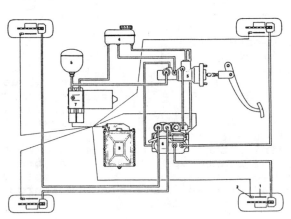

Fig.1: WABCO-ABS booster, modular system

position and closed when the booster pressure rises.

The quantity of fluid which is displaced by the master cylinder is determined by the annular area in the servo-cylinder and the stroke of the piston. By choosing appropriate piston diameters for the servo-cylinder, equal or different brake pressures can be obtained in the hydrostatically boosted circuit and the hydrodynamic circuit.

If, in the case of a high pedal force and full accumulator pressure being supplied by the power valve, the pressure in the annular chamber is higher than the pressure in the brake wheel-cylinder, the fluid from the master cylinder flows through the opening check valve and increases the brake pressure in the brake wheel cylinders which are supplied by the servo-cylinders. This represents an important element of safety in the event of the power pressure being low.

BEHAVIOUR IN THE EVENT OF BRAKE SYSTEM FAULTS - Possible faults which may arise in the hydraulic booster system and their effect on the braking behaviour are explained in the following examples of a total power pressure failure and a master cylinder failure.

It is justified to describe only these two cases, as the most adverse fault situation occurs when the power pressure fails (Fig. 3).

LOSS OF MASTER-CYLINDER PRESSURE - If there is a loss of master-cylinder pressure, the hydrodynamic pressure is modulated, after greater lost motion than in normal operation, via a spring, causing a short active pedal travel to be simulated. The driver is able to achieve a level of vehicle deceleration at almost the same pedal force as in normal operation, but at a longer pedal travel (Fig. 3a and 3b).

If the circuit in connection with the servo-cylinder is supplied as normal, the braking takes place without the fraction of brake force provided by the master cylinder.

LOSS OF BOOSTER PRESSURE - If there is a loss of booster pressure the check valve of the servo-cylinder does not close and the master cylinder pressure is transmitted directly to the brakes. In contrast to an application with the system functioning correctly, the internal ratio comes into action. This makes an effective deceleration of the vehicle possible with comparatively low pedal force but with a greater pedal travel.

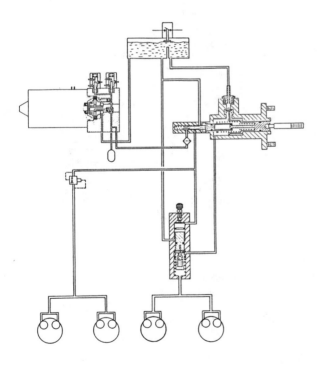

Fig.2:  Booster brake system

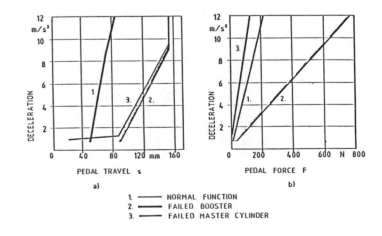

Fig.3:  Pedal characteristics

WARNING DEVICES - Both the dynamic circuit supplied from the hydraulic accumulator, and the static circuit are protected by warning devices, which activate a warning lamp when there is a loss in supply pressure or the fluid level in the reservoir falls below a certain level. Since the ABS-regulation takes place on the dynamic side of the booster, the ABS system is switched off when the accumulator pressure falls below a certain level, this also being indicated by the ABS-warning lamp.

MODULES FOR VARIOUS BRAKE CIRCUITS - The emergency braking effect which is required in the event of brake system faults can be achieved with different circuit arrangements. One factor which determines the system to be used is the laden to unladen vehicle weight ratio (7).

The II brake circuit configuration which has already been described, Figs. 2 and 4a, can be modified to an IH brake circuit configuration, Fig.4b, by applying the rear brakes and one half of the front brakes from the dynamic circuit, the remaining half front braking being supplied from the booster circuit.

The extension to a three-circuit system is achieved by using an additional servo-cylinder. The servo-cylinders are distributed either separately on the front wheels, Fig. 5 and 6a, on one axle of the vehicle or diagonally. An extension of this system to the H- or the HH- brake circuit configurations is possible, Fig. 4c, 4d, thus maintaining the brake force distribution when there is a failure in a brake circuit.

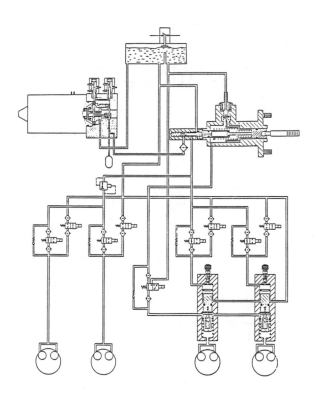

Fig.5: Booster brake system with ABS-modulation

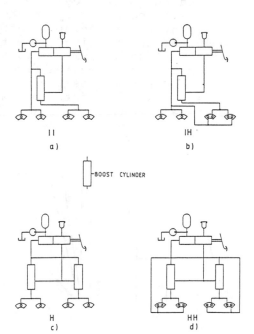

Fig.4: Booster applied to alternative brake system splits

ABS-BOOSTER ARRANGEMENTS - The basis of the WABCO-ABS is a unit with two servo-cylinders which can be used in II-, IH- and diagonal brake circuit configurations.

The regulating valves required for a 4-channel system with individual wheel control are incorporated in the dynamic brake circuit, Fig. 5. In order to avoid the pressure developed in the master-cylinder from over-riding the ABS control, an additional isolating valve is required - its function being described below. The regulating valves are situated before the servo-cylinders on the hydrodynamic side of the booster. Pressure control is transmitted through the floating pistons into the hydrostatic circuits and therefore to the brake wheel cylinders. The advantage of this arrangement is that the unavoidable loss of fluid which occurs during ABS control (as a result of repeated pressure drops) takes place in the booster section, where it can be rapidly replenished by the pump or accumulator. It would not be feasible to control ABS by the master cylinder, since its limited fluid volume would be quickly exhausted.

If a horizontal brake split is provided in the vehicle, two additional servo-cylinders are required for the individual wheel control function.

Leiber described in (8) the ABS-components required in the BOSCH-ABS for different circuit arrangements. It should be pointed out that for the system dealt with in this paper, the number of solenoid valves remains the same regardless of the circuit arrangement. The hardware of the electronic control unit thus needs no additional modification, Fig. 6.

In principle it is possible to modify the 4-channel system dealt with in this text to a 3-channel system, with select-low-regulation on the rear axle corresponding to the systems described in (9) and (6). In this case a pair of solenoid valves and one servo-cylinder could be omitted from the systems shown in Figures 6c and 6d.

ISOLATING VALVE-CONTROL AND PEDAL FEEL - The function of the isolating valve is to unload the annular areas of the servo-cylinders during ABS operation, in order to enable the pressure in the brake wheel cylinders to be reduced to zero despite high pedal forces.

This is especially necessary when the ABS is controlling the braking on low coefficients of friction. The isolating valve is switched over by the electronic control unit as soon as there is ABS-regulation on at least one wheel. By controlling the isolating valve the volume of the master cylinder is locked in. Due to this the brake pedal cannot be moved any further, it resists movement.

This occurs at a pedal travel which depends upon the coefficient of friction between the tyre and the road surface, independently of the elasticity of the brake, so that the tyre to road friction can be reduced from the fixed pedal position, Fig. 7.

Separating the master-cylinder (during ABS control) by means of the isolating valve prevents the pressure variations which are unavoidable in any ABS-control sequence (pressure build-up and reduction) from being transmitted to the master cylinder, so that only a slight movement is noticeable at the foot pedal, caused by the opening and closing of the power valve.

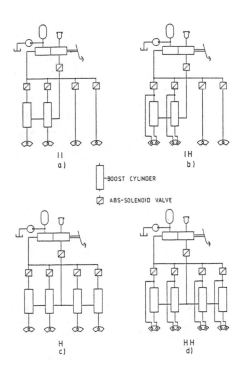

Fig.6:  ABS-booster applied to alternative brake system splits

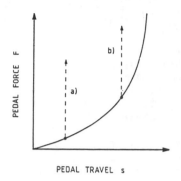

a) MODULATION ON LOW COEFFICIENT SURFACE

b) MODULATION ON HIGH COEFFICIENT SURFACE

Fig.7: Pedal characteristics during ABS-operation

## ABS-COMPONENTS

SENSORS - As with the other anti-lock braking systems (10) the rotational behaviour of all wheels is monitored by sensors. The installation of the inductive rod sensors and the pole wheels depends on the particular vehicle design.

REGULATING VALVES - The four pairs of solenoid valves consist of separately controlled pressure maintaining inlet valves and pressure reducing outlet valves. In this way the three individual ABS operations -
. increasing pressure
. maintaining pressure
. reducing pressure
are achieved.
The electrical reaction time (energizing and de-energizing) of the solenoid valves is less than four milliseconds. The various pressure stages required for comfortable and flexible regulation can thus be achieved by varying the energized and de-energized periods. Controlling an outlet valve without controlling an inlet valve is not possible.

Each inlet valve incorporates a non-return valve which permits pressure release in the brake wheel cylinder if the brake effort is reduced during ABS regulation, or should the inlet valve jam.

In order to protect the solenoid valves from dirt, a filter mesh is placed in each inlet and outlet.

## ABS-CONTROL AND SAFETY CONCEPT -
The WABCO hydraulic ABS has been designed with the following aims in view:
- to ensure stability and steerability
- to ensure shortest stopping distances
- to ensure fail-safe operation of the system.

This has been achieved by the following characteristics:
- high quality, adaptive control is achieved on all road surface coefficients by a special combination of acceleration, deceleration and slip theresholds;
- finely graduated adjusting of brake pressure and the avoidance of high slip values;
- stable control is ensured even on wet ice with coefficients of friction down to less than 0.1;
- control is effective almost to a standstill;
- special control techniques used in calculating wheel speeds and computing the vehicle's reference speed make it possible for the ABS to function even in all-wheel drive vehicles with side-to-side and/or inter-axle differential locks in operation.

MODIFIED INDIVIDUAL CONTROL (MIR) - Most of the existing anti-lock braking systems for passenger cars have a 3- or 4- channel system with individual control (IR) of the front wheels and select low control (SLR) of the rear wheels. The select-low-control was considered to be necessary to ensure stability on road surfaces with laterally different friction coefficients (split-$\mu$) (6, 8, 11). Individual control ensures short braking distances under the same circumstances.

The concept which has suceeded for commercial vehicles has been designed by WABCO and DAIMLER-BENZ. It consists of modified individual control (MIR) of the front wheels and individual control (IR) of the rear wheels (10,4, 12, 13). The same concept is used for the hydraulic ABS described in this paper for light commercial vehicles and passenger cars, Fig. 8. It ensures not only the necessary reduction in yawing moment when braking on split-coefficient road surfaces but also an extraordinary stability and steerability behaviour during emergency braking on surfaces with a varying coefficient of friction ($\mu$-jump).

When ABS signals occur on the low-$\mu$-wheel, MIR effects a pressure modulation on the high-$\mu$ wheel which allows a difference in braking effort between the front wheels to built up gradually, Fig. 8.

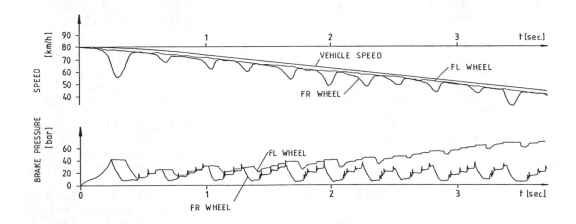

Fig.8: ABS MIR-control on split-μ
surface

Due to the delay in this developing, the driver is able to adjust to
the increasing steering and yawing moments. Further details can be obtained
from the literature (12).

SAFETY CONCEPT - The WABCO-ABS
safety concept already realised in air-
braked commercial vehicles has been
transferred to this hydraulic ABS. If
the hydraulic accumulator pressure
drops below the danger level, see
WARNING DEVICES, or if the voltage to
the solenoid valves drops below an adequate value, the ABS-warning lamp comes
on and ABS is switched off. This is
also true if the central plug is not
connected or the solenoid valve plug
is not connected to the ABS servo-
cylinder unit.

Other faults within the ABS
wiring, the ABS components or the dual
electric circuit electronic control
unit (12) lead to a diagonal switch
off. This results in normal service
braking of the defective diagonal and
ABS controlled braking of the other
diagonal.

THE FUTURE
A COMBINATION OF BRAKE FORCE AND
DRIVE SLIP CONTROL - Since the ABS
leads to an impressive improvement in
the braking control of vehicles, it is
logical to control the process of acceleration, in order to prevent dangerous
wheel-spin of the driving wheels on one
or both sides of the vehicle.

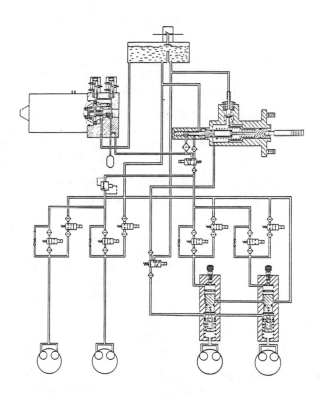

Fig.9: ABS brake system incorporating Drive Slip Control

By means of the ABS sensors which monitor the rotation of the wheels it is easy to get information about wheel-spin. With the aid of an additional solenoid valve, the voltage supply of the WABCO-ABS enables brake pressure to be applied to the wheel cylinders of one or more spinning driving wheels.

The brake pressure can be finely controlled by the ABS solenoid valves, and the brakes of wheels which are not spinning can be prevented from being applied by energizing the appropriate inlet valves (Fig. 9).

As an extension of this function, which may be described as a "differential" brake, Drive Slip Control also requires the engine power to be modulated by controlling the fuel mixture.

REFERENCES

1. D. Otte. Unfälle mit Beteiligung von LKW und Bussen - Einfluß von ABS. MHH, July 1984.

2. M. Burckhardt. Erfahrungen bei der Konzeption und Entwicklung des Mercedes-Benz/Bosch-Anti-Blockier-Systems (ABS). ATZ 81 (1979) 5

3. A. Mischke. Aufbau und Wirkungsweise des Antiblockiersystems ABS für Nutzfahrzeuge. ATZ 83 (1981) 9.

4. E. Petersen. K.G. Quicke. New anti-lock system for commercial vehicles realised with single-chip micro computers. I Mech E, 1981. C 205/81.

5. Continental gets anti-lock brakes, 10 Automotive Engineering 92. (1984) 10.

6. H.W. Bleckmann. J. Burgdorf. H.E. Gruenberg. K. von Timtner. The first compact 4-wheel antiskid system with integral hydraulic booster. 20 SAE Paper 830483, 1980.

7. H. Strien. Auslegung und Berechnung von PKW-Bremsanlagen. Alfred Teves GmbH, 1981.

8. H. Leiber. Bremskreisaufteilungen und mögliche Antiblockiersysteme. FISITA Paper, 82125, 1982.

9. H. Leiber. Antiblockiersysteme für Personenwagen mit digitaler Elektronik-Aufbau und Funktion. ATZ 81 (1979) 11

10. E. Petersen. Anti-Blockier-Systeme in Nutzfahrzeugkombinationen. KIVI-RAI Symposium, Amsterdam 1984.

11. P. Wiegner. Über den Einfluß von Blockierverhinderern auf das Fahr-verhalten von Personenkraftwagen bei Panikbremsungen. Diss. TU Braunschweig. 1974.

12. E. Reinecke. Neues Anti-Blockier-System mit erweiterten Sicherheits- und Systemfunktionen. VDI-Berichte Nr. 418, 1981.

13. A. Mischke. Antiblockiersystem (ABS) für Nutzfahrzeuge. VDI-Berichte Nr. 418, 1981.

14. WABCO-ABS für hydraulisch gebremste PKW, leichte Omnibusse und Nutz-fahrzeuge, WABCO Westinghouse Fahrzeugbremsen GmbH, 1984.

# ADVANCED ABS TECHNOLOGY

*Theory*

# Digital Algorithm Design for Wheel Lock Control System*

**Syed F. Hussain**
Allied Automotive
Bendix Chassis & Brake Components Division - Engineering

*Paper 860509 presented at the International Congress and Exposition, Detroit, Michigan, February, 1986.

## ABSTRACT

THE APPLICATION OF HIGH SPEED, HIGHLY INTEGRATED DIGITAL MICROCONTROLLERS IN MODERN ANTI-LOCK SYSTEMS allows increased computational capabilities with the potential for improved control and performance over previously developed analog designs.

The success of a microcontroller based ECU in demonstrating its performance capabilities lies in the basic control algorithm design and development. This paper will discuss basic algorithm design considerations and alternatives in digitally calculating basic parameters such as wheel and vehicle speeds, and deceleration.

In addition, several microcontrollers are investigated and evaluated for suitability for anti-lock control application by review of their architecture and instruction sets.

## INTRODUCTION

The overall performance capabilities of wheel lock control systems have been limited in the past primarily by the unavailability of low cost, flexible, high speed electronic technology.

Analog based electronic control units (ECU's) limited the early planning efforts in control system design mainly because of their inability to handle computations cost-effectively.

The availability of 8-bit digital microcontrollers allows complex, high speed, cost-effective computations by incorporating the functions of analog hardware circuits in digital software. Wheel lock control system designs, however, are still limited somewhat in response time, flexibility, and accuracy of both measured and calculated variables with an 8-bit machine. These limitations are due mainly to bit/input speed resolution and clock speeds. Additionally, significant portions of real time (scan time) are allocated to software modules such as speed processing, thus shortening the time available for control functions. The digital algorithm designer is limited in control philosophy by the architecture and instruction set of the 8-bit microcontroller used. The designer may use dual microcontrollers in either parallel or serial configurations to distribute real time overhead.

The impact of a 16-bit microcontroller design with very powerful instruction sets, architecture and input/output functions, has been to improve accuracy, flexibility and performance at reduced response time and overall cost.

The purpose of this paper will be to outline some considerations in the design and development of hydraulic anti-lock brake control systems including a review of microprocessor selection criteria and the design of some basic digital algorithms.

## MICROPROCESSOR DESIGN ALTERNATIVES

There is a wide array of microcontrollers available on the market ranging from 8-bit to 16-bit CPU's, and an assortment of on-board peripherals with differing architectures and instruction sets. A set of criteria was formulated to evaluate various microcontroller designs for application to wheel lock control systems, the overall objective being to maximize performance at the lowest cost with the best reliability. These criteria considered the following issues:

Maximum I/O integration to reduce off chip circuits and components to

effectively accommodate the following I/O requirements:
- Four wheel speed inputs
- Multiple solenoid driver/outputs
- Additional digital input/output ports for expandability
- Power down RAM
- Serial communication to outside world for diagnostics
- Watchdog timer

CPU and instruction set flexibility:
- Real time computational capability including clock speed and double byte data manipulation
- Available RAM
- High level instruction set for minimum software design and development time

Overall cost per function analysis. Environmental and packaging considerations. Expandability and adaptability to other wheel control applications such as traction control. Development support equipment availability. Manufacturer support and future commitment.

The microcontroller families evaluated were:

| | |
|---|---|
| MCS 8096 (MCS 96) | HD6301 |
| MC 68HC11 | MCS 8051 |
| MK 68200 | MC 6801 |

A comparison chart is attached in Figure I.

The use of 6801, 6301 or 8051 devices would entail the use of dual chips either in parallel or serial configuration to meet desired RAM, ROM, input/output line, and input capture speed requirements. To provide increased performance at reduced cost, the 8096, 68HC11 and MK 68200 were evaluated in a single chip configuration. Both 68HC11 and MC 68200 have only three high speed inputs, and thus require additional hardware to handle four wheel speed sensors.

The MC 68200 has only a 4K ROM, which was judged insufficient for wheel-lock control applications, especially when future expansions were considered. It would require off board ROM with address/data decode hardware making it less cost effective.

The MCS 8096 offered advantages over the MC 68HC11 in real time computational capability, ability to handle four wheel sensor inputs without peripheral hardware, and a 16-bit CPU with no accumulators. The design and development time for the MCS 8096 was considered shorter, due to the instruction set and highly intensive input/output architecture.

## ALGORITHM DESIGN CONSIDERATIONS

The following discussion outlines the software associated with the main control loop of a sample, viable anti-lock system as shown in Figure 2. The flow of events in software modules is as follows.

RESET: Immediately after power-up, the RAM is cleared, the stack pointer is initialized, the input/output ports and registers are set, the internal watchdog is initialized, and the interrupts are enabled.

MAIN: This is the start of the main program loop; the program ensures that the main loop is executed at a prescribed scan time (i.e. every 5 msec). The various wheel lock control software modules and subroutines are initialized, indexing structure is reset, control flags are reset, addresses and parameters are configured.

FAILSAFE: The module detects system failures and initiates either a fault-tolerant, adaptive, or system-shutdown mode, with appropriate driver warning. The action taken is dependent on type, timing and duration of the failure.

MODE SELECT: This program segment determines if anti-lock action is permissible based on failure status and low vehicle speed cut-off considerations. The decision is based upon vehicle velocity, wheel speed information and system failure mode status.

ANTI-LOCK MODE OFF: The software module shuts down the modulator (if it was activated) in a specified sequence, dependent on system failure modes or normal system conditions. It also treats each channel sequentially utilizing the "Wheel Speed" subroutine only.

ANTI-LOCK MODE ON: The software module treats each channel sequentially left front, right front and rears, respectively. For each of the channels, the following subroutines are called.

Wheel Speed
Analysis
Valve action

These three subroutines will now be discussed in more detail.

Wheel Speed: A magnetic pickup and tonewheel assembly generate an analog signal the frequency of which is proportional to the equivalent wheel translational speed. The electronic control unit treats the sensor output converting the typical sinewave signal into a cleaner square wave form.

The actual wheel speed can be expressed as:

$$\text{Wheel speed} = VWHL = \frac{SEN\ FREQ}{SENHZ} \qquad EQ\ 1$$

Where: SENHZ = The frequency generated at the wheel speed sensor at 1 mph. (This is a reference point determined by the tire rolling radius and number of tonewheel teeth).

SEN FREQ = The sensor output frequency at any given wheel speed.

The MCS-96 microcontroller high speed input is programmed to capture the value of the free running timer into the first-in,

## Microcontroller Comparison Chart

| Considerations | 8096 (Intel) [4] | 68HC11 (Motorola) [3] | MK68200 (Mostek) [3] | 6301 (Hitachi) [1] | 8051 (Intel) [2] | 6801 (Motorola) [1] | Comments |
|---|---|---|---|---|---|---|---|
| Wheel speed inputs | | | | | | | All except 8096 would either require use of two uP's or additional front end hardware to accommodate four inputs |
| Multiple solenoid drivers Multiple digital I/O's | 32 (48 pin pkg.) 40 (68 pin pkg.) | 40 | 40 | 29 | 29 | 29 | A single 8096, 68HC11 or MK68200 would be able to handle the system I/O requirements. However, the rest would either require dual processors or off-chip circuitry for memory mapped I/O |
| Watchdog timer (internal) | Yes | Yes | No | No | No | No | |
| Power down RAM, NVRAM, E²PROM | Power down RAM | E²PROM Power down RAM | Power down RAM | Power down RAM | Power down RAM | Power down RAM | The E²PROM offers the best option but presently limited in the number of write cycles |
| Serial communication | Yes | Yes | Yes | Yes | Yes | Yes | |
| Real time computational capability | Very high | High | Very high | Low | Low | Low | |
| CPU architecture for real time I/O | Very high | High | High | High | High | High | The 8096 has no accumulators/data registers; all 232 bytes of RAM form the source or destination of an instruction |
| Available RAM (bytes) | 232 | 256 | 256 | 128 | 128 | 128 | 128 Bytes not sufficient for application; need for additional RAM may be accomplished by dual configurations or off board RAM |
| Software design and development | Short | Medium | Short | Long | Long | Long | Application and instruction set dependent |
| 32 by 16 bit divide | 6.5-12.5 usec | 150 usec | 4 usec | 700 usec | 700 usec | 700 usec | |
| Double byte data manipulation instruction | Yes | Yes | Yes | Yes | None | Yes | |
| Accuracy and response time for processing wheel speed sensor inputs | Very high | Medium | Very high | Low | Low | Low | A single 6301, 8051 or 6801 would not be able to handle 4-sensor inputs to meet the response time or hardware requirements. A single 68HC11 or MK68200 would need peripheral hardware to process 4-wheel speeds with loss of input information and accuracy |
| Environmental and packaging considerations: | | | | | | | |
| 1) Automotive temperature | Yes | Yes | No | Yes | Yes | Yes | EPROM version is closer to a masked ROM version and can be used for fleet testing, EMI, RFI and fairly representative of production mask design. 4K of ROM marginal for wheel lock control application |
| 2) CMOS | Planned | Yes | Not known | Yes | Yes | Not Known | |
| 3) SMD | Yes | Yes | Yes | Yes | Yes | Yes | |
| 4) EPROM version | Planned | No | No | Yes | Yes | Yes | |
| 5) On board ROM (bytes) | 8K | 8K | 4K | 4K | 2K/4K | 2K/4K | |

NOTE: The above objective data is what was available to the author at the time of printing. Subjective comments and evaluations reflect the opinions of the author.

FIGURE 1.

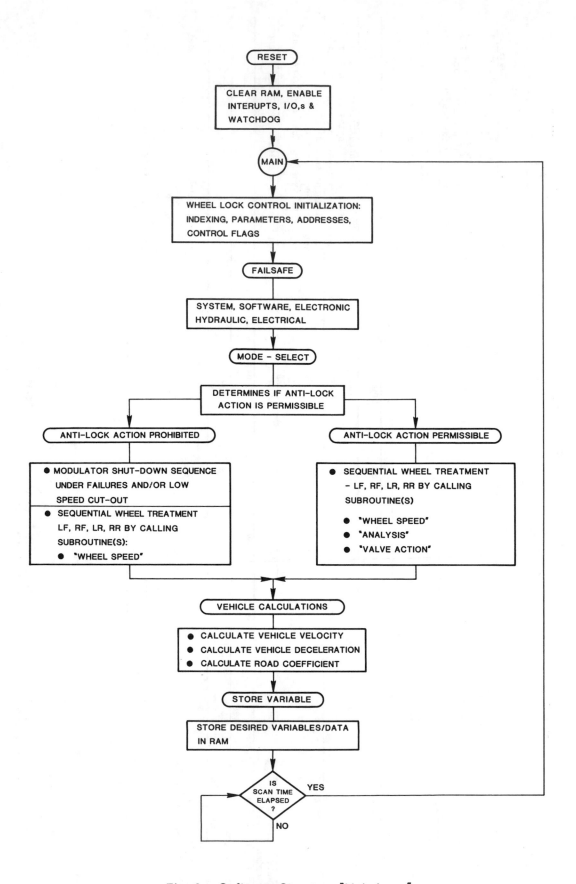

Fig. 2 – Software Structure "Main Loop"

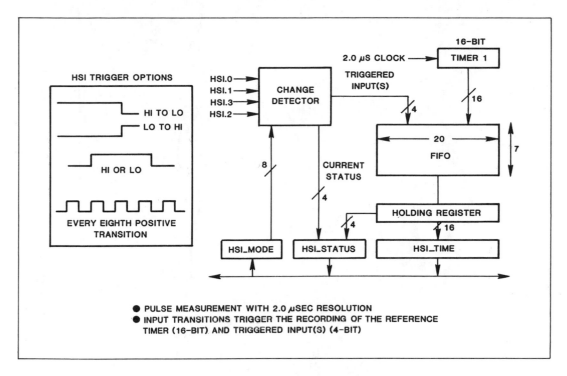

Fig. 3 – High Speed Input Unit

(Courtesy of Intel Corporation)

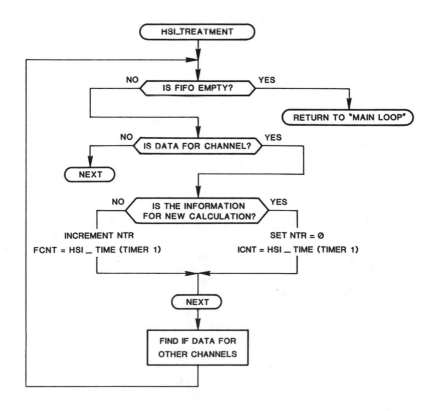

Fig. 4 – High Speed Input Service Routine

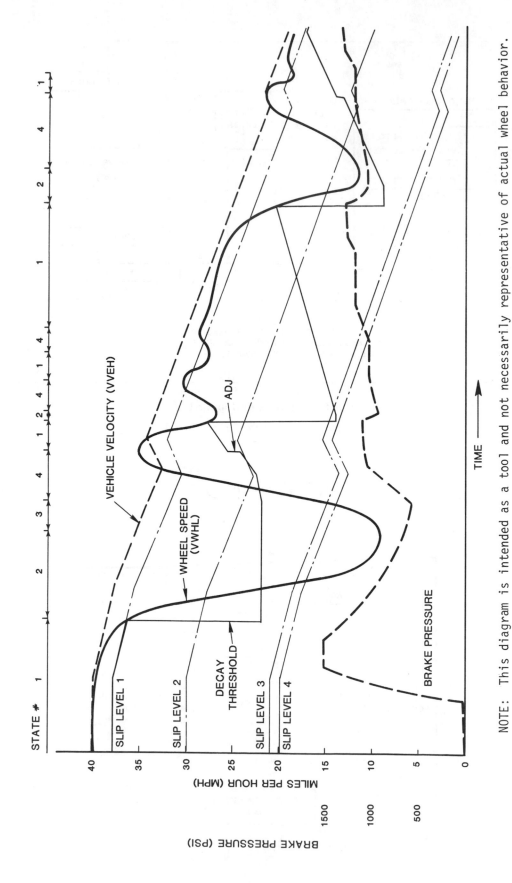

**Fig. 5 – Wheel State Diagram**

NOTE: This diagram is intended as a tool and not necessarily representative of actual wheel behavior.

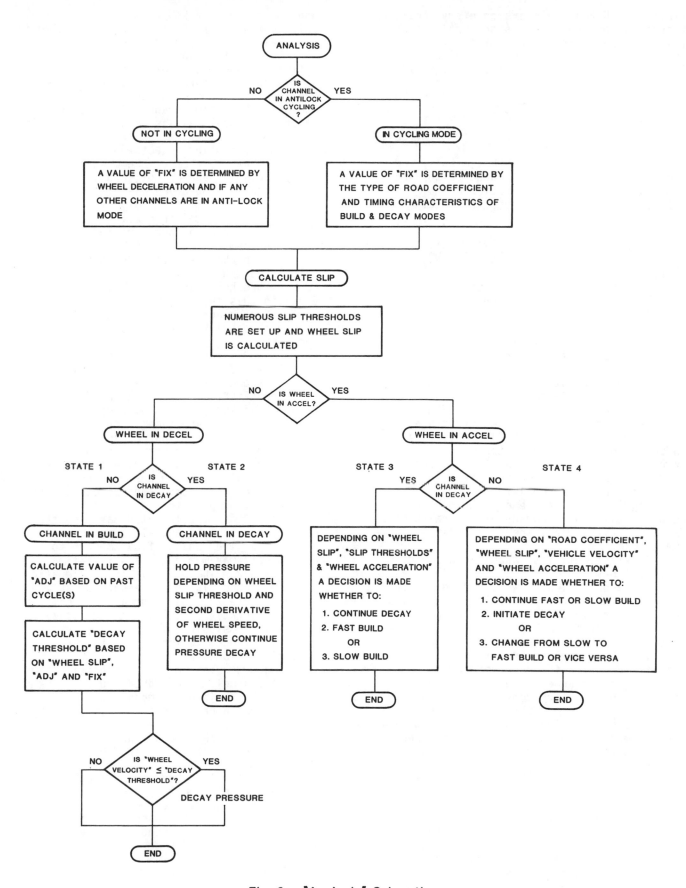

Fig. 6 – "Analysis" Subroutine

343

first out (FIFO) registers on every negative transition of the sensor signal pulse train (Figure 3). If the number of negative transitions (NTR), the initial value of timer (ICNT), and final value of "TIMER1" (FCNT) are known, then

$$\text{SEN FREQ} = \frac{NTR}{(FCNT - ICNT)} \qquad \text{EQ 2}$$

and

$$\text{VWHL} = \frac{NTR}{(FCNT - ICNT) * SENHZ} \qquad \text{EQ 3}$$

If both past (NTR1, FCNT1, ICNT1) and present (NTR, FCNT, ICNT) speed values are stored, then wheel deceleration or acceleration (GWHL) can be calculated.

$$\text{GWHL} = \frac{VWHL - VWHL1}{Ts} \qquad \text{EQ 4}$$

$$\text{GWHL} = \frac{NTR * TCNT1 - NTR1 * TCNT}{SENHZ * TS * TCNT1 * TCNT} \qquad \text{EQ 5}$$

Where:
VWHL = Present wheel speed
VWHL1 = Past wheel speed
Ts = Program sample time
TCNT = FCNT - ICNT
TCNT1 = FCNT1 - ICNT1

This speed processing approach offers the following:

Speed calculations for different sensors and applications require only a parametric change in the program (SEN HZ).

The high speed input RAM structure (first-in, first-out [FIFO]) enables the software to process every tonewheel transition and a larger sample size can be used to offset manufacturing variation of sensor/tonewheel assembly.

Significant improvement in both response time and accuracy of the calculated variables (i.e. wheel speeds, decel, etc.) due to:

An efficient, fast architecture and instruction set.

The acquisition of wheel speed related variables (NTR, FCNT and ICNT) is performed in a separate input interrupt routine asynchronously to the main program timing loop.

The wheel speed calculation in the main loop is performed for each channel prior to analyzing the channel ("Analysis") for appropriate anti-lock control commands.

The wheel speed subroutine takes approximately 60 microseconds. See Figure 4 for a summary flow chart of the high speed input interrupt service routine. This interrupt service routine is entered from the main program loop when any negative transition is detected on the high speed input. The program segment acquires transition and timer count (NTR) at each transition at each input. After interrupt service, the main loop program is resumed.

Analysis: The module forms the core of the wheel lock control system. In it, wheel characteristics (wheel velocity, acceleration, deceleration and jerk) are compared to a calculated vehicle velocity and appropriate brake pressure modulator valve commands are formulated and handled in the "Valve Action" subroutine.

Note that in addition to summary software flowchart references, it will be useful to use Figure 5 to get an overview of the functions on a more recognizable speed/function/time graph. A flowchart describing the "Analysis" sub-routine is in Figure 6.

The subroutine detects one of four states; that is, whether the channel is in a decay pressure mode or not, and if the wheel is in acceleration or deceleration (see Figures 5 and 6).

State 1 - Wheel in deceleration and channel not in decay pressure mode: This module determines whether to initiate anti-lock brake pressure modulation. The calculated "wheel velocity" is compared to a "decay threshold" and, if found less than or equal to this threshold, decay is commanded in the "Valve Action" subroutine. The "decay threshold" is composed of three components.

"Slip" variable is a percentage of vehicle velocity which is more significant at high speed than at low speeds.

The "Fix" is calculated based on the road coefficient, status of other brake channel activity and time in pressure build after the end of the last pressure decay. The "fix" provides adjustment of parameters for different road surfaces, and is intended to reduce false cycling and improves performance under conditions of oscillations and chatter bumps, etc.

The "ADJ" is calculated based on the performance of past anti-lock cycles; it provides a fine tuning of parameters to accommodate maximum surface utilization.

State 2 - The wheel is in deceleration and the channel is in a pressure decay mode. In this state, a "hold" command is initiated if wheel jerk is calculated to be positive indicating a slowing of wheel deceleration. This is intended to improve performance and possibly fluid usage.

State 3 - The wheel is in acceleration and the channel is in a decay mode. In this module, the decision is made when to initiate build pressure mode. It also determines the build rate so as to keep the wheel close to the top of the mu-slip curve. The build rates are determined by the road coefficient.

State 4 - The wheel is in acceleration and the channel is not in a decay mode. The module predominantly controls the build rates as the wheel approaches vehicle velocity. It sets commands for the "Valve Action" subroutine for desired brake pressure build rate(s).

Immediately after "Analysis," the "Valve Action" subroutine is called.

"Valve Action": This software module translates "Analysis" module commands to

344

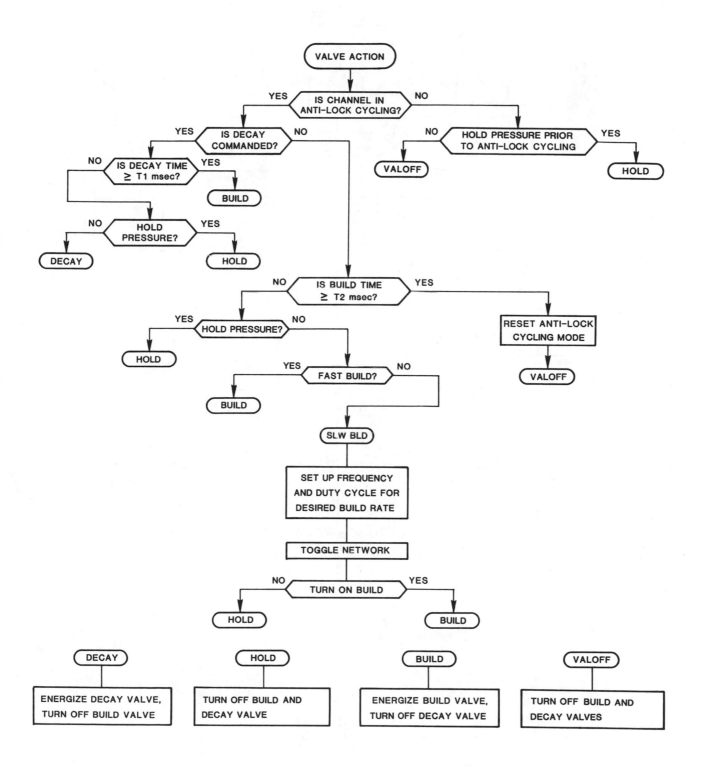

Fig. 7 – "Valve Action" Subroutine

345

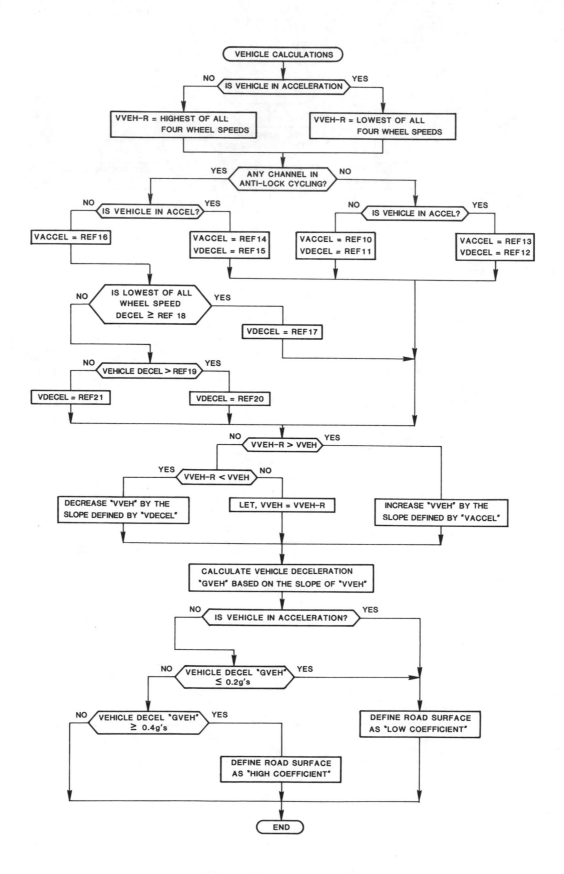

**Fig. 8 – "Vehicle Calculation" Subroutine**

action at the pressure modulator solenoid valves. The different requirements of brake pressure modulation are met by four output command modules as shown in Figure 7.

"Valve Off" - The build and decay valves are in a de-energized state. This is a normal state when the channel is not in an anti-lock cycling mode.

"Hold" - The decay and build valves are de-energized. Thus, the pressure is neither increasing nor decreasing; the channel is "holding" the desired pressure.

"Decay" - The decay valves are energized but the build valve is de-energized. The effect is the decrease (decay) brake pressure.

"Build" - The build valves are energized but the decay valve is de-energized, the effect being to increase (build) brake pressure.

Some of the main features are:

If build time exceeds a fixed time (T1), the module commands the channel to discontinue the anti-lock cycling mode.

If decay time exceeds a fixed time (T2), fast build is initiated and a system fault is recognized.

The slow build is achieved by a subroutine which sets the frequency and duty cycle to achieve the desired build rates.

Referring back to Figure 2, the program loop continues.

VEHICLE CALCULATION: The vehicle velocity is calculated from the four wheel speeds (see Figure 8). If the vehicle is in acceleration, the lowest of all wheel speeds is used for calculating the vehicle velocity. If the vehicle is in deceleration or if anti-lock action is occurring, then the highest of all wheel speeds is used for vehicle velocity calculations. This referenced variable is referred to as vehicle "raw" speed (VVEH-R).

The "actual" vehicle velocity (VVEH) is computed by comparing previous scan time vehicle velocity (VVEH) to present vehicle "raw" speed (VVEH-R) if:

VVEH-R is greater than VVEH, then vehicle velocity (VVEH) is allowed to increase at a rate defined by another variable, vehicle acceleration (VACCEL).

VVEH-R = VVEH, then vehicle velocity is the same as vehicle raw speed.

VVEH-R is less than VVEH, then vehicle velocity (VVEH) is allowed to decrease at a rate defined by the variable, vehicle decrease (VDECEL).

The variable "VACCEL" and "VDECEL" are dynamically set depending upon vehicle decel, road coefficient, all four wheel speed decelerations, and if the system is in anti-lock cycling (Note: Defined as "Ref 10", "Ref 11," etc.).

The vehicle decel (GVEH) is calculated based on the slope of the "VVEH."

The vehicle decel (GVEH) is then used to determine overall road coefficient, for example if vehicle decel is calculated to be less than .2 GS, then the surface is defined as a lo-co surface.

At the end of the "main loop," (Figure 2), various parameters like wheel speeds, acceleration or deceleration of wheel, vehicle velocity, etc. are stored in RAM for future reference. Finally, a check is made if the scan time for the "main loop" has elapsed. If it has, then the program execution is initiated all over again.

## SUMMARY AND CONCLUSIONS

The wheel lock control design, as realized by the software structure and algorithms, is flexible to meet the future requirements relating to wheel lock control applications. The modular design permits addition of other software modules for traction control and related applications. The main portion of scan time is utilized by the "analysis" subroutine, as opposed to other functions.

The on-chip input/output hardware like 10-bit analog inputs, pulse width modulated outputs, high speed inputs, serial communication, etc. allow interface with various sensors and actuators. Communication with other on-board computer systems, and additional wheel lock control functions are also possible.

The software structure, instruction set, 16-bit CPU, and on-chip input/output peripherals have resulted in improved accuracy, system response time, flexibility, and performance at a reduced overall cost.

## REFERENCES

1.  T. C. Schafer, Bendix Corporation, J. W. Douglas, Chrysler. The Chrysler "Sure-Brake" - The First Production Four-Wheel Anti-Skid System, 1971.
2.  HCMOS Single Chip Microcomputer "MC68C11A8" Motorola Semi-Conductor Technical Data, 1985 (AD11207).
3.  8-bit Single Chip Microcomputer Data Book, Hitachi America Ltd. (Hitachi #UTI), 1985.
4.  Single Chip Microcomputers, 68200 16-Bit United Technologies Mostek, 1984 (Pub #4420466).
5.  Microcontroller Handbook 1985, Intel Corporation, USA (ISBNO-917017-15-3).

# 4-Sensor 2-Channel Anti-Lock System for FWD Cars*

**Yasuo Kita, Masato Yoshino, and Hideaki Higashimura**
Sumitomo Electric Ind. Ltd.

*Paper 860511 presented at the International
Congress and Exposition, Detroit, Michigan,
February, 1986.

## ABSTRACT

The possibility of 2 Channel anti-lock system, which controls each of two independent hydraulic circuits of diagonal split braking system of FWD car seperately, were studied.

Theoretical investigation suggested two out of four possible control logics to be promising and they were proved to be practically satisfactory through vehicle test.

This system is almost as effective as expensive 3-channel or 4-channel system, when the braking force distribution between front and rear axles is correct as required by EEC Braking regulation. Under extreme condition that rear wheels lock earlier than fronts, the compromise between stopping distance and stability is necessary.

## BACKGROUND OF THIS STUDY

Anti-lock systems have been applied to aircraft and high-speed trains for more than 30 years to optimize braking characteristics and reduce tire wear.

Application to the automobile field, where large numbers of vehicles move under independent control in widely varying conditions, has proved more difficult. Rear axle control systems for cars were first introduced in the late '60s. They never gained wide acceptance, however, because their limited contribution to road safety failed to justify the extra cost. Fig.1 shows a subjective figure how anti-lock systems reduce trafic accidents.

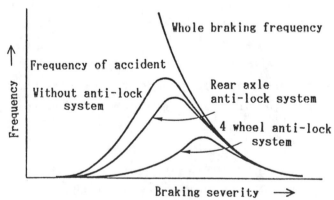

Fig.1 Subjective Figure on Anti-Lock Systems Contribution to Road Safety

Since that time, rapid progress in electronics technology has made cheaper and more reliable control systems available, enabling the deve-

lopment of more effective four-wheel
anti-lock systems. Many of these rely
on independent front-wheel and rear-
axle control for braking (3-channel).
Due to their high cost these systems
were used at first only on European
luxury cars in the late '70s, and even
now their use appears limited to higher
priced, high performance cars. Because
FWD is now used on most high-volume,
lower-cost cars, a more economical
anti-lock system is appropriate for
this market.

The weight distribution of these
vehicles, however, usually dictates the
use of diagonal, rather than front/rear
split brake circuits, to ensure that an
adequate secondary braking capability
is maintained in the event of hydraulic
circuit-failure. Consequently, a fourth
control channel is now required to
modulate pressure when an anti-lock
system is fitted to individual rear
brake lines. This additional channel
makes anti-lock systems for FWD cars
more expensive than those previously
developed for RWD cars. Several efforts
have already been made to redress this
cost problem and meet the demand for a
cheaper system for FWD cars.(1),(2),(3)*

This report describes the results
of recent studies to determine which
control principles are most suitable
for use with a 2-channel, diagonal split
anti-lock system. Vehicle tests have
proved that progress is being made toward
the practical application of anti-lock
systems to FWD automobiles.

INVESTIGATION INTO THE PRACTICAL
APPLICATION OF 4-SENSOR, 2-CHANNEL,
DIAGONAL SPLIT ANTI-LOCK BRAKING SYSTEM

BASIC SYSTEMS - Fig.2 shows the

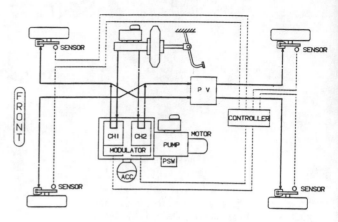

Fig.2  2-channel Diagonal Control System

proposed 2-channel anti-lock system,
which controls brake pressure in each of
the diagonal circuits in response to
information gathered from speed sensors
on all four wheels. The front-to-rear
braking ratio is controlled by propotioning valves to ensure that the front
wheels lock before the rear wheels.

The decision to initiate anti-lock
control depends on the "logic" built
into the software of an electronic
controller. Four different control
logics have been studied to determine
their effect on performance. Note that
with this system layout, modulation of
brake pressure affects the braking
force acting on both the front brake
and the diagonally opposite rear brake.

<u>Logic A</u> -- Diagonal "Select-Low"
The anti-lock system will be activated
in either of the diagonal circuits,
whenever the front or rear wheel controlled by that circuit tends to lock
first.

<u>Logic B</u> -- Rear Override
The anti-lock system will be activated
in either of the diagonal circuits whenever the front wheel controlled by that

_____

* Numbers in parentheses designate
  references at end of paper

circuit tends to lock. No action will be taken if one rear wheel locks, but should a second rear wheel lock, the anti-lock system will be activated in the corresponding second diagonal circuit.

Logic C -- Front Override
The anti-lock system will be activated in either of the diagonal circuits whenever the rear wheel controlled by that circuit tends to lock. No action will be taken if one front wheel locks, but should a second front wheel lock, the anti-lock system will be activated in the corresponding second diagonal circuit.

Logic D -- Double Override
No action will be taken if one front wheel locks, but should a second front wheel lock, the anti-lock system will be activated in the corresponding second diagonal circuit. Similarly, no action will be taken if one rear wheel locks, but should a second rear wheel lock, the anti-lock system will be activated in the corresponding diagonal circuit.

INVESTIGATION BY THEORETICAL CALCULATION - To examine the practical feasibility of these logic systems, we first estimated the performance by the method shown below. For purposes of comparison, estimates were also calcu-

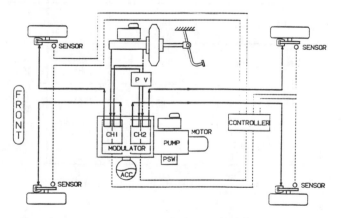

Fig.3  2-channel axle-control system

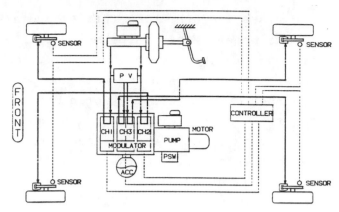

Fig.4  3-channel system

lated for 2-channel axle-control systems and for 3-channel systems as shown in Fig.3 and Fig.4.

The conditions in the comparison are:
1)  A symmetrical road surface
2)  An asymmetrical road surface
            (Split-$\mu$)
3)  The special case where rear wheels tend to lock before the front wheels.

The third condition is included to assess the consequences of unbalanced braking caused, for example, by variations in brake lining friction.

The subjects of this comparison are:
(1)  Wheel condition under anti-lock
(2)  Stopping distance
(3)  Directional stability
(4)  Steerability
(5)  Tire wear

In practice, these items are greatly influenced by the timing of wheel-lock detection, the timing of reduction/reapplication of pressure, etc. For the purposes of this study we assumed that these factors were effectively the same for all systems, and only the differences due to control logics were compared.

Briefly describing, upon application of the brake pedal, brake pressur rises and a braking torque develops at the wheel. This value has a limitation due to the amount of friction between the tire and the road surface, governed by the reactive force between the two and the corresponding adhesion coefficient, which varies according to the degree of slip. Cornering and side forces may also vary depending on the nature of the skid and how the weight is transferred to the tire.

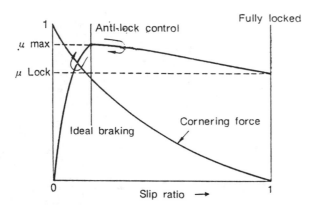

Fig.5 Adhesion coefficient of road surface

In this study, we used the following equations and values to calculate an approximate way of judging the systems:

Dynamic Weight Distributed on Wheels (wi)* - This ratio is the effective distribution of the vehicle's weight while braking. The ratio of distribution is assumed as:

w1=w2=0.35W        for front wheels
w3=w4=0.15W        for rear wheels
where        W=total vehicle weight

Braking Torque Efficiency (Wheel-Restraining Torque Efficiency)(Ebi) - Braking torque is taken relative to the torque at which a wheel locks. We assumed the following values:

---

* suffix i means individual wheels
  i = 1, 2 --- front wheels
  i = 3, 4 --- rear wheels

Ebi=1.0  for a locked wheel
Ebi=0.9  for an anti-lock controlled wheel
Ebi=0.9  for a rear wheel in a front/ rear split braking system under anti-lock control related to the other rear wheel
Ebi=0.8  for a rear wheel in a diagonal split braking system under anti-lock control related to the front wheel
Ebi=0.8 x 0.3 / 0.8 =0.3  for a wheel on a high-$\mu$ surface, when the other wheel in the same circuit is on a low-$\mu$ surface and is under anti-lock control

Effective Adhesion Utilization At The Contact Area Between Tire and Road Surface (Eti,Esi) - Compared with braking torque, the effective braking force utilized at the contact area between tire and road surface cannot achieve the theoretical value calculated with the maximum adhesion coefficient. In this study braking force, side force, and cornering force are derived from values calculated using the maximum-$\mu$, and the assumed efficiency shown in Table 1 below.

Table 1. The proportion of the force transfered from the road to the tire

| Braking condition | Usage proportion of braking efficiency (%) | Side force Cornering force (%) | Symbol Used in Table 3 |
|---|---|---|---|
| Light braking | 100 | 100 | □ |
| Ideal braking | 100 | 60 | ⊠ |
| Anti-lock braking | 95 | 60 | ▨ |
| lock braking | 80 | 0 | ■ |

Braking Force (Bf) - The braking force of each wheel can be expressed by the formula using items written above as:

$$bfi = wi \times K \times Ebi \times Eti$$

For the investigation of stopping distance, we compared the value of Bf.

$$Bf = \sum bfi$$

Directional Stability (St) - The side force of each rear wheel, which is most important for directional stability, can be expressed as follows.

$$sti = wi \times k \times Esi$$

For the investigation of directnal stability, we compared the value of St, the sum of sti calculated for each rear wheel.

$$St = st3 + st4$$

Steerability(Co) - The side force of each front wheel, which is the most essential component of cornering force and steerability, can be expressed by the formula

$$coi = wi \times k \times Esi$$

For the investigation of steerability, we compared the value of Co, the sum of coi calculated for each front wheel.

$$Co = co1 + co2$$

Amplitude of Braking Force Modulation (Av) - Av is defined as the percentage of braking force lost by the entire vehicle when braking torque efficiency, identified above, for a single braked wheel is 50% (representative of braking torque efficiency during a momentary loss of pressure). Calclations were based on that equal pressure is lost at the anti-lock controlled wheel and the other wheel coupled with the same braking circuit of the front/rear split system. In vehicle with diagonal split brake system, braking torque efficiency at the rear wheel when the front wheel of the same braking circuit is under anti-lock control was assumed to be 80%, and braking torque efficiency at the front wheel when the rear wheel of the same braking circuit is under anti-lock control was assumed to be 30% due to proportioning valve characteristics.

RESULTS OF THE INVESTIGATION - Table 3 shows the results of the theoritical calculations. As a base-line for each item, excluding tire wear, we use the performance of the 3-channel system (front independent, rear select low) which is the most widely-used system in the industry today. For ease of comparison, we divided the value calculated for each logic system by the value of the baseline system, and ranked them accordingly as shown below in Table 2.

Table 2. Theoreitical standards

| Symbol | Braking force Directional stability Steerability | Amplitude of braking force | Tire wear |
|--------|--------------------------------------------------|----------------------------|-----------|
| ◉ | More than 90 % | Less than 110 % | GOOD |
| ○ | More than 75 % | Less than 125 % | — |
| △ | More than 50 % | Less than 150 % | POOR |
| × | Less than 50 % | More than 200 % | VERY POOR |

Consequently, as shown in Table 3, we found that control logics C, D and E not only have a tendency to cause detrimental tire wear when the wheel is fully locked, but are also less acceptable in areas of directional stability and steerability, compared with the 3-channel control system.

Logic A, on the contrary, is expected to perform at almost the same level as the 3-channel system, except for relatively long stopping distances on split-$\mu$ road surfaces.

Logic B is also expected to provide about the same performance as the 3-channel system except for its relatively poor stability when a rear wheel locks earlier than a front.

## TEST PROGRAM DESIGN AND EXECUTION

CONTROL PROGRAM FOR VEHICLE TESTS - Based on the results of tests with the above basic variations, the following three anti-lock programs were considered and selected as candidate systems offering better performance.

Logic AB - Logic AB was proposed to compensate for the shortcomings of logics A and B. It is a system in which anti-lock control switches alternately between logic A and logic B on a time basis according to the specific vehicle design. A:B cotrol time was varied from 64 msec to 256 msec for differrent tests, but remained constant during each individual test.

Logic B - This system is identical to logic B described before.

Logic AX - This system was also proposed to compensate for the shortcomings of logics A and B. Logic AX differs from logic AB in that the anti-lock control is biased towards logic A. Split-$\mu$ road surfaces are determined according to several factors including the combination

Table 3.  * Reference for comparison

| System variation | | | 2-channel, diagonal split circuit | | | | * 2-channel, axle split circuit | * 3-channel |
|---|---|---|---|---|---|---|---|---|
| | | | A | B | C | D | E | F |
| | | | Front+rear | Front + higher rear | Higher front + rear | Higher front + Higher rear | Front high select rear low select | Front Independent rear low select |
| Wheel condition under anti-lock | | | | | | | | |
| Braking force B | 1st wheel | | 29.9 $\mu$ | 29.9 $\mu$ | 28.0 $\mu$ | 28.0 $\mu$ | 28.0 $\mu$ | 29.9 $\mu$ |
| | 2nd wheel | | 29.9 $\mu$ | 29.9 $\mu$ | 29.9 $\mu$ | 29.9 $\mu$ | 29.9 $\mu$ | 29.9 $\mu$ |
| | 3rd wheel | | 12.0 $\mu$ | 12.0 $\mu$ | 12.0 $\mu$ | 12.0 $\mu$ | 12.8 $\mu$ | 12.8 $\mu$ |
| | 4th wheel | | 12.0 $\mu$ | 12.0 $\mu$ | 12.8 $\mu$ | 12.0 $\mu$ | 13.5 $\mu$ | 13.5 $\mu$ |
| | Total | | 83.8 $\mu$ | 83.8 $\mu$ | 82.7 $\mu$ | 81.9 $\mu$ | 84.2 $\mu$ | 86.1 $\mu$ |
| | Ratio to 3ch | | 97.3 % O | 97.3 % O | 96.1 % O | 95.1 % O | 97.8 % O | 100 % (Standard) |
| Stability S | 3rd wheel | | 9.0 $\mu$ | 9.0 $\mu$ | 9.0 $\mu$ | 9.0 $\mu$ | 9.0 $\mu$ | 9.0 $\mu$ |
| | 4th wheel | | 9.0 $\mu$ | 9.0 $\mu$ | 9.0 $\mu$ | 0 $\mu$ | 9.0 $\mu$ | 9.0 $\mu$ |
| | Total | | 18.0 $\mu$ | 18.0 $\mu$ | 18.0 $\mu$ | 9.0 $\mu$ | 18.0 $\mu$ | 18.0 $\mu$ |
| | Ratio to 3ch | | 100 % O | 100 % O | 100 % O | 50 % △ | 100 % O | 100 % (Standard) |
| Steerability C | 1st wheel | | 21.0 $\mu$ | 21.0 $\mu$ | 0 $\mu$ | 0 $\mu$ | 0 $\mu$ | 21.0 $\mu$ |
| | 2nd wheel | | 21.0 $\mu$ | 21.0 $\mu$ | 21.0 $\mu$ | 21.0 $\mu$ | 21.0 $\mu$ | 21.0 $\mu$ |
| | Total | | 42.0 $\mu$ | 42.0 $\mu$ | 21.0 $\mu$ | 21.0 $\mu$ | 21.0 $\mu$ | 42.0 $\mu$ |
| | Ratio to 3ch | | 100 % O | 100 % O | 50 % △ | 50 % △ | 50 % △ | 100 % (Standard) |
| Amplitude of braking force modulation | When the front wheel | | 21 % | 21 % | 21 % | 21 % | 34 % | 17 % |
| | When the rear wheel | | — | — | 31 % | — | 16 % | 16 % |
| | Maximum value | | 21 % | 21 % | 31 % | 21 % | 34 % | 17 % |
| | Ratio to 3ch | | 123 % O | 123 % O | 182 % △ | 123 % O | 200 % × | 100 % (Standard) |
| Tire wear | | | GOOD O | GOOD O | POOR △ | VERY POOR × | POOR △ | GOOD O |

A/L performance on a uniform $\mu$ road surface

| System variation | | | 2-channel, diagonal split circuit | | | | \# 2-channel, axle split circuit | \# 3-channel |
|---|---|---|---|---|---|---|---|---|
| | | | A | B | C | D | E | F |
| | | | Front+rear | Front + higher rear | Higher front + rear | Higher front + Higher rear | Front high select rear low select | Front Independent rear low select |
| | Wheel condition under anti-lock | | 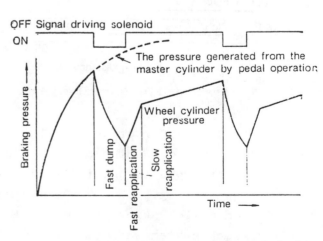 low↑high 1⊠2 / 3⊠4 | (diagram) | (diagram) | (diagram) | (diagram) | (diagram) |
| Performance on an asymmetrical (SPLIT) μ road surface | Braking force B | 1st wheel | 9.0 μ | 9.0 μ | 8.4 μ | 8.4 μ | 8.4 μ | 9.0 μ |
| | | 2nd wheel | 10.5 μ | 23.9 μ | 10.5 μ | 23.9 μ | 23.9 μ | 23.0 μ |
| | | 3rd wheel | 3.8 μ | 3.6 μ | 3.8 μ | 3.6 μ | 3.8 μ | 3.8 μ |
| | | 4th wheel | 3.6 μ | 3.6 μ | 10.3 μ | 10.3 μ | 4.1 μ | 4.1 μ |
| | | Total | 26.9 μ | 40.1 μ | 33.0 μ | 46.2 μ | 40.2 μ | 40.8 μ |
| | | Ratio to 3ch | 66 % ○ | 98 % ◎ | 81 % ○ | 113 % ◎ | 99 % ◎ | 100 % (Standard) |
| | Stability S | 3rd wheel | 2.7 μ | 0 μ | 2.7 μ | 0 μ | 2.7 μ | 2.7 μ |
| | | 4th wheel | 12.0 μ | 12.0 μ | 7.2 μ | 7.2 μ | 12.0 μ | 12.0 μ |
| | | Total | 14.7 μ | 12.0 μ | 9.9 μ | 7.2 μ | 14.7 μ | 14.7 μ |
| | | Ratio to 3ch | 100 % ◎ | 82 % ○ | 67 % △ | 49 % × | 100 % ◎ | 100 % (Standard) |
| | Steerability C | 1st wheel | 6.3 μ | 6.3 μ | 0 μ | 0 μ | 0 μ | 6.3 μ |
| | | 2nd wheel | 28.0 μ | 16.8 μ | 28.0 μ | 16.8 μ | 16.8 μ | 16.8 μ |
| | | Total | 34.3 μ | 23.1 μ | 28.0 μ | 16.8 μ | 16.8 μ | 23.1 μ |
| | | Ratio to 3ch | 148 % ◎ | 100 % ◎ | 121 % ◎ | 73 % △ | 73 % △ | 100 % (Standard) |
| When rear wheels lock first | Wheel condition under anti-lock | | (diagram) | (diagram) | (diagram) | (diagram) | (diagram) | (diagram) |
| | Stability S | 3rd wheel | 9.0 μ | 0 μ | 9.0 μ | 9.0 μ | 9.0 μ | 9.0 μ |
| | | 4th wheel | 9.0 μ | 9.0 μ | 9.0 μ | 0 μ | 9.0 μ | 9.0 μ |
| | | Total | 18.0 μ | 9.0 μ | 18.0 μ | 9.0 μ | 18.0 μ | 18.0 μ |
| | | Ratio to 3ch | 100 % ◎ | 50 % △ | 100 % ◎ | 50 % × | 100 % ◎ | 100 % (Standard) |

of wheels which the anti-lock system has to dead with. When a split-μ road surface is detected anti-lock control switches logic B.

TESTS –

Test Vehicle Selected – The test vehicle was a 2000 cc, front wheel drive vehicle.

Anti-Lock Actuator – The actuator used one solenoid per channel and three cylinder pressurization patterns, fast dump fast reapply, and slow reapply.

Fig.6 Typical brake pressuer modulation of the actuator used for the tests

Test Schedule - To examine practial usage, tests were conducted under the following conditions.

1) Straight line braking on a symmetrical road surface (high-$\mu$ and low-$\mu$)
2) Lane change performance
3) Straight-line braking on an asymmetrical road surface (high and low-$\mu$ split)
4) Braking while cornering (cornering performance)
5) Braking when the rear brake locks prior to the front brake

TEST RESULTS -

Straight Line Braking on A Symmetrical Road Surface. - Fig.7 shows the results of braking on a high-$\mu$ surface (dry asphalt) at a speed of 80 km/h, and braking on a low-$\mu$ surface (wet epoxy) at a speed of 50 km/h.

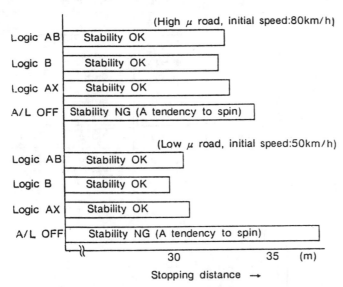

Fig.7 Straight line braking performance

Lane Change Performance - The lane change test was conducted on a medium-u surface (wet asphalt) under the conditions shown in Fig.8. The maximum entry speed when full braking is applied with an anti-lock system was compared with the value obtained without braking. The re-

sults were just over 70 km/h in the no braking condition, and over 80 km/h with the anti-lock system, disregarding logics. Needless to say, when braking fully without anti-lock, there is not enough steerability to pass through the course.

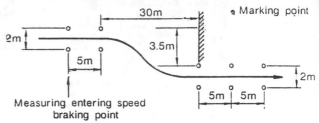

Fig.8 Lane change test conditions

Straight Line Braking on An Asymmetrical (Split-$\mu$) Surface - The split-$\mu$ test was conducted under the conditions shown in Fig.9. The road surface is asymmetrical with a large relative difference between the adhesion coefficients on each side of the vehicle. Fig.10 shows the stopping distance and the final yaw angle after braking from 50 km/h with the steering wheel held fixed, and braking from

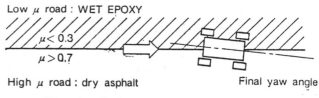

Fig.9 Split $\mu$ test

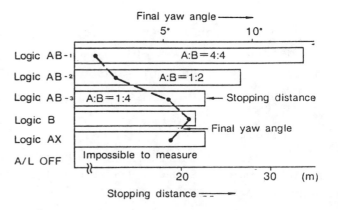

Fig.10 Split $\mu$ test result from 50 km/h with the steering wheel held fixed

356

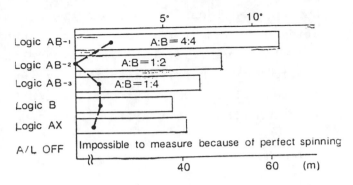

Fig.11　Split $\mu$ test result from 70 km/h
　　　　with steering wheel corrections

70 km/h with steering wheel corrections
to keep the vehicle straight.

<u>Braking While Cornering</u> － Fig.13
shows the results of braking while cornering (the wavy lane shown in Fig.12).
Full braking was applied from 50 km/h,
almost the critical speed for this 25 m
redius curved lane. Cornering performance
"SLIGHT OVERSTEER" is the same as the
original characteristics of the vehicle

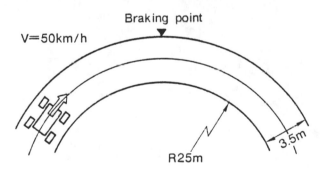

Fig.12　Braking test while cornering

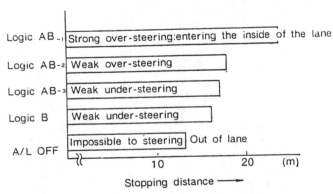

Fig.13　Braking performance while
　　　　cornering

used in the test for normal braking when
cornering. With 'LOGIC B' this is shown
to be 'SLIGHT UNDERSTEER',but in the test
with a 30 m radius and an initial speed
of 50 km/h (not so critical), the characteristics varied towards 'SLIGHT OVER-
STEER' as with the other logics.

<u>Braking When The Rear Wheel Locks</u>
<u>Before The Front</u> - Even if the braking
force is distributed by the proportioning
valve to lock the front wheels earlier
than the rear,  so long as there is no
special surface condition such as a
split-$\mu$, the possibility remains that the
rear wheels will lock earlier than the
fronts due to the variation of the $\mu$-level
of the brake pads or the "fade" phenomenon. To examine performance under these
conditions, tests were conducted using
the special proportioning valve shown in
Fig.14(B), the results of which are shown
in Fig.15.

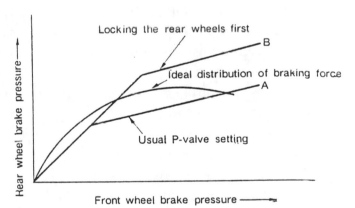

Fig.14　Propotioning valve setting

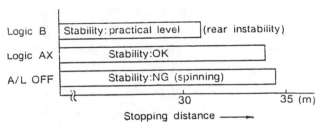

Fig.15　Braking performance when rear
　　　　wheels lock earlier than fronts

## CONCLUSION

From the vehicle tests results shown above, we can conclude:

1) Using 'LOGIC AX', rear wheel locking is completely prevented on symmetrical road surfaces. On the other hand, the stopping distance is kept short on asymmetrical road surfaces, allowing the rear wheel to lock on lower-$\mu$ surfaces. This means we attempted to discriminate road surfaces only by the behavior of four wheels.

   To follow up all the situations on various road surface conditions by proper anti-lock control, more verification tests will be carried out.

2) With 'LOGICS B', the stopping distance is shortest of all the tests. Vehicle stability and steerability can be considered as practical.

3) With 'LOGIC AB', the split-$\mu$ stopping distance is longer in proportion to the ratio of switching time A:B than with 'LOGIC B'. But vehicle stability and steerability are clearly superior. In this series of tests, the ratio of period A:B was fixed during the test. We believe that if we introduce the idea of an in-stop flexible A:B ratio to exfract the optimum performance for each situation (based on data such as vehicle speed, vehicle deceleration, road surface-$\mu$ or tire slip ratio), we should be able to greatly improve the total balance of the braking performance.

As noted above, we have proved that by a simpler 2-channel control of each circuit for a FWD vehicle, the anti-lock system can provide satisfactory performance for practical use. We are now carrying out improvements in the control Logics (mentioned above), and are conducting endurance tests for each component as well as vehicle tests to prove reliability. We look forward to the day when our LOW-COST, 2-CHANNEL, ANTI-LOCK SYSTEM will contribute to traffic safety.

## REFERENCSES

(1) Makoto Satoh, Shuji Shiraishi
"Performance of Antilock Brakes with Simplified Control Technique"
SAE Paper, No. 830484

(2) W.R.Newton, F.T.Riddy
"Evaluation Criteria for Low Cost Anti-Lock Brake Systems for FWD Pasenger Cars"
SAE Paper, No. 840464

(3) Peter Hattwig
"Cost Benefit Analysis of Simplified ABS"
SAE Paper, No. 850053

*Systems*

# Upgrade Levels of the Bosch ABS*

**Wolf-Dieter Jonner and Armin Czinczel**
Robert Bosch GmbH

*Paper 860508 presented at the International
Congress and Exposition, Detroit, Michigan,
February, 1986.

ABSTRACT

The Bosch ABS for passenger cars which has
been in production since 1978 has been
described in numerous publications.
Following the gathering of extensive experience
with the Bosch ABS and its installation in the
different models of passenger car, the concept
has been revised with various upgrade levels
in order to further optimize braking performance on μ-split road surfaces with different
right/left adhesion coefficients, in order
further to improve the operation of the system
when braking on very slippery road surfaces
and also to adapt the control algorithm to
four-wheel-drive vehicles with differential
locks.

## YAWING MOMENT BUILDUP DELAY

When Bosch ABS was introduced to series
production, only vehicles with a relatively
large wheel base and a large moment of inertia
about the vertical axis were equipped with it.
These required no yawing moment buildup delay
because the large wheel base resulted in a
large stabilizing torque thanks to the lateral
forces. This meant that the vehicle yawing
caused by the unequal braking forces on the
front wheels due to the large moment of inertia
about the vehicle's vertical axis was so slow,
that the driver had enough time to compensate
for these unequal forces by turning the
steering wheel accordingly.

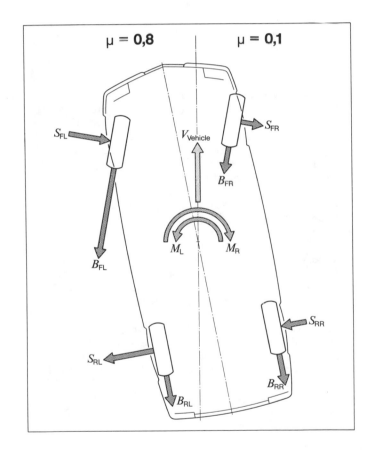

Fig. 1: Steady-state braking with ABS on a
μ-split road surface without yawing
moment buildup delay

Figure 1 shows a vehicle with ABS on a µ-split road surface with different right/left adhesion coefficients during a steady-state braking operation without yawing moment buildup delay. During emergency braking, differing brake forces are built up on the front wheels corresponding to the differing right/left adhesion coefficients. This produces a left-handed torque $M_L$ about the vertical axis which leads to counterclockwise yawing of the vehicle. Due to the ABS-select-low principle, equal braking forces arise on the rear wheels which correspond to the lower adhesion coefficient. By steering the vehicle, the driver has to turn the front wheels in clockwise direction to compensate for the left-handed torque $M_L$. Lateral forces orientated to the right thus come into being on the front wheels and, together with the left-orientated lateral forces acting on the rear wheels, generate the right-orientated torque $M_R$ which then compensates for the left-handed torque $M_L$.

It must be emphasized that steady-state braking on split coefficient road surfaces is only possible with a compensating motion of the steering wheel (closed loop). Experiments, and above all also computer simulations, with the steering wheel held tight (open loop) do not suffice to arrive at an assessment of braking response on µ-split road surfaces. This applies also to braking when cornering.

As Bosch ABS was increasingly installed in smaller vehicles with a smaller wheel base and moment of inertia about the vertical axis, it increasingly became necessary to install a yawing moment buildup delay to keep these vehicles under control during panic braking operations on µ-split road surfaces. A smaller wheel base results in higher lateral forces and thus increased slip angles leading to increased yawing. Lower moments of inertia result in more rapid yawing.

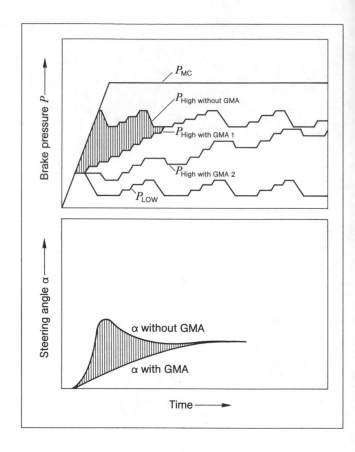

Fig. 2: Yawing moment buildup delay (GMA) for passenger cars

The yawing moment buildup delay introduced in 1984 produces a delay in buildup of the brake pressure applied to the front wheel running on the side of the vehicle with the higher adhesion coefficient (high wheel). This principle is shown in Figure 2. The top line shows the pressure $p_{MC}$ generated in the master cylinder.

If no yawing moment buildup delay (i.e. individual control) is employed, the pressure $p_{high}$ is produced which very rapidly leads to a high braking force difference on the front axle due to the low pressure applied to the other front wheel $p_{low}$.

In less critical vehicles GMA1 is used which builds up the brake pressure on the high wheel in steps during the initial braking phase as soon as a pressure decrease indicating instability of the wheel occurs at the low wheel. Once the brake pressure on the high wheel has reached the lock-up pressure level, its wheel pressure is no longer influenced by the low wheel, but is autonomously controlled with full utilization of the adhesion force. In less critical vehicles this measure ensures statisfactory steering response during panic braking on µ-split road surfaces. Since the lock-up pressure level is always reached relatively quickly (in approximately 750 ms), the increase in the stopping distance is small compared to individual control.

A further trace on the graph shows the brake pressure on the high wheel when using the GMA2 in particularly critical vehicles. As soon as the brake pressure on the low wheel is decreased, the ABS solenoid valve of the high wheel is controlled with a specific pressure-holding and pressure-reduction time. The renewed pressure buildup on the low wheel then triggers off a gradual pressure buildup on the high wheel whereby the pressure buildup times are longer by a specific factor than on the low wheel. This metering of pressure takes place not only during the first control cycle, but continuously during the entire panic braking operation.

The greater the initial braking speed, the more critical is the effect of the yawing moment on steering response. This is why the speed of the vehicle is subdivided into 4 sectors in which yawing moment buildup delays of differing magnitude are active. In the sectors with high vehicle speeds, pressure buildup times on the high wheel are increasingly shortened while the pressure decrease times are increasingly extended to achieve a reduced buildup speed of the yawing moment as the initial braking speed increases.

While the GMA1 is contained in the large scale integrated circuits of the electronic ABS controller, the GMA2 is realized in two additional microprocessors operating redundantly for safety. Faulty operation is indicated if both microprocessors emit differing signals during several clock pulse cycles and the ABS is deactivated.

The lower part of the figure qualitatively shows the steering angle required with GMA during braking on µ-split road surfaces in comparison with individual control.

In choosing the optimum yawing moment buildup delay, a compromise must always be made between a good steering response and an adequately short stopping distance. This parameter can only be defined in agreement with vehicle manufacturers.

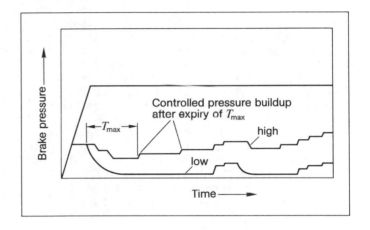

Fig. 3: Controlled pressure buildup on the high wheel with long pressure reduction phases on the low wheel

Particularly when braking on µ-split road surfaces with the clutch engaged, it may happen that no pressure buildup occurs at all for long periods of time due to the braking effect of the engine drag torque on the low wheel of front-wheel-drive vehicles. Figure 3 shows that, in this case, the brake pressure on the high wheel is built up in steps after expiry of a specific time Tmax to ensure a satisfactory braking effect in this case also.

Braking behaviour while cornering must also be examined for introduction of the GMA. It is known that, when the initial braking phase is very slow, starting from the cornering limit speed, vehicles tend to turn in towards the curve and are then not easy to control by steering (oversteering occurs).

During panic braking with ABS and individual control of the front wheels under these conditions, a relatively high braking force is built up on the front wheel on the outer side of the corner which compensates for the tendency to turn in towards the curve. When GMA2 is employed, the braking force on the front wheel on the outer side of the curve is built up so slowly that, analogously to very slow initial braking phases, turning into the curve occurs in many vehicles because there is no high stabilizing braking force on the front wheel on the outer side of the curve. This may lead to a critical oversteering tendency which can no longer be mastered by the average driver. To prevent this critical situation Bosch offers a lateral acceleration switch which can be employed in an interaction with the GMA2 to deactivate the yawing moment buildup delay when braking, when cornering close to the cornering limit speed with high adhesion coefficients. In this way, the braking force acting on the front wheel on the outer side of the curve is built up rapidly, compensating for the vehicle's oversteering tendency. Depending on the vehicle type, driving conditions at the cornering limit speed can also be derived from the differences in speed between the inner front and rear wheels, and then used to deactivate the GMA. Figure 4 shows the braking behaviour when cornering without and with an active yawing moment buildup delay.

The top part of the figure shows the braking behaviour when cornering without yawing moment buildup delay. Corresponding to the high normal wheel force, a high lateral force and, due to individual control, a high braking force, act on the outer front wheel. The high braking force leads to an only slight tendency to understeer with the result that the vehicle remains easy to control. The bottom part of the figure shows the vehicle with the yawing moment buildup delay active. Only a slight braking force is now built up on the outer front wheel which cannot maintain equilibrium with the high lateral force acting on this wheel. This results in a torque turning in towards the curve, and thus to oversteering of the vehicle. The vehicle is then so difficult to control that often the demands placed on the normal driver's abilities are too high.

| Vehicle | Steering angle reductions | | Stopping distance increase |
| --- | --- | --- | --- |
| | In the initial braking phase | During control | |
| Upper middle class | – 72 % | – 40 % | + 14 % |
| Middle class | – 41 % | – 47 % | + 10 % |
| Lower middle class | – 33 % | – 7 % | + 5,2 % |

Fig. 5: Steering angle reductions and stopping distance increases during braking on μ-split road surfaces (μ = 0,8/0,2) with GMA2 in comparison with individual control

Figure 5 shows the results of measurement of the reduction in steering angle and the enlargement of stopping distances by the GMA2 for 3 middle class vehicles in contrast to individual control. A comparison of the measured data shows that considerable reductions in the required steering angle are achieved along with justifiable extension of the stopping distance. The comparison of the figures also shows that the demands for as great a reduction in steering angle as possible and minimum extension of the stopping distance stand in mutual opposition.

Yawing moment buildup delay is an effective measure of improving control ability during panic braking operations on μ-split road surfaces.

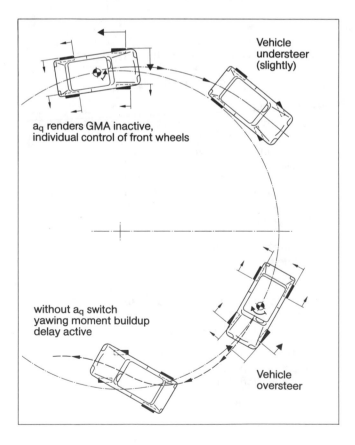

a_q renders GMA inactive, individual control of front wheels

Vehicle understeer (slightly)

without a_q switch yawing moment buildup delay active

Vehicle oversteer

Fig. 4: Cornering behaviour at limit speed with GMA and without/with a_q switch

## ENGINE DRAG TORQUE CONTROL (MSR)

Particularly on very slippery road surfaces, the drag torque of the engine with the clutch engaged can have an adverse affect on braking performance. Control of the braking torque alone by the ABS is not able to compensate for the drag torque of the engine. In rear-wheel-drive vehicles the drag torque of the engine can lead to instability of the vehicle, while in the case of front-wheel-drive vehicles steerability may be considerably affected. As a first stage, it is possible to reduce the drag torque of the engine by suppressing the overrun fuel cutoff during ABS operation. The second step might be to raise the engine speed if the drive wheels show signs of locking. As a third step, electronically controllable automatic transmissions can be shifted automatically to neutral, while, in the case of manual transmission, the clutch can be automatically disengaged. In such cases, attention must be paid to the problems of re-engaging the clutch.

The possible fourth step is to control the drag torque of the engine with the aid of the actuator of the electronic accelerator. This component, which is nearing the start of series production, permits optimum control of the engine drag torque. If a specific wheel slip occurs on the drive wheels, the throttle valve is opened with the aid of the electronic accelerator to such an extent that the engine drag torque is compensated.

Without MSR, a slip of up to 80 % can occur on the drive wheels when braking with ABS on very slippery road surfaces. In rear-wheel-drive vehicles this critically endangers driving stability, while it considerably reduces the steerability of the front-wheel-drive vehicles. Driving stability and steerability can be crucially improved with MSR.

On slippery road surfaces, the MSR is not only beneficial during braking, but also when shifting down to a lower gear when cornering without any engine drive. In this case also, the MSR ensures compensation by controlled opening of the throttle valve, thus maintaining driving stability und steerability, when slip occurs due to the engine drag torque.

## ABS FOR FOUR-WHEEL-DRIVE VEHICLES

Currently, more and more models of passenger car are being offered with four-wheel drive. Automobile manufacturers have designed their four-wheel-drive vehicles in accordance with various criteria, where traction, driving dynamics and braking performance play an essential role. Previously known concepts call for differing upgrade levels in comparison with the conventional ABS. With the differential locks selected and under certain road surface conditions, ABS operation encounters problems which necessitate special action. When the rear axle differential lock is selected, the rear wheels are rigidly coupled, i.e. they run at the same speed and behave like a rigid body in respect of the encountered braking moments and road surface friction moments. When the central lock is selected, the average speed of the front wheels is then equal to the speed of the rear wheels. All wheels are thus dynamically coupled and the engine drag torque and moment of inertia act on all wheels as shown in Figure 6.

---

1. On μ-split road surfaces, the differential lock on the rear axle produces an increased yawing moment by cancelling select-low mode.

2. The central lock couples all wheels to each other, to the inert engine mass and to the engine drag torque. This makes reference speed calculation (generation) difficult.

---

Fig. 6: Problems encountered in four-wheel-drive concepts with ABS

Figure 7 shows the dynamic conditions prevailing when braking with ABS on μ-split road surfaces with and without the rear axle lock selected and with differing right/left adhesion coefficients.

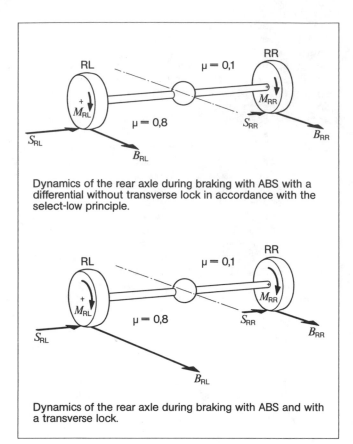

Dynamics of the rear axle during braking with ABS with a differential without transverse lock in accordance with the select-low principle.

Dynamics of the rear axle during braking with ABS and with a transverse lock.

Fig. 7: Dynamics of rear-axle lock

Bosch ABS operates on the rear axle in accordance with the select-low principle. The rear wheel braking with the lower adhesion coefficient determines the joint brake pressure of both rear wheels. As shown in the top part of the figure, only a low braking force $B_{RL} = B_{RR}$ corresponding to the adhesion coefficient $\mu = 0,1$ is developed on the left-hand rear wheel in the system without transverse lock. The available cornering force $S_{RL}$ is accordingly high, ensuring the stability required for retardation on $\mu$-split road surfaces.

The bottom part of the figure shows the dynamics of the system with transverse lock. Since the axle with the wheels represents a rigid system, such high braking torques are introduced that the maximum possible braking forces corresponding to the differing adhesion coefficients are transmitted by both wheels. This results in a tangible decrease in the cornering force reserve $S_{RL}$ on the left-hand rear wheel. In turn, this considerably reduces driving stability when braking on $\mu$-split road surfaces.

As known from numerous publications in the trade press, Bosch ABS makes use of the two controlled variables of the angular wheel acceleration and wheel slip for brake control. With conventional rear-wheel or front-wheel drive vehicles, the non-driven wheels which are never coupled to the engine's moment of inertia are particularly suitable for generating a wheel angle deceleration signal -a in due time in the event of a tendency to lock-up in order to eliminate this tendency by decreasing the brake pressure. In four-wheel-drive vehicles with the transverse and longitudinal locks engaged, all wheels are coupled to the inert mass of the engine with the result that the wheel's sensitivity to braking torque changes is reduced considerably. Thus, the wheel's suitability for registering a tendency to lock-up by generating a wheel angle deceleration signal is critically impaired.

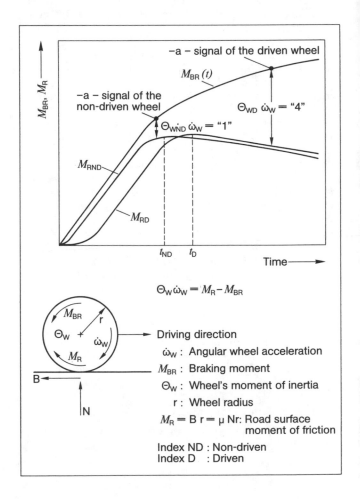

Fig. 8: Generation of wheel angle deceleration signals for driven and non-driven wheels

Figure 8 indicates this problem with reference to an initial braking operation with one non-driven wheel and one wheel coupled to the engine. The top part of the Figure shows the course of the braking torque $M_{BR}(t)$ generated in the brake and the road surface friction moments for the non-driven and driven wheels $M_{RND}$, $M_{RD}$.

The bottom part of the Figure elucidates the designations used. This Figure is based on the assumption that the inertia of the driven wheel is increased by a factor of 4 due to the engine inertia, $\theta_{WD} = 4\,\theta_{WND}$.

If a specific rise of the braking torque $M_{BR}(t)$ in time is chosen, the road surface moment of friction $M_R = \mu \times N \times r$ follows with a short delay in the case of the non-driven wheel, or a large delay in the case of the driven wheel. In accordance with the spin equation $\theta_W\,\dot{\omega}_W = M_R - M_{BR}$, the difference between the $M_{BR}(t)$ curve and the $M_{RND}$ -, $M_{RD}$ curves corresponds to the term $\theta_W\,\dot{\omega}_W$. The road surface friction moments $M_{RND}$, $M_{RD}$ rise in accordance with an assumed $\mu$-slip-curve, reach their maxima at the times $t_{ND}$, $t_D$ and drop again as the time continues to increase.

Based on the assumption $\theta_{WD} = 4\,\theta_{WND}$, the difference between $M_{BR}(t)$ and $M_{WND}$ in the case of the driven wheel must assume a 4-fold value in comparison with the non-driven wheel before the wheel angle deceleration control signal -a is triggered off ($\theta_{WND}\,\dot{\omega}_W$ has the standardized value of "1" while $\theta_{WD}\,\dot{\omega}_W$ has the standardized value of "4").

While the -a signal is generated in the non-driven wheel when slip is optimum, it is not triggered off in the driven wheel until slip is relatively high.

Under a precisely metered brake pressure increase, particularly on slippery road surfaces, wheels of four-wheel-drive vehicles with permanently acting locks may lock-up unless special remedies are found.

| Four-wheel-drive concept | Measures for ensuring ABS functioning |
|---|---|
| 1. Permanent or connectable locks in the longitudinal section and rear axle | – Yawing moment buildup delay<br>– Auxiliary signal $-a_L$ for support of the reference speed and halving the wheel angle deceleration response threshold<br>– Increase in the idling speed (alternative: Engine drag torque control, automatic disengagement) |
| 2. Viscous clutch with freewheel in the longitudinal section, no rear axle lock | – No measures required |
| 3. Controllable/automatically connectable locks | – Disengagement of locking effect during braking |

Fig. 9: ABS upgrade levels for various four-wheel-drive concepts

Figure 9 shows four-wheel-drive concepts which are known up to now and the additional measures required to ensure ABS functioning.

In system 1 with permanently acting locks in the longitudinal section and on the rear axle (viscous locks), the rear wheels are quasi-rigidly coupled and, due to the longitudinal lock, the average speed of the front wheels is equal to that of the rear wheels. As already explained, the result of the permanently acting rear axle lock is that "select-low" mode at the rear wheels is no longer effective and, instead, the maximum possible braking force is utilized on each rear wheel on $\mu$-split road surfaces. The braking force difference on the rear wheels results in a yawing moment on the vehicle which critically impairs its stability. If the maximum possible braking force difference were built up rapidly at the front axle by individual control of the front wheels, it would no longer be possible to maintain driving stability. In this four-wheel-drive concept, yawing moment buildup delay on the front wheels must be introduced to maintain driving stability and steerability on $\mu$-split road surfaces.

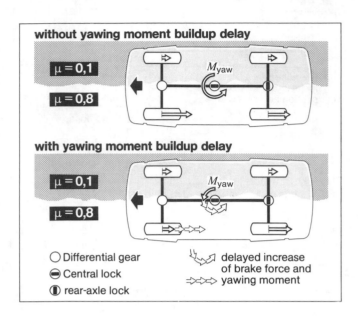

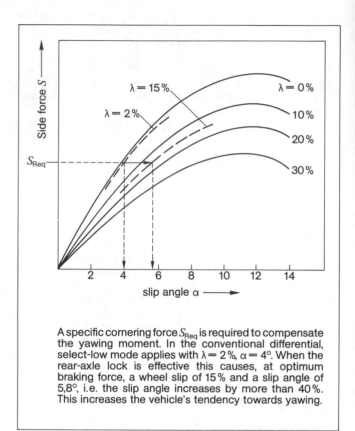

A specific cornering force $S_{Req}$ is required to compensate the yawing moment. In the conventional differential, select-low mode applies with $\lambda = 2\%$, $\alpha = 4°$. When the rear-axle lock is effective this causes, at optimum braking force, a wheel slip of 15% and a slip angle of 5,8°, i.e. the slip angle increases by more than 40%. This increases the vehicle's tendency towards yawing.

Fig. 10: Increased yawing moment caused by
         rear-axle lock on a split coefficient
         roadway

Fig. 11: Increase in the slip angle on rear
         wheel on a high adhesion coefficient
         during the transition from select-low
         ( $\lambda$ = 2 %) to optimum braking force
         ( $\lambda$ = 15 %)

Figure 10 shows the braking forces and yawing moments in a four-wheel-drive vehicle with the rear axle lock effective, both with and without yawing moment buildup delay when braking on a $\mu$-split road surface. Without yawing moment buildup delay, the maximum possible braking forces act on the left-hand wheels after a short pressure buildup time during panic braking with the result that a high yawing moment very rapidly occurs. With yawing moment buildup delay, the braking force on the left-hand front wheel arises slowly. Starting from its initial amount predetermined by the braking force difference on the rear axle, the yawing moment also increases slowly, giving the driver enough time to compensate for the yawing moment by operating the steering wheel.

Nevertheless, greater attentiveness in compensation of the yawing moment by corresponding operation of the steering wheel is necessary when braking this four-wheel-drive vehicle compared with a conventional vehicle such as one with a rear-wheel drive. Due to full utilization of the braking force on the rear wheel braking on the road surface side with the high adhesion coefficient, the cornering force reserve on this wheel is reduced in comparison with vehicles having rear axle select-low. Thus, the slip angle $\alpha$ occurring at the rear axle is increased as shown in Figure 11.

This thus increases the vehicle's tendency to yaw, with the result that the driver has to react faster and more exactly in his steering response.

To maintain ABS functioning on slippery road surfaces, the engine drag torque acting on all wheels in this four-wheel-drive concept must be reduced. The above-described engine drag torque control (MSR) is used for this purpose.

However, in this four-wheel-drive concept, coupling of the wheels to each other and to the inert engine mass necessitates additional remedies as regards signal shaping and the logic of the electronic controller when braking on slippery surfaces. The reduced wheel sensitivity in respect of changes in the road surface friction moment has to be compensated for by refinement of the brake contol on slippery surfaces to prevent wheel lock-up during finely metered braking.

Bosch makes use of a longitudinal acceleration switch in order to recognize slippery road surfaces with adhesion coefficient μ < 0,3. It would also be possible to record the vehicle deceleration dependent upon the friction coefficient by using an analog acceleration sensor, a pressure switch or an analog pressure sensor. The possibility of recording friction coefficient-dependent compression of the axles contrary to the vehicle's direction of motion on the basis of the wheel's speed, and thus of arriving at conclusions as to the road surface's friction coefficient, appears to be capable of realization.

If the vehicle is braked on a slippery road surface, the angular acceleration response threshold is halved and the downward pitch of the reference speed is limited to specific, relatively low values. This registers the wheel's tendency to lock-up more sensitively and at an earlier time. Reduction in the reference speed is better adapted to the maximum possible vehicle deceleration on slippery road surfaces. This measure also ensures ABS functioning when braking on slippery surfaces.

In a four-wheel-drive vehicle, all wheels may spin if the vehicle is accelerated excessively on road surfaces with low coefficients of friction, thus eliminating the usual relationship between the speeds of the wheels and vehicle.

To prevent an increase in the reference speed required for wheel-slip control above the vehicle's speed due to the spinning wheels (which would lead to incorrect wheel-slip signals using usual calculations, thus reducing the braking effect), the positive pitch of the reference speed is limited outside the ABS control in all four-wheel-drive ABS controllers to a value which corresponds to the maximum possible vehicle acceleration on a road surface with a low coefficient of friction. On the other hand, the result of this limiting is that, when braking directly after large-scale vehicle acceleration of long duration, the reference speed is considerably less than the vehicle's. For this reason, the pressure increase is not initiated during the first control cycle via the -a signal und a wheel-slip signal calculated in reference to the reference speed, as is usually the case, but by way of the -a signal and an additional wheel speed difference.

In system 2 as shown in Figure 9 (viscous clutch with freewheel in the longitudinal section, no rear axle lock), all wheels may spin, as in the case of system 1, when the vehicle is accelerated excessively on slippery road surfaces. Therefore, as in the case of system 1, the reference speed pitch must be reduced to small values and the first ABS control cycle must be initiated by a -a signal and an additional wheel speed difference if the reference speed is far less than the vehicle's speed after considerable acceleration of the vehicle of long duration.

Further measures of ensuring ABS functioning are not necessary. During braking the over-running clutch uncouples the rear axle from the front axle with the result that the wheels are mutually uncoupled. In this system also, ABS functioning on very slippery road surfaces can be improved considerably by means of engine drag torque control.

In system 3 (controllable or automatically switchable locks), the reference speed pitch must be limited, as in the case of systems 1 and 2, and the pressure decrease in the first control cycle must be initiated, if necessary, by means of a -a signal and an additional wheel speed difference. Automatic lock disengagement appears appropriate in this system when initiating each braking operation. This thus dispenses with any need for additional measures for ensuring ABS function.

## OUTLOOK

Now that Bosch ABS is being introduced to an
increased extent in middle class vehicles with
conventional single-axle drives, and also in
four-wheel-drive vehicles, it has become
necessary to adapt the ABS to their driving
and braking characteristics. Thanks to a
"controlled" yawing moment buildup delay, it
has now become possible to master the generally
increased tendency of middle class vehicles
to yaw in comparison with higher-class vehicles.
In future, modern microelectronics in
conjunction with favorably priced sensors will
permit a further optimization of driving
dynamics in conjunction with ABS, and a
transition to genuine control systems.

It has now been possible to make the ABS fully
suitable of use in four-wheel-drive vehicles
also thanks to special refinements of the
ABS control algorithm, the yawing moment buildup delay and engine drag torque control.
Further optimization particularly in the case
of four-wheel-drive vehicles with permanently
acting longitudinal and transverse locks, are
possible and desirable.

## REFERENCES

(1) Heinz Leiber and Armin Czinczel:
    Antiskid System for Passenger Cars with
    a Digital Electronic Control Unit,
    SAE 790458, 1979

(2) Heinz Leiber and Armin Czinczel:
    Four Years of Experience with 4-Wheel
    Antiskid Brake Systems (ABS)
    SAE 830481, 1983

# Traction Control System with Teves ABS Mark II*

**Hans-W. Bleckmann, Helmut Fennel, Johannes Gräber, and Wolfram W. Seibert**

Ate/Alfred Teves GmbH
Frankfurt, West Germany

*Paper 860506 presented at the International Congress and Exposition, Detroit, Michigan, February, 1986.

## ABSTRACT

After having increased the safety of braking by means of production line installed Antilock Brake Systems (ABS), the next problem to be solved was to optimize the acceleration procedure. In this context, standard ABS—components should be used extensively to improve the forward traction especially on split—friction surfaces and to increase the safety during acceleration in general.

This paper shows the possibilities to fulfil the demands made on an intelligent Traction Control System (TCS) with the aid of an already existing, fully integrated ABS, which has a modular built-up in hydraulics and electronic hard— and software.

SIMILAR TO THE PROBLEMS caused by locked wheels in braking, the spinning power wheels lead to highly reduced steerability with front—wheel drive and respectively, a loss of stability with rear—wheel drive. In some situations (e. g. overtaking processes) an optimal acceleration, especially on low—$\mu$ surfaces can be an important factor for safety. Beyond this, a spinning wheel on asymmetrical road conditions can prevent any forward movement at all.

The purpose of the paper is to define the main functions of an intelligent move-off and acceleration control and to describe a solution concept based on the TEVES ABS Mark II.

To clearly illustrate the relatively complex relationship between Antilock and Traction Control this paper covers only Traction Control for front— and rear—wheel driven cars. The presentation of the very interesting application "Traction Control for all—wheel drive" would be beyond the scope of the paper.

## MAIN FUNCTIONAL REQUIREMENTS

DIFFERENTIAL GEAR LOCKING — On nearly all motor vehicles the drive for both wheels of an axle is through a differential gear. This transmission, conceived for cornering only, is not necessary for straight line driving and can in some cases become a disadvantage.

Therefore, the first range of duties for an intelligent Traction Control is given: The differential must be controllable with variable locking ratio. Thus the speed difference of spinning wheels must be reduced to the value, defined by the vehicle geometrie during cornering. This task is particularly important for traction on split-friction slopes. Fig. 1 shows the maximum climbing ability of front- and rear-wheel driven cars with and without Traction Control for various split—$\mu$ conditions. As a comparison, the achievable gradient for an uncontrolled and unlocked all—wheel drive is delineated in addition.

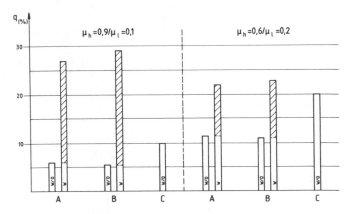

Fig. 1: Climbing ability of vehicles with and without Traction Control
A : front—wheel driven car
B : rear—wheel driven car
C : all—wheel driven car

The same problem appears with high acceleration on split-μ surfaces. In this case the term "maximum gradient q [%]" in Fig. 1 can be replaced by "maximum acceleration in % of 'g'".

HANDLING OF EXCESSIVE ENGINE TORQUE – Beyond this, a further important task of Traction Control is to maintain steerability, or stability respectively, with optimal traction. Therefore a sophisticated control also must be able to handle excessive engine torque preset by the driver.

ACCELERATING IN CORNERS – The third and decisive task for driving safety is to keep the car steerable and stable while accelerating in corners.

In Fig. 2, two different traction and cornering force characteristics for the left and right wheel (left high-μ, right low-μ) are plotted. The values of the maximum forces shown on the right side of the diagram are results of indiviual wheel control for traction optimum. In comparison, the values for the uncontrolled case show that a loss of cornering forces stands opposite to the distinct benefit of tractive power.

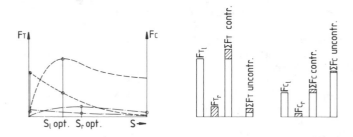

Fig. 2: Tractive and cornering forces resulting from individual wheel control for traction optimum

This must be compared with Traction Control according to the select-low principle shown in Fig. 3. A significant increase of cornering forces is evident in comparison to Fig. 2. Despite a reduction, the remaining tractive power is still larger than in the uncontrolled mode.

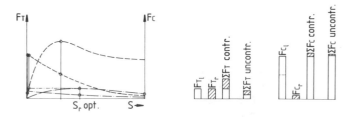

Fig. 3: Tractive and cornering forces resulting from select-low Traction Control

For Antilock Control, optimal deceleration is a general requirement. Contrary to this, with Traction Control under special conditions such as accelerating in corners, maximum tractive power can be compromised to increase cornering forces.

Thus intelligent Traction Control strategies must be applied to decide between these two control situations.

COMPARISON OF DIFFERENT TRACTION OPTIMIZING SYSTEMS

According to the above-mentioned main functions, an intelligent Traction Control System has to meet the following requirements:
- Variable and adaptive differential locking from 0% to 100%
- Effectiveness with transmitted engine torque less or equal to sum of transmittable road torques
- Effectiveness with transmitted engine torque higher than sum of transmittable road torques
- Short time lag in reducing excessive engine torque
- Ability to decide between optimal traction and increased cornering force

To optimize move-off and acceleration maneuvers, various systems were developed in the past, which did not fulfil all essential requirements completely. In the matrix of Table 1, the most important developements in this field are shown and compared with advanced Traction Control Systems.

Table 1: Requirement matrix of different traction optimizing systems

| Requirements / Systems | Variable and adaptive locking ratio | Effectiveness with $T_{eng} \leq \Sigma T_{road}$ | Effectiveness with $T_{eng} > \Sigma T_{road}$ | Short delay in reducing excessive engine torque | Ability to decide between optimal traction and increased cornering force |
|---|---|---|---|---|---|
| Mechanical differential lock | | YES, but unfavourable in cornering because of fixed locking ratio | | | |
| Visco-locking differential | WITH RESTRICTIONS: - locking ratio 0% is not available - time dependent locking ratio characteristic is not predictable | | | | |
| Simple engine control systems | | YES, but no optimal traction because of select-low control | YES, but no optimal traction because of select-low control | - SHORT delay with intervention in ignition or injection | |
| Simple engine control systems combined with a visco-locking differential | WITH RESTRICTIONS: see above | | YES | - LONG delay with actuating the throttle-flap | |
| Traction control only controlling wheel brakes | YES | YES | YES, but brake temperature problems in long-time controlling | SHORT delay by reducing the excess torque with the brakes | YES |
| Combined traction control with brake modulation and engine control | YES | YES | YES | SEE ABOVE and long-time reduction of excess torque by actuating throttle-flap | YES |

372

This matrix makes evident that only a combined brake and engine controlling system is able to fulfil all requirements in a suitable way. The advantage of this system is, that it can work with a slowly responding engine control compensated by the fast responding brake modulation. Therefore, no intervention in ignition or electronic injection, with the well-known problems involving the exhaust gas composition, is necessary. The engine torque is simply controlled by a superimposed throttle-flap actuation.

Since the TEVES ABS Mk II is the first system equipped with an auxiliary energy source and being extendable because of its modular design in hard- and software, the TEVES Traction Control System has the advantage to use existing standard modules enhanced by a few additional components.

## THEORY

DIFFERENTIAL GEAR LOCKING — A normal unlocked differential gear balances the wheel torques:

$$T_{W1} = T_{W2} \qquad (1)$$

(with $T_{W(1,2)}$ = wheel torques of the driven wheels 1 and 2)

However, on asymmetrical friction surfaces unequal torques for optimal traction are needed. The Traction Control System (TCS) locks partially the differential by an external brake torque. This additional locking torque should be proportional to the friction coefficient difference of the driven wheels (Fig. 4):

$$T_{W1} + T_{B1} = T_{W2} \qquad (2)$$

(with $T_{B1}$ = necessary brake torque of wheel 1 running on low friction coefficient)

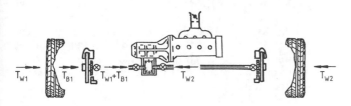

Fig. 4: Additional brake torque used as locking torque

The TCS has to determine the correct locking torque value by automatic feedback control of wheel slip.

BALANCE BETWEEN DRIVE TORQUE AND WHEEL TORQUE — During steady-state driving the transmitted engine torque is balanced to the torque sum of the driven wheels:

$$T_{W1} + T_{W2} = T_E \cdot C_D \qquad (3)$$

(with $C_D = \eta_D \cdot i_D$
$\quad \eta_D$ = efficiency of drive train
$\quad i_D$ = gear ratio between engine and driven wheels
$\quad T_E$ = engine torque for steady-state driving)

With increased engine torque, and no automatic brake intervention, the engine torque must be reduced by the driver or by automatic engine intervention:

$$T_{W1} + T_{W2} = (T_{Ee} - \Delta T_E) \cdot C_D \qquad (4)$$

(with $T_{Ee}$ = excessive engine torque
$\quad \Delta T_E$ = engine torque reduction)

With partial differential lock by brake intervention, the engine torque reduction $\Delta T_E'$ is smaller due to the additional brake torque $T_{B1}$:

$$T_{W1} + T_{B1} + T_{W2} = (T_{Ee} - \Delta T_E') \cdot C_D \qquad (5)$$

A combined TCS with engine and brake intervention compensates the wheel torque difference by one-side braking and achieves the engine-to-wheel torque balance by engine torque reduction. In case of fast dynamic load changes, the response of brake and engine intervention must also be fast. This is not realistic considering the time lag typical for the engine control response.

Fast response is possible if the Traction Control compensates the excessive engine torque by adding a torque on both brakes of the driven axle instead of reducing the engine torque:

$$T_{W1} + T_{B1} + T_{BE} + T_{W2} + T_{BE} = T_{Ee} \cdot C_D \qquad (6)$$

$$\text{(with } T_{BE} = 1/2 \cdot C_D \cdot \Delta T_E')$$
or

$$T_{W1} + T_{B1}' + T_{W2} + T_{B2}' = T_{Ee} \cdot C_D \qquad (7)$$

(with $T_{B1}' = T_{B1} + T_{BE}$ and $T_{B2}' = T_{BE}$)

Not considering increased brake power dissipation, this equation shows that a TCS in principle can work without engine intervention.

POWER DISSIPATION IN THE BRAKES — Theoretical calculations for an average European front-wheel driven car show, that the power dissipated in the brakes during Traction Control without engine intervention is comparable to the power dissipated during normal braking.

However, contrary to the braking process which ends at a complete stop, the duration of Traction Control can be unlimited in special situations. On an inclined road with low friction surface it is possible to maintain Traction Control by fully pressing the accelerator pedal without moving of the car.

To avoid brake overheating in such situations a combined Traction Control with brake and engine intervention is necessary.

This combined control becomes effective in the steady-state case as per Eq. (5).

TOTAL CONCEPT - Due to this analysis the total concept of the TEVES Traction Control and Antilock Brake System is defined as follows (Fig. 5):
— One-side brake intervention to achieve the external locking function of differential gear
— Additional both-side brake intervention to achieve the balance between transmitted engine torque and drive-wheel torques
— Engine torque reduction via throttle-flap intervention
— Four wheel speed sensors

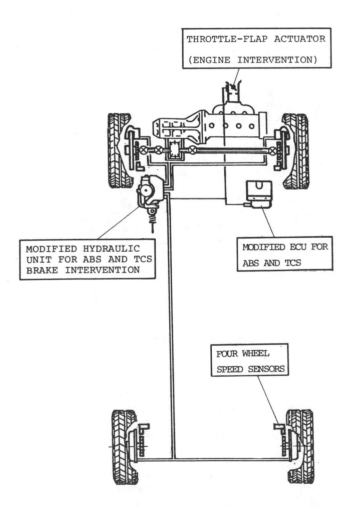

Fig. 5: Total concept of a TCS

HYDRAULIC CONCEPT

FUNDAMENTALS — To achieve a brake intervention for Traction Control it is necessary to design a hydraulic brake actuation unit, that, in Traction Control mode delivers pressure to the brakes without actuating the brake pedal. The pressure must be individually modulated by the controller for each drive wheel.

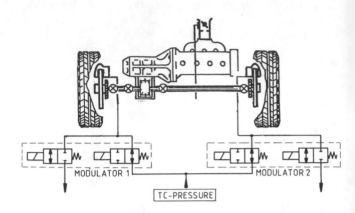

Fig. 6: Principle of Traction Control brake actuation

To keep expenses low the integration of the Traction Control brake actuation into the ABS hydraulic unit is necessary. Beyond this, for the benefit of driving safety, a TCS should only be installed together with an ABS.

For the concept of the TCS hydraulic unit the principle of the antilock pressure modulation is very important. Two types of pressure modulation are known in mass produced Antilock Brake Systems:
— The power recharge principle (Fig. 7) in which the fluid volume of the controlled brake circuit is pumped back into the brake actuation unit for pressure decrease. To increase pressure again, volume is taken from the actuation unit.
— The flow-in principle (Fig. 8) in which the brake fluid of the controlled brake circuit flows back to the reservoir caused by the pressure difference. The lost volume is refilled by an energy supply unit with pedal force proportional pressure via a flow-in valve (dynamic flow-in).

The TEVES ABS Mk II works according to the flow-in principle with an energy supply unit including pump and accumulator. The flow-in valve is situated between hydraulic booster and master cylinder. During pressure modulation at the front wheels any fluid loss through the outlet valve is refilled by a dynamic flow from the booster circuit to the wheel brakes via the switched flow-in valve and the master cylinder piston seals (Fig. 8).

Contrary to the front wheels, the rear wheel brakes are powered directly by the dynamic booster circuit.

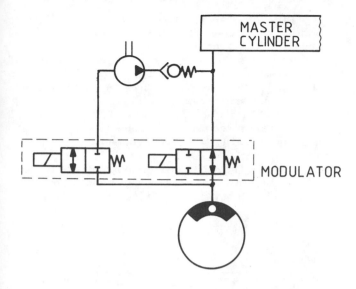

Fig. 7: Basic draft of the power recharge principle

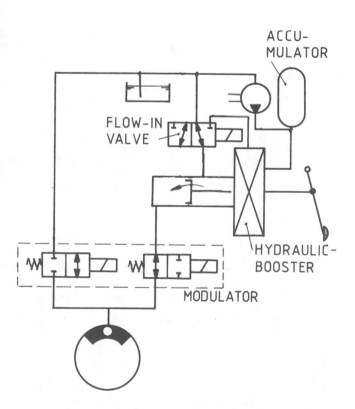

Fig. 8: Basic draft of the flow-in principle

HYDRAULIC UNIT FOR TRACTION CONTROL — Both the existing energy supply unit and the wheel pressure modulation valves of the ABS Mk II are used for controlled brake pressure supply during Traction Control in the integrated ABS and TCS.

By a specially designed flow-in valve principle, the accumulator is connected to the pressure modulation valves during Traction Control function.

Thus a configuration according to the principle shown in Fig. 6 is achieved by using most common parts.

In the ABS Mk II, the rear axle pressure is controlled by one pair of modulation valves. To achieve individual wheel pressure modulation during rear axle Traction Control it is necessary to integrate a further pair of modulation valves to the rear axle. In Antilock Control mode, the rear axle valves are actuated jointly by the electronic controller (according to the select-low principle) whereas in Traction Control mode they are actuated individually.

ADVANTAGES OF A MODULAR HYDRAULIC UNIT — Due to the modular concept of TEVES ABS Mk II it is very easy to extend the hydraulic unit for Traction Control.

To attain a combined unit it is only necessary to exchange the ABS pressure modulator completely with an integrated version, containing the normal modulation valves and the Traction Control components (Fig. 9).

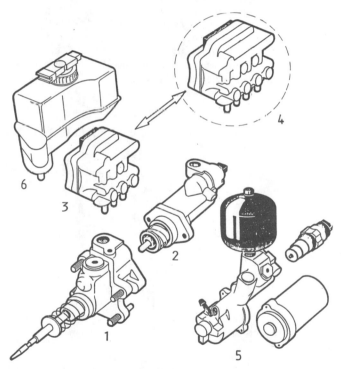

Fig. 9: Modular designed hydraulic unit
1: Hydraulic booster
2: ABS and TCS master cylinder
3: ABS valve block
4: ABS and TCS valve block
5: Energy supply unit
6: Reservoir

HYDRAULIC SAFETY CONCEPT — The ABS Mk II energy supply and fluid level monitoring is also used for the combined ABS and TCS unit.

Due to the hydraulic Traction Control concept, any mechanical failure of the additional components leads to a warning by the energy supply monitoring.

This warning signal forces the system into a safe state. According to this and the electronic failsafe concept, no additional failure monitoring is necessary for the combined ABS and TCS hydraulic unit.

## ENGINE CONTROL INTERVENTION

In the Traction Control concept described above the excessive engine torque can be compensated immediately by means of the wheel brakes. Therefore the engine torque need not be reduced by intervention in ignition or electronic injection but can be modulated by superimposed actuation of the throttle-flap.

For safety reasons two very important requirements are fulfilled by the throttle-flap actuator:

1. The functional safety of the normal, mechanical throttle control is maintained.
2. The actuator is designed in a way, that the throttle-flap angle can be reduced by the driver in any case, even in Traction Control mode.

Since the throttle-flap actuator is designed as a pure add-on system, it is located within the Bowden cable connection or the accelerator linkage, without causing any modification to the throttle housing. Especially important is, that the coherence between pedal travel and throttle-flap angle is not changed for normal actuation.

## MODULAR HARD- AND SOFTWARE CONCEPT OF THE ELECTRONIC CONTROLLER

During its development the TC-controller ran through several stages of evolution. This evolution is shown in Fig. 10 to Fig. 13.

Fig. 10 shows two separate electronic controllers, one for the Antilock, and the other for the Traction Control System.

In this version a double amount of electro-mechanic, electric, and electronic components related to controller hardware and connections are necessary.

With the step from Fig. 10 to Fig. 11 many electro-mechanic components are saved. Only one housing with one connector is necessary.

Inside this controller box only one voltage regulator and half the number of components for input signal filtering are needed.

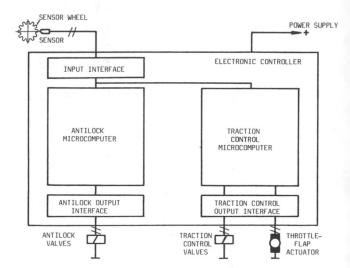

Fig. 11: One electronic controller with two microcomputers for Antilock and Traction Control

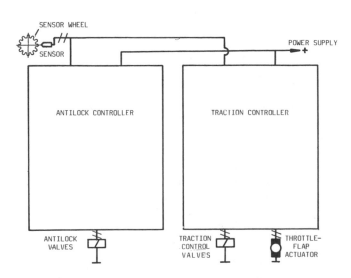

Fig. 10: Two separate electronic controllers for Antilock and Traction control

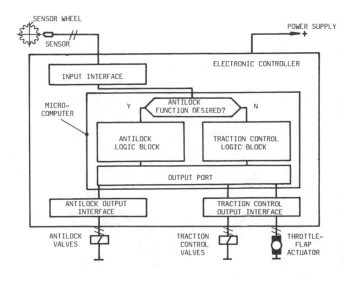

Fig. 12: One electronic controller with one microcomputer and two separate logic blocks for Antilock and Traction Control

376

The amount of large scale integrated circuitry is still high because for either task, Antilock and Traction Control, one microcomputer is used.

Fig. 12 shows a design which reduces this high amount of electronics by using only one microcomputer for both tasks.

The CPU (Central Processing Unit) of the microcomputer can process one of the two implemented software tasks at a time. Both of the independently executable software blocks reside in the ROM section of the microcomputer. By using only one microcomputer instead of two, the interfacing is much easier than the design shown in Fig. 11. So the number of peripheral parts can be reduced further. Only the amount of memory is still high in this architecture.

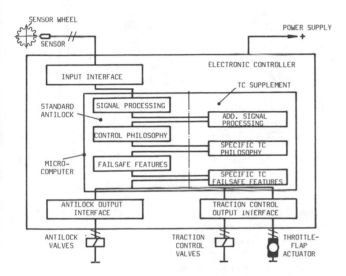

Fig. 13: One electronic controller with one microcomputer and one combined Antilock and Traction Control logic block

The next step shown in Fig. 13 represents the most advanced approach: the memory size is drastically reduced by using the same software structures for both tasks whenever possible.

Since there are many similarities in both applications, which base on identical sensor inputs, many software modules like signal processing, pressure modulation and most of the failsafe features are commonly used for both tasks.

The additional software modules to be developed are related to the automatic throttle-flap actuation and special Traction Control and failsafe features.

As a conclusion, a modular hard- and software concept helps to minimize the controller size and the electromechanic, electric, and electronic part count for cost and reliability reasons and for reduced electromagnetic interference.

## BASIC TASKS OF TRACTION CONTROL PHILOSOPHY

At a first glance the mechanisms for Antilock and Traction Control are very similar. In both cases the physical laws of slip and friction which decelerate or accelerate a wheel are equivalent. The sum of torques influencing the wheel are generated by the same aggregates like tire, engine, transmission, differential gear, and wheel brake.

Strongly simplified, only the signs for friction, slip, and pressure modulation gradient have to be inverted to get from Antilock to Traction Control and vice versa.

In fact, at the beginning of Traction Control development the first steps were made with such an "inverted" Antilock controller structure. This type of control immediately brought reasonable results.

In practice, however, contrary to many similarities, there are influences producing several differing effects in both applications. These effects have their reason in the variing importance that specific aggregates and their

Table 2: Differing problems for Antilock and Traction Control

| Antilock System | Traction Control System |
|---|---|
| - Both axles of the car are controlled | Only the driven axle is controlled |
| - The control of the driven wheels is done normally at idle operation of the engine or in declutched drive mode | The control is always done in the clutched drive mode at high engine inertia |
| - The problems of oscillations caused by the power train are only of secondary importance | The oscillations caused by the power train are always present |
| - The influence from one wheel to the other via differential gear is only of secondary importance | The influence from the differential gear is always present (in combination with the high inertia of the engine) |
| - Only one control loop (brake control) with nearly constant response time of the system | Two control loops with different response times (factor 10): one brake control loop and one engine control loop; multi loop control system |

Table 3: Similar tasks for Antilock and Traction Control

| |
|---|
| - Optimal utilization of the friction between tire and road |
| - Steerability to be achieved by precise control of the front axle (generally valid for Antilock function and especially for Traction Control on front-wheel driven cars) |
| - Stability to be achieved by precise control of the rear axle (generally valid for Antilock function and especially for Traction Control on rear-wheel driven cars) |
| - Fast control adaptation at different friction coefficients |
| - Minimum control deviations to avoid oscillations and to minimize the energy consumption |
| - Powerful suppresion of mechanical disturbances which can influence the control behaviour (caused by road, suspension, and power train) |

interactions assume in a multi loop control system which changes its structure depending on acceleration or braking.

In Table 2 the differing problems are explained.

Despite the distinctive problems described in Table 2, similar tasks listed in Table 3 have to be fulfilled for Antilock and Traction Control.

Comparing the differing problems with the similar tasks it is evident that dedicated strategies for optimal performance of Traction Control are necessary.

A powerful software package residing in the ROM of the mask programmed single-chip microcomputer is able to handle all the necessary tasks with high accuracy in a very short time.

The software structure is divided in several partitions which
— calculate the wheel speeds and their derivatives
— calculate the reference speed at optimum friction
— select the control strategies with their highly sophisticated pressure modulation algorithms
— determine the throttle-flap actuation
— and perform the failsafe philosophy.

## FAILSAFE CONCEPT

As mentioned above, for the presented Antilock / Traction Control System nearly all functional blocks like signal processing, control philosophy, and failsafe system are realized by software. On account of the modular software structure the already existing and proven failsafe methods of the Antilock System have been fully transferred to the Traction Control application:
— the redundancy concept
— the plausibilty criteria
— the test stimuli
In addition some new features had to be worked out for the Traction Control application and included in the software package.

One powerful and convenient tool to recognize defects in the electronic system is the redundancy principle. Two microcomputers are calculating their data independently, transferring the data to the other one and comparing their own calculated data with the received data from the other microcomputer (e. g. the acceleration and deceleration signals, the reference speed).

The components of the system, which are not existing redundantly like sensors and tooth wheels, are monitored by plausibility criteria.

For monitoring peripheral elements like the solenoid valves, their connections, and their driving circuits, special test procedures are provided which are part of the complete system, working all the time the controller is switched on. For this task the microcomputers are capable to transmit stimuli to the com-

ponents to be monitored, and to receive the stimulated reaction.

This method is very smart if included in the redundancy concept. In Fig. 14 one example is shown for monitoring the actuators. For simplification, only the essential parts and connections are shown.

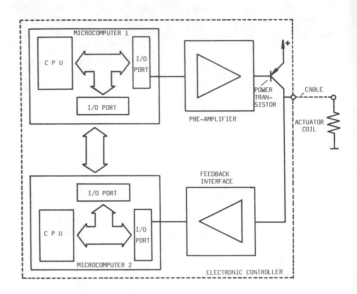

Fig. 14: Signal loop for monitoring the actuators

Assume that microcomputer 1 sends test pulses which are so short, that the actuator cannot react mechanically.

Microcomputer 2 receives the stimulated answer fed back from the actuator coil and compares this answer with its own calculated value. Upon short circuit or disconnection at any part of the chain, microprocessor 2 detects this failure and forces the controller into a safe state.

With this simple method nearly all electric or electronic parts in the chain can be monitored:
— the CPU of microcomputer 1
— the connection between CPU and I/O port
— the I/O port function
— the connection between the I/O port and the pre-amplifier
— the connection between the pre-amplifier and the driving power transistor
— the power tranistor
— the connector on the electronic controller
— the connecting wire
— the actuator coil and its connections
— the feedback connection
— the feedback interface
— the I/O port of microcomputer 2
— the internal connection to the CPU
— the CPU of microcomputer 2
In addition to these three failsafe methods (redundancy, plausibility, stimuli) it is necessary for the Traction Control application

to apply additional methods to prevent the engine control actuator and the wheel brakes from damage or early abrasion.

A correct solution of this problem would require additional sensors for brake temperature and mechanical stops of the throttle actuator which, however, should be omitted for cost and reliability reasons.

Thus, instead of using actually measured data the controller uses simulated data derived from software models of the throttle actuator and the brake temperature.

The realization of these sophisticated control strategies at reasonable cost became possible only with widespread use of microcomputers.


SUMMARY

Based on general requirements of Traction Control various technical approaches are discussed resulting in a system which uses brake and engine control intervention to achieve optimal traction under all driving conditions.

The unique concept of the TEVES ABS Mark II hydraulics covers the straight-forward and cost effective Traction Control extension.

To cope with the additional Traction Control task the electronic controller utilizes optimal partitioning of jointly shared functional blocks. The safety aspect in Traction Control has been supplemented with the proven failsafe features in ABS.

This combination of highly advanced brake actuation technology together with effective use of microcomputers leads to a well balanced system being prepared for mass production.

BIBLIOGRAPHY

1    H.-W. Bleckmann, J. Burgdorf, H.-E. von Grünberg, K. Timtner, L. Weise, "The first compact 4-wheel Anti-Skid System with integral hydraulic booster". SAE technical paper series 830483,Feb./March 1983

2    A. Czinczel, H. Leiber, "The potential and the problems involved in integrated Anti-Lock braking systems". Paper C192/85 read at the conference on Anti-Lock Braking Systems at the Institution of Mechanical Engineers IMechE, London on 11-12 Sept. 1985

3    H.-W. Bleckmann, L. Weise,"The new four-wheel Anti-Lock generation - a compact Anti-Lock and booster aggregate and an avdvanced electronic safety concept". Paper read at the conference on Anti-Lock Braking Systems at the Institution of Mechanical Engineers IMechE, London on 11-12 Sept. 1985

4    J. Gerstenmeier, R. Emig, "ABS electronics, current status and future prospects". Paper C239/85 read at the conference on automotive electronics of the Institution of Mechanical Engineers IMechE, Nov. 1985

# REFERENCES

The compilers wish to thank the SAE staff for their support in gathering information on SAE Papers on the broad subject of ABS. We sincerely hope we have not omitted any deserving authors. Likewise, we acknowledge the tremendous volume of papers on this subject which have been published by sister organizations, not least of which are the Institution of Mechanical Engineers and the equivalent bodies in the other major vehicle producing nations.

For further reference we provide below a list of SAE Papers which are **not** included in this publication, covering ABS applications on other vehicle categories. The papers are listed by number and date order (Prefix C ' I. Mech E. Papers)

## Commercial Vehicle

**R. H. Madison and Hugh E. Riordan, Evolution of Sure-Track Brake System, SAE Paper 690213, 1969.**

The history, system philosophy, design evolution, and performance of the Sure-Track anti-lock automotive braking system are presented and discussed. Considerations of performance, driver skill, reliability, and commercial acceptance resulted in the choice of a vacuum-electronic rear wheel anti-lock system that incorporates individual wheel speed sensing and control of braking as a pair.

The system provides superior directional stability under "panic" braking conditions while maintaining stopping distance equal to or shorter than those for locked wheels under most road conditions.

**John L. Harned, L. E. Johnston, and Glen Scharpf, Measurement of Tire Brake Force Characteristics as Related to Wheel Slip (antilock) Control System Design, SAE Paper 690214, 1969.**

Tire brake force characteristic data are presented that should be helpful in the design of wheel slip control systems. Correlation of these data has been established with antilock system performance. Experimentally measured mu-slip curves are given for a large number of tire/road pairings. These measurements cover a wide range of commercial tire types on dry and wet road surfaces and glare ice. It is shown how wet road characteristics are affected by road construction, water cover depth, and tread wear. The measuring system used to obtain these data is described and variability of the experimental measurements is discussed.

**Edwin E. Stewart and Lauren L. Bowler, Road Testing of Wheel Slip Control Systems in the Laboratory, SAE Paper 690215, 1969.**

The use of a laboratory simulator to evaluate the performance of wheel slip control systems under controlled operating conditions is reported. It is shown how an analog computer can be interconnected with a hydraulic brake system and wheel slip control hardware to form a hybrid simulation of a vehicle installation. An analog computer can also be used to simulate vehicle dynamics and tire-to-road friction characteristics. Simulator accuracy is established by correlating laboratory results with road data. Advanatges and disadvantages of using the simulator in lieu of experimental road testing are pointed out.

**Bruce E. Latvala and R. J. Morse, Adaptive Anti-Lock Braking - A Reality for Air Braked Vehicles, SAE Paper 700112, 1970.**

A braking control system for air braked vehicles has been developed which prevents wheel lock under any braking condition and thereby improves vehicle stability and reduces stopping distance.

The theory of operation and the specifications of the system are discussed and performance under several typical conditions is shown.

**Gordon W. Yarber and Franklin B. Airheart, Controlled Rotation of Braked Wheels, SAE Paper 700113, 1970.**

Automatic skid control provides the rotation control needed to stop a wheeled vehicle safely through a control system which senses wheel speed, has control logic, and controls braking torque. This paper describes how these essential control functions are used on the vehicle.

Skid or rotation control can be provided for all braked wheels, either on an individual or paired wheel control basis, if maximum safety of vehicle operation is the primary consideration. If the major objective is to prevent complete loss of vehicle control, rotation control on the rear wheels only, either individual or paired, will be satisfactory, realizing, however, that steering control is lost when the front wheels lock up. To determine the final configuration, the performance as related to vehicle stability, directional control, stopping distance, tire damage, and relative cost must be evaluated. Depending on the requirements, cost can vary almost 300%; therefore, a compromise in the specified requirements may be necessary.

**R.C. Bueler and E. J. Falk, A Practical Approach to the Selection and Sizing of Brakes to Meet FMVSS-121, SAE Paper 730198, 1973.**

A rational, practical method has been developed to size and select brakes which will meet the requirements of FMVSS-121 for a wide variety of truck models and sizes, with a minimum number of brakes. This paper presents the technique of utilizing Newtonian mechanics to categorize vehicle sizes and allow selection of common brake packages for each category. Actual sizing and selection was based on dynamometer test data.

**Clarke F. Thornton, Operation Redesign: Axles and Brakes for MVSS 121, SAE Paper 730697, 1973.**

The purpose of this paper is to discuss the detail known to affect the design of axles, brakes, wheels, and related equipment that will be used on air-braked vehicles under Federal Motor Vehicle Safety Standard (FMVSS) 121. In specific, this paper shows the change that is expected to occur on redesigned axle and braking equipment compatible with the higher levels of vehicle deceleration and controllability to satisfy the

Standard. Variables affecting vehicle brake performance and design and application problems related to MVSS 121 qualification are presented.

**Paul A. Turvill, Comparative Analysis of Air Brake Antiskid Systems - A Truck Builder's View, SAE Paper 730698, 1973.**

Numerous air brake antiskid systems, each representing a unique approach to achieving a common goal, are being promoted to fill the market created by the performance criteria contained in Federal Motor Vehicle Safety Standard (MVSS) 121. Nine of these systems were examined in order to determine the range of available features and performance, and to discover what new factors will have to be taken into account by vehicle manufacturers to accommodate antiskid systems as standard equipment.

The early effective date of MVSS 121 has forced introduction of antiskid systems with only minimal field exposure. The next generation of systems designs is already being considered, and is expected to incorporate refinements based on current experiences of skid control suppliers, vehicle manufacturers, and users.

**G. W. Stearns, Intermixing of Tractors and Trailers Equipped with Existing and FMVSS 121 Braking Systems, SAE Paper 730699, 1973.**

There are a number of factors affecting the braking stability on tractor-semitrailer combination vehicles when intermixing new and existing braking systems. The changes in the stability of combination vehicles caused by the differences in brake performance levels, brake torque utilization, and brake system response of new and existing systems will be evaluated. The new and the existing systems will be pictorially compared and discussed to clarify the features and performance parameters required on the new equipment which must work in conjunction with pre-FMVSS 121 brake systems.

**James M. Lewis, A Fleet Operator's Comments on FMVSS 121 Braking System Compatibility, SAE Paper 740049, 1974.**

Federal Motor Vehicle Safety Standard (FMVSS) 121 has introduced many new and varied problems for vehicle manufacturers, component manufacturers, and vehicle operators. An area of great concern to vehicle operators, particularly fleets, is the question of compatibility associated with antilock system intermix, old and new vehicle intermix, and control standardization. This paper discusses the compatibility related problems observed during limited fleet tests and evaluation of FMVSS 121 braking systems. Test data, observations, and possible solutions are presented with emphasis on the need for more extensive investigation in this area to ensure that the goal of FMVSS 121 - increased truck safety - is achieved.

**T.D. Gillespie, Front Brake Interactions with Heavy Vehicle Steering and Handling During Braking, SAE Paper 760025, 1976.**

The increased braking performance required on air-brake equipped commercial vehicles by the Federal Motor Vehicle Safety Standard 121 results in vehicles with higher front brake torque capacity and greater deceleration capability. Using a simple analytical model, certain mechanisms by which handling during braking is influenced by tire characteristics, load transfer during braking, steering system characteristics, brake imbalance, and other factors are demonstrated. In addition, analysis of the steering system shows how steer angle deviations arise from braking and lateral forces acting against compliance of the steering linkage, and the influence of caster geometry on these deviations.

To investigate certain quantifiable characteristics of handling performance, the HSRI Directional Response Computer Program for predicting the longitudinal and directional response behavior of trucks was modified to include the effects of a compliant steering system subject to the force and moment inputs of the front tires. Measurement of bias ply truck tire force and moment characteristics for use in the computer simulation revealed that tire aligning torque characteristics reverse in direction at high braking levels and may dominate the effect of geometric caster built into the steering system.

Studies utilizing the modified program indicate that (a) no vehicle will stop perfectly straight without driver steering corrections because of steer angle deviations and (b) the steering reactions fed back to the steering wheel during braking may reverse direction with anti-lock brake cycling, largely because of the reversal of tire aligning torques. A relationship between these steering reactions and front brake torque level is shown.

**Maurice H. Cardon, George B. Hickner, and Ralph W. Rothfusz, Development and Evaluation of Anti-Lock Brake Systems, SAE Paper 760348, 1976.**

Anti-lock systems which effectively prevent wheel lock have been developed for passenger cars, trucks, articulated vehicles and buses. Six anti-lock system configurations involving individual wheel and axle control are discussed. Also discussed are techniques for evaluating the performance anti-systems; included are straight line braking, the use of a split coefficient surface and braking in a turn. The results of computer simulation studies and vehicle tests conducted to evaluate the performance of the various anti-lock system configurations are presented. It is concluded that the best anti-lock system configuration for a particular vehicle requires a trade off among vehicle design characteristics, desired level of braking, and vehicle handling performance and cost.

**J. P. Koenig and R. D. Kreider, Air Brake System Trends For The 80's, SAE Paper 760587, 1976.**

The air brake system of the 1980's will evolve from the present design because of forces acting on the vehicle manufacturer through the government, the trucking industry and the component industry. Changes will occur to improve safety/performance, maintenance, vehicle efficiency, reduce noise and maintain vehicle compatibility. All of these factors will dictate the cost effective designs of the 1980's.

**John W. Kourik, An Equipment Supplier's View on Regulations, SAE Paper 760593, 1976.**

Equipment suppliers have both generated and accommodated to evolutionary changes as regulations evolved with advances in the state-of-the art.

Federal Motor Vehicle Safety Standards have revolutionized the equipment suppliers' interest in Regulations. The scope and activity of groups within the NHTSA is reviewed to show the impact of the performance, design and quality assurance aspects of such Standards at the equipment supplier level. Compliance and defect responsibilities are discussed. Specific examples have been selected from experience with FMVSS 121-Air Brake Systems.

### Joseph J. Winderlin, Thin-Lift Recycling of Asphalt Pavement, SAE Paper 760662, 1976.

Five basic steps of efficient and economical continuous thin-lift asphalt pavement recycling are listed and the design features of the Cutler Repaver are described. The huge, mobile, self-propelled asphalt paving plant utilizes the five steps with integrated components as follows:

1. Softening the old road surface with radiant heat produced by ceramic emitters.
2. Loosening a thin layer of old material with air-bag mounted scarifiers.
3. Mixing new and old material with liquid asphalt in a pug mill.
4. Laying and compacting the new road surface with a vibratory screen.
5. Completing the softening, loosening, mixing, laying and compacting operations in one continuous operation.

### C. P. Lam, R. R. Guntur, and J. Y. Wong, Evaluation of the Braking Performance of a Tractor-Semitrailer Equipped with Two Different Types of Anti-Lock Systems, SAE Paper 791046, 1979.

In this paper, a digital computer model for studying the braking performance of an articulated vehicle equipped with anti-lock devices is presented. Using this computer model, the braking characteristics of a tractor-semitrailer fitted with a commercially available system (System A) is compared with that of the same vehicle equipped with a proposed system (system B). The deficiencies of system A are identified. The merits and disadvantages of system B are also examined. Based on the results of the simulation study, guiding principles for the development of the control logic of anti-lock brake systems are suggested.

### C. D. Cernes, The Design of a Range of Buses for Economic Operation and Maintenance, SAE Paper 825070, 1982.

The layout of a bus, its mechanics, systems, components and procedures of maintenance influence the operation and the maintenance of city buses and coaches. The bus must operate reliably and continually throughout its long life. Diesel buses have achieved wide acceptance; one considers the reasons for its emergence, the alternatives available and the probable future developments. The influence of the gearbox, its efficiency and economical operation, require careful choice. One reviews the 'demands' for public transport, and describes several opportunities to introduce product innovations to save costs of operation. The influence of the auxiliary systems of the bus over this economy of operation is significant and developments to increase economy and efficiency can be achieved. The layout of the bus influences greatly the first costs, the costs of ownership and the intrusion into city traffic, and the alternatives are considered.

### H. Goebels, H. Moller, and H. Schramm, ABS: Highly Flexible Modular System for Commercial Vehicles, SAE Paper 852326, 1985.

This paper provides an overview of the different Bosch Anti-Lock Braking Systems (ABS). Basically there are three types of ABS available providing a suitable design for the various types of single-axle and multiple-axle vehicles. A semitrailer for example can be equipped with a single-axle ABS, a truck, trailer or bus mostly with a 2-axle ABS and an articulated bus will be equipped with a 3-axle ABS. In order to comply with all requirements, and particularly to take into account the installation conditions imposed by the vehicle, Bosch has 2 different types of pressure-modulation valve available, a single-channel and a dual-channel valve. With the different anti-lock systems discussed, Bosch can equip the wide variety of vehicle types which range from volume-production vehicles to special-purpose vehicles.

### E. G. Gohring, Tractor/Semitrailer Anti-Lock Performance and Compatibility as Seen by the Commercial Vehicle Manufacturer, I. Mech. E. C184/85.

Since the beginning of the 1980's anti-lock systems have been available for trucks, tractors and buses, too. Looking at the present vehicle population and the resulting transition period - surely quite a number of years - until all tractor/semitrailer units of a fleet will be fitted with anti-lock systems, operators of commercial vehicles will inevitably be confronted with questions about the performance and the compatibility of tractor semitrailer systems and the answers to these are of special importance in the present introductory phase of these systems.

For this reason, it is indispensible to know the tractor/semitrailer braking performance and their compatibility, especially if only one part of the rig, e.g. either tractor or semitrailer, is equipped with an anti-lock system. The following study describes the problems arising.

### P. S. Fancher, Integrating Anti-Lock Braking Systems with the Directional Control Properties of Heavy Trucks, I. Mech. E. C185/85.

Existing computerized models of articulated heavy trucks are used to investigate driver/vehicle system performance in (a) braking in a turn and (b) straight-line braking on a roadway with sidewise differences in friction level. Results from simulated maneuvers are employed in examining directional control difficulties that may be experienced during antilock braking of empty and fully laden commercial vehicles operating on poor, wet road surfaces. The findings of these analyses are used in discussing (1) implications of various antilock control strategies (worst-wheel or independent-wheel arrangements, for example) on directional control during braking, (2) vehicle-performance testing, and (3) type-approval calculations.

**G. S. Bowker and Tanguy, C., Brief Presentation of an Adaptive Brake System, I. Mech. E. C186/85.**

An adaptive brake system is described which meets the currently proposed E.E.C. requirements for a Category 1 anti-lock brake system. Selection of vehicle control philosophy is outlined and major system components described. One aspect of wheel control logic, namely, when to recommend the pressure build phase of operation, is selected for analysis. Derivation of the formula used by this system for this calculation is detailed. The remaining wheel control logic is presented in simplified form. A traction control system, which is currently under development, is also outlined. This system will work in conjunction with the adaptive brake system.

**E. Reinecke, An Anti-Lock System with Extended Safety and Control System Functions, I. Mech. E. C187/85.**

Anti-lock braking systems for commercial vehicles and passenger cars must provide directional stability, steerability and short stopping distance requirements. The ABS dualing-circuit anti-lock system for commercial vehicles having Modified Individual Regulation (MIR) at the front wheels and with integrated drive slip control is examined. Due to the independent diagonal arrangement of the control circuits and safety circuits, a dual circuit function can be extended to the total braking system and applies at the same time to the directional stability and steerability functions.

**W. Maisch and H. Schramm, Further Development of the Anti-Lock Braking Systems for Commercial Vehicles with Compressed Air Brakes, I. Mech. E. C191/85.**

Bosch has been producing and delivering the ABS Anti-Lock Braking System for passenger cars since 1978 and for commercial vehicles since 1982. The system is subject to permanent ongoing development. In the following, the actual status of the development of the ABS for commercial vehicles is described.

First of all, the wheel-speed sensor is briefly dealt with. Next, the basic functions of the electronic control module are discussed, as well as its self-testing facility and the monitoring of its periphery. Furthermore, the function of the so-called single-channel and 2-channel pressure-modulation valves is described. Either of these valve types can be installed depending upon the requirements. In the case of tractor-trailer combinations in which only the tractor or the trailer is equipped with ABS, Bosch has retrofit kits available for completing the installation.

**W. S. Broome, Three Years Experience of Volume Produced Anti-Lock for Trailers, I. Mech. E. C193/85.**

This paper deals with the introduction of antilock on trailers as the sole means of controlling over braking and meeting legislative requirements. It describes the design choices which were made and analizes their effect in retrospect. Electrical connections, the effects of suspension and the choice and disposition of sensors is discussed. In particular, the principles involved in 2 and 3 axle bogie control including lift axles is explored with particular reference to service experience. The special service conditions to which trailers are subjected have particular bearing on the requirements of antilock and the effect of the experience gained on future design specifications are considered.

# Other Cars

**G. B. Hickner, J. G. Elliott, and G. A. Cornell, Hybrid Computer Simulation of the Dynamic Response of a Vehicle with Four-Wheel Adaptive Brakes, SAE Paper 710225, 1971.**

An improved version of the 17-degree-of-freedom hybrid computer simulation is being modified to include a 4-wheel adaptive braking system (ABS). The derivation and verification of the ABS model, the form of the integrated vehicle/ABS model, and future plans for validation and utilization of the integrated hybrid simulation are presented.

**Robert A. Grimm, Wheel Lock Control Braking System, SAE Paper 741083, 1974.**

Automobile and truck manufacturers have given increasing attention to electronic wheel lock control brake systems during the last few years. These systems prevent continuous wheel lock-up during maximum braking stops, thus aiding the driver in retaining lateral stability and generally improving stopping distances.

This presentation discusses a system for preventing continuous rear wheel lock-up of an automobile during maximum braking stops. Included is a description of the control system components, tire and road characteristics, brake and vehicle dynamics, and an analysis leading to the requirements for optimum control.

**David Brown and Colin Harrington, Brake Fluid Functionability in Conventional and Anti-Skid Systems in Arctic Conditions, SAE Paper 750383, 1975.**

The effect and description of vehicle braketests performed under conditions of low ambient temperature and the evaluation of the effect of brake fluid viscosity, under these conditions, on brake system performance. Vehicle braking tests were performed within the arctic circle.

**T. O. Jones, T. R. Schlax, and R. L. Colling, Application of Microprocessors to the Automobile, SAE Paper 750432, 1975.**

This paper describes microprocessor technology as it may be applied to the automobiles of the future.

The microprocessor requirements described in this paper were generated as a result of the evaluation of the Alpha IV vehicle system which utilizes a solid-state, digital 4-bit, microprocessor to perform several vehicle control and display functions. The development of the Alpha IV system encompassed not only the interface circuit design and microprocessor programming, but also, the derivation of the digital algorithms and control laws for the functions which have traditionally been performed in an analog fashion.

The control functions performed include: Cruise Control, Four-Wheel-Lock Control, Traction Control, Speed Warning, Speed Limiting, Ignition Spark Advance and Dwell, Automatic Door Locks, and Anti-

Theft System. The display functions include: Speedometer (both analog and digital), Odometer, Trip Odometer, Tachometer, Clock, and Elapsed Time.

## MacAdam, C.C., Computer Model Predictions of the Directional Response and Stability of Driver Vehicle Systems During Anti-Skid Braking, I. Mech. E. C175/85.

Findings specific to this work and applicable to moderate turning maneuvers during which full brake torque applications occur are: (1) passenger car antiskid systems operating under high friction conditions are effective in assisting drivers maintain directional and path control during full braking demand, only if, a) the rear axle is equipped with a select low ("worst wheel") axle control system, and/or, b) conservative prediction rules are employed in the antiskid logic, particularly at the rear axle location; and (2) under moderate friction conditions (most wet asphalt or concrete pavements), antiskid systems appear to have greater effectiveness in assisting drivers maintain directional and path control of vehicles during limit braking maneuvers. Since the principal mechanism responsible for severe alteration of the directional dynamics of the vehicle during hard braking is the fore/aft load transfer, the level of deceleration and the ability of drivers to adapt quickly to an altered dynamical system, together determine the level of closed-loop braking performance that can be achieved.

## G. P. R. Farr, Low Cost Anti-Lock for Hydrostatic Braking System, I. Mech. E. C176/85.

The average private motor vehicle has its braking ratio arranged so that the front wheels lock before the rears for all surface conditions, and related vehicle decelerations, taking account of the dynamic weight transfer from the rear to the front wheels. For normal driving conditions this front bias is sufficient to provide a stable stop and avoids the possiblity of a sideways skid due to a premature rear wheel lock.

For the experienced driver, therefore, the shortest stop can be achieved by pressing the brake pedal just hard enough to prevent locking of the front wheels, but this impending lock is not easily sensed by the average driver. Also, if the road surface friction level varies during the stop it is even more difficult to sense an impending skid and prevent loss of steering, especially under emergency conditions.

The object of this paper is to show that it is possible to fit a simple anti-lock device to the front wheels of a front wheel drive car, to permit the average driver to steer the car, no matter how hard the brake pedal is pressed, although warning will be given by a slight pedal pulsation, that the optimum front adhesion level has been attained. Also by connecting a front to its diagonally opposite rear, via an improved apportioning valve, it is possible to provide directional stability and achieve a creditable stopping distance.

## J. H. McLoughlin, Limited Slip Braking, I. Mech. E. C177/85.

Limited slip braking is an advanced form of anti-lock braking which limits tire to road slip. During braking, car and wheel velocities are compared with the aid of dedicated electronics and a highly damped accelero-

meter. Braking force is reduced if tire slip exceeds preset limits. The system virtually removes the poor performance aspects of most anti-lock braking systems on rough road surfaces and gravel.

## H. Y. Yuasa, M. T. Tani, and T. F. Funakoshi, Anti-skid Brake System for Four-Wheel Drive Cars, I. Mech. E. C06/86.

A simple two channel ASB system for 4WD cars with two sensors and two diagonally connected brake lines has shown almost similar improvement in braking distance and directional stability as compared with a three channel ASB system for 2WD cars.

## J. R. W. Mansfield, Design Development and Testing of the Sierra XRx4, I.Mech. E. C11/86.

This paper discusses the background of the design development and testing of the Sierra XRx4, including becoming the leading design for a full Sierra 4 wheel drive program, and selection of the Ferguson Formula high performance high specification concept.

# Aircraft

## Edgar A. Hirzel, Real-Time Microprocessor Technology Applied to Automatic Braking Systems, SAE Paper 801194, 1980.

Since the advent of the jet engine, aircraft stopping capability has been dependent primarily on brakes and tires. To provide high performance control after landing while conserving tires and brakes, efficient hydraulic pressure systems have been developed. Associated electronic controls have historically relied on analog technology; but with the availability of high speed microprocessors, real-time digital control is now possible for automatic braking functions. This paper reviews the role of digital technology and the microprocessor in the braking control loops.

## John A. Tanner, Review of NASA Antiskid Braking Research, SAE Paper 821393, 1982.

NASA antiskid braking system research programs are reviewed. These programs include experimental studies of four antiskid systems on the Langley Landing Loads Track, flight tests with a DC-9 airplane, and computer simulation studies. Results from these research efforts include identification of factors contributing to degraded antiskid performance under adverse weather conditions, tire tread temperature measurements during antiskid braking on dry runway surfaces, and an assessment of the accuracy of various brake pressure-torque computer models. This information should lead to the development of better antiskid systems in the future.

# Radar Braking

## John B. Flannery, Automatic Braking by Radar, SAE Paper 740094, 1974.

Braking system, developed by AutoStop Corp., uses Doppler-generated signals to reduce stopping distance of vehicle in danger of collision. Safety device acts on conventional controls of vehicle, provides backup where driver's reaction time is not adequate, and rapid deceleration is required to avoid collision. Actuation of controls to achieve deceleration is accomplished by two vacuum bellows, one mechanically linked to the brake pedal, another linked to throttle and accelerator. Braking action is initiated by driver. System discriminates between objects which present danger of collision, and false targets such as traffic in adjacent lanes and signposts.

### William C. Troll, Automotive Radar Brake, SAE Paper 740095, 1974.

An automatic braking system for automotive vehicles is described. The system employs an onboard radar sensor to measure distance and relative closing velocity to obstacles in the vehicle path. This range and range-rate information is processed to generate a control signal which is a measure of the critical braking level existing in the dynamic environment. In response to selected control signal thresholds, the system provides the driver with advance warning of potential collision situations and can subsequently automatically apply vehicle braking if the driver response to the warning is judged inadequate. The critical threshold at which automatic braking is activated is selected to be well beyond that of a normal alert driver, thereby allowing him time to exercise his own options.

Problem areas associated with practical implementation of the automatic braking system on the production automobile are discussed.

Approaches to the problem of similarly equipped vehicles mutually interfering with each other consider controlled radiation, modulation, and signal processing techniques. All-weather performance is discussed in terms of radar operating frequency and road surface conditions. The problem of false alarming on off-path nonhazardous objects typified by signs, bridges, and other lane traffic is treated with respect to suppressing false alarms while maintaining adequate detectability and performance response on true obstacles. System-driver interface considerations and related human factors are discussed in terms of impact on system design and operational philosophy.

### R. A. Chandler, L. E. Wood, and W. A. Lemeshewsky, A Review of Philosophical Considerations in the Development of Radar Brake Systems, SAE Paper 750086, 1975.

The National Highway Traffic Safety Administration (NHTSA) has been involved in an investigation into the economic and technical feasibility of applying radar devices as sensors for automatic braking systems. Several different system application philosophies have been defined and discussed with consideration being given to the expected economic and safety benefits afforded by each.

The technical feasibility study, performed for NHTSA by the Institute for Telecommunication Sciences of the U.S. Department of Commerce, included such topics as radiation hazards, intersystem blinding effects, performance restrictions imposed by common highway geometries, effects of precipitation on signal propagation, and analysis of vehicular radar cross sections.

### M. Kiyoto, Y. Takeuchi, H. Iizuka, Y. Fukumori, and M. Katsumata, Skid Control System with Doppler Radar Speed Sensor, SAE Paper 765059, 1976.

A two-wheel anti-skid system incorporating a 24 GHz microwave doppler radar has been developed. The radar and electronic control circuit of this system are housed in a waterproof case, part of which forms a radar horn antenna. In this system, wheel slip is kept at an appropriate level by comparing the true ground speed measured by the radar with the circumferential speed of the rear wheels. The parameters of the control system have been optimized by using a simulator consisting of a hydraulic braking system and a hybrid computer. The road-test data agreed well with the simulated results, and showed that the new anti-skid system is superior in performance to the existing one.

### William C. Troll, Richard E. Wong, and Yung Kuang Wu, Results from a Collisions Avoidance Radar Braking System Investigation, SAE Paper 770265, 1977.

Results of previous studies have indicated that an automatic/noncooperative radar braking system may provide a significant benefit in preventing accidents that may otherwise be caused by driver inattention or tardy driver response. However, one of the major technical problem areas in implementing the automatic/noncooperative radar brake system is in achieving sufficient target discrimination. This is necessary to allow rejection of non-hazardous objects and to maintain a sufficiently low false alarm rate while retaining recognition capability on all potential hazards.

This paper presents the results of an experimental and computer simulation study conducted to resolve the effects of the various system parameters which may be significant to the target recognition problem. The target discrimination experimental study was conducted using an instrumented test vehicle equipped with an automatic/noncooperative radar braking system to gather parametric data under typical traffic conditions. The test courses selected for the experiments typify much of the high density, high speed, urban and suburban driving in the United States. The sensitivities of the various radar brake system parameters are also discussed. This work was sponsored by National Highway Traffic Safety Administration of the U. S. Department of Transportation.

### Nicholas S. Tumbas, John R. Treat, and Stephen T. McDonald, An Assessment of the Accident Avoidance and Severity Reduction Potential of Radar Warning, Radar Actuated, and Anti-Lock Braking Systems, SAE Paper 770266, 1977.

A group of 215 in-depth accident reports prepared as part of a tri-level accident causation study by a multidisciplinary team was examined to assess the benefit derived from the hypothetical application of various combinations of radar warning, radar actuated, and anti-lock braking systems. The approach was to have an accident analyst evaluate post hoc the benefit which would have been derived if one or more of the vehicles involved in each accident had been equipped with various types and combinations of these hypothetical systems; ten system types or combinations were defined. On one extreme, it was found that two-wheel anti-lock systems, by themselves, had relatively little accident prevention potential; only one of the 215 accidents (0.5%) would definitely have been prevented by such a system, although with less assurance there was some possibility or prevention of up to eight accidents (3.7%). On the other extreme, the most complex of the systems defined, comprised of a non-cooperative radar system with both actuation and warning potential, coupled with a four-wheel anti-lock system, would definitely have prevented 39 of these accidents (18.0%), with some possibility of prevention of up to 90 accidents (41.9% of those examined).

### Kohsaku Baba, Yukitsugu Fukumori, Yoichi Kaneko, Kenji Sekine, and Akira Endo, Doppler Radar Speed Sensor for Anti-Skid Control System, SAE Paper 780857, 1978.

A 24 GHz doppler speed sensor for skid controls has been developed. The microwave sensor is designed using both waveguide and thin-film technologies and assembled into a small integrated unit measuring 27#*#10#*#9 mm. The radar unit and the control circuitry are housed in a waterproof module of 94#*#140#*#78 mm. Part of the casing forms a horn antenna, which radiates a vertically polarized beam incident at 45" on the road surface, when mounted on the vehicle. The error in speed measuring is usually less than 10 percent.

### Gerald F. Ross, A Baseband Radar System for Auto Braking Application SAE Paper 780262, 1978.

This paper describes a Baseband Radar (BAR) sensor for radar braking application; an early version of the BAR concept was reported previously as a precollision sensor for air bag activation. In this paper we show how the normally wide effective beamwidth of the BAR is narrowed by using interferometry in conjunction with a novel delay line digital processor scheme. The beamwidth of the breadboard system spans a traffic lane width at 45 meters. The paper describes the details of the BAR sensor front-end and preliminary test results sponsored by the U.S. Department of Transportation and the Institute for Telecommunication Sciences.

# Testing/Legal

### E. P. Williams, Testing Anti-Lock Braking Systems - The Early Years, I. Mech. E. C180/85.

When the authorities in the United Kingdom were first required to test anti-lock braking systems there were many facets of ECE Regulation 13 that were ambiguous. The test procedures and equipment that were evolved during the first year or two now allow an objective and repeatable assessment to be made of an anti-lock braking system on a road vehicle. The particular aspects of the tests which are examined in some detail are the measurement of tire to road adhesion, the measurement of energy consumption, the testing of semi-trailer systems which use 'slave controlled' wheels and the assessment of the system susceptibility to interference from electro-magnetic fields.

### H. C. Allsopp, Road Surface Friction Measurement, Methods and Machines, I. Mech. E. C181/85.

Various methods of obtaining a road surface coefficient of friction are discussed together with six different machines in use world wide for obtaining same. Three of these are British viz. SCRIM, Mu Meter and The Portable Skid Resistance Tester; two are Swedish viz. Saab Friction Tester and the BV11; one is American viz. the ASIM Trailer.

Results of some correlation tests are presented and mention is made of ISO efforts to promulgate a standard measuring method and machine.

### A. Grimm, The Development of Anti-Lock Braking Regulations in the UK, I. Mech. E. C182/85.

The development leading up to the introduction of the UN ECE Regulation 13 antilock requirements in UK law are described. Further adaptation and application to motorcycles are discussed.

### J. Todorovic, The Influence of Anti-Lock Devices on the Reliability of Braking Systems, I. Mech. E. C183/85.

The reliability of a braking system actually represents the probability of successful braking, under all conditions in which the vehicle can be used. Anti-lock devices, which contribute to a higher safety during braking, even on slippery roads, may contribute, therefore, to a higher braking system reliability. A means of assessing the influence of anti-lock devices on the braking system reliability is presented in this paper.

### T. Willis and M. P. Kaplan, The Rear Wheel Lock-Up Problem and the American Legal System, I. Mech. E. C189/85.

The American legal system recognizes the concept of strict liability, wherein a manufacturer can be held liable for accidents caused by his product if in general it is shown to be (among other things) unreasonably dangerous.

Currently, several major vehicle manufacturers have specific models on the road which have been the subject of customer complaints, Consumer Group investigation and tests, governmental inquiries, and several lawsuits. Some of these involve design problems with respect to rear wheel locking before the front, alleging loss of control.

The parameters governing this phenomenon are well known to the thoroughly experienced engineers called upon to testify as experts in these matters. Control of these parameters at the vehicle design stage is not only possible, but is reasonably achievable. Modern design and manufacturing economics appear to require that one base vehicle be used with dozens of variations to create consumer demand.

This paper discusses the impact of this litigation on vehicle brake system design and its cost to the manufacturer. Through case studies, the mechanism whereby litigation has become a driving force to achieve major changes in vehicle design is explained in the context of the American legal system.

# Motorcycles

**Raymond J. Miennert, Antilock Brake System Application to a Motorcycle Front Wheel, SAE Paper 740630, 1974.**

Antilock brake systems suitable for use on a motorcycle were investigated under an experimental safety motorcycle program.

Due to program constraints, the only systems investigated were those already perfected by brake system manufacturers. This enabled utilization of the latest state-of-the-art principles to accomplish the desired result in the shortest time.

Various types of systems and power sources were investigated. The system chosen for the motorcycle application was a fluid powered system in conjunction with a production hydraulic disc brake. Laboratory and field tests have been conducted with results exceeding expectations.

**John W. Zellner, Advanced Motorcycle Brake Systems - Recent Results, SAE Paper 830153, 1983.**

Results of an evaluation of possible advanced brake components and systems for motorcycles are reviewed. Potential improved conventional brake components included: friction materials aimed at improving wet brake performance; and components affecting brake system feel properties. A prototype all-mechanical antilock brake system was evaluated. Results showed improvements in performance may be realized via all three of these areas, based on prototype results that might apply to future designs.

**G. L. Donne, The Development of Anti-Locking Brakes for Motorcycles at the Transport and Road Research Laboratory, I. Mech. E. C179/85.**

This paper describes development work done at the Transport and Road Research Laboratory on various designs of anti-lock brake systems for use on motorcycles. The need for such systems and possible future developments are briefly discussed.

**J. Cart, An Anti-Lock Braking System for Motorcycles, I. Mech. E. C188/85.**

If the wheel of a motorcycle is overbraked, lock-up and vehicle instability rapidly occurs. Capsize with vehicle damage and rider injury is likely to follow. Anti-lock braking has been available for many years on more stable vehicles such as cars and trucks but nothing for motorcycles even though TRRL demonstrated its worth in 1964. The specification for a motorcycle system is particularly stringent with severe constraints on cost, size, appearance, environment etc. A non-electronic system which meets the necessary parameters is described together with results from a wide range of road surface conditions.

# ABS + Four-Wheel Drive Vehicle Handling

**Jean Odier, Road-Holding: Braking and Traction, SAE Paper 700367, 1970.**

Every car-driver - even an experienced one - is perfectly aware of the steering instability during the braking process. These instability phenomena will often appear with unpleasant treachery and in a spectacular way that may lead to serious consequences. It proves useful to divide them into two groups - often complementary - to clarify the argument: 1. Defects of braking stability that show up without any wheel-lock-or before locking if locking results. These are the so-called driving stresses, pulls and deviation. When subject to these defects, the vehicle swerves, out by itself from its trajectory, without any driver's action on the steering wheel - except to compensate suitably for the defect, is possible. 2. Defects in braking stability with locked wheels that show up under three possible ways: locking of rear wheels only (with turning round of the car: "slew-round"); locking of front wheels only and losing steering control; and locking of all four wheels simultaneously.

Every car-driver is, out of experience, perfectly aware of these phenomena, but he generally does not know how to cope with them when they show up -unless he is very experienced. It is the engineer's craft to look for the causes of these defects, by means of studies and tests, then to make the necessary arrangements for curing them, if possible with an automatic device.

Each group of defects is examined, along with recent advances, conclusions, and prospects.

This survey is limited to the problems connected with Europe and takes its stand on the authority of a bibliography that, though not exhaustive, is thorough enough on the last 10 or 15 years.

**T. Kapitaniak, The Influence of Vehicle Suspension Displacements on the Working of Anti-Lock Braking Systems, I. Mech. E. C178/85.**

During braking the displacements of the pilot elements of a vehicle suspension take place and these displacements have a serious influence on the rigidity of a vehicle suspension. The model and experimental researches of these phenomena are presented in the paper. The characteristics of vehicle suspensions were made based on the real vibrations of the sprung and unsprung mass. The tests were carried out on several models of cars having different types of suspensions. During braking with an anti-lock system the displacements of pilot elements of the suspension are changing as a function of the braking moment, so the suspension rigidity is also changing. The variable vertical rigidity has a significant influence on vertical vibration of the vehicle body during braking - it moves a resonant area and in some cases causes the striking of suspension elements. The variable torsional rigidity causes the disturbance of the measurement of the rotational speed of the driving wheels. The horizontal rigidity makes the disturbance of the linear velocity of the centre of the driving wheels.

# Index

**A**

**Acceleration**
    vehicle, 7, 8, 26, 30, 96, 372
    wheel, 38, 63, 76-77, 234, 236, 239
    wheel tread, 8
**Accumulators,** 95, 97-103, 108-110, 186
**Actuators,** 22, 95-103, 168-170, 186-188, 355
    testing, 188-190
    throttle-flap, 376
**Adaptive brake control systems,** 35-51, 75, 151-159, 163, 241, 251
    effect on
        stability, 157-158
        steering, 157-158
        stopping distances, 157-158
    hardware, 36-37
    models, 154
    software, 37-38, 49-51
    tests, 156-157
**Adhesion**
    antiskid systems, 138-139
**Air brakes**
    electronic control standards, 131-133
**Amplifiers,** 307-308
**Antilock systems;** *see Antiskid Systems*
**Antiskid systems;** *see also Skid Control*
    15-22, 25-33, 36, 53-63, 65-72, 73-85, 87-91, 101-103, 107-111, 113-116, 197-205, 223-230, 303-310, 311-323, 337-347
    aircraft, 161-162
    Audi, costs, 303-304
    Bosch, 207-213, 215-221, 233-239, 241-248, 315-316, 361-370
    Chrysler, 175-184
    comparisons, 311-323
    costs, 278, 303-310
    effect of
        directional dynamics, 65
        road surfaces, 90-91, 119, 197-205, 280, 365, 369
    effect on
        cornering, 70, 366-368
        stability, 242-243, 272-273, 280, 307, 353-355
        steering, 65, 69, 176, 200, 242-245, 269-270, 306-307, 353-355, 362
        stopping distances, 20-21, 25-28, 162, 171-173, 176, 181-182, 190-191, 197-205, 242-243, 254-255, 269-273, 280-281, 297, 363-365
        suspension systems, 87-91
    electrical control, 57

electromagnetic interference, 138, 183, 255
energy consumption, 138-139
energy storage, 313
energy supply, 313, 326
FAG, 289-301, 316-317
failure, 207-213, 218-220, 229, 246, 254, 265, 291, 327
Ford Motor Co., 161-174
forecasts, 320-321
four channel
    Bosch, 242
    FAG, 290
    Lucas-Girling, 277
    WABCO, 325-332
four wheel drive vehicles, 364-370
front wheel drive vehicles, 53, 277-287, 350, 365
Girling, 185-192
history, 113-116, 259, 349
Honda, 317-318
hydraulic control, 56-57, 95, 97, 101-103, 233, 236, 242-247, 261-264, 289-301, 304, 318-320, 325-329, 374-376
integrated, 207-213, 215-221
Kelsey-Hayes, 162
Lucas-Girling, 277-287, 317
Mitsubishi, 249-255, 318
models, 88-89
motorcycle, 73
radio frequency interference, 114-115
regulations, Europe, 135-146
safety, 208, 210, 215-221, 226-227, 245, 254, 264-265, 331, 378-379
single channel, 111
    Audi, 307
software, 338-347, 378-379
state of the art, 311-323
Teldix, 27-28, 31-33
tests, 119, 135-146, 180-181, 183, 188-190, 355-358
Teves, 107-111, 257-267, 318-320, 371-379
three channel
    Audi, 304-305
    Bosch, 242
    Teves, 110-111
tractor, 142
trailers, 42, 113, 131-133
truck, 21-22, 42, 113, 131-133
two channel, 111, 349-358
    Audi, 305-306
    Bosch, 243
    Lucas-Girling, 279
WABCO, 325-332

**Audi,** 303-304
**Axles**
  adhesion, 139
  load, 15
    effect on stopping distance, 25
  rear
    control, 27
    dynamics in braking, 366
  sensors, 165-167, 177
**Amplifiers,** 60

**B**
**Bode plots,** 76-77
**Bosch,** 207-213, 215-221, 233-239, 241-248,
  315-316, 361-370
**Brake boosters,** 105, 223-230
  hydraulic, 97-100, 105, 210, 247, 257, 258,
    325-329
  vacuum, 210
**Brake pedal effort,** 96-97, 105, 108, 208,
  210, 226-229, 270-271, 301, 327,
  329
**Brakes**
  air, 113
    standards, 131-133
  control systems, 5-13, 15
  efficiency, 96, 104-105, 353
  force, 26, 30, 96, 99, 104, 294, 353-355
    amplification, 307-308, 354
    effect of
      braking, 362-364, 368
      wheel slip, 35, 58-59, 234
    effect on steering, 244-245, 271,
      361-362
  frequency response, 75-80, 154, 175-180
  friction, 7
  front wheel drive vehicles, 279
  history, 258
  moment, 300-301
  motorcycle, 73-81
  power, 95-105
  power dissipation in, 373
  pressure, 6, 8, 10, 35-37, 107-108, 114,
    208-209, 215-221, 224-226, 294,
    300, 342, 363
    effect of
      road surfaces, 40-43, 45-47
      valves, 234-235
      vehicle speed, 363
    effect on
      brake torque, 154, 294, 352
      steering, 244-245
      vehicle deceleration, 39-41, 45-46,
        59
      wheel slip, 38, 41-42, 45, 47, 367

torque, 6, 8, 10, 16, 57-59, 87-89, 198-199,
  269-270, 352-353, 367
  effect of
    brake pressure, 154, 294, 352
    road surfaces, 89-90, 362, 369
    suspension systems, 87
    wheel cylinder pressure, 36
  wet pavement performance testing,
    119-129
**Braking,** 89, 373
  effect of
    cornering, 67, 70, 191, 357, 363-364
    engine drag torque, 365
    road surfaces, 362, 369
    wheel slip, 58-59
    wheels, 6, 8
  effect on
    brake force, 362-364, 368
    steering, 65-66, 244-245, 269, 271-274,
      361-364
    yawing, 67, 171-173, 244-245, 361-362,
      365-370
  efficiency, 16-17, 280
    effect of
      tire-road-coefficient of friction, 16,
        25, 362, 369
      vehicle loads, 17
    truck, 17
    van, 16, 17
  four wheel drive vehicles, 365-370
  lane changing, 282, 356
  tests, 138, 356
  turns, 25-26, 28, 200, 205, 273
  wet pavement performance test, 119-
    129, 356

**C**
**Climbing ability**
  of vehicles, 371-372
**Computers**
  analog in simulation of wheel slip, 5-10
**Constant velocity mode,** 10
**Control systems**
  adaptive brake, 35-51, 151-159, 163, 241,
    257
    tests, 156-157
  amplifiers, 307-308
  antiskid, 25-33, 53-63, 75, 161-174, 185-
    192, 197-198, 207-208, 236, 242-247,
    257-267, 269-276, 280-281, 289-301,
    303-310, 350-351, 374-376
    Audi, 303
    costs, 303-310
    Honda, 269-276

hydraulic 56-57, 95, 97, 101-103, 233,
236, 242-247, 261-264, 289-301, 304,
318-320, 325-329, 374-376
Mitsubishi, 252
brake, 5-13, 15
digital, 233, 337-347
electrical, 57
feedback, 75-76, 79-80, 163
pneumatic, 22-23, 57, 119
power brake, 97-101
vacuum, 210, 250, 318
wheel lock, 53-63, 75
wheel slip, 5-13
**Controllers,** 55, 60-61, 75, 78, 178-179,
252-253
electronic, 60-61, 259, 264-265, 320,
376-379
quality control, 246
**Cornering,** 66-68, 70
effect of
antiskid systems, 70, 366-368
braking, 67, 70, 191, 357, 363-364
tires, 66, 191
effect on
stopping distances, 282, 357
traction, 372

**D**
**Deceleration**
effect on threshold values, 44
locked wheel, 37, 40-41
vans, 17
vehicle, 15, 60-61
effect of
brake pressure, 39-41, 45-46, 59
loads, 17
effect on
weight transfer, 16, 153, 352
wheel slip, 46-47
sensors, 251
wheel, 60-61, 234, 236, 251-252, 294, 366
**Deformation**
suspension systems, 88-89
**Directional dynamics**
effect of antiskid systems, 65
**Disc brakes,** 99
**Dynamics**
motorcycle, 76
vehicle braking in a turn, 25-26

**E**
**Efficiency factor,** 96, 104-105
**Electrical control**
antiskid systems, 57
**Electromagnetic interference**
antiskid systems, 138, 183, 255

**Energy consumption**
antiskid systems testing, 138-139
**Energy storage,** 313
**Energy supply,** 313, 326
**Engine drag torque**
control, 365
**Engines**
torque, 365, 372-373

**F**
**FAG,** 289-301, 316-317
**Failure**
antiskid systems, 207-213, 218-220, 229,
246, 254, 265, 291, 327
**Feedback control,** 75-76, 79-80, 163
**Filters**
noise, 38
**Forces**
brake, 26, 30, 96, 99, 294, 353-355
amplification, 307-308, 354
effect of
braking, 362-364, 368
wheel slip, 35, 45, 58-59, 234
effect on steering, 244-245, 271,
361-362
tire, 9, 176
wheel, 9-10
**Ford Motor Co.,** 161-174
**Four phase modulators,** 37, 45-48, 51
**Four wheel control systems,** 26, 175-184,
197-205, 241-248, 257-267, 269-276, 277
**Four wheel drive vehicles,** 361, 365-370
antiskid systems, 365-370
**Frequency response,** 75-80, 154
**Friction**
brake, 7
effect of road surfaces, 7
tire-road
effect on braking efficiency, 16, 25,
362, 369
**Front wheel drive vehicles,** 53, 279, 350,
365

**G**
**Girling;** *see also Lucas-Girling* 185-192

**H**
**Hardware**
adaptive brake control systems, 36-37
traction control systems, 377-378
**Honda,** 317-318
**Hydraulic brake boosters,** 210, 258
Teves, 258
WABCO, 325-329
**Hydraulic control**
analog, 289-301

antiskid systems, 56-57, 95, 97, 101-103, 6, 242-247, 261-264, 289-301, 304, 318-320, 325-329, 374-376
Bosch, 207-213, 215-221, 233-239, 241-248, 315-316, 361-370
FAG, 289-301, 316-317
Ford, 161-174
Girling, 185-192
Honda, 318
Lucas-Girling, 277-287
power brakes, 95, 97
Teves, 318, 374-376
traction control systems, 371-379
valves, 17-18
WABCO, 325-332
**Hydraulic valves,** 15, 17-18
effect on brake pressure, 234-235

**I**
**Inertia; see also Moment of Inertia**
wheel, 10

**J**
**Jackknifing,** 21

**K**
**Kelsey-Hayes,** 162

**L**
**Lane changing**
effect on stopping distances, 282, 356
**Load sensitive proportioning valves,** 15, 18-19
effect on stopping distances, 17-18
**Loads**
axle, 15
effect on stopping distances, 25
front-axle-to-rear-axle, 15
vehicle
effect on deceleration, 17
stopping distances, 17
**Lucas-Girling; see also Girling** 277-287, 317

**M**
**Maneuverability,** 205, 274
**Microprocessors,** 264-265, 267, 320, 337-339, 377-378
**MIRA Test Facility,** 119-129
**Mitsubishi,** 249-255, 318
**Model-predictive self-adaptive brake control system,** 163
**Models**
adaptive brake systems, 154
antiskid systems, 88-89
half-car, 8

modulators, 83-85
motorcycle, 76, 81-85
quarter-car, 8
road surfaces, 88, 201
suspension systems, 200
tire, 201
undriven wheel, 298-299
vehicle, 8
wheel, 8
**Modulation**
brake pressure, 107-108, 207-208, 215-221, 223-224, 300-301, 317
**Modulators,** 55-56, 77-78, 155
brake pressure, 36-37, 55-56, 179-180, 253, 275, 283-286
four-phase, 45-48, 51
models, 83-85
motorcycle, 83-85
on-off, 37, 41-44, 49-51
vacuum powered, 164, 183
**Moment of inertia,** 235, 367
polar, 153
**Motorcyles**
brakes, 73-85
dynamics, 76
models, 76, 81-85

**N**
**Noise**
filters, 38

**O**
**On-off modulators,** 37, 41-44, 49-51

**P**
**Pedal effort; see Brake Pedal Effort**
**Pneumatic control**
antiskid systems, 19, 22-23, 57
valves, 15, 19
**Polar moment of inertia**
wheels, 153-154
**Power brakes,** 97-101
hydraulic control, 95, 97
**Power spectrum density**
road surfaces, 202
**Power steering,** 110
pumps, 112, 116
**Pressure**
brake, 6, 8, 10, 35-37, 107-108, 114, 207-208, 215-222, 224-226, 294, 300, 342, 363
effect of
road surfaces, 40-43, 45-47
valves, 234-235, 304
vehicle speed, 363

effect on
   brake torque, 154, 294, 352
   steering, 244-245
   vehicle deceleration, 39-41, 45-46, 59
   wheel slip, 38, 41-42
modulation, 36-37, 41-51, 55-56, 77-78, 83-85, 107-108, 164, 179-180, 183, 207-208, 215-221, 223-224, 253, 275, 283-286, 317
proportioning valve, 45
wheel cylinder, 36-37
**Propellor shafts**
   sensors, 251
**Proportioning valves,** 15, 17
**Pulses circuit,** 61
**Pumps,** 261
   power steering, 112, 186

**Q**
**Quality control**
   Bosch, 246

**R**
**Radio frequency interference**
   antiskid systems, 114-115
**Rear brake lock-up control systems**
   Mitsubishi, 249-255
**Rear wheel brake control;** *see also Two Wheel Brake Control,* 20, 53
**Recovery time,** 41, 43-44
**Regulations**
   antiskid systems Europe, 135-146
**Road surfaces**
   effect on
      antiskid systems, 90-91, 119, 197-205, 280, 365, 369
      brake friction, 7
      brake torque, 89-90, 362, 369
      stability, 272-273
      stopping distances, 11-12, 20, 25-27, 58, 181-182, 190-191, 197-205, 242-243, 254-255, 269-273, 280-281, 297, 363-365, 369
      traction, 371-372
      wheel slip, 44, 46, 48, 59
      wheel speed, 198, 270
   models, 88, 201
   power spectrum density, 202
   torque, 372
**Road tests**
   of wheel slip control systems, 10-11
**Rolling radius,** 57, 153

**S**
**Sensors,** 107-110, 166-167
   axle shaft, 165-167, 177
   propeller shaft, 251
   speed, 53-54, 165, 251
   tests, 167
   transmission speed, 54, 165
   vehicle deceleration, 251
   wear, 166-168
   wheel speed, 54-55, 165, 186, 236, 242, 251, 264-266, 313-314
   wheel velocity, 20-21, 36-37
**Shock absorbers,** 88
**Simulation**
   of wheel slip control systems, 5-13
**Signal processing unit,** 38
**Software**
   adaptive brake control systems, 37-38, 49-51
   antiskid systems, 338-347, 378-379
**Solenoid valves,** 55, 114, 236
   effect on brake pressure, 234-235, 330
**Speed**
   sensors, 53-54, 251
   transmission sensors, 42, 165
   vehicle, 61, 74-75, 199, 204
      effect on
         brake pressure, 363
         tire-road friction coefficient, 153
   wheel, 61, 74-76, 89, 155, 198-199, 202-204, 251, 270, 338, 342
      effect of suspension systems, 87-89, 198, 270
      sensors, 54-55, 165, 186, 236, 242, 251, 264-266, 313-314
**Split friction test,** 143-145
**Stability,** 15, 19
   effect of
      adaptive brake control systems, 157-158
      antiskid systems, 242-243, 272-273, 280, 307, 353-355
      engine drag torque control, 365
      road surfaces, 272-273
**Standards**
   air brake, 131-133
   antiskid systems Europe, 135-146
**Steering**
   effect of
      adaptive brake control systems, 157-158
      antiskid systems, 65, 69, 76, 200, 242-245, 269-270, 306-307, 353-355, 362
      braking, 65-66, 244-245, 269, 271-274, 361-364
      engine drag torque control, 365

wheel lock, 26, 280, 308
power, 110, 112, 186
**Stop control system,** 277-282
**Stopping distances,** 8, 11-12, 15, 17, 20, 25, 62
    effect of
        adaptive brake control systems, 157-158
        antiskid systems, 20-21, 25-28, 162, 171-173, 176, 181-182, 190-191, 199-200, 242-243, 254-255, 269-271, 280-281, 297, 365
        axle loads, 25
        cornering, 282, 357
        lane changing, 282, 356
        load sensitive proportioning valves, 17-18
        road surfaces, 11-12, 20, 25-27, 58, 181-182, 190-191, 197-205, 242-243, 254-255, 269-273, 280-281, 297, 363-365, 369
    effect on
        dry surfaces, 11-12, 20, 26-27, 181, 255
        gravel roads, 182, 197, 281
        ice, 182, 255, 281, 297
        rumble strips, 181-182
        snow, 182
        wet surfaces, 11-12, 20, 26-27, 119-129, 181, 245, 255, 272-274, 281, 297, 356
    truck, 17
**Sure-track brake systems,** 161
**Suspension systems**
    deformation, 88-89
    effect of antiskid systems, 87-91
    effect on
        brake torque, 87
        wheel speeds, 87-89, 198, 270
    models, 200

**T**
**Teldix antiskid system,** 27-28, 31-33
**Teves,** 107-111, 257-267, 318-320, 371-379
**Threshold values**
    effect of deceleration, 44
**Tire-road friction coefficient,** 5, 16, 20, 25-26, 57, 59, 82, 152-153, 352, 362, 367, 369
**Tires**
    cornering, 66, 191
    forces, 9, 176
    load, 8
    models, 201
    rolling radius, 57, 153
    slip, 29, 30, 82
    tests, 119-129
    torque, 6-10, 57-58, 60

wear, 353-355
wet pavement performance, 119-129
**Toggles,** 79-81
**Torque**
    brake, 6, 8, 10, 16, 57-59, 87-89, 198-199, 269-270, 352-353, 367, 369
    effect of
        brake pressure, 154, 352
        road surfaces, 89-90
        suspension systems, 87
        wheel cylinder pressure, 36
    engine, 365, 372-373
    road, 372
    tire, 6-10, 57-58, 60
    wheel, 8, 16
**Traction control system,** 371-379
**Tractors**
    antiskid systems, 142
        effect of cornering, 372
**Trailers**
    antiskid systems, 42, 113, 131-133
**Trucks**
    antiskid systems, 21-22, 42, 113, 131-132
    braking, 17
    stopping distances, 17
**Two wheel brake control,** 53, 161-174

**V**
**Vacuum brake booster,** 210
**Vacuum control**
    antiskid systems, 154, 210, 250
        Mitsubishi, 318
**Valves**
    control, 101, 114, 179, 207-208, 215-221, 291-293
        effect on brake pressure, 234-235, 304
    hydraulic, 15, 17-18
        effect on brake pressure, 234-235
    load sensitive proportioning, 15, 17-19
    pneumatic, 15, 19
    solenoid, 55, 114, 236
        effect on brake pressure, 235
**Vans**
    braking, 17
    deceleration, 17
**Vehicles**
    acceleration, 7-8, 26, 30, 96, 372
    climbing ability, 371-372
    deceleration, 15, 60-61
        effect of
            brake pressure, 39-41, 45-46, 59
            weight transfer, 16, 153, 352
            wheel slip, 46-47
        sensors, 251
    dynamics braking in a turn, 25-26
    loads, effect on braking, 17

maneuverability, 205, 274
models, 8
speed, 61
   effect on brake pressure, 363
velocity, 7-8, 58, 342
**Velocity**
   sensors, 19-21, 57
   vehicle, 7-8, 58, 342
   wheel sensors, 20-21, 36-37
   wheel tread, 8

**W**
**WABCO,** 325-332
**Warning systems**
   brake failure, 60, 220, 229, 264, 328
   regulations, 143
**Wear**
   of sensors tests, 166-168
   of tires, 353-355
**Weight transfer**
   vehicle, effect of deceleration, 16, 153,
   352
**Wet pavement performance,** 11-12, 20,
26-27
   tests, 119-129, 181, 245, 255, 272, 274,
   281, 356
**Wheel lock,** 20, 38-42, 44, 46, 53, 59, 71,
197, 227-228, 249, 273, 280, 306, 308, 357,
363, 366
   effect of
      brake pressure, 38, 42, 227-228
      load-sensitive proportioning valves, 19
      tire road friction coefficient, 25, 270
   effect on
      deceleration, 39-41
      steering, 26, 280, 308
      stopping distances, 11, 357
   four wheel drive vehicles, 367
   motorcycles, 74
**Wheel slide protection**
   Girling, 185-192
   tests, 188-190
**Wheel slip,** 7, 20, 25, 29, 59, 82, 151-153,
199, 271, 295, 309, 342, 352, 365
   control systems, 5-13
   effect of
      brake pressure, 38, 41-43, 45, 47, 367
      deceleration, 46-47
      road surfaces, 44, 46, 48, 59
      wheel velocity, 44
   effect on
      brake force, 35, 58-59, 234
      braking, 58-59
      tire force, 176
   four wheel drive vehicles, 367-368
   road tests, 10-11

simulation, 5-13
**Wheel spin,** 371
**Wheel tread,** 8
**Wheels**
   acceleration, 38, 63, 76-77, 234, 236, 239,
   252
   cylinder pressure, 36
   deceleration, 60-61, 234, 236, 251-252,
   294, 366
   effect of braking, 6, 8
   effect on brake torque, 5
   forces, 9-10
   inertia, 10
   models, 8
   polar moment of inertia, 153-154
   radius, 10
   speed, 61, 74-76, 89, 155, 198-199, 202-
   204, 251, 270, 338, 342
      effect of
         road surfaces, 198, 270
         suspension systems, 87-89, 198, 270
      sensors, 54-55, 165, 186, 236, 242, 251,
      264-266, 313-314
   torque, 8, 16
   undriven models, 298-299
   velocity
      effect on slip, 44
      sensors, 19-20, 36-37

**Y**
**Yawing,** 65-67, 71, 272
   effect of braking, 67, 171-173, 244-245,
   361-362, 365-370
   effect on
      stability, 367-368
      steering, 363, 367-369
   four wheel drive vehicles, 367-370